寰宇文献 Universal Library | SINOLOGY 系列

SELECTED WORKS OF BERTHOLD LAUFER

劳费尔著作集

第十三卷

[美] 劳费尔 著

黄曙辉 编

中西书局

ZHONGXI BOOK COMPANY

图书在版编目(CIP)数据

劳费尔著作集 / (美) 劳费尔著；黄曙辉编. —上
海：中西书局，2022
　(寰宇文献)
　ISBN 978-7-5475-2015-4

　Ⅰ.①劳… Ⅱ.①劳… ②黄… Ⅲ.①劳费尔 – 人类
学 – 文集 Ⅳ.①Q98-53

中国版本图书馆CIP数据核字（2022）第207067号

第 13 卷

200

食土癖

FIELD MUSEUM OF NATURAL HISTORY

FOUNDED BY MARSHALL FIELD, 1893

PUBLICATION 280

ANTHROPOLOGICAL SERIES   VOLUME XVIII, No. 2

GEOPHAGY

BY

BERTHOLD LAUFER

CURATOR, DEPARTMENT OF ANTHROPOLOGY

CHICAGO, U. S. A.

1930

CONTENTS

99

GEOPHAGY

INTRODUCTION

The bibliography appended to this study may appear impressive at first sight, and a glance at it may even convey the impression as though a novel investigation of the subject were superfluous, but such an impression would be a delusion. The only really profound and serious research is represented by the fundamental work of Ehrenberg, which has unfortunately been forgotten or overlooked by the majority of those who have subsequently written on the subject. Ehrenberg, a geologist by profession, has studied and analyzed many hundreds of specimens of edible earths from all parts of the globe, and has had a wider and deeper knowledge of the subject than all his successors combined. Science does not always progress consistently in a straight line. Many articles cited in the bibliography are informative on special lines and useful, particularly the work of Hooper and Mann, which is important as far as India is concerned. The whole subject, however, is deserving of a new treatment in the light of fresh material and from the standpoint of the universal history of mankind.

In this article is given for the first time a correct exposition of the facts concerning geophagy, as revealed by Chinese records which are abundant. It will be noticed that these are very instructive and contribute important material toward the evaluation of the whole question of geophagy. For this reason China opens this investigation. The days are gone when the discussion of a problem started with the Greeks and Romans whose importance in the history of civilization is not much greater than and in many respects inferior to that of the Asiatic nations. Next to China the relevant conditions in Indo-China, Malaysia and Polynesia, Melanesia and Australia, India, Burma, and Siam, Central Asia and Siberia, among Persians and Arabs, in Africa, Europe, and America will be reviewed and discussed. In all these sections a great many new data unknown to previous investigators will be found. America especially has never before been adequately treated.

Geophagy is a convenient term which comprises a series of most varied phenomena resulting from entirely different causes and moving along different psychological lines.

101

In regard to the various earths and clays used by mankind Ehrenberg's work gives the best possible information, and any new geological and chemical researches should continue where he left his task. As a rule, not every kind of earth is eaten, but only those kinds which recommend themselves through certain qualities, such as color, odor, flavor, softness, and plasticity. The most important from the standpoint of edibility is what is called diatomaceous earth or kieselguhr, popularly known as "mountain meal" or "fossil meal" (in Chinese "stone meal" or "earth-rice"), which is a very light, porous earth resembling chalk or clay and consisting of the siliceous remains of very minute aquatic organisms or diatoms in several thousand varieties (hence, also styled "infusorial earth"). It varies in color from white to different shades of gray to black. Earths used as medicines or for enjoyment are almost without exception fine, fat, and usually ferruginous clays. They are consumed either in their natural state or lightly baked. Diatomaceous earth is at present of great industrial importance (cf. N. Goodwin, Diatomaceous Earth, *Chemical and Metallurgical Engineering*, 1920, pp. 1158-1160; R. B. Ladoo, Non-metallic Minerals, 1925, p. 190).

Geophagy has been characterized by previous authors as an "evil" or a "vice," while others have qualified it with such attributes as "disgusting" or "depraved appetite." Such characterizations are subjective and meaningless, and do not help us in understanding the phenomenon. Man, at the outset, will taste and test everything offered to him by nature; and consuming earth, mud, or clay is no more surprising than eating salt, pepper, bark, insects, snakes, or monkeys, or chewing gum, coca leaves, betel, or tobacco.

Earth or clay is nowhere used as an ordinary and regular article of diet, on a par with vegetal and animal food-stuffs; as it essentially consists of inorganic matter, it is naturally indigestible. It was used, however, and may still be used by many peoples in times of scarcity and famine as a food substitute to allay the pangs of hunger, giving as it does a sensation of fullness to the stomach; as a sort of condiment or relish, usually in combination with articles of food; mixed with acrid tubers or acorns as a corrective of taste; as a dainty or delicacy for its own sake; as a remedy for certain diseases; as a part of religious rites and ceremonies. These are the normal applications of clay and earth. There is, further, an abnormal or morbid use produced by or accompanying certain diseases, or due to nervous conditions.

Most writers have indulged in the sweeping assertion that geophagy is a universal phenomenon and was practised in times of antiquity. Neither of these statements is true. Generalization is the worst of all setbacks in scientific research and unfortunately an only too common sin in ethnological studies. This or that custom is observed in a single or a few individuals or in a single settlement, and it is at once fastened on the whole community or tribe or country. A traveler may have seen a certain person lick or chew a bit of earth, and the nation to which this individual belongs will go down in history as one of geophagists. Lasch (p. 216) asserts that earth-eating is exceedingly diffused over Africa, but this notion of a wide diffusion is merely fortified by a total of seven references. What is needed in ethnology is application of statistical methods or judicious restriction to really observed cases. Geophagy is not universal; it is unknown, for example, in Japan ancient and modern, Korea, Polynesia excepting New Zealand (while it occurs in Malaysia and Melanesia), Madagascar, as well as in many parts of Africa and Europe, and in the southern part of South America. It was likewise unknown in ancient Egypt and Babylonia as well as among the ancient Semites in general. It was equally foreign to the Greeks and Romans of classical times, while in the Hellenistic period the use of clay was confined to that of a medicine; neither Greeks nor Romans were geophagists. In China, Indo-China, India, and Persia earth-eating was practised to a certain extent, and is still widely practised in India and Persia, but in none of these countries is there an ancient record of this custom preserved; at least I have found none that would antedate our era. Maybe, this is fortuitous; maybe, it is not; the coincidence of the lack of ancient records in all great civilizations of Asia, at any rate, is suggestive.

While geophagy is not a universal phenomenon, yet it occurs sporadically almost anywhere. It has nothing to do with climate, race, creed, culture areas, or a higher or lesser degree of culture. It is found among the most civilized nations, even in our own midst, as well as among primitive tribes. It occurs in the Old and New Worlds alike. On the other hand, the habit is not general in any particular tribal or social group, and none can positively be labeled with a clear distinction as geophagists or non-geophagists. There are individuals who eat earth, and there are other members of the same tribe who abstain from it and even disapprove of the habit or may even see fit to dissuade their countrymen from indulging in it. In other words, the habit is more or less individual, not typically tribal; and this is exactly the point which has aroused my interest in the subject. We are wont to look upon the life and thoughts of a primitive people as something typical and collective, as a standard adopted and followed by all members of the community. This in general is true, but there are also features in primitive cultures which are left to individual decision and which require careful study. One of these is geophagy, the causes of which lie chiefly in the physical and mental constitution of the individual. Imitation, as in all human habits, has, of course, been a powerful factor in contributing toward the expansion of the custom. It could not have been diffused so widely all over India in all classes of the population unless by contamination of example. Again, if women during the period of pregnancy are especially devoted to clay-eating in a continuous area—Persia, India, Malaysia, and Melanesia—while this is not the case in China, Indo-China, Europe, and America, we must believe in an historical dissemination over the aforementioned area. In other words—while, on the one hand, geophagy may spring up anywhere spontaneously and independently, it has, on the other hand, assumed certain forms which can be explained only through contact and diffusion.

Clay-eating, consequently, cannot be interpreted as a racial characteristic or as a peculiar trait of this or that group of peoples, as has been done by Sarat Chandra Mitra, who expressed the opinion, "It seems that the use of clay for food is more confined to the Indian branch of the Aryan race, some Dravidian races and the various peoples belonging to the Mongolian stock, than to any other offshoot of the Aryan family or to any other race." This conclusion has also been antagonized by Hooper and Mann. In fact, the custom is not more characteristic of one tribe than of the other and pervades all classes of Indian society without distinction.

Clay-eating is not exclusively a poor man's habit either. In the Panjab "the very rich and the very poor are not free from it" (Hooper and Mann, p. 253). In Assam "the best working classes are affected by it" (*ibid.*, p. 252). It is likewise as common in the cities of India as among the peasantry; it prevails among all castes, regardless of race and creed.

There is a medical angle to this subject which is beyond the scope of this article. It has been suggested that geophagy is a symptom of ankylostomiasis and can be subdued together with this disease (H. Prowe, *Zeitschrift für Ethnologie*, 1900, p. (354)). Which is the cause and which is the effect seems not to be certain.

Certain it is that cases of ankylostomiasis do occur without being accompanied by geophagy; this disease, for instance, is widely diffused throughout China and Formosa (J. L. Maxwell, Diseases of China, pp. 174-182; G. Olpp, Beiträge zur Medizin in China, pp. 86-87), but neither Maxwell nor Olpp mentions any clay-eating on the part of patients. There is a disease known as cachexia africana, which is a disorder of the nutritive functions among Negroes and in certain kinds of disturbances of health among women, in which there is a morbid craving to eat clay (for details see p. 159). This so-called pathological geophagy is of limited interest to the ethnologist, but belongs properly to the domain of the physician. That inordinate and indiscriminate clay-eating is injurious to health and may lead to untimely death is obvious; even a Chinese author of the seventeenth century has plainly pointed it out. On the other hand, the perils of the indulgence have been overstated, and there is no doubt that occasional consumption of diatomaceous earth or a sprinkling of earth or clay over ordinary food is harmless. Again, the situation is not the same everywhere. In India, where the habit perhaps is more widely spread than in any other country and where it has developed into a veritable passion with many individuals, especially women, the appalling effects have grown proportionately. Here again, however, experiences recorded as to ill effects of clay-eating vary a great deal. One observer in India who made wide inquiries from women habitually eating clay was invariably informed that they experienced no ill effect whatever. Another correspondent who has known numerous instances of earth-addicts in Mysore reports that "the habit once contracted by women is rarely, if ever, abandoned by them, and is invariably followed by fatal results" (Thurston, p. 553).

"Reports are almost unanimous in stating that the habit when indulged in causes anaemia. Cases of intense anaemia are recorded with the history that the patients were perfectly well until they took to mud-eating. It is, however, almost certain that anaemia gives rise to the habit, and most probable that the habit is both the cause and the consequence of anaemia. Clay is eaten by people who are already anaemic, and the more they eat it, the more anaemic they become.

"Earth-eaters are frequently troubled by worms, but whether they are caused by earth-eating, or their presence is a contributory cause of the habit, is not quite decided. The most general idea among medical men who have had to deal with large numbers of cases is that anaemia accompanied by morbid gastric sensations is most often due to the commencement of the habit. The anaemia due to the ankylostoma worm is particularly accompanied by gastric cravings. Dr. Brooks says it may or may not cause ankylostomiasis of which anaemia is in his districts nearly always a symptom" (Hooper and Mann, p. 264).

Clay-eating is seldom openly practised and does not belong to the obvious things lying at the surface that would come within the ordinary traveler's observation. Many natives feel that the habit displeases the white man and will keep it secret or are loath to talk about it. It is reported that the female coolies of the Cochin hills "seem to be ashamed of the habit and, if other people see them eating clay, try to hide it" (Thurston, p. 525).

While the effects of geophagy are comparatively easy to recognize, it is more difficult to account for its causes.

Deniker is inclined to ascribe the habit of eating earthy substances to the need of supplying the deficiency of mineral substances (calcareous or alkaline salts), which induces the use of salt. F. W. Krickeberg (in Buschan, Vergl. Völkerkunde, I, 1922, p. 146) likewise regards the craving for salt as the cause leading to geophagy.

This theory is most improbable. In the first place, the clays consumed by man, as a rule, contain no salts, or if so, only a negligible quantity. Second, if Deniker's opinion were correct, we should justly expect that the maximum of clay-eating would be reached by people who command little or no salt and that with the growth of the salt supply the habit of clay-eating would proportionately decrease. This, however, is not the case. To cite but one example— the Iroquois and related tribes formerly did not make use of salt, but nothing is known about clay-eating on their part (cf. F. W. Waugh, Iroquois Foods and Food Preparation, pp. 150-153, Canada Geol. Survey, Memoir 86, 1916). The fact remains that all geophagists have access to salt, and probably more easily than to clay. Hooper and Mann (p. 263) point out that among children of India the salty nature of the ingredients of some earths is the recommendation for their use, but add judiciously, "This, however, can be the reason in but few cases of the habit." In another passage (p. 258) they dissociate completely the use of salt earths from the habit of earth-eating, contending that the use of the former can only be referred to as occurring commonly in districts where salt is expensive. In India, accordingly, earth-eating and the use of salt earth are two distinct and unrelated phenomena. The same is the case in China. In ancient China a great amount of salt was obtained from saline earth (J. O. von Buschman, Das Salz, II, 1906, pp. 4, 9; F. von Richthofen, China, I, 1877, p. 102), but such saline earth was never consumed, while other kinds of earth free from salt were eaten by the people. The same situation, again, is met with in some parts of Africa (Buschman, II, p. 278). A seeming exception occurs in South America. Brazil is very deficient in salt (Buschman, II, p. 413), and the Indians take recourse to various substitutes for salt in preparing their food, usually by burning saline plants and using the salty ashes; sometimes a reddish earth which has the appearance of salt ashes is resorted to for the same purpose. F. d'Azara, who traveled in South America from 1781 to 1801 (German translation by C. Weyland, 1810, p. 19), has some interesting notes on a salty clay (called by the Spaniards *barrero*) craved by the grazing cattle which cannot be kept away from it even by blows and which frequently feed on it to such an excess that they will die. A few travelers in South America report the consumption of salty clay on the part of Indians in lieu of salt, but the notes assembled in the chapter on South America (p. 184) demonstrate abundantly, that the widespread habit of eating non-salty clays throughout South America springs from causes which are entirely independent of the hunger for salt.

H. Schurtz (Katechismus der Völkerkunde, 1893, p. 21) believes that the original object of earth-eating was to silence the hungry stomach for a short while with an indigestible morsel.

Hooper and Mann (p. 270) are inclined to attribute the cause of geophagy "primarily to the purely mechanical effect it seems to have in comforting gastric or intestinal irritation. This may or may not be due to disease; if it is so due, the result is quickly to aggravate the disease it is taken to alleviate; if not, it rapidly produces effects which bring on disease. Gastric or similar irritation is inseparable from certain periods in a woman's life, and these are precisely the periods when the earth-eating habit is contracted. Once indulged in, the wish for similar alleviation becomes a craving; and the habit, as is usually the case with similar ones, strengthens itself, and brings on disease of the digestive canal. In the cases where men indulge, probably the habit has some similar origin."

The two last statements quoted assuredly contain some elements of truth, but do not explain all the phenomena connected with geophagy, and a formula applicable to the subject in its entire range can hardly be found, as geophagy appears in so many widely varying forms. It is best to emphasize a few specific cases. When we hear that the Pomo Indians of California mix clay with acorn-meal, their staple food, we may at first be inclined to dismiss this case as an unusual or queer practice; but when we further read that exactly the same thing is done by the peasants of Sardinia, we pause and think. As an historical contact between the Pomo and Sardinians is out of the question, the cause for this practice can only be physiological. The Zuñi swallow a bit of white clay with the tubers of *Solanum fendleri*, and it has been suggested that this is done to counteract or reduce the acridity and astringency of the tuber; this explanation may be correct as far as it goes, although it remains unexplained why it is just clay that is resorted to as a corrective. This is a matter that awaits the investigation of a physiologist.

Chemical analyses of edible clays are all right as far as they go, but are of no great utility to the ethnologist in understanding the problem. Moreover, most of the analyses made date a considerable time back when chemistry was not yet so perfected as it is at present, and when the usual conclusion of the investigators has been that the clays consumed by mankind contain neither nutritive nor medicinal properties. Maybe this is true, maybe it is but partially correct; but we need more solid and renewed information from a biochemist and physiologist, in the light of modern science, especially as to the effects of clays on the human organism. If these pages should have the good fortune to attract the attention of a biochemist and physiologist and to stimulate them to a fresh investigation of the problems involved, I should feel amply rewarded for the trouble and time I have taken in gathering this material from all parts of the world; but it must be studied comparatively. It cannot be fortuitous, for instance, that the identical phenomena appear in the most diverse regions and peoples, as the example of the Pomo and Sardinians just mentioned, or the craving for the bucaro pottery made of a reddish, odoriferous clay on the part of Peruvian and Portuguese women alike.

When we read again and again that to people living widely apart certain clays have an agreeable and spicy flavor and that they are attracted to them irresistibly and experience a pleasant and beneficial effect on their systems, we cannot simply brand such folks as maniacs, but there must be a physiological cause for such behavior.

For the geophagy of the pregnant Lasch (p. 219) has tried to give an explanation which does not satisfy me. According to him, the stomach does not bear substances like earth and clay, which will result in more or less violent vomitings which will cause, especially during the last months of pregnancy, contractions of the uterus and may facilitate delivery. This is theoretical speculation, but is not based on really observed facts. None of the authors who reports the craving of the pregnant for clay (and this is chiefly the case in Melanesia, India, and certain parts of Africa) says a word about vomiting, while the majority of women addicted to clay-eating take it habitually, whether pregnant or not; it is only during the periods of menstruation and pregnancy that the habit appears more intensified. It is clear, moreover, that a woman would not enjoy clay-eating and continue the habit if it really operated as an emetic. It is curious that Lasch himself cites Modigliani, who refers to the clay eaten by the Toba-Batak of Sumatra, as saying that it has the property of stopping the vomiting of women during pregnancy—the opposite of his theory—and this is far more probable. The Greeks used the earth of Samos as a means of stopping the vomiting of blood (Dioscorides), and the Arabic pharmacologists recommended the clay of Nishapur as a good remedy to relieve or stop nausea and vomiting (L. Leclerc, Traité des simples, II, p. 426).

The craving for earth so universally displayed by infants and young children, even in our midst, is presumably not pathological, but is simply due to insufficient roughage or insufficient mineral matter in their regular diet, and to an instinctive desire for roughage, which is usually supplied by wheat bran, potato-skins, green vegetables, and cereals. The case is known to me of a man (American) who for a few years swallowed two teaspoonfuls of white sand twice a day and declared that it kept him feeling fine in every respect; then he developed sarcoma of the intestine and died; whether the sarcoma was caused by his sand-eating habit has not been determined.

Explanations given by natives for earth-eating must be taken, of course, with a grain of salt. How many of us are able, if the question were put to us unceremoniously, to give an intelligent answer as to why we use salt and have a more intense craving for salt at one time than another. The common explanation given by primitives is that they believe earth or clay is good for them, that it benefits the stomach and promotes digestion. Others are satisfied with the notion that it has a pleasant odor and taste, that it tickles the palate and gratifies the stomach; others are merely attracted by the peculiar bright colors of some clays.

It seems that in its origin geophagy is not allied with religious ideas, in particular, as one might think, with the worship of earth as a deity or the notion of mother-earth. China, as will be seen, affords the best example to this effect (p. 125).

It is curious that tribes which make an extensive use of clays ceremonially, for instance, in body painting, do not take to eating it; for example, the Andamans (A. R. Brown, The Andaman Islanders, 1922, pp. 90, 99, 102, 106, 111, 122, etc.) and the Cheyenne (G. B. Grinnell, The Cheyenne Indians, 1923, II, pp. 235-236, 242).

On the other hand, geophagy frequently enters into religious ceremonies, notably in ancient Mexico and among some Malayans, who consume earth in ordeals, or among the Chins of Burma and the Negroes of Barbados who swallow it in affirmation of an oath. In China, diatomaceous earth was regarded as being of super-natural origin, as the food of dragons and immortals; and the discovery of such earth was hailed as a happy omen, and its consumption could not fail to have a beneficial effect on the health and welfare of pious believers.

Earth is also eaten by animals. Ehrenberg (II p. 19) mentions a case of earth-eating horses from Africa. Examples are known of wolves eating earth. Yet Wilken's theory (Handleiding van der vergel. volkenkunde van Ned.-Indië, p. 21), that man hit upon the idea of earth-eating in imitation of animals, is not convincing and must be rejected. The physiological causes driving both animal and man to earth-eating possibly are identical, and if so, the assumption of a mutual imitation is superfluous.

I wish to express my thanks to Dr. S. A. Barrett, Mr. Elsdon Best, Mr. C. Daryll Forde, Mr. I. Lopatin, Mr. Marshall H. Saville, Dr. Frank G. Speck and Mr. J. Eric Thompson for specific information. Their contributions are quoted verbatim under their names and may be easily traced by consulting the index.

CHINA

As regards geophagy in China, three different ways of using earth must be distinguished: (1) the magical method of the Taoists, (2) the medicinal employment, (3) earth as a famine food.

In European literature we meet only a few casual references to the subject with reference to China. As far as I know, Edouard Biot, who took a profound interest in all scientific questions, first called attention to this singular phenomenon. In his "Etudes sur les montagnes et les cavernes de la Chine, d'après les géographies chinoises" (*Journal asiatique*, 1840, p. 290), he has the following observations:—

"On Mount Lo-pao, department of Lin-ngan fu (Yün-nan), the mountain-people make the earth of this mountain into balls; it is fat and soft, and according to the text of the *Kwang yü ki*, they feed on it habitually."

This translation, as will be seen presently, is not exact. The name of the mountain is Lo-jung 樂粲, not Lo-pao. Biot adds the remark, "This is a new example of the depravation of taste observed for the first time by de Humboldt among the Ottomac."

The account of the *Kwang yü ki*, alluded to by Biot, which is a geographical description of China, is as follows: "Mount Lo-jung (or yung) 樂粲 山 is south of the prefectural city (Lin-ngan fu 臨 安 府 in Yün-nan). The earth of this locality has a fine odor, and is made into cakes used for purposes of cauterizing. When cooked (or heated), it can be eaten. The women of the P'o 焚 are fond of it." This text occurs in the original edition of the work, published in 1600 (chap. 21, p. 6b), as well as in the subsequent reprints of 1686 and 1744 (chap. 21, p. 11b). The question is of an edible clay; but the point emphasized by Biot, that the people feed on it habitually, is not directly brought out by the text, while he makes no reference to the P'o tribe; and this is an important feature. It is, accordingly, not the Chinese, but an aboriginal tribe of T'ai stock, which indulges in the habit; and, again, it is especially their women, who have developed this appetite. A similar reference to an aboriginal tribe is made in the *Nan chao ye shi*: "When the Li-su suffer hunger, they swallow earth mixed with honey" (C. Sainson, Histoire particulière du Nan-tchao, p. 181).

111

According to the *King chou ki* 荆 州 記, written by Sheng Hung-chi 盛 弘 之 in the fifth century A.D., there is in the district Wu-tang 武 當 a ravine on the banks of which there is a clay of fresh-yellow color; also it is eatable (*T'ai p'ing yü lan*, chap. 37, p. 8). It is not stated, however, that this clay was actually eaten, although this probably was the case.

The *Shen sien chuan* 神 仙 傳, attributed to Ko Hung of the fourth century, contains the following story:—

"Wang Lie 王 烈 lived solitary in the T'ai-hang Mountains 太 行 山 when all of a sudden he heard a crash on the east side of the mountain and the earth rolling like thunder. Lie proceeded to find out what had happened. He noticed that the mountain was cracked, and that the rocks were split over a distance of a thousand feet. Both sides of the road were covered with green stones exhibiting holes more than a foot in diameter. These holes were filled with a green mud which flew out like marrow. Lie took a sample of this mud, examined it, and formed it into a pill. Instantaneously it became hard like stone, as if hot wax were formed, and hardened immediately. It had an odor like boiled rice; and when he chewed it, it also tasted like rice. Lie collected several such pills of the size of peaches. He took these along and returned to Ki Shu-ye 嵇 叔 夜, with the report that he had found a strange object. Shu-ye, very pleased, took one of the pills and examined it; it changed into a green stone, and when struck, gave a sound like copper. Shu-ye then went along with Lie to inspect the spot, but the mountain which was previously torn asunder had resumed its normal shape."

There are several mountains bearing the above name—two in Shan-si (in P'ing-yang fu and Tse chou) and three in Ho-nan (in Chang-te fu, Wei-hui fu, and Hwai-k'ing fu). As follows from a notice in the *Kwang yü ki* (chap. 6, p. 26; original edition of 1600), the T'ai-hang of the prefecture of Hwai-k'ing is hinted at in the above story; for an abstract of it is given under the name of this mountain.

A landslip 山 崩 in the T'ai-hang is reported in the year A.D. 265 under the Emperor Yüan 元 帝 of the Wei (*T'ung chi*, chap. 74, p. 29b), and it is plausible that this catastrophe forms the historical background of Ko Hung's story.

The Gazetteer of Yi-hing 宜 興 縣 志 has this story:—

"As to Yao Sheng 姚 生, it is unknown from what place he came. Once he traveled to the Chang-kung Grotto 張 公 洞 and, a torch in his hand, entered it. There he met two Taoists 道士 seated opposite each other and engaged in a game of *wei-k'i*. Sheng expressed the wish to obtain some food. The Taoists pointed to several lumps of blue (or dark) clay or mud 青泥. He chewed a morsel of it, and found it very fragrant. The Taoists then bade him go and not speak to mortals about his adventure. Sheng bowed and thanked them, and carried away in his bosom the remains of the clay. He left the grotto and met Kia Hu 賈胡, who became frightened and said, 'This is the food of dragons. Clay is produced in grottoes, in the same manner as rocks.' " In a Chinese tale, entitled "The Nine-headed Bird," a youth meets a dragon in its cave and notices it lick a stone; the youth, tortured by the pangs of hunger, follows the dragon's example and no longer experiences hunger (R. Wilhelm, Chinesische Volksmärchen, 1927, p. 14).

Under the heading *t'u fan* 土飯 ("earth-rice"), a fundamental document, hitherto not indicated, is contained in the *K'ien shu* 黔書 ("Records of Kwei-chou Province"), written by T'ien Wen 田雯 (*hao* Mung-chai 蒙齋). In the edition of the *Yüe ya t'ang ts'ung shu* (chap. 4, pp. 25b-26b) it is as follows:—

"During the period Wan-li (1573-1620) of the Ming dynasty, the district Tse-yang 滋陽 (in the prefecture of Yen-chou, Shantung) was struck by a great famine. Suddenly appeared there a Taoist monk with a star-cap, gourd, and sword, and pointing to a lot of waste-land, said, 'Beneath this spot there is earth-rice, which may serve as food.' He vanished at once, and the crowd regarded him as a strange apparition. The people dug the soil more than a foot deep, and found earth of a bluish color, which somewhat had a flavor like grain. The famished people swallowed it eagerly, and as they greatly enjoyed it, quarrelled about the same piece. Several thousand men took so much of this earth away that it resulted in a pit several acres wide and about twenty feet deep. The following year, when wheat had matured, the Taoist monk came down to the same spot, as if he had something to fill out the pit. All of a sudden it was full, and again the people began to dig; however, they found nothing but sandy earth which could not be eaten; for the fairies 仙家 are crafty and make such earth only to help men. Further, in the year *ping-tse* 丙子 of the period Tsung-cheng (1636), there was an intense drought north of the Yang-tse, and in the Fung-yang mountains 鳳陽山 this earth was produced. Many people depended on it to keep themselves alive. In examining the records of K'ien 黔志, I find that for a number of years and in former times people used to dig earth on the occasion of great famines and to subsist on it. People unable to procure food, even when there was no drought, continually consumed such earth; nor is this astounding in view of the poverty of the populace of K'ien. When I heard of this, I was moved to sympathy with the people. Then I searched for this earth in order to examine it: it is white and unctuous like rice or meat-cakes. I tried it and found that it is flat of taste, but has no special characteristic. It is swallowed with some difficulty; when it has reached the stomach, however, one is satiated, but with a feeling of depression. Excessive eating of earth will cause obstructions and evil effects, and will ultimately lead to death. Ordinarily, people doomed to death from starvation have no leisure to select wherewith to fill their stomachs; anything is appetizing to them, and their thoughts are occupied day and night with devising new means of subsistence. Those who escape death owe it to the fact that they had mixed other things with the clay. This earth, therefore, is not to be regarded very highly, and does not even satisfy as much as chaff."

It is obvious that the specimen of white clay examined by T'ien Wen is not identical with the earth-rice of bluish color eaten by the people at the instigation of a Taoist monk. The former was a common inorganic clay, the latter a kind of kieselguhr containing organic substances and in principle identical with the "stone flour" to be discussed presently.

A substance shi mien 石麵 ("stone meal" or "mineral flour") is mentioned by Li Shi-chen in his Pen ts'ao kang mu (chap. 9, p. 22b) published at the end of the sixteenth century. Apparently, it is not pointed out in any previous Pen ts'ao. "Shi mien is not a substance of ordinary growth, but is an object of good augury 瑞物. According to some, it is produced only in times of famine. In the third year of the period T'ien-pao 天寶 (A.D. 744), under the reign of Hüan Tsung of the T'ang dynasty, in the districts Wu-wei 武威 and P'an-ho 番禾縣 [in Liang-chou fu, Kan-su], a sweet spring suddenly arose and brought forth stones, which were transformed into flour.[1] This was taken and eaten by the poor. In the fourth year of

---

[1] Inexact and incomplete translations of the text of the Pen ts'ao have been given by Biot (p. 216), Schott (in Ehrenberg I p. 145), and F. de Mély (Lapidaire chinois, p. 101). Biot translates, "Une source miraculeuse sortit de terre," omitting the geographical names entirely. Schott renders, "A source in Wu-jin (now Liang-chou fu) threw stones out." De Mély has, "La source de Li produisit une pierre"; in his translation, based on the unreliable text of the San ts'ai t'u hui (also utilized by Biot), all geographical names are eliminated, which renders

the period Yüan-ho 元 和 (A.D. 810), in the mountain valleys of the three chou—Yün, Wei, and Tai—of Shan-si 山 西 雲 蔚 代 三 州, stones were transformed into flour, which was consumed by the people. In the fourth month of the fifth year of the period Siang-fu 祥 符 (A.D. 1012), under the reign of the Emperor Chen Tsung of the Sung dynasty, there was a famine in the populace of Ts'e chou 慈 州 [now Ki chou 吉 州 in P'ing-yang fu, Shan-si]; the mountains in the district Hiang-ning 鄉 寧 縣 [in P'ing-yang fu] produced a greasy substance on stones like flour, which could be made into cakes and eaten. In the third month of the seventh year of the period Kia-yu 嘉 祐 (A.D. 1062),[1] under the Emperor Jen Tsung, the soil around P'eng-ch'eng 彭 城 [in Chi-li] produced flour; in the fifth month [of the same year] the soil in the district of Chung-li 鍾 離 [in the prefecture of Fung-yang, An-hwi] produced flour. In

the information valueless for scientific purposes, and the Chinese dates are not even correlated with those of our chronology. As the above quotation relates to the T'ang period, it is necessary to consult the geographical section of the T'ang Annals in order to understand this terminology. There we find (*T'ang shu*, chap. 40, pp. 7b-8a) that the district Wu-wei 武 威 郡 in Liang chou 涼 州 was divided into six *fu* 府; namely, Ming-wei 明 威, Hung-ch'i 洪 池, P'an-ho 番 禾, Wu-ngan 武 安, Li-shwi 麗 水, and Ku-ts'ang 姑 臧; in the year A.D. 744, a hill came forth from under a sweet spring (*li ts'üan*), and in consequence of this event the name was changed into T'ung-hua 通 化.—The natural event, as described above, was doubtless caused by a landslip. In 48 B.C., we read (*T'ung chi*, chap. 74, p. 29), mountains collapsed in Lung-si (Kan-su), and water-springs burst forth to the surface (山 崩 水 泉 涌 出). The same phrase (水 涌 出) occurs in two passages of the *Hou Han shu* (chap. 26, pp. 3b, 4) in connection with landslips.—The "sweet spring" (*li ts'üan* 體 泉) was prominent among the phenomena of good augury. It was regarded as the essence of water, of sweet and fine taste, and was believed to come forth only at a time when the sovereign practised righteous principles. This first happened in A.D. 25 under the Emperor Kwang Wu of the Han, when those suffering from chronic diseases and partaking of this water were all cured. It appeared again in the beginning of the reign of the Emperor Wen of the Wei and in A.D. 435 under Wen of the Liu Sung dynasty (*Sung shu*, chap. 29, p. 41). In A.D. 1008 a sweet spring came forth on the T'ai-shan (*Shan-tung t'ung chi*, chap. 63, p. 8); and in the same year, the same event is reported in Ju chou 迦 州, Ho-nan (*Ju chou ts'üan chi*, chap. 9, p. 63). A *li ts'üan* with a wine-like aroma exists on the sacred Hwa-shan in Shen-si (*Hwa yo ts'üan tsi*, ed. 1597, chap. 2, p. 3). *Li* is not the name of a river, as conceived in de Mély's work, but *li ts'üan* designates only "a spring of sweet water of miraculous origin."

[1] The date is erroneous. The passage is copied from the *Sung shi* (chap. 66, p. 18), where the date is given as the first year of Kia-yu (A.D. 1056). Moreover, the locality is more exactly defined as the village Pai-hao 白 鶴 in the district P'eng-ch'eng; and it is added, "The soothsayers stated, 'When the earth produces flour, the people will be stricken by hunger.'"

the fifth month of the third year of the period Yüan-fung 元 豐 (A.D. 1080), under the reign of Shen Tsung, all stones in Lin-k'ü 臨 朐 and Yi-tu 益 都, in the prefecture of Ts'ing-chou 青 州 [Shan-tung] were transformed into flour, gathered and eaten by the people. Inquiring into this phenomenon, it must be accounted for by the desire to secure food. As to the taste of this substance, it is sweet and non-poisonous. As to its healing powers, it benefits the breath; and eaten, when mixed with other things, it stops hunger."

Li Shi-chen does not state that he has ever seen or examined this substance; and in view of his assertion that it does not ordinarily occur in nature, but appears in a prodigious or miraculous manner, this is not even probable. It is no longer known in China under this name, and is not given, for instance, in the "List of Medicines," published by the Imperial Maritime Customs. Li Shi-chen seems to be the only author who has reference to this matter, for the *T'u shu tsi ch'eng* cites no other text under this heading. There is no description of the substance preserved; and what it was, must remain more or less a matter of guesswork. Read and Pak (Minerals and Stones, *Peking Soc. of Nat. Hist. Bull.*, III, pt. 2, 1928, No. 72) also give *shi mien* as unidentified.

We may positively state, however, what it was not: it was not a famine-food. The intimation that it only appears in famine-times is a gratuitous speculation; for under the dates recorded there were no famines, nor is it said that the people were driven by hunger to eat this substance; they ate it, simply because it was found and thought to be eatable. On the other hand, in the numerous records of famines under the Sung dynasty, it is not stated in a single case that people subsisted on this mineral flour. On the contrary, whenever food-substitutes are mentioned in such cases, they are given as leaves, wood, roots, chaff, ferns, mosses, rats, and human flesh.

A. J. C. Geerts (Produits de la nature japonaise et chinoise, 1883, p. 388) has a brief notice on *shi mien* (Japanese *seki-men*), saying that he has in his collection under this name a grayish white friable clay coming from Iwakimura in the province of Kaga and not containing organic matters. "Mixed with flour," he adds, "this is eaten in China in times of famine as a supplement of an insufficient nutrition, but it appears that in Japan where bad harvests are fortunately much more seldom than in China geophagy is not practised." This statement lacks sense and logic. If the Japanese abstain from eating earth, how is any one to know that a clay specimen from Japan is edible and how is it possible to assert that this Japanese specimen is identical with the Chinese "stone meal"? As a matter of fact, the former has nothing to do with the latter, and Geerts' note is no contribution to the problem.

In my opinion the "stone meal" of the Chinese is a fossil earth or kieselguhr, akin to the "mountain meal" of Germany (p. 168).

The last of the events mentioned by Li Shi-chen is also referred to by a contemporary writer, Wang P'i-chi 王 闢 之, in his *Sheng shwi yen t'an lu* 澠 水 燕 談 錄 (chap. 9, p. 19, and chap. 10, p. 9b), written toward the end of the eleventh century. This author reports a famine which took place in Lin-tse 臨 淄, in the prefecture of Ts'ing-chou 青 州 (Shan-tung), during the period Yüan-fung (A.D. 1078-86); it then happened that in the mountains and plains grew everywhere a white flour and white stone 白 麵 白 石 like lime 灰, but unctuous; the people obtained several tens of *hu* 斛 of this substance and mixed it with flour made into gruels and cakes, which could be eaten and proved very helpful. The author assures us that he made this observation with his own eyes.

Under the heading Kwan-yin fen 觀 音 粉 ("powder or flour of Kwan-yin," Avalokiteçvara), the *Pen ts'ao kang mu shi i* (chap. 2, pp. 28b-29b; written by Chao Hio-min in 1650), which is a supplement to the *Pen ts'ao kang mu*, gives the following additional information on edible clays:—

"According to the Gazetteer of Ch'u-chou fu 處 州 府 志 [in Che-kiang], there is a white clay of muddy appearance in the Yün-ho Mountains 雲 和 山. It is mixed with water and beaten on a stone; flour of glutinous rice is added, the proportions being half and half. This compound is steamed and consumed. It is capable of appeasing hunger, and is called Kwan-yin flour.

"There is an earth or clay produced in mountains, which in its interior is as white as flour, very fine and glossy. In years of dearth the villagers hastily dig it up, mix it with wheaten flour, and bake the mass into cakes which they use as food. But moderation must be observed; in case too much is eaten, there is danger of the belly being closed, as the natural properties of this clay are apt to obstruct the stomach and bowels. Earth produced in caves must not be administered for fear lest it might be poisoned with the saliva of venomous snakes.

"Cheng Chung-k'wei 鄭 仲 夔, in his *Leng ch'ang tai* 冷 賞 載, tells this story: In the year *ping-tse* 丙 子 there was a dearth in the villages I-yang and Shi-wo. The Buddhist monk in charge of the temple there had a dream in which the Mahāsatva 大 上 announced that in the soil at the foot of the mountain there was a mineral flour (*shi fen* 石 粉), which might be taken to satisfy hunger. In accordance with these words he set out to dig and obtained this mineral flour, which was very much like fern flour (*küe fen* 蕨 粉).[1] He ground it finely and made it into cakes which were steamed until well cooked and of pleasant taste, quite unusual. The villagers, as soon as they received the news, vied with one another to gather this flour. Some placed it in cabbage oil which made it so bitter that it was unfit to eat. This substance was what is called 'flour of the Mahā-satva' (*ta shi fen* 大 士 粉).

"The mineral flour discussed in the section 'Stones' of the *Pen ts'ao kang mu* is exactly the same. It is regarded as something extra-ordinary and grows imperceptibly. Now everywhere in mountains there are lakes on the banks of which is found a kind of earth that has curative properties, inasmuch as it stops hunger, benefits the breath, and adjusts the inner organs. When eaten, it stops hunger unconsciously. It has the merit of removing moisture, and in this respect is superior to *ts'ang shu* 蒼 朮 (*Atractylis* sp.), for even earth may perform the function of the element water. Its taste is a bit sweet and bitter; its nature is even, it neutralizes poison caused by insects, it cures dropsy, clears the eyes and heals jaundice caused by moisture."

In the Gazetteer of the district of Hwa-yang, which with Ch'eng-tu forms the prefectural city of Ch'eng-tu and capital of Se-ch'wan Province (*Hwa-yang hien chi*, chap. 43, p. 3), it is reported that "in the forty-ninth year of K'ien-lung (1784) an ochre-colored earth was produced in the town of Hwa-yang and that the people picked it up and ate it, as it was as fine as flour." There was no famine at that time, and there was no necessity of consuming this earth. It simply appealed to the people for the reason that the appearance of this earth was an unusual natural occurrence and that it was dis-tinguished as to color, fineness, and possibly flavor.

An allusion to "mineral flour" is perhaps contained in the following tradition which is pointed out by J. F. Davis (On the Poetry of the Chinese, p. 95, Macao, 1834), but which I have not been able to verify from Chinese records. "When Yung-lo usurped the whole empire (A.D. 1403), one of his nephews, the proper heir, shaved his head, and assuming the habit of a priest, retired to the

---

[1] The young shoots of some kinds of fern are eaten, and a kind of arrow-root is made from the rhizomes, which, after proper washing and cooking, are also eaten, in spite of their bitterness—only as substitutes in times of famine (G. A. Stuart, Chinese Materia Medica, p. 173).

depths of the mountains. The living rock there opened, and poured out a constant supply of grain for the support of the royal refugee. After his death, the miracle still went on, until a covetous priest, not satisfied with the quantity of grain thus obtained, enlarged the hole or fissure in the stone through which it flowed—when the supply immediately stopped altogether, as the proper reward of his cupidity."

Rockhill (*J. R. A. S.*, 1891, p. 267) was informed that an eatable clay is found in holes in the low ground near the river at Wu-tai shan in Shan-si.

From the notices of the *Pen ts'ao kang mu* and *Pen ts'ao kang mu shi i* it follows that medicinal properties also were attributed to edible clays. Li Shi-chen has devoted chapter VII of his work to earthy and clayish substances, discussing sixty-one species and their administration in the pharmacopoeia. This subject belongs to the history of pharmacology and has no direct bearing on geophagy; these medicinal clays were administered in small quantities in the form of pills, and were usually blended with other ingredients. Such pills surely were not capable of leading one into a habit of or passion for earth-eating. In this connection, however, attention must be drawn to the fact that it was the Taoists again who inaugurated the employment of earth as a remedy against disease. There is a story of a Taoist, Ch'en Nan 陳楠 by name, who was possessed of the power of curing disease with a medicine which he made by kneading earth and charmed water together into a bolus. In consequence he was nicknamed by his contemporaries Ch'en Ni-wan 陳泥丸; that is, Mud-pill Ch'en (cf. W. P. Yetts, *New China Review*, I, 1919, p. 17).

Rains of earth are also recorded in Chinese chronicles, thus in 1098 B.C. in the Bamboo Annals (E. Biot, Tchou chou ki nien, 1842, p. 29); others during the period Shi-yüan (86-80 B.C.), in A.D. 503, 535 ("yellow dust"), 536, 550 ("yellow sand"), 580 ("yellow earth"), and 582 ("earth"); see *T'ung chi* 通志, chap. 74, p. 4. In no case is it recorded, however, that such earth was consumed, presumably because an earth rain was considered an evil augury.

Ehrenberg (I p. 144) has analyzed two specimens of edible earth from China. One of these, a white earth, he received from A. von Humboldt, while the latter resided in Paris, and forwarded from China to Paris by French missionaries. The other specimen was a yellow earth which Ehrenberg obtained in 1847 from one of the large geological collections of London and which proved to be a sort of loam.

The text quoted above from the *K'ien shu* demonstrates clearly that clay also served occasionally as a famine food and that it was a Taoist monk who pointed it out to the populace. It must be emphasized, however, that, comparatively speaking, geophagy has been a very rare occurrence in China in times of famine.

Famines, droughts, inundations, and other similar catastrophes, to which the country has so frequently been subject, are listed with minute care in the chapters of the Annals, entitled *Wu hing chi* 五 行 志 ("Records relating to the Five Elements"). In the majority of cases, merely the fact of a famine is recorded under a given year (see, for instance, *Sung shu*, chap. 34, p. 31; and *T'ang shu*, chap. 34, p. 14), while food-substitutes used in famines are but seldom mentioned in the Annals. The gruesome phrase 人 相 食 ("men ate one another") recurs constantly. In A.D. 939, when locusts ravaged the fields of Chu chou 諸 州 (Shan-tung), the people were forced to subsist on grass and leaves (*Kiu Wu tai shi*, chap. 141, p. 6 b).

In A.D. 1127 when the city of Pien-liang (now K'ai-fung, capital of Ho-nan Province) was stricken by a great famine, the price of a pint of rice soared to three hundred copper coins, a single rat reached a high mark of several hundred copper coins, and people subsisted on aquatic plants and leaves of trees like *Sophora japonica* (*Sung shi*, Annals of the Sung Dynasty, chap. 67, p. 2). In A.D. 1148 when the eastern part of Che-kiang Province was visited by a famine, food was reduced to distillers' grains, chaff, grass, and wood (*ibid.*). In A.D. 1640 there prevailed a drought, locust-plague, and in consequence a famine in Ju chou 沕 州 (Ho-nan) when leaves of cotton-trees and other plants sold for a hundred copper coins the catty; crows and magpies deserted the country and flew southward, leaving their nests empty (*Ju chou ts'üan chi*, chap. 9, p. 63). We read also that people driven by hunger gnawed at crossbows or even boiled shoes, armor, leather, or sinews.

Grass, foliage, weeds, wild herbs, and tree-bark have always been the principal food-substitutes in famine times up to the present day. The best known historical example of recent times is the so-called "sweet dew" (*kan lu*) consumed by the T'ai-p'ing rebels during the siege of Nanking in 1863. Li Siu-ch'eng 李 秀 成, the so-called Chung Wang 忠 王 ("King of Loyalty"), as he tells in his memoirs, induced the T'ien Wang ("The Heavenly King"), the leader of the

T'ai-p'ings (Hung Siu-ts'üan 洪秀全), to issue a decree with suggestions to meet the distress of the famished population. "The decree was that they should eat 'sweet dew' in order to support themselves, whereupon I asked, 'How can they subsist on sweet dew?' The T'ien Wang replied, 'Let them take of the things which the earth brings forth'—this, it appears, was what he called 'sweet dew.' In concert with others I then represented that such was not a fit article for food, whereupon the T'ien Wang observed, 'Bring some here, and after preparing it, I will partake of some first.' As no one complied with his request, he gathered several herbs from his own palace garden and, having made them up into a ball, he sent the ball outside with orders to the people to prepare their food in like manner . . . . Three or four years prior to the present crisis orders had been issued to each household to collect ten piculs of 'sweet dew,' and deliver it into the treasury. Some obeyed and contributed their quota, others did not. The T'ien Wang for many days ate this stuff in his palace, and if my chief could do so, there was no reason why I should not do the same" (The Autobiography of the Chung Wang, translated from the Chinese by W. T. Lay, p. 62, Shanghai, 1865).

Some examples of geophagy in times of famine, which have come to my notice, may now follow. Such cases have occurred indeed, though rarely, and clay has been the last resort of the people when all other means of subsistence were exhausted.

The *Kiu hwang hwo min shu* 求荒活民書 is a monograph dealing with famines, droughts, and other catastrophes and the means employed on such occasions in saving human life. This work was written under the Sung by Tung Wei 董煟 (title Ki-hing 季興), who graduated as *tsin shi* in A.D. 1194, and has been reprinted in the collection *Ch'ang en shu shi* 長恩書室 published in 1854. Several examples of geophagy during times of famine are cited in this book (chap. 拾遺, pp. 2 and 11). Thus, in A.D. 618, at a time of scarcity, people gathered bark and leaves of trees, or pounded straw into a powder, or baked earth and ate it.

It once happened under the T'ang (A.D. 618-906), when military forces besieged Lo-yang and supplies were exhausted in the city, that people ate grass, roots, and leaves. When all this was finished, they subsisted on cakes made from pulverized rice dipped in floating mud. All fell ill, their bodies swelled, and their feet weakened until finally they died. In the period K'ien-tao (A.D. 1165-74), when a great famine prevailed in Kiang-si, there were people who ate white clay (*pai shan t'u* 白 墡 土) and choked to death.

It is on record in the *Wu Tai shi* 五 代 史 (Annals of the Five Dynasties of the tenth century): "When the town Ts'ang chou 滄 州 was besieged by Liu Shou-kwang 劉 守 光 [he died in A.D. 912], the inhabitants ate pieces of clay mixed with their food" 雜 食 菫 塊 (cf. Couvreur, Dictionnaire classique chinois, p. 172).

In the Gazetteer of the District of Wen shwi 文 水 (*Wen shwi hien chi*, chap. 1, p. 7b), in the prefecture of T'ai-yüan in Shan-si Province, it is on record that in A.D. 1586 there was no rain during the entire year, so that a huge famine prevailed and people ate grass, roots, and white clay or kaolin (*pai t'u* 白 土), with a very large number of dead in consequence.

In 1834 the Chinese missionary Mathieu-Ly, stationed in the province of Kiang-si, reported in the *Annales de la Propagation de la Foi* (No. XLVIII, 1836, p. 85), "Several of our Christians will surely die of starvation this year [1834]. God only can remedy so many and so great needs. All crops have been swept away by the inundation of the rivers. For three years numerous people feed on the bark of a tree which grows here; others eat *a light, white earth* discovered in a mountain. This earth can only be bought for silver, so that not every one is able to procure it. The people first sold their wives, then sons and daughters, then their utensils and furniture; finally they demolished their houses in order to dispose of the timber. Many of them were wealthy four years ago." Reporting on the great famine which overtook Shen-si Province in 1900-01, F. H. Nichols (Through Hidden Shensi, p. 232) states, "In order to buy food the farmers sold first their scanty stock of furniture and farming tools, then the roofs of their houses, and, lastly, their children."

"Regarding the straits as to food to which the sufferers by famine were put, various details are given. As a general rule, when famine was at its height, the sufferers from it, as long as they were able to do so, were in the habit of gathering grass, weeds, and other herbage they could find in the fields, and of eating these alone, or with such scanty supplies of better food as they were able to get. Others betook themselves to a soft *clayey slate*, which for a time allayed the pangs of hunger, but had a very injurious effect upon them. Those who had bean-cake, cotton seeds, and grass seeds swept from the roadsides, or bark and dried leaves, were considered fortunate. In Shan-si *stone-cakes* were somewhat extensively made use of as food, and were exposed for sale. The stone of which they chiefly consisted was the same as that of which English soft slate pencils are made. This was pounded to dust and mixed with millet husks, in greater or less proportions according to the poverty of the people, and then baked. It did not look bad, but tasted like what it was—dust. Elsewhere the people made use, as food, of a kind of *white earth* brought from the mountains, and which has much the appearance of corn-flour. Many of the people, for want of other sustenance, supported themselves upon this 'mountain meal.' In many places it was impossible to see any trees with the bark upon them; it had all been stripped off to be reduced and so consumed as food. Of another locality it is recorded that the most common food of the people consisted of leaves, mainly willow-leaves, weeds, and elm-bark; that the trees in summer were so stripped of foliage as to look bare as in early spring; the very weeds fast getting used up. Near T'ai-yüan fu, at the extreme northern limit of the famine in Shan-si, the roots of rushes were all eaten up; there were no trees left to bark except the poisonous ones, and hunger made the people often try these. In the same locality every family lived on the seeds of thorn-bushes or wild herbs, which they ground and mixed with a little corn-flour. In the southern part of that province every tree whose bark was not actually bare was stripped bare, and the dead trunks were cut up as firewood; in one district there some fine persimmon (*Dyospyros kaki*) orchards were left nearly uninjured, from which circumstances it was concluded that the bark of that tree could not be eaten, notwithstanding the excellent quality of its fruit. Elsewhere in that province, the root of the flag-rush (*Typha?*), stems of wheat, millet, maize, etc., and leaves of the willow, peach, plum, apricot, mulberry, and persimmon were eaten; also wild herbs, too numerous to name, *oily earth*, and many other articles not usually consumed. In some instances it was recorded that by means of small sums of money given by the several agencies of relief, those who were living on straw and reeds ground up with a little mud or chaff or boiled bark, were able by the addition of more substantial food thus put within their reach to tide over the time pretty well until the autumn harvest was cut" (Surgeon-General C. A. Gordon, An Epitome of the Reports of the Medical Officers to the Chinese Imperial Maritime Customs Service, from 1871 to 1882, pp. 387, 388, London, 1884).

In the report of the great famine in northern China during 1920 and 1921, mention is made of "flour made of ground leaves, fuller's earth, flower seed, etc." used in the daily diet of the famine-stricken (W. H. Mallory, China: Land of Famine, p. 2, New York, 1926).

Speaking of steatite or soapstone found in the environment of Lai-chou, Shan-tung Province, A. A. Fauvel (La Province chinoise du Chan-toung, p. 163, Bruxelles, 1892) remarks that steatite in a pulverized state is still employed in Shan-tung for the purpose of rendering wheat flour white and heavy; during the famine of 1876-77 many unfortunate people ate such flour in the hope of deceiving their stomachs and appeasing their hunger; the result was a terrible constipation which entailed death.

The Chinese and also the Japanese have a class of literature styled "treatises of eatable things" and devoted to a discussion of vegetal and animal foods for human consumption. None of these books makes any reference to earth or clay as an article of diet or as a relish; nor have I ever heard or read of an habitual earth-eater in China. The Chinese, although they regarded diatomaceous earth as a marvel of nature and occasionally ate it and although the destitute when driven by starvation occasionally resorted to earth-eating, cannot be classified as geophagists.

Finally I deem it my duty to refute a few of the many errors and misrepresentations from which this subject has suffered on the part of previous writers. Ehrenberg (I p. 144) asserts that clay-eating goes back in China to ancient times. There is no evidence for this generalization. Ancient Chinese literature contains no reference to such a practice. In this case, negative evidence may claim some degree of validity; for the Chinese have always been keen observers of the soil, its formation, color, and other properties, for purposes of agriculture and industry. The chapter Yü kung of the *Shu king* is the best witness thereof: the nature of the soil in each of the Nine Provinces is briefly characterized; for instance, as "whitish and rich," as "red, clayish, and rich," as "yellow and mellow," etc. In no passage, however, is any mention made of geophagy. In the *Chou li*, the various qualities of soils are set forth, and five classes are assumed according to aptitude for cultivation, productions, and physical characteristics of the inhabitants (E. Biot, Tcheou-li, I, pp. 194, 276). There are, further, numerous references to earths and clays in technical literature, which, however, maintains complete silence as to edible sorts (cf. Beginnings of Porcelain in China, Field Museum Anthr. Series, XV, No. 2, pp. 111-117).

Earth colored and plain played a great role in the worship of the god of the Soil and in the ceremony of investiture with a fief when a clod of earth enveloped by the white herb *mao* 白 矛 was bestowed upon the vassal by the liege-lord (cf. Chavannes, Le T'ai Chan, 1910, pp. 450-459; Le royaume de Wou et de Yue, *T'oung Pao*, 1916, p. 187; J. Przyluski, *Bull. de l'Ecole française*, X, 1910, p. 347).

A clod of earth was the symbol of the land and sovereign power over it. In 643 B.C. when Ch'ung-er 重 耳 left Wei, he begged some food from a villager, who handed him a clod of earth. The prince became irritated and was about to whip him, but Tse-fan 子 犯 restrained him, saying that this is a gift of Heaven. Ch'ung-er then touched the ground with his forehead, received the clod, and took it with him in his carriage (*Tso chwan*, V, Hi kung, 23d year; cf. Legge, Classics, V, p. 186; Couvreur, Tch'ouen Ts'iou et Tso Tchouan, I, p. 342). These examples are instructive in demonstrating that the sacred character of earth did not lead to earth-eating.

D. Hooper and H. H. Mann (p. 251) assert that "the Chinese are addicted to the habit and eat a white clay free from all organic remains." No authority is cited for this bold generalization,[1] but reference is made to D. Hanbury's "Science Papers" (p. 219), where an aluminous and an argillaceous earth, used for medicinal purposes, are described; but Hanbury does not state that they are ever taken as food. Hooper and Mann, further, remark that the Chinese, in many parts, mix gypsum with pulse, and thus form a jelly, which they greatly relish. What is meant here is doubtless traceable to F. Porter Smith (Contributions toward the Materia Medica of China, p. 108), who says, "The mineral gypsum is largely used as an ingredient in the bean-curd of ordinary diet. It enters into the composition of some sorts of putty, and is used to give rice a whiter face, after hulling and preparing it for sale." This phenomenon, however, is radically different from clay-eating. The question is here merely of an adulteration of food-stuffs, but the Chinese certainly have no craving or appetite for gypsum.

R. Lasch (p. 216) states, "In China, earth-eating is widely diffused. Pater Du Halde mentions a clay from the province of Shen-si utilized by Chinese women in order to render their complexion pale. Such clays are also found in many other places of China, and as in

---

[1] The sentence is evidently taken from the article of Sarat Chandra Mitra, who says (p. 288), "The Chinese, the Annamites, etc., are also addicted to this habit." Almost all data in the first chapter of Hooper's and Mann's treatise are derived from Mitra's article without acknowledgment. Who has ever observed an earth-addict among the Chinese?

Persia and Java, are publicly sold." He quotes Du Halde's work, but gives no exact page-reference. In fact, Du Halde says nothing of the kind; at least he does not say that Chinese women eat clay to bring this effect about; he does say (Description of the Empire of China, I, p. 281), "It is affirmed that they rub their faces every morning with a kind of paint to make them look fair and give them a complexion, but that it soon spoils their skin and makes it full of wrinkles." It is an old story that Chinese women, besides rice powder, use pulverized clay as a face powder, but they never took it internally. The *Ling piao lu i* (chap. A, p. 4, ed. of *Wu ying tien*), written at the end of the ninth century by Liu Sün (Sino-Iranica, p. 268), for instance, points out a pit of white clay north of the city of Fu-chou 富 州 (in the province of Hu-pei), the material being dug and traded by the people of the district and being used as a face powder by women.

INDO-CHINA

The brief communication of E. T. Hamy (see Bibliography) is based on information received by him from G. Dumoutier at Hanoi, who sent him specimens of earth cakes dried or cooked and consumed in four provinces of Tonking—Nam-Dinh, Thai-Binh, Hai-Duong, and Sontay. These cakes are said to be regarded rather as dainties than as articles of food, but their consumption is not connected with any superstitious idea or any belief in medicinal virtues of the substance; it is, according to Dumoutier, a simple depravation of taste maintained by local tradition. There are two kinds of these cakes; one consisting of thin shavings cut off from a compact block and rather dried than cooked over bricks made red hot by fire. The natives call them "cat-ears tiles" (*ngoe tai mèo*). They sell on the market on an average at 18 silver dollars for 600 grams. The other specimen looks like a thin tile, and has a beautiful red color in consequence of a rather strong roasting; its price is the same as for the preceding one. At the end of Hamy's notice a few chemical observations are made by E. Demoussy. The cakes in question have the physical properties of clay, unctuous to the touch, almost completely free from grains of sand, sticking to the tongue like kaolin and having the same flavor as the latter or rather lack of flavor. The clay includes a bit of iron and lime without an appreciable proportion of limestone, a little phosphoric acid, and a quantity of azote in that proportion generally found in a good soil; that is, about 15 per cent. The only characteristic that distinguishes these specimens from ordinary earth is that they contain a bit of combined ammonia, but in a quantity not sufficient to convey to them the slightest flavor. In short, they do not contain any ingredients that would justify their use as an article of food.

As the information given by Dumoutier seemed little satisfactory to me, I applied to the Ecole française d'Extrême-Orient of Hanoi, and the then secretary, Noel Péri, whose premature death is much to be deplored, was good enough to transmit to me in 1919 the following precise information which had been communicated to him by Dr. med. Paucot after the latter's own observations. "Cases of geophagy were observed only among the Annamese, not among the Muong. There is in Tonking no fossil edible clay. The cases known date more than twenty years back, the last being recorded in 1899-1900. The question was of eaters of an alluvial potter's clay observed in only two villages, one located a few kilometers south-

127

east of Hanoi on the right bank of the Red River, the other on the same bank opposite the town Yên-bay. There was but a small number of such persons, all in a wretched condition, who seemed to have acquired this habit in consequence of famines. How they got this idea could not be determined. The habit of eating a few mouthfuls of earth at their meals persisted even when it was possible for them to return to a normal state of nutrition, and they consumed this earth jointly with other foods. The clay was cut up into the shape of thin tiles of small size and simply dried in the sun. This consumption of clay resulted in the following symptoms: increase in volume of the intestines; extensive dilatation of the stomach which in some cases dropped to a point beneath the umbilicus; frequent helminthiasis; ankylostomiasis in all cases; state of emaciation and cachexy within the lapse of one or two years. From time to time cases of morbid geophagy are observed among children in the Annamese population; the parents are generally annoyed and alarmed and consult a physician. These cases are of interest only from a medical point of view, but seem to be devoid of interest to the ethnographer."

Monsieur Péri added that these earth cakes were never regarded as dainties, that they occurred until a few years ago not far from Hanoi in the provinces of Hà-dông and Son-tây, but that this custom appears to have almost vanished at present owing to the cessation of famines, as he was assured by a high Annamese functionary. Mitra (p. 288), without citing an authority, asserts that "the Annamese look upon the pasty and tasteless clay as a great delicacy."

MALAYSIA AND POLYNESIA

The earliest mention of geophagy with reference to Java is made by Labillardière (Relation du voyage à la recherche de la Pérouse fait en 1791-92 et 1798, II, p. 322 or Account of a Voyage in Search of La Pérouse, II, p. 338, London, 1800). In the villages between Surabaya and Samarang he noticed with surprise in the markets of several villages shops filled with little square, flat loaves of a reddish potter's earth which the inhabitants called *tana ampo*. This term means "clay earth." In his Malay vocabulary appended to his work (p. 376) the author defines it as "potter's clay which the Javanese eat." "I had at first imagined," Labillardière writes, "that they might probably employ these cakes for scouring their clothes; but presently I saw the natives chew them in small quantities, and they assured me that they made no other use of them." A specimen of this loam was sent in 1847 by Mohnike to Berlin, where it was analyzed by Ehrenberg (Bericht über die Verhandlungen der Berliner Akademie, 1848, pp. 222-225). Dutch scholars have since done considerable work in studying the edible clays used in Malaysia, above all J. J. Altheer, a chemist, who has examined and analyzed eleven specimens from Java and Borneo, and J. Heringa, who has investigated a clay coming from the west coast of Sumatra.

The word *ampo* is explained by J. Rigg (Dictionary of the Sunda Language, p. 13, Batavia, 1862) as follows: "Said of animals, particularly buffalo and deer, which lick the places where salt has been deposited, or are in the habit of licking the ground or rocks which contain some saline matter. *Batu ampo* is ampo stone which is found in many parts of Java and eaten by the natives. It is either a rock in a high state of decomposition, from having undergone a sort of *caries in situ*, or in other cases may be an aggregation of minute animal exuviae."

In a letter addressed to A. von Humboldt, Leschenault has given the following information: "The earth sometimes eaten by the Javanese is a sort of reddish ferruginous clay. It is spread out on rather thin leaves and then rolled into the shape of small tubes (almost in the form of the cinnamon of commerce) which are toasted over a fire. In this state the clay is called ampo, and is sold in the markets. The ampo has an insipid and empyreumatic flavor. It is rather absorbing, sticking to the tongue, and dries it up. Only women will eat it,

129

especially during the period of maternity or when attacked by the malady known in Europe as pica. Some men also eat ampo, for the purpose of checking obesity. I believe that ampo only acts on the stomach as a substance which absorbs the gastric juices" (Camilli, p. 188).

Hekmeyer, who was an officer in charge of the distribution of drugs in the Dutch Indies, stated that the Javanese first remove sand and other hard substances from the edible clay, and then reduce it to a paste by kneading it with water. The dressed clay is then molded into small cakes or tablets of about the thickness of lead pencils. The latter are baked in an iron sauce-pan, and when thoroughly roasted, look like pieces of dried pork. The Javanese often partake of small figures roughly made from clay in the form of animals or little men like those made by pastry-cooks (Mitra, p. 288). E. Ferrand gives illustrations of such clay figures representing a girl astride a dog, a woman holding a child, and a dancing girl. It is reported also that the women of Java eat pieces of a red pottery made at Samarang (Heringa, p. 186) and that at Batavia red pieces of clay wrapped in dried leaves of pisang or other plants are sold in the market (Altheer, p. 84).

The women of Java are also said to eat earth when attacked by chlorosis or pica. Others resort to it as an alleged means of reducing weight, because a slender figure is regarded as beautiful.

The preparation of ampo in Java forms an industry of its own which is practised by professionals, called *tukang ampo* (A. Maass, Durch Zentral-Sumatra, II, p. 252).

A red-brown earth is eaten by the Batak women on the west coast of Sumatra (Heringa, p. 186).

In the highlands of Padang in Sumatra earth is eaten, especially by pregnant women. To bring about abortion, a pap made of leaves and eatable clay is heated and applied to the abdomen. In Nias women put hot slices of clay on the abdomen to the same end (A. Maass, *op. cit.*).

The Encyclopaedie van Nederlandsch-Indië (2d ed., I, p. 3) gives the following brief summary under Eetbare aarde:—

"Eating earth is a custom encountered throughout the Archipelago, both in Java and Sumatra among Malayans and Batak, among the Dayak of Borneo, in Sumbawa, and even in New Guinea. The earth which is eaten, called *ampo* in Java, consists of a fat clay white, yellow, reddish, yellow brown, or gray green in color, and which besides the common components of clay contains bituminous and organic substances. It is carefully cleaned; when it has settled after a night, it is rubbed and formed into disks or tubes. The cakes are often covered with a solution of salt, smeared with coconut oil, and are then roasted. The earth is usually eaten as a delicacy, sometimes also by pregnant women, that the unborn infant may be fond of it. Its use leads to constipation and illness."

Other writers say that Javanese pregnant women eat clay in the belief that their foetus is fond of it.

Aside from this realistic geophagy, there is a ceremonial form of it. Eating of earth features in the ordeals of the Javanese: when a dispute arises about a boundary, it is believed that a bit of the controversial earth swallowed will swell the wrong-doer or burst him (P. J. Veth, Java, IV, 1907, p. 146). This custom may be traceable to India (below, p. 141).

In the island of Timor earth-eating played a role in ordeals. When the oath was sworn, a bit of rice was scattered, and some earth was eaten while the Mistress of the Earth was invoked (Riedel, Die Landschaft Dawan oder West-Timor, Deutsche Geogr. Blätter, X, p. 280; and A. H. Post, Grundriss der ethnologischen Jurisprudenz, I, 1895, pp. 482-483).

H. L. Roth (The Natives of Sarawak and British North Borneo, I, 1896, p. 385) quotes from Sir Spencer St. John (1862) that "in their boat expeditions Borneo people take a supply of red ochre to eat, in case of becoming short of other provisions; and we once found in some deserted Seribas' prahus many packets of a white oleaginous clay used for the same purpose"; and from Bishop McDougall (1863) that "there is a certain slimy clay which the Sakarran Dyaks always provide themselves with when they make their excursions in their boats, and which they suck when their stock of rice is exhausted: they say it is very nutritious." Roth was informed that the Undup occasionally eat a clay much resembling fuller's earth; they did not like it, but thought it a healthy thing to do—they seemed to think it acted as a purifier.

A. W. Nieuwenhuis (Quer durch Borneo, I, 1904, p. 83) informs us, "The fact is noteworthy that the natives of central Borneo sometimes crave a peculiar relish; thus, I observed that men and women, particularly pregnant women, sought in the soil of the banks for a yellowish or reddish loam consisting of weathered slate."

O. Beccari (Wanderings in the Great Forests of Borneo, 1904, pp. 335, 337) tells of Dayak of Borneo hunting among the pebbles of a torrent for a peculiar stone and nibbling it greedily as if it were a sweetmeat. It was a kind of clayey schist, soft and brittle and greasy to the touch. At Ruma Sale he saw again some Dayak eating clay schist with evident relish and observes, "It certainly was not eaten to appease hunger, but as a delicacy or perhaps to assuage an instinctive craving of the stomach for some alkaline substance."

H. W. Walker (Wanderings among South Sea Savages, 1909, p. 220) writes, "I made the discovery that some of my Dayak friends were addicted to the horrible[!] habit of eating clay, and actually found a regular little digging in the side of a hill where they worked to get these lumps of reddish gray clay, and soon caught some of the old men eating it. They declared that they enjoyed it." Clay-eating seems to be quite general among the Dayak (see also Altheer, pp. 85-87).

Among the Kayan of Borneo "it frequently happens that the woman begins to crave to eat a peculiar soapy earth (batu krap), and this is generally supplied to her" (C. Hose and W. McDougall, Pagan Tribes of Borneo, II, 1912, p. 153).

I. H. N. Evans (Among Primitive Peoples in Borneo, 1922, p. 114) writes, "At Tuaran the women have the abnormal habit of eating earth, which is also found in other parts of Borneo, in Java, and the Federated Malay States. Not far from the Chinese shops at this station there is a gully, which at the time of heavy rains has a small stream running at the bottom of it. The sides of the gully are made of a bluish gray clay with one or two bands of a hard dark purplish red clay running through it. At about six o'clock in the evening it is usual to see anything up to about a dozen women digging out this red clay with pointed sticks or small knives, and putting it into baskets. I have been told that the clay is roasted before being eaten, and that some women consume very large quantities. It is said to be a good medicine for women who are enceinte. I have several times dug out a sample and eaten it myself; it has rather the consistency of chocolate, but is almost tasteless."

With reference to the same locality O. Rutter (The Pagans of North Borneo, 1929, p. 72) supplies the following interesting information: "The women of the Tuaran group have a habit of eating a dark red clay which is found near the Chinese shops at the Tuaran Government Station, and tastes something like unsweetened chocolate. Mr. E. A. Pearson, who was stationed at Tuaran for some time, tells me that this earth is eaten by women who wish to bear children, since it is supposed to have particular effect at or about the time of the menstrual periods. That is, it is eaten as a means of securing pregnancy and not as a medicine during pregnancy, as Mr. I. H. N. Evans states. It seems to be rather a stealthy habit and the women (naturally enough) are shy about admitting that they eat it; they dig it out of the ground quite openly, but it is always 'for someone else.' Some women undoubtedly become addicts and cannot give up the habit, even when they are long past childbearing. One elderly Dusun crone told Mr. Pearson that she would rather give up her betel-nut than her daily whack of clay."

The analysis of an edible clay from Borneo is given in *Zeitschrift für Ethnologie*, III, 1871, p. 273.

In the Moluccas, a grayish white clay is eaten at Abubu in Nusalaut and in Saparua, notably by women during the period of pregnancy, and as stated by one informant "for the purpose of giving birth to white children" (K. Martin, Reisen in den Molukken, 1894, p. 55).

No case of genuine geophagy has become known to me from the Philippines. The following instances in which earth is used ceremonially and medicinally by the Tinguians have been kindly communicated to me by Dr. F. C. Cole.

The second day following a marriage is known as *sipsipot* ("the beginning or the start"). The couple go with their parents to the fields, and after the boy has cut grass along the edge of the land, he takes a little of the soil on his headaxe. Both bride and groom taste of this, "so that the ground will yield good harvests for them."

As a cure for dysentery and cholera, leaves of the sobosob (*Blumea balsamifera*) are placed in a jar of water. Above this a ball of clay is suspended, and banana leaves are placed over the mouth of the jar to prevent escape of the steam. The leaves are boiled for a time, and then the ball of clay is crushed and mixed with water, and this is given the patient to drink.

At the beginning of the rice harvest, the woman of the family goes alone to the fields until she has cut a hundred bundles of rice. During this time she uses no salt in her food, but sand is used as a substitute.

Throughout the Islands it is a common thing to mix the earth taken from nests of "white ants" with water and give it to patients troubled with bowel complaints. It is also mixed with water and applied to sores.

The nest of the *nido* (a small cave bird) is mixed with water, and is used as a cure for coughs and consumption.

As regards Polynesia, some forms of geophagy are reported from New Zealand, and possibly it was anciently known in Tahiti.

A. S. Thomson (The Story of New Zealand, I, 1859, p. 157) refers to "a clay called *kotou*, with an alkaline taste and an unctuous feel, which was eaten by the New Zealanders when pressed by hunger."

E. Best (The Maori, I, p. 432, Wellington, 1924) writes that "in times of great scarcity a kind of clay (*uku*) was eaten, as during the long siege of Kura-a-renga at Te Mahia; hence that fortified village was afterwards known as Kai-uku ('clay eating')."

The same scholar, a well-known authority on Maori agriculture and life, has been good enough to favor me with the following notes:

"In 1824 the Puke-karoro fortified village at Te Mahia was occupied by some hundreds of the Ngati-Kahu-ngunu tribe, when it was surrounded and besieged by a large force of raiders of the Tuhoe, Ngati-Maru, and other tribes. The siege continued for some months, until the besieged were reduced to cannibalism, families exchanging children so as to be guiltless of eating their own. Other non-combatants were also eaten, also quantities of the bluish diatomaceous clay called *uku*. Hence the siege and fort are often referred to as Kai-uku ('clay eating'). Cf. *Journ. Polyn. Soc.*, X, 1901, p. 26; XIII, 1904, p. 2; XVI, 1907, p. 20; *Transactions New Zealand Institute*, XXXV, p. 81.

"Some form of mud or clay was eaten in the Rotorua District in times of scarcity; a favored deposit of it was at Rotomahana.

"The Rev. R. Taylor mentions an unctuous clay or earth of a yellowish color that was eaten under similar circumstances."

The Maoris living around Taupo Lake are said to have eaten a fine, gray yellow ooze ejected by the volcanoes of the north island and called "native porridge" by the English settlers (Lasch, p. 217).

J. C. Crawford (Recollections of Travel in New Zealand, 1880, pp. 135, 139), who visited Lake Taupo, mentions mud springs the deposit from which is chiefly siliceous, and writes that the Maoris employ steam and mud springs for stewing food and the boiling springs for boiling it and for scalding pigs, but he does not say that this substance is eaten.

No accounts of edible earth are available for the other Polynesian islands, but from a legend given by W. Ellis (Polynesian Researches, I, 1831, p. 68) it would appear that a kind of red earth was formerly consumed in Tahiti. The tradition in question is an attempt at explaining the origin of the breadfruit. Under the reign of a certain king, when the people ate red earth (*araea*), there were a husband and wife who had an only son whom they tenderly loved. The youth was weak and delicate; and one day the husband said to his wife, "I compassionate our son, he is unable to eat the red earth. I will die and become food for our son." He died, and from his organs planted in the ground sprang a breadfruit tree. The mother directed her son to gather a number of fruits, to take the first to the family god and to the king; to eat no more red earth, but to roast and eat the fruit of the tree growing before them.

Earth is used by the Polynesians for industrial purposes. Red ochre is found in several islands, and in Rurutu and some others its color is so strong as to enable the people to form a bright red pigment for staining or painting their doors, window-shutters, canoes, and mixed with lime, the walls of their houses (Ellis, I, p. 24). This presents another example for the fact that industrial utilization of earth does not necessarily lead to earth-eating.

MELANESIA AND AUSTRALIA

R. Bruce (Annual Report on British New Guinea from 1899 to 1900, p. 102, Brisbane, 1901) saw white clay eaten in New Guinea. The cakes looked like white sausages, with a string running through their center which joined a lot together. After many inquiries as to the use of this clay he found that it was scraped down with a shell and used as a relish to food. He tasted it and fancied that it contained arsenic. He adds that "many natives of Torres Straits and New Guinea eat red-fat earth which contains iron; the women of the Straits eat it when pregnant so as to make the child light-skinned, etc." W. N. Beaver (Unexplored New Guinea, p. 144, Philadelphia, 1920), alluding to the report of Bruce (he locates the edible white clay at Tapamone on the Bituru River), writes that this clay is also found near Sui, a small village near the mouth of the estuary; there are one or two villages located on the northwestern side on Mount Lamington in the valley of the Kumusi River, the inhabitants of which are clay-eaters and invariably carry supplies of this "food" about them. In fact, from all accounts they pine away when deprived of it.

R. Neuhauss (Deutsch Neu-Guinea, I, 1911, p. 275) informs us, "Everywhere in Kaiser-Wilhelmsland [former German New Guinea] the blacks eat earth; it is an exceedingly fine-grained gray, yellow, or reddish material. In Bukaua the gray white clay comes from the mouth of the Bulesom; it is eaten, without special preparation, mainly by pregnant women. At the Sattelberg it is a reddish, ferruginous clay which is taken in a dried state. At Sissanu this delicacy has a gray yellow color, and is swallowed without further preparation. These clays are devoid of any nutritive values, but are agreeable in taste, especially the clay from the Sattelberg."

L. M. d'Albertis (New Guinea: What I did and what I saw, II, 1881, p. 89) writes that a red clay is chewed and even eaten by some of the people of Hall Sound and that he found this red clay.

Edible clays were located by P. Wirz (Die Marind-anim, pt. 1, 1922, p. 96) in Dutch Southern New Guinea. He describes them as gray or yellowish white and of acid taste. According to appearance, flavor, or origin various sorts are distinguished; they are partially much appreciated and used for barter. A white clay found and dug near Senajo is especially popular. It serves both as a cosmetic for painting face and body and as a relish. When the people of Senajo visit the coast, they will bring this clay along and exchange it with

136

the people of the coast. Another gray, recent marine clay, called *dave*, occurs in many places on the beach; it has likewise an acid flavor and is said to be good for the stomach; it is particularly eaten by expectant mothers. At Mevi Wirz saw a pregnant woman fashion this clay into loaves and dry these in the sun; she stated that consumed they are good for the foetus and must be eaten daily till the day of delivery. Wirz also refers to Bali where an edible clay found in the western part of the island is offered for sale in the bazars and is likewise enjoyed by the pregnant.

O. Finsch (Samoafahrten, 1888, pp. 295, 346) observed edible clay on the north coast of what then was Kaiser-Wilhelmsland (now Australian mandated territory); it was offered in the shape of flat cakes 20 cm wide and perforated in the center for the passage of a cord.

The title of the brief article of Meigen (see Bibliography) is misleading, for the edible earth analyzed by him did not come from New Guinea, but from New Mecklenburg. According to a communication of Dr. Hahl, then governor of German New Guinea, this sample came from Lakurefange on the east side of New Mecklenburg, and the natives ascribe to it healing powers in stomach and intestinal troubles. It is a fat clay of ochre yellow color, a terra rossa, of a camphor-like odor and of a not disagreeable spicy flavor.

E. Stephan and F. Graebner (Neu-Mecklenburg, 1907, p. 10) mention the eating of earth in New Mecklenburg with reference to the Gazelle Expedition of 1874-76, but offer no more recent information.

In New Caledonia geophagy was formerly widely practised, and partially it is still in vogue. As early as the eighteenth century it is reported by Labillardière (Account of a Voyage in Search of La Pérouse, II, p. 213), who visited New Caledonia in 1793. The natives approached the ship's landing-place and received bits of biscuit for which they asked. He then gives the following interesting account: "I saw, however, one of them come up who already had his stomach well filled, but who nevertheless ate in our presence a lump of a very soft steatite of a greenish color and as big as his two fists. We afterwards saw a number of others eat quantities of the same sort of earth. It serves to deaden the sense of hunger by filling their stomach, thus supporting the viscera attached to the diaphragm; and although this substance does not afford any nutritious juice, it is yet very useful to these people, who must be often exposed to be long in want of food, for they apply themselves little to the culture of their lands, which besides are very sterile.

It is to be remarked that undoubtedly the inhabitants of New Caledonia have made choice of the steatite only because from its great friability it does not remain long in their stomach and intestines. I should never have imagined that cannibals would have recourse to such an expedient when pressed by hunger."

Vauquelin, the chemist, found in this steatite from New Caledonia a not inconsiderable proportion of oxid of copper. In the northern parts of the island steatite occurs abundantly in the ancient slate formation. According to some authors, earth is merely eaten in times of scarcity to appease hunger; according to others, only women take it in doses of the size of a hazel-nut, and children imitate the practice. Among the people of Tiari, near Baladea, Garnier found a few geophagists, but only women who he says were prompted by a morbid craving to eat but a little earth, which is insipid in taste and is called by them *pagute* (*Globus*, XIII, 1868, p. 102). Lemire mentions balls of steatite which are dissolved in the saliva and have a somewhat sweetish flavor (F. Sarrasin, Ethnologie der Neu-Caledonier, 1929, p. 64).

According to Glaumont (*Revue d'ethnographie*, VII, 1888, pp. 85-86, not cited by Sarrasin), the inhabitants of New Caledonia chew a friable grayish earth found on the sides of the mountains. This author holds that the custom of earth-eating is on the same level as betel-chewing or opium and tobacco smoking. Sarrasin was informed by a native of the isle of Baaba in the north of Caledonia that baskets full of gray soft earth were collected there. He refers to another account that women on the march finished a whole basketful of earth, giving preference to it to real food. As steatite is not found everywhere in the island, many tribes must be content with clayish and marly minerals. This is also the case in the Loyalty Islands which consist merely of chalk. In a cave near La Roche on Maré, Sarrasin found weathered yellow marl of which the natives told him that it is crushed and eaten, particularly by women, as a dainty; red earth, too, they said, is eaten there after it has been burnt. This seems to refer to the weathered product of chalk which is colored red by iron.

V. de Rochas (La Nouvelle Calédonie, 1862, p. 140) reports that in the Loyalty Islands people eat an aluminous earth full of organic detritus, which is gathered in caves abounding in humus and which is kneaded into hard balls; these are dissolved in the saliva without leaving a bad taste. Sarrasin thinks that this substance may contain a trace of nutritive value, which is not the case with steatite, marl, and clay.

Earth is eaten in North Santo and Malekula in the New Hebrides. This is a tough, dark brown clay apparently mixed with organic substances and particularly coveted by pregnant women. In East Santo it is said to be flattened out like a biscuit and dried in the smoke. In Malekula the earth is shaped into small balls which are dried and sucked like a sweet-meat; the clay has indeed a sweetish flavor (F. Speiser, Ethnographische Materialien aus den Neuen Hebriden, 1923, p. 133).

Some authors, quite in general, assign geophagy to aboriginal Australia. It seems certain that it occurs among some tribes, but not among others. The following specific cases have come to my notice.

R. Brough Smyth (The Aborigines of Victoria, I, 1878, p. XXXIV) writes, "There is nothing in the records relating to Victoria respecting the use of any earth for the purpose of appeasing hunger; but Grey mentions that one kind of earth, pounded and mixed with the root of the *mene* (a species of *Haemodorum*), is eaten by the natives of West Australia." Seven or eight species of this genus occur in Australia, all of them furnishing roots which are eaten by the natives; they are acrid when raw, but mild when roasted (E. L. Sturtevant, Notes on Edible Plants, p. 297). The case therefore is analogous to what is found among the Ainu, Pomo, and Hopi.

The aborigines of Queensland use huge clay or mud pills, one or two of which at a time are prescribed for diarrhoea (W. E. Roth, Ethnological Studies among the North-West-Central Queensland Aborigines, 1897, p. 163).

E. Eylmann (Die Eingeborenen der Kolonie Südaustralien, 1908, p. 448) mentions medicinal employment of earth, ashes, and sand; women rub their breasts with a pap made of gypsum for the purpose of causing a secretion of milk.

INDIA, BURMA, AND SIAM

In India clay is generally eaten by women and children, rarely by men; by women usually during the period of menstruation and pregnancy, by others habitually at all times.

Examinations and analyses of Indic edible clays have been conducted by Ehrenberg (I pp. 116-177) and by Hooper and Mann (pp. 260-263).

The fact that clay is eaten in India was known in Europe early in the nineteenth century. Curiously enough, the edible clay of India was then designated "clay of the Mogol." G. I. Molina (Saggio sulla storia naturale del Chili, 1810, p. 50), therefore, wrote at that time that the Peruvian women are in the habit of eating pottery sherds as the Mogol women eat the dishes of Patna (como le Mogolesi mangiano il vasellame di Patna). This Indic pottery is described as being gray in color with a yellow tinge, known under the name "earth of Patna" and found principally in the environment of Seringapatnam. From this clay were manufactured vases so light in weight and so delicate in shape that "a breath from one's mouth was sufficient to turn them upside down on the table." Water poured into these vessels assumed a pleasant flavor and odor; and the ladies of India when they had emptied them would break them to pieces, swallowing the sherds with pleasure, especially in the period of maternity (Camilli, p. 188).

The clay consumed by the women of Bengal is a fine, light ochreous-colored specimen fashioned into thin cups with a perforation in the center and then baked in a kiln. In other words, it is ready-made pottery which they consume and which emits a curious smoky odor. It is this particular odor which makes it such a favorite with delicate women. The cups are strung on a cord and sold by the potters at so many pieces for one pice. Formerly these cups were hawked about in the streets of Calcutta, but this is no longer customary. Such a street vendor of baked clay cups once figured in a Bengali play staged in a Calcutta theatre; she recommended her ware in a song, pointing out that her cups are well baked, crisp to eat and yet cheap, and that delicate ladies about to become mothers should buy them without delay, as eating them would bless them with sons (Mitra, p. 286).

Saucer-shaped chips of partially baked clay are sold in the Calcutta bazar for eating (G. Watt, Commercial Products of India,

140

1908, p. 330). Burnt earth is considered less injurious in India than fresh earth.

The habit of clay-eating, though at present universal in India, cannot be proved to be of ancient date in that country. The earliest literary references to it, first pointed out by Mitra, occur in Kālidāsa's *Raghuvamça*. In one case, the question is of a queen who partakes of baked clay to render her breath fragrant and pleasing to her lord. In another case, the queen of Ayodhya, before giving birth to Raghu, feels a hankering for baked clay (Sanskrit *katikā*, Hindi *khariyā*). Mallinātha, in his commentary to the poem, observes that it is well known that pregnant women eat earth. These allusions contain nothing that would warrant the belief that clay-eating then (fifth or sixth century A.D.) was a general and habitual practice. In Vedic literature, no reference is made to it, nor in the Arthaçāstra. In such encyclopaedic works, as Varāhamihira's *Bṛhat-Samhitā*, where we might expect to find a trace of it, it is not mentioned either. Likewise in the literature on alchemy it appears to be absent, as evidenced at least by Ray's "History of Hindu Chemistry." Notably the Chinese pilgrims who traveled in India have not recorded the practice. Also so keen an observer as Garcia da Orta maintains silence about it, and W. Ainslie, in his "Materia Indica" (1826), ignores it; no reference to it is made in early Portuguese and English accounts of India. While this negative evidence is not in any way conclusive, it must be admitted that the wide diffusion of geophagy, though sporadic cases are on record for earlier periods, is only the result of more or less recent times.

In ancient prescriptions occurs earth from the roots of Jambū trees. This is a vegetable mold or black soil formed with decaying vegetal matter, such as is found in ponds and round the foot of trees (Hoernle, The Bower Manuscript, p. 149). A baked clod of clay with other ingredients was kept in water to relieve morbid thirst (*ibid.*, p. 137).

Indian physicians mention a kind of chlorosis (*pāṇḍuroga*) as being caused by the consumption of earth (Jolly, Indische Medicin, p. 86, who unfortunately does not say which physicians, the older or more recent ones).

The symptoms which appear in confirmed and habitual geophagists in India are usually reported as the face being unnaturally swollen or puffed, the abdomen distended, the limbs shrunk except at the joints which appear enlarged and are said to be painful.

Swelling-up of the face and abdomen may result from clay-eating for a period of twelve months.

White-ants' nests constructed of soft, fine earth, generally of a reddish black color, are consumed in India in the same manner as in Africa. Coolies of Assam are disposed toward white-ant soil taken from the center of the nest, white ants themselves being included as a delicacy (Hooper and Mann, p. 257). Among the mountain tribes of Travancore the men, not the women, eat this earth with the ants inside the cells, sometimes adding honey to it. It is taken, not in small medicinal doses, but in rather large quantities. No evil effects have been noticed to follow its use (*ibid.*, p. 259).

Steatite or soapstone ground to powder and mixed with flour has served in India as a regular famine food, in the same manner as in China (above, p. 124).

Consumption of small quantities of earth from holy places is prevalent throughout India. Such sacred earth is supposed to have healing properties. The followers of the Vaishnava sect keep in their houses the earth of the sacred river Jumna. At the close of their daily worship, a pinch of this earth is placed on the tip of the tongue and swallowed. There is a hill a few miles from Madras; and one particular spot in it is considered sacred, and the earth found there is credited with miraculous, curative properties. Those who visit the hill on a pilgrimage take a handful of this earth along, making it into pills used for various internal disorders as occasion arises (Hooper and Mann, p. 259).

He who is especially interested in the subject should not fail to read the valuable monograph of Hooper and Mann who have dealt with geophagy in India almost exhaustively.

The following interesting case is reported by E. Thurston (Omens and Superstitions of Southern India, 1912, p. 38): "Some years ago Mr. H. D. Taylor was called on to settle a boundary dispute between two villages in Jeypore under the following circumstances. As the result of a *panchāyat* ('council meeting'), the men of one village had agreed to accept the boundary claimed by the other party if the head of their village walked round the boundary and eat earth at intervals, provided that no harm came to him within six months. The man accordingly perambulated the boundary eating earth, and a conditional order of possession was given. Shortly afterwards the man's cattle died, one of his children died of smallpox, and finally he himself died within three months. The other party then claimed the land on the ground that the earth-goddess had proved him to have perjured himself. It was urged in defence that the man had been made to eat earth at such frequent intervals that he contracted dysentery, and died from the effects of earth-eating."

According to W. C. Smith (The Ao Naga Tribe of Assam, 1925, p. 33), "the Ao eat a whitish clay which they say is salty. The women use it more than the men. The Lakhers eat it and declare it can sustain a man without food for thirty-six hours, and women soon to become mothers are very fond of it."

L. and C. Scherman (Im Stromgebiet des Irrawaddy, p. 55, München, 1922), visiting a bazar at Yawnghwe in the Southern Shan States, found among the articles offered for sale also edible earth or more exactly gray, yellow and reddish clays.

Among the Chin of Upper Burma it is customary to eat earth as a sign of swearing to tell the truth, and earth is administered to witnesses giving evidence in a criminal case. This is considered a very binding oath and more likely to extract the truth from a Chin than anything else (Gazetteer of Upper Burma and the Shan States, I, pt. 1, p. 472, Rangoon, 1900). In a similar manner it was formerly customary among the Angami Naga tribe in rendering an oath to snatch up a handful of grass and earth, and after placing it on the head, to shove it into the mouth, chewing it and pretending to eat it (J. H. Hutton, The Angami Nagas, 1921, p. 146; cf. also J. P. Mills, The Lhota Nagas, 1922, p. 103).

In Siam, it is said, people consume steatite which consists of 65.6 per cent silic acid, 30.8 per cent magnesia, and 3.6 per cent oxid of iron (Altheer, p. 90).

N. Annandale (Fasciculi Malayenses, Anthr., pt. II, p. 62) has observed that both Malay and Siamese women eat a kind of earth dug out of the banks of a river and roasted; this is administered as a tonic.

CENTRAL ASIA AND SIBERIA

The Tibetan Kanjur contains a translation of the Buddhistic work *Vinayavastu* in which is embodied a curious story concerning the earlier periods of the world. In the course of these supposed periods a gradual deterioration of man and his foodstuffs is believed to have taken place. First there was the "sap of the earth" (Tibetan *sa-i bčud*, Sanskrit *pṛthivīrasa*) of excellent color, fragrance, and flavor, in color resembling butter, in taste like honey. The bodies of the spiritual beings who partook of this substance waxed hard and heavy and lost their fine luster, whereupon darkness arose in the world. Then originated sun, moon, and stars, and in consequence day, night, months, and years. Men subsisted on that earthly food and reached a high old age. Those who consumed but little were beautiful in appearance, but those who ate too much of it were ugly. The former grew haughty and despised the ugly. The sap of the earth vanished in the wake of this quarrel, and was replaced with an "earth grease or oil" (Tibetan *sa-i žag*, Sanskrit *pṛthivī-pārvataka*, Mongol *gadzar-un tosun*, "earth oil or butter"), which served as food. The same happens as previously, and the earth oil disappears to give way to vegetable foods. This legend has first been excerpted from the Kanjur by A. Schiefner (Über die Verschlechterungsperioden der Menschheit nach buddhistischer Anschauungsweise. Bull. histor.-philol. de l'Académie de St.-Pétersbourg, IX, No. 1, 1851).

A kind of eatable clay is reported from Tibet by Ma Shao-yün and Sheng Mei-k'i in their *Wei Ts'ang t'u shi*, an account of Tibet written in 1792. Near the monastery rDo-rje-'dra, not far from the celebrated temples of bSam-yas (southeast of Lhasa), there is a mountain with a cavern containing an eatable white clay, which has a taste like *tsamba* ("roasted barley-flour," the staple food of the Tibetans). Whenever clay is removed, it will grow again. The cavern must be entered with candles. Behind it there is a large lake (Bitchourin and Klaproth, Description du Tubet, pp. 131-132; Rockhill, *J.R.A.S.*, 1891, p. 267). According to Rockhill, this clay is styled *sa rtsam-pa* ("earth tsamba").

Earth is also used as a medicine in Tibet; *sa smug* is a dark red earth employed medicinally.

The Mongol chronicler Sanang Setsen relates in regard to Öljäi Ilduchi, who lived toward the end of the sixteenth century,

144

that he and his army, while on a warlike expedition, suffered from want of food, and were compelled to sustain their lives by eating of a stone, called *barkilda* (I. J. Schmidt, Geschichte der Ost-Mongolen, p. 217). The editor and translator of Sanang Setsen's work remarks (p. 413) that he does not feel certain whether this eatable stone or earth is identical with the Siberian "stone butter" described by Pallas. He also alludes to the earth eaten by certain tribes of South America. Nothing can directly be inferred from the Mongol term, which is isolated in this passage and is not known otherwise. Golstunski, at least, with reference to this word in his Mongol-Russian Dictionary, cites solely the text here in question. The word *barkilda*, which cannot be derived from any known Mongol stem, and which does not occur in Turkish, means also "aerolith," and is correlated with Tibetan *ka-tu* or *ke-tu* (that is, Sanskrit *ketu*). It may be, therefore, that the stone mentioned by Sanang Setsen was believed to be of celestial origin. It certainly is not identical with the "stone butter" of Siberia, which is a substance of vitriolic origin, first described, as far as I know, by P. J. von Strahlenberg (Das nord- und östliche Theil von Europa und Asia, 1730, p. 384).

P. S. Pallas (Reise durch verschiedene Provinzen des russischen Reiches, II, 1771, pp. 88, 656, 697; III, 1776, p. 258) found this "stone butter" in the Ural, near Tomsk, on the Yenisei and the Chilok. He explains it as plume alum or stone alum, a white yellowish substance of vitriolic origin flowing out of slate. Some inhabitants of Tomsk boiled from it an impure yellow vitriol which assumed a sand-like hardened shape and which was sold on the market—for industrial purposes only, as, for example, for dyeing leather black (Strahlenberg). According to Pallas, the "proper natural stone butter" is not so frequently gathered that Tobolsk and other Siberian towns could be supplied with it. At Krasnoyarsk only it was offered for sale in abundant quantity, being collected in the neighborhood of the town. It is described by him as very white and light in weight; when burnt at a flame, it flows easily, and when boiled, it emits red vitriolic fumes, while a light, very white and savory earth remains. Several puds of this earth were annually collected and sent to Krasnoyarsk, where the pound sold at from fifteen to twenty kopeks. The common people used this substance chiefly as a remedy in cases of diarrhoea and dysentery or for copious bleeding of lying-in women (cf. also Pallas, Neue nordische Beiträge, V, 1793, p. 290).

J. B. Müller (Les moeurs et usages des Ostiackes, in Nouveaux Memoires sur l'état présent de la Grande Russie ou Moscovie, II, p. 160, Amsterdam, 1725) writes, "On the highest mountains and rocks of Siberia is found an extraordinary mineral called by the inhabitants of the country *kamine masla* or stone butter. The heat of the sun causes it to flow down the rocks to which it is attached as chalk to walls. It is dissolved in water like salt, and is as strong as vitriol. They attribute to it many virtues and use it in several diseases, especially in dysentery. I believe that we ought not to get accustomed to this remedy, and I know of no one who has ever made use of it."

According to J. G. Georgi (Bemerkungen einer Reise im russischen Reiche im Jahre 1772, St. Petersburg, 1775), the stone or rock butter served also as a specific against syphilis. He reports also that there is in Kamchatka near the river Olontora and in several other localities a lithomarge clay which both the Tungusian tribes and the Russians eat, either alone or dissolved in water or milk. This substance, he concludes, produces in those people merely a light constipation which perhaps is wholesome to them in the spring when they eat an abundance of fish, which will cause diarrhoea. Georgi informs us also that in the countries located between the Volga, Kama, and Ural there is a sort of powdered plaster termed by the inhabitants "rock flour" or "celestial flour." This substance was mixed with flour in times of scarcity, but those who ate such bread almost always experienced fatal effects.

The Sungar picked up earth during earthquakes which do not infrequently occur around the Altai mountains, and placed it on the tongue of a parturient woman, believing that it was a good means of expediting birth and expelling the after-birth (P. S. Pallas, Samlungen histor. Nachrichten über die mongolischen Völkerschaften, I, p. 166).

G. W. Steller, in his famous "Beschreibung von dem Lande Kamtschatka" (1774, pp. 72, 324), speaks of *Sory officinarum* or so-called Siberian *kamenna masla* ("stone butter") and a soft bolus earth which tastes like cream and which is eaten; the latter he calls *semlanoi smetana* ("earth sour cream"). Like the Tungus around Okhotsk, he continues, the Itälmen and Koryak eat a kind of fine white clay which looks like cream and which is not devoid of an agreeable flavor, but is at the same time astringent. According to A. Erman (*Zeitschrift für Ethnologie*, III, 1871, p. 150), who himself visited Kamchatka, the so-called flowing clay or earth cream (i.e. the gelatinous detritus of a trachytic rock) was eaten there but exceptionally and only in certain places.

Mr. I. Lopatin, who has devoted many years of his life to investigations of the native tribes of eastern and northeastern Siberia, kindly informs me that in his experience clay-eating is not practised by any of these, but that some Tungusian tribes, such as the Oroche, Udekhe, and Olcha make use of clay as a medicine. He did not observe this practice, however, among the Golde with whom he is particularly familiar and to whom he has devoted a very interesting monograph. "Many times during my expeditions into the countries of these peoples," Mr. Lopatin writes me, "I saw small pieces of clay fashioned into cakes, about one and a half inch square and a quarter of an inch thick, and suspended from the roofs of their huts. On two occasions I watched the preparation of these clay cakes. An Udekhe woman made a sort of dough of the clay, and after having kneaded it well, she turned out two or three dozens of cakes somewhat resembling American crackers. When these clay crackers were sufficiently dried, she perforated each piece in the center and on both sides made four rows of cavities by pressing, about four or five in a row, whereupon she strung the cakes through the perforations in the center and suspended them under the roof of the hut. On another occasion I saw a man of the same tribe make such cakes which were of the same size and shape as previously. He said that a particular kind of clay, which is yellowish gray in color, must be used for this purpose. The clay cakes must be thoroughly dried before being consumed. They are kept under the roof for at least five or six months and in fact for two or three years before they are ready for use. Udekhe, Oroche, and Olcha believe that these cakes are very helpful in stomachic troubles and diarrhoea. In the event of such complaint these cakes are taken internally for a period of six or seven days. I wish to stress the point that these clay cakes are but seldom eaten by these people and exclusively as a remedy in case of illness."

It is certainly possible that this remedy is apt to stop diarrhoea; it is so employed elsewhere, for instance, in Sumatra (Heringa, p. 186), and as has been stated, by the natives of Queensland in Australia (above, p. 139). In our own time powdered clay has been recommended as a remedy for cholera (*Berliner Klinische Wochenschrift*, 1905, p. 750), and clay pills have been used for hemorrhoids (Hahneman, Chronic Diseases, II; Hooper and Mann, p. 269).

During his excavations conducted in Kamchatka in 1910-11 W. Jochelson (Archaeological Investigations in Kamchatka, p. 66, Carnegie Institution, 1928) found pieces of white clay in some of the excavations of dwellings, which he is inclined to think was eaten by the inhabitants. He refers to Krasheninnikow's "Description of Kamchatka" as mentioning white clay as a remedy for diarrhoea.

The edition consulted by him is the third in Russian, published at St. Petersburg, 1818. I have a German edition of this work (Lemgo, 1766) in which this passage is not contained, neither in the chapter on Diseases and Remedies nor in the chapter on Food and Drinks of the Kamchadal; in discussing the different kinds of earth found in Kamchatka (p. 97) no reference is made either to clay-eating. Of course I do not doubt that in the edition consulted by Jochelson the passage in question is contained, but what I venture to call into doubt is that the clay pieces found by him in the deserted dwellings were really intended for internal medicinal use. Unfortunately, Jochelson has neglected to state the essential point, and this is, of what shape these clay pieces were, whether they were shaped into a certain form by human hand or just odd pieces in their natural state. If, for instance, they were like the clay cakes described by Lopatin, this would constitute sufficient evidence for his conclusion; but if not artificially fashioned in some way or other, the hypothesis is not convincing, or the case remains at least doubtful.

According to W. Bogoras (The Chukchee, p. 200, Jesup North Pacific Expedition, VII, 1904), "the Reindeer Chukchee as well as the Lamut and the Koryak in Kamchatka occasionally use as food a kind of white clay, which is called 'earth fat' (*nute-echen*). This, of course, is eaten only in moderate quantities, mixed with broth or with reindeer-milk." If this be true, the clay consumed cannot, of course, be designated as a food, but is rather a condiment added to articles of food.

L. J. Sternberg (The Gilyak, *Ethnograficheskie Obozränie*, 1905, p. 17) mentions a dish of the Gilyak consisting of the gluey broth of fish-skins, seal's fat, berries, rice, and sometimes of minced dried fish, being mixed with dissolved white clay; this dish is favorite for treating guests.

H. von Siebold (Ethnol. Studien über die Ainos, 1881, p. 37, Suppl. Z. Ethn.) was told that the Ainu occasionally eat a clay mixed with herbs and roots; he had no occasion to see this himself. Hooper and Mann (p. 251), without citing any source, assert, "Among the Ainu, the aborigines of Japan, there is a kind of clay which is eaten to a considerable extent, mixed with fragments of the leaves of a plant and used as an ingredient in the preparation of soup. The clay occurs in a bed in the valley of Tsie-tonai ('eat-earth valley') on the north of the coast of Yezo. It is of light-gray color and fine consistency, and is consumed, not as a matter of necessity, but because it is believed to contain some beneficial ingredient." The above name should be written Chi-e-tonai; *chi* means "earth," *e* "to eat"; but a word *tonai* is not given in the Ainu Dictionaries of Batchelor and Dobrotvorski. In the works of J. Batchelor, the best informed authority on the Yezo Ainu, no reference is made to consumption of earth in a pure state, nor have I learned anything to this effect among the Saghalin Ainu. This, of course, does not mean that the habit does not exist, or might not formerly have existed. There is, however, an Ainu practice recorded by Batchelor which offers a striking parallel with what is found among the Pomo and Hopi, as well as among the natives of western Australia (above, p. 138).

The bulbs of *Corydalis ambigua* (Ainu *toma*, Japanese *engosaku*) are extensively eaten by the Ainu, especially those in the Ishikari valley of Saghalin Island and in the southern Kuriles. The bulb has a slightly bitter taste which is removed by repeated boilings in water. In Etorup, the Ainu boil the bulbs with a certain kind of earth to remove its bitterness. They are eaten either simply boiled or mixed with rice. In Saghalin, it is said, they are cooked generally with the fat of seals (J. Batchelor and K. Miyabe, Ainu Economic Plants, No. 48, *Transactions of the Asiatic Society of Japan*, XXI, 1893, p. 215). The Ainu also feed on acorns (*ibid.*, No. 108) which are usually boiled and occasionally roasted, but earth is not applied to these as by the Pomo and the peasants of Sardinia.

The Ainu have traditions of famines in early times when people were dying from want of food, and this seems to be one of their typical forms of legend two of which are recorded by J. Batchelor (Specimens of Ainu Folk-lore, *Transactions of the Asiatic Society of Japan*, XVI, 1888, pp. 112-122). In these no allusion is made to earth-eating; in fact, no famine food is mentioned. It is known, however, that the ancient Ainu subsisted a great deal upon the stem and leaves of the mugwort (*Artemisia vulgaris*) which has been the means of keeping them alive throughout more than one famine (J. Batchelor and K. Miyabe, Ainu Economic Plants, No. 78).

In mixing earth with certain foodstuffs there is agreement between the Ainu, Gilyak, and Chukchi; and this perhaps may be regarded as an ancient feature of the culture of the Palaeo-Asiatic tribes. Considering the further fact that earth is still eaten by Tungusian tribes and that a certain kind was consumed by the ancient Kamchadal, there is a continuous area in northeastern Siberia for the practice of earth-eating. It will be seen that this continues in points of the far north of North America.

PERSIANS AND ARABS

One of the infernal punishments of the Parsis was that a man, who used false measure and weight and who adulterated his merchandize, was compelled to eat dust and earth meted out to him on a scale (M. Haug, Über das Ardāi Vīrāf nāmeh, 1870, p. 25).

Ibn al-Baiṭār (1197-1248), an Arabic scholar born at Malaga, Spain, and author of a famous work on pharmacology, discusses eight kinds of medicinal earth (L. Leclerc, Traité des simples, II, 1881, pp. 421-427; for a general appreciation of this work see Baron Carra de Vaux, Les penseurs de l'Islam, II, 1921, pp. 289-296). The eight kinds are the terra sigillata, Egyptian earth, Samian earth, earth of Chios, Cimolean earth or pure clay (cimolite), earth of vines called *ampelītis* (Pliny XXXV, 56) or *pharmakītis* from Seleucia in Syria, Armenian earth, and earth of Nishapur. A great deal of the information given by the Arabic scholar is derived from Dioscorides and Galen. Earths used in medicine were but rarely taken internally, but usually applied locally; these cases, therefore, do not come within the subject of this monograph. Reference is made here only to Ibn al-Baiṭār's notes as far as they relate to earths administered internally. It appears that the sigillated earth, which will be more fully discussed under the heading "Europe," was regarded by Avicenna as an antidote and having the tendency to eject poisons from the system when taken before or after the act of poisoning. Under Cimolean Earth, Ali Ibn Mohammed is quoted as saying that this soft earth, called *al-hurr*, green in color like verdigris, is smoked together with almond bark to serve as food when it will turn red and assume a good flavor and that it is but rarely eaten without being smoked. The Cimolean earth is named for Cimolus (Greek Kimolos), one of the Cyclades, also called Argentiera (cf. Dioscorides V, 175; Pliny XXXV, 57; E. Seidel, Mechithar, 1908, No. 204).

The Armenian earth (boļe armenic), according to Ishak Ibn Amrān, was salutary in cases of bubonic plague, being administered both externally and internally. The same is affirmed by Leo Africanus (in Ramusio, 4th ed., 1588, fol. 10b; French ed. by Schefer, I, p. 114) with reference to Barbary, save that there the Armenian earth was applied externally to the bubos. At present no longer used, this article (Latin *bolus armena*) was renowned in ancient times and extensively traded from Armenia, where it is abundant. It was introduced into medical practice by Galen (Seidel, Mechithar,

150

No. 132). It is a soft earth, greasy to the touch, strongly adhering to the tongue, very fragile, generally of a yellowish brown color, sometimes of a fine flesh red. According to J. Chardin (Travels in Persia, ed. Sykes, p. 164), it also occurred abundantly in Persia, where it was especially used by women in washing their heads. According to W. Ainslie (Materia Indica, I, 1826, p. 43), it was brought from the Persian Gulf to India, where the Tamul practitioners prescribed it as an astringent in fluxes of long standing and supposed it to have considerable efficacy in correcting the state of the humors in cases of malignant fever. Its constituent parts, according to Ainslie, are silica 47 per cent, alumina 19 per cent, magnesia 6.20 per cent, lime 5.40 per cent, iron 5.40 per cent, water 7.50 per cent.

The most celebrated of all edible clays was that found near Nishapur in Persia. The Arabic historian al-Ta'alibī (A.D. 961-1038), who calls it al-naql, writes that it occurred exclusively at Nishapur and was exported from Zauzan into all quarters of the globe to places near and distant; a raṭl of this clay was sometimes valued at a dinar in Egypt and in the Maghreb (E. Wiedemann, Zur Mineralogie im Islam, Sitzber. phys.-med. Soz. Erlangen, 1912, p. 242). According to Edrīsī (Jaubert, Géographie, I, pp. 452, 454), there was two days' journey from Caneïn, or Caïn, on the road leading to Nishapur, a kind of brilliant white clay, called ṭīn el-mehāji and exported for purposes of consumption to distant regions. The same fact is mentioned by Ibn Haukal (W. Ouseley, Oriental Geography of Ebn Haukal, 1800, p. 223).

Ibn al-Baiṭār devotes much space to the clay of Nishapur, chiefly relying upon Ali Ibn Mohammed and the celebrated physician Mohammed Ibn Zakkariyā al-Rāzī (i.e. born at Rei, the ancient Rhages), known as Razes, of the tenth century. This clay is described as being white, of an agreeable taste, taken either in its natural state or roasted. It is sweet to the taste, and soils the lips on account of its great softness; on the other hand, it is said that its flavor is somewhat saline, but that exposed to a fire it will lose this saline property and grow sweet. There are people who pound it and soften it with rose-water and a little camphor and who then shape this compound into bread loaves, tablets, or other forms. Others scent the clay with musk, camphor, or some other aromatic, and thus take it after having indulged in wine to perfume their breath and to assuage the heat of the stomach.

According to Razes, the clay of Nishapur fortifies the heart and combats nausea. It stops vomiting (or is used as an anti-emetic) and especially counteracts nausea provoked by sugared and greasy foods. Razes holds that the Nishapur clay is not apt to cause obstructions in the reins and bladder, as it happens with other clays. In his "Treatise on Clays" Razes tells an interesting story of how he cured an individual seized by a very grave choleric affection accentuated by violent fits of vomiting and cramps. The usual remedies were of no avail; he administered to the patient powdered Nishapur clay in doses of thirty drams, three times, twice in a decoction of sweet apples and once in a decoction of sweet rush (*Andropogon schoenanthus*), and his nausea and indigestion were immediately relieved. What was still more marvelous was that the patient found himself stronger and merrier than before as though the medicine had nourished him.

Razes further maintains that he employed Nishapur clay in treating affections of the stomach, as well as in cases of nausea and indigestion caused immediately after a meal. This convinced him that it was necessary to administer a small dose of clay after a meal, which relieved the indigestion, chills in the abdomen, and the tendency to vomit. He considers Nishapur clay as a capital remedy for the treatment of affections of the stomach especially with patients who apparently have no obstruction of the liver or contraction of the bowels. In these cases this remedy is rarely harmful; on the contrary, it seems that the body gains weight. He administered this clay also to individuals who suffered from a considerable secretion of saliva and to all patients seized by a ravenous appetite—all these were radically cured.

The modest and unadorned report of Razes inspires confidence and merits full credence.

At present, the habit of clay-eating is widely diffused over Persia. It has developed into a passion among those people who have taken to it, and these swallow considerable quantities of clay. The habit extends to both sexes, notably to women, and is said to be restricted to common people, while it is rare among the better classes. The reasons advanced by the people are that "it tastes well" and "satiates their hunger." The clay fiends are characterized by leanness and sallow, earth-like complexion. Edible clays form a not unimportant article of trade, and are sold in the bazars of most cities. Two edible clays are especially reputed—one traded from Kirman and called *ghel-i-giveh*, and another from Kum under the name *ghel-mahallat*. The two sorts have been analyzed and described by Goebel (see also Ehrenberg I p. 184; II p. 36). According to Goebel, these clays contain no nutritive substances, but some agents which have an effect on the nervous system. Their action is mechanical, not chemical. They leave the organism without exerting a disturbing influence on the composition of the blood in case indulgence has not been excessive. Tietze (Die Mineralreichtümer Persiens, Jahrbuch der k.k. geol. Reichsanstalt Wien, 1879, p. 654) gives an analysis of three kinds of earth from Persia.

J. L. Schlimmer (Terminologie médico-pharmaceutique français-persane, p. 299, Teheran, 1874) writes that "geophagy is a general habit among the women of Persia, even when they are not pregnant. The Persian physicians attribute this 'idiosyncrasy' to the presence of intestinal worms, which for the rest is far from being proved. Among young children, however, this particular habit is often connected with the existence of intestinal worms, and in this case vermifuges administered in small doses continued for a long time and the simultaneous use of wine will overcome this 'depraved appetite'; but a cure becomes difficult in cases where geophagy is the concomitant symptom of scrophulous diathesis when the young patients assume a cachectic appearance, which is quite characteristic of their condition."

Polak (Persien, II, p. 273) observes that the Persians have trained their taste to such an extent that they discriminate between various kinds of clay without hesitation.

An earthy, soap-like substance that the natives term *chunniah* is obtained from lakes not far from Halla. It is largely eaten by the women of Sind (J. Wood, Journey to the Source of the River Oxus, 1872, p. 19). In Lasch's article (p. 220) this *chunniah* has been transformed into *tschamiah*.

Hajaj, a military officer, who served under the Caliph Abdul Malik (A.D. 685-705), was in the habit of eating clay. Determined to wean himself from this habit, he consulted Theodocus (Theodunus or Tiaduq), a renowned physician, as to the proper remedy. "The will of a man of your mold," Theodocus responded. Hajaj then ceased to eat clay (L. Leclerc, Histoire de la médecine arabe, I, 1876, p. 83).

As in China, earth-eating was also connected with religious beliefs among the Arabs and the Mohammedans of India. Hooper and Mann (p. 259) inform us that dust from the tomb of the prophet is an auspicious article, said to be a cure for every disease. According to E. W. Lane (Manners and Customs of the Modern Egyptians, 5th ed., I, p. 323), who received such specimens from a Mecca pilgrim, they come in oblong, flat cakes of a grayish earth, each about an inch in length and stamped with an Arabic inscription, "In the name of Allah! Dust of our land [mixed] with the saliva of some of us." They are alleged to be composed of earth obtained from the surface of the grave of the Prophet and to be a cure for every disease, and are sold at Mecca. A cake of this kind is sometimes worn as an amulet in a leather case. It is also formed into lumps of the size and shape of a pear, and is suspended from the railing which surrounds the monument erected over the grave of a saint.

Sir Richard Burton (Pilgrimage to Al-Madineh and Mecca, I, p. 415) found in Arabia a yellow loam or bole being eaten by anaemic women. It was used as a soap in some parts of the East, and was supposed to have some miraculous properties owing to the Prophet having employed it with success as a medical agent.

In 1612 William Lithgow visited the cave near Bethlehem in which the Virgin Mary, at the time of the persecution of Herodes, took refuge, and gives this account (Totall Discourse of the Rare Adventures and Painefull Peregrinations, p. 247 of the edition reprinted at Glasgow, 1906): "The earth of the cave is white as snow, and hath this miraculous operation, that a little of it drunke in any liquor, to a woman, that after her childbirth is barren of milke, shall forthwith give abundance: which is not onely availeable to Christians, but likewise to Turkish, Moorish, and Arabianish women, who will come from farre countries, to fetch of this earth. I have seene the nature of this dust practised, wherefore I may boldly affirme it, to have the force of a strange vertue: Of the which earth I brought with me a pound weight, and presented the halfe of it to our sometimes Gracious Queene Anne of blessed memory, with divers other rare relicts also, as a girdle, and a paire of garters of the Holy Grave, all richly wrought in silke and gold, having this inscription at every end of them in golden letters, Sancto Sepulchro, and the word Jerusalem, etc."

The legend goes that the milk of the Virgin when she took refuge in that grotto spurted against the rock, and ever since this earth has been capable of increasing the milk of both women and animals. In the first place, of course, the question is here of earth-eating (cf. the analogous custom in Australia of using earth externally as a means of promoting lactation, above, p. 139).

The Italian designation for diatomaceous earth, *latte di luna* ("lunar milk"), may be connected with this belief (other Italian terms for it are *agarico minerale* and *farina fossile*).

In an interesting study entitled "Mohammedan Saints and Sanctuaries in Palestine" (*Journal of the Palestine Oriental Society*,

V, 1925, p. 188), T. Canaan writes, "Christians as well as Moham-medans use the soft whitish stones of the milk-grotto in Bethlehem to increase mothers' milk. The stones are rubbed in water and given to the nursing women. It is supposed that the holy family took refuge in this cave where a drop of Mary's milk fell to the floor."

The same author reports that plaster, stones, and sweepings of many shrines are used medicinally. Some of the earth of a certain locality made with oil into a paste cures sores of the head. Earth gathered from another holy place is dissolved in water, and given to cattle will guard them against disease. Everything that belongs to or comes in contact with a saint or his shrine is believed to receive some of his power which may be transmitted to others. Thus the earth of a saint's tomb (likewise stones, water, grass and trees) is believed to possess supernatural power.

AFRICA

T. F. Ehrmann (Geschichte der merkwürdigsten Reisen, VII, 1793, p. 70; after J. Matthew's Journey to Sierra Leone 1785-87) speaks of a white, soap-like earth found here and there in Sierra Leone and so fat that the Negroes frequently eat it with rice, because it melts like butter; it is also used for white-washing their houses. Ehrmann adds, "A curiosity which merits a closer investigation." The same clay was also reported by Golberry (1785-87) from Senegambia and described by him as a white, soap-like earth as soft as butter and so fat that the Negroes add it to their rice and other foods which thus become very savory. This clay is said not to injure the stomach (Lasch, p. 216).

In the third edition of his "Ansichten der Natur" (1849, I, p. 167), A. von Humboldt writes, "In Guinea the Negroes eat a yellowish earth which they call *caouac*. When carried as slaves to the West Indies, they try to procure there a similar earth. They affirm that earth-eating is quite harmless in their home country. The *caouac* of the American islands, however, makes the slaves sick. Therefore, earth-eating was forbidden there, though in 1751 earth was secretly sold in the markets of Martinique. The Negroes of Guinea assert that in their country they eat habitually a certain clay whose flavor gratifies them without being harmed by it. Those addicted to eating *caouac* are so fond of it that no punishment can prevent them from swallowing earth." Humboldt's information is derived from Thibault de Chanvallon (Voyage à la Martinique, p. 85).

Ehrenberg (II pp. 15, 53) has refuted the idea propounded by Thibault de Chanvallon that the Guinea Negroes generally and habitually eat a red earth, without endangering their health. Ehrenberg's conclusions are based on the observations of many missionaries stationed at many points of the Gold and Slave Coasts during more than thirty years. On the whole, earth-eating occurs there but seldom, chiefly on the part of children and thoughtless persons. Ehrenberg (p. 19) has also analyzed a clay specimen from Cuba and arrived at the conclusion that the *caouac* substitute of the West Indies alleged to be so harmful does not appear to be more harmful than the earth of Guinea.

In regard to the Congo region we are well informed by Catholic missionaries who have paid special attention to this subject. It is noteworthy that geophagy prevails among some tribes of the Congo

156

and is absent among others. F. Gaud (Les Mandja, p. 151, Brussels, 1911) writes, "At the present time it is only during famines that the Mandja (in the French Congo) gather the earth of termites'-nests and consume it mixed with water and powdered tree-bark. This compound is said to assuage the tortures of hunger in a singular manner. We think that this effect must be attributed not only to the physical action resulting from the filling of the stomach, but also to the absorption of organic products existing in the clay. It is in fact known that the walls of the termites'-nests are built by the female workers with tiny clay balls kneaded by them by means of their saliva. It would not be surprising that this saliva contains formic acid."

The buildings of the great ants (*Termes bellicosus*) are constructed from red ferruginous clays in the shape of mushrooms (see illustration in G. Schweinfurth, The Heart of Africa, I, p. 349).

C. van Overbergh (Les Basonge, p. 151) gives the following information: "The Baluba frequently eat *pembe* or white earth. Result: appalling leanness and swelling of the abdomen. Pregnant women do not eat white earth. In general women eat earth; I have never seen men eat it, but I do not guarantee that men will not eat it. Another observer, Michaud, states that he saw men and women alike eat earth. It appears that a person who has once tasted this earth becomes infatuated with it, but dies in consequence."

Another missionary among the Baluba, R. P. Colle (Les Baluba, Congo belge, I, p. 131), states, "A certain number of children display a very lively desire to eat the embers of the hearth and clay. It is that firm, fat, white and unctuous clay which serves for the manufacture of pottery. Perhaps they are driven to this by the need of salt. The embers in fact contain potash; and the clay in question, a slight quantity of magnesia. The result of this habit is the disease called *le carreau*."

While among the Baluba, as stated, pregnant women do not consume earth, it is eaten by pregnant women of the Mayombe (C. van Overbergh, Les Mayombe, p. 121).

"At Nouvelle-Anvers in the Belgian Congo, it is reported by eye-witnesses, can be procured for five Centimes the kilo a sort of clay of which the natives are very fond. This is a yellow earth of agreeable odor which contains silicic acid, oxid of aluminum, sodium, and a little iron" (C. van Oberbergh, Les Bangala, état indigène du Congo, p. 123).

R. Schmitz (Les Baholoholo, Congo belge, p. 65) writes, "Earth is not alimentary. Once in a while one encounters a case of geo-mania, a sick person who has a passion for the wall of a hut or an ants'-nest and who eats of it till he dies."

On the other hand we read, "Not the slightest indication of geophagy among the Mangbetu, Mangbellet, and Mobadi" to which another observer adds, "Save among a few sick" (C. van Overbergh, Les Mangbetu, p. 181).

"The Ababua does not eat any species of earth" (J. Halkin, Les Ababua, Congo belge, p. 151). "No case of geophagy exists among the Warega" (Delhaise, Les Warega, Congo belge, p. 87).

According to Winwood Reade, the famous author of "The Martyrdom of Man," a white clay is frequently chewed or drunk in solution on the Gold Coast, the young people taking it as a sweetmeat, and the old people as a medicine (*Journal Anthrop. Institute*, X, 1881, p. 467).

The following interesting account of geophagy with reference to the people of Batanga is given by W. L. Distant (*Journal Anthrop. Institute*, X, 1881, p. 467):—

"A somewhat curious instance of this custom came before me at Batanga in May, 1880; and subsequent inquiry has enabled me to throw some light upon it. From what I could gather while at Small Batanga, the custom seems to prevail all along the coast as far as the island of Corisco, where I believe it is also known, and perhaps it extends farther south. I met with it first at Babani, where there occurs a deposit of yellowish red clay, containing about 15 per cent of iron and a considerable quantity of mica and some quartz particles, but there is evidently a large quantity of organic matter in it. This clay is made up into balls of about five inches in diameter, and baked over a slow fire. When quite dry and ready for use, a small portion is broken off, and placed in the hollow of any smooth leaf and reduced to powder between the finger and thumb. The leaf is then gently shaken in order to cause the harder and more gritty particles to fall aside. These are carefully removed, and the residue, consisting of a fine powder, is transferred to the mouth, masticated, and swallowed. I was informed that the men use it while on a long journey, when they do not wish to stop in order to cook food. As, however, they travel far without carrying something in the way of provisions that can be eaten readily, this scarcely accounts adequately for the origin of the custom. Some inquiries made at Camaroons elicited the following additional information. The custom is known there, but does not exist to the same extent, or in the same manner as at Batanga. The material used is a very dirty earthy clay, with but little iron and no mica, and is derived from a deposit on the banks of the rivers. When baked in the sun, it becomes very hard; and, indeed, is sometimes used in the construction of houses. The men sometimes, but seldom, eat it; but I am told the women, during the time of pregnancy, when they are supposed to be assailed by very unnatural appetites, use it largely. Is it the result of inheritance, or merely from the force of imitation, that the custom is almost universal among the Camaroons children? I am told that all of them eat it, even those belonging to the mission, who are well fed, and are strangers to the sensation of hunger. By way of test, I showed some of them a small piece of the Batanga earth. They looked at it for a moment as if to make sure of it, then eagerly besought me to give them some. I gave them what I had in my hand, and they greedily swallowed it, afterwards expressing a desire that, as the kind I had given them was so nice, they would like some more. These children had just supped, and their evident appreciation of the clay could, therefore, hardly be connected with hunger, and would seem to indicate an appetite, or at least a liking, however unnatural, not much related to the desire for food. One of those children, I was informed, usually took a piece of the clay to bed with her, but this child, though well-fed, was always hungry."

The Negro slaves imported from West Africa to America continued the habit of earth-eating, especially in the West Indies. P. Browne (Civil and Natural History of Jamaica, 1756, p. 64), who estimated the number of Negroes living in the island at that time at 120,000 (p. 24), describes a peculiar sort of earth that runs in veins, and is chiefly found in marly beds. "It is of different colors," he writes, "but these generally answer to that of the layer wherein it is found; it is apparently smooth, and greasy, and somewhat cohesive in its nature; but dissolves easily in the mouth. The Negroes, who make frequent use of this substance, say, that it is sweetish; and many get a habit of eating it to such excess, that it often proves fatal to them. It is the most certain poison I have known, when used for any length of time; and often enters so abundantly into the course of the circulation, as to obstruct all the minute capillaries of the body; nay, has been often found concreted in the glands, and smaller vessels of the lungs, so far as to become sensibly perceptible to the touch. It breaks the texture of the blood entirely; and for many months before they die, a general languor affects the machine, and all the internal parts, lips, gums, and tongue, are quite pale, insomuch, that the whole mass of their juices seems to be no better than a waterish lymph. It is probable they are first induced to the use of this substance (which is generally well known among them) to allay some sharp cravings of the stomach; either from hunger, worms, or an unnatural habit of body."

It is even suggested that Negro and Indian slaves took to earth in despair as a means of slow suicide and that the Carib slaves ate earth whenever they were punished or mistreated (Ehrenberg II p. 16, after W. Irving, Columbus). What is more interesting to us is the ceremonial use of earth on the part of American Negroes.

G. Hughes (Natural History of Barbados, 1750, p. 15), speaking of the ordeals of the Negroes of Barbados, writes, "They take a piece of earth from the grave of their nearest relations, or parents, if it can be had; if not, from any other grave. This being mingled with water, they drink it, imprecating the divine vengeance to inflict an immediate punishment upon them; but in particular, that the water and mingled grave-dust which they have drunk (if they are guilty of the crime) may cause them to swell, and burst their bellies. Most of them are so firmly persuaded that it will have this effect upon the guilty, that few, if any (provided they are conscious of the imputed crime), will put the proof of their innocency upon the experiment."

In the disease known as cachexia africana (*mal d'estomac* of the French), common among the Negroes of the West Indies and Guiana, an essential symptom is a generally depraved appetite and an ungovernable determination to the eating of dirt. According to Cragin (p. 358), "the only appreciable signs of mental activity during the course of this disease are the crafty and cunning plans which the patient most subtly matures and as stealthily executes to procure his desired repast. This consists usually of charcoal, chalk, dried mortar, mud, clay, sand, shells, rotten wood, shreds of cloth or paper, hair, or occasionally some other unnatural substance. The patient, when accused of dirt-eating, which is too often urged as a voluntary crime rather than an irresistible disease, invariably denies the charge. As curative means, neither promises nor threats (even when put in execution), nor yet the confinement of the legs and hands in stocks and manacles exert the least influence and their preventive effect is as temporary as their employment; so great is the depravity of the appetite, and so strongly are the unfortunate sufferers under this complaint subjected to its irresistible dominion. A metallic mask or mouthpiece secured by a lock is the principal means of security for providing against their indulging in dirt-eating, if left for a moment to themselves, nor does this effect a cure or save the life of the patient."

Cragin quotes from a work "Practical Rules for the management and medical treatment of Negro slaves in the sugar colonies, by a professional planter" (London, 1811) the statement, "We find that Negroes laboring under any great depression of mind, from the rigorous treatment of their masters, or from any other cause, addict themselves singularly to the eating of dirt."

Cragin is inclined to think that the disposition to eat chalk, clay, and earth arises from a purely physiological cause, an acidity of the stomach, not from a melancholic or any other affection of the mind. He concludes that the effect has been mistaken for the cause. As one of the facts to prove his position he cites the following: persons living on the same plantation, perhaps on the identical section of the same plantation, on which they were born and reared, with all their friends around them, and by indulgent masters and owners, who are themselves the real slaves, while the owned are only nominally so, provided with ample food, raiment, and if necessary, medical aid, are also subject to this malady.

Dr. Melville J. Herskovits of Northwestern University informs me that the Bush Negroes and the Negroes of the coastal region of Surinam eat earth only on ceremonial occasions. Many times he saw women who were possessed by the spirit rolling lumps of a white sacred clay (*pemba doti*) in their hands during the time of possession and repeatedly licking their hands or the clay.

According to Major J. O. Browne (The Vanishing Tribes of Kenya, 1925, p. 104), "instances occur from time to time of earth-eating, but they are always associated with an outbreak of ankylostomiasis, of which, of course, it is a well-known symptom."

F. Fülleborn (Das deutsche Njassa- und Ruwuma-Gebiet, 1906, p. 115) has the following notice: "In the south of German East Africa earth is eaten, although not so generally as it is reported with reference to Asiatic and American peoples. The fact that pregnant women among the Wakissi are said to eat earth once in a while would mean nothing, since the pregnant often have desires for strange things. I was witness of how at Wiedhafen on the Nyassa relatives brought to a prisoner together with his daily ration a piece of loam (not a special kind, but a quite common one apparently detached from the wall of a hut) of which he ate with seeming enjoyment. This, it is true, was the only case observed by myself, but Elton reports in regard to the Wassangu that he saw there young children and women emaciated into skeletons who had contracted a disease from earth-eating, and Johnston reports similar cases from British Central Africa."

Ehrenberg (II p. 19) received also a clay from Abyssinia with the remark that it was eagerly eaten by women.

In Morocco the earth from the tombs of saints is used in the healing of disease. It is called the *hanna* or *henne* of the saint. It is made into plasters to be applied to the skin or into amulets. It is also moistened with the water of the sanctuary, and then becomes a potion which will cure the most obstinate evils. It is known that the objects concealed in a sanctuary are never stolen, thanks to the protection of the saint who would blind, paralyze or instantly slay thievish intruders. By making a bag containing earth from a saint's tomb and suspending it in a tree, on the walls surrounding a garden, in the flour-chest, or in a shop which remains unguarded at night, the saint is obliged to protect these places; he is transformed into a veritable guardian and is compelled to punish the thief as though his own sanctuary had been violated (Legey, Essai de folklore marocain, 1926, p. 10).

During her stay in Taourirth Abdallah, which is one of the Kabyl towns in the foothills of the Atlas, in Algeria, in 1928, Miss Georgiana B. Such, as she kindly informs me, noticed numerous cases of clay-eating and always in individuals obviously suffering from some more or less obvious polyglandular disturbance or insufficiency—many had goiter; all those examined by her were suffering from intestinal parasites, many had tapeworms, and all were undernourished.

L. Rauwolf (Beschreibung der Raiss inn die Morgenländer, 1583, p. 32) writes that in Tripoli an ash-colored earth called *malun* was used for washing the head and that another earth called *iusabor* was frequently eaten by women as among us the pregnant eat coal and other things.

R. F. Burton (Lake Regions of Central Africa, II, p. 28) writes that clay of ant-hills, called "sweet earth," is commonly eaten on both coasts of Africa. According to Major Tremearne (The Ban of the Bori, p. 80), the women of Nigeria eat white earth during the first three months of pregnancy to insure a successful delivery, but earth is not used as food during a famine.

EUROPE

In his *Naturalis Historia* (XVIII, 29) Pliny discusses *alica*, a preparation or a kind of porridge made from peeled spelt for which Italy was famed. It was manufactured in several localities, for instance, in the territories of Verona and Pisae, but the product of Campania was most renowned. Pliny describes in detail how the grain was dealt with in Campania for this purpose and that three kinds of alica, the finest, the seconds, and the coarse were distinguished; none of these, however, had as yet the white gloss for which they were reputed. For this purpose, Pliny continues—and expresses his surprise by adding a *mirum dictu* ("strange to relate") —a white marl or chalk (*creta*) is mixed with the grain, and this chalk well embodied in the mass lends it color and tenderness (postea, mirum dictu, admiscetur creta, quae transit in corpus coloremque et teneritatem adfert). This chalk, he writes, is found between Puteoli and Neapolis upon a hill called Leucogaeum (a Greek name meaning "white earth"). He refers to a decree, then still in existence, of the emperor Augustus, in which the latter ordered an annual allotment of twenty thousand sesterces to be paid from his exchequer to the Neapolitans for the lease of this hill. The reason for this contribution, the emperor stated, was that the people of Campania alleged that their alica could not be made without this mineral.

It must be emphasized that what Pliny reports with reference to the alica of Campania was not a regular, but an exceptional practice at which Pliny himself marvels as a very singular fact. In other places of Italy as well as in Egypt the alica was prepared without the addition of creta. Accordingly we face here a purely local custom whose principal object was to whiten the meal or to intensify its whiteness. Nothing like improving its flavor or pleasure in eating a clayish substance is mentioned by Pliny. This passage is not conclusive in attributing to the ancients the habit of earth-eating, as has rashly been done by Ehrenberg (II p. 2).

Pliny further mentions an adulterated kind of alica produced in Africa, over which gypsum, in the proportion of one fourth, is sprinkled. No reason therefor is given. Fée, one of Pliny's commentators, wonders how the African mixture accommodated itself to the stomachs of those who ate it. I believe, very well, and that Fée himself with millions of others has numerous times consumed flour adulterated with gypsum and perhaps worse ingredients.

163

I know of no passage in Greek or Roman literature to warrant the opinion that earth, clay, or chalk was occasionally or habitually consumed, either for pleasure or as a necessity. Various renowned clays like those of Samos, Chios, and Selinos were only employed medicinally or for industrial purposes (Pliny XXXV, 16, 53-56).

Galen (A.D. 129-199) has left an interesting account of his journeying back and forth between Rome and Pergamum in order to stop at Lemnos and procure a supply of the famous terra sigillata, a reddish clay stamped into pellets with the sacred seal of Diana. He describes the solemn procedure by which the priestess from the neighboring city gathered the red earth from the hill where it was found, sacrificing no animals, but wheat and barley to the earth. He brought away with him some twenty thousand of the little disks or seals which were supposed to cure even lethal poisons and the bite of mad dogs. Berthelot believed that this earth was an oxid of iron more or less hydrated and impure. During the middle ages and later Greek monks replaced the priestess of Diana, and the religious ceremony was performed in the presence of Turkish officials (L. Thorndike, I, p. 130).

The learned Dr. Covel, in his Diary (1670-79), gives us an interesting account of what he saw in connection with the terra sigillata of Lemnos, the sacred earth with supposed curative properties:

"On the side hills, on the contrary side of the valley, directly over against the middle point betwixt this hill and Panagiá kotzinátz is the place where they dig the terra sigillata. At the foot of a hard rock of gray hard freestone enclining to marble is a little clear spring of most excellent water, which, falling down a little lower, looseth its water in a kind of milky bogge; on the East side of this spring, within a foot or my hand's breadth of it, they every year take out the earth on the 6th of August, about three hours after the sun. Several papas, as well as others, would fain have persuaded me that, at the time of our Saviour's transfiguration, this place was sanctifyed to have his virtuous earth, and that it is never to be found soft and unctuous, but always perfect rock, unlesse only that day, which they keep holy in remembrance of the Metamorphosis, and at that time when the priest hath said his liturgy; but I believe they take it onely that day, and set the greater price upon it by its scarcenesse. Either it was the Venetian, or perhaps Turkish policy for the Grand Signor to engrosse it all to himself, unless some little, which the Greeks steal; and they prefer no poor Greek to take any for his own occasions, for they count it an infallible cure of all agues taken in the beginning of the fit with water, and drank so two or three times.

Their women drink it to hasten childbirth, and to stop the fluxes that are extraordinary; and they count it an excellent counter-poyson, and have got a story that no vessel made of it will hold poison, but immediately splinter in a thousand pieces. I have seen several finganes (Turkish cups) made of it in Stamboul; we had a good store of it presented to us by Agathone and others, all incomparably good. We had some such as it is naturally dig'd out and not wash'd . . . Thus they take it out: before day they begin and digge a well about 1 ½ yards wide, and a little above a man's height deep; and then the earth is taken out soft and loomy, some of it like butter, which the Greeks say, and the Turks believe, is turned out of rocky stone into soft clay by virtues of their mass. When they have taken out some 20 or 30 kintals for the Greeks' use, they fill it up again, and so leave it stop't without any guard in the world . . .

"We came down to a town called Hagiapate, where there is a great large fountain, where they wash and prepare the hagion choma (sacred earth) for the Turkish seal. They first dissolve it in water, well working it with their hands; then let the water pass through a sive, and what remains they throw away. They let the water stand till settled, then take of the clear, and, when dry enough, they mould in their hands; and most of this we have is shaped from thence. It is all here white, yet I had some given me flesh-coloured. I enquired diligently about it, and they all told me it came out of the same pit; but I expect some of these fellows have found some other place which they conceal. We had some little quantity given us of several people, but very privately, for fear of the *Avaniás*. Agathone, being the Pasha's favourite, feared nothing, but gave us at least 20 okes before 20 people. They tell a story that the earth is hollow from the holy well, when dig'd, to the fountain, where they wash it; and that a duck once dived in the water there and was taken up here; but it seemed an impossible thing to me, there being not water enough in the first place to cover a duck, and the water in the bogge so very shallow, and the earth not sinuous."

J. T. Bent, editor of Covel's Diary (Early Voyages in the Levant, 1913, p. 285), adds the following comments: "Dr. Covel's remarks on the sacred earth of Lemnos are particularly valuable, as this is one of the clearest instances of a pagan superstition being carried on through the influence of Christianity down to our own times. Pliny mentions it (XXIX, 5); also Dioscorides (V, 113); and Galen made an expedition to Lemnos on purpose to see it, and gives us an account of it (De simpl. med., IX, 2). He mentions the disorders

for which it was considered beneficial; he also gives us the ceremonies and mode of operation; on certain occasions a priestess of
Artemis came, and after certain rites carried off a cartload to the
city; she mixed it with water, kneaded it, and strained off both the
moisture and gritty particles, and when it was like wax, she impressed it with the seal of Artemis. During the middle ages, the
reputed virtues of this earth remained unimpaired as a remedy for
the plague."

Pierre Belon witnessed the ceremony on August 6, 1533. When
Tozer visited Lemnos in 1890, the ceremony was still performed
annually on August 6, and was to be completed before sunrise, or
the earth would lose its efficacy. Mohammedan Khojas then shared
in the religious ceremony, sacrificing a lamb. In the twentieth century the entire ceremony was abandoned. In western Europe the
terra sigillata continued to be held in high esteem, and was included
in pharmacopoeias as late as 1833 and 1848. C. J. S. Thompson
has given a chemical analysis of a sixteenth-century tablet of the
Lemnian earth, with the result that no evidence therein of its possessing any medicinal property could be found (L. Thorndike, II,
p. 131).

Hegiage Ben Josef al-Thakefi, governor of Arabia at the time of
the Caliphs, is said to have died of phthisis caused by overeating of
terra sigillata, called by the Arabs *ṭin makhtum, lutum* and *lutum
sigillatum* (D'Herbelot, Bibliothèque orientale, II, 1777, p. 229).
This earth is also mentioned in the pharmacological literature of the
Arabs, for instance, in Serapion's Liber de semplici medicina (P.
Guigues, Les noms arabes dans Sérapion, *Journal asiatique*, 1905,
p. 85) and by Ibn al-Baiṭār (L. Leclerc, Traité des simples, II,
p. 421).

Peter of Abano, in his Treatise on Poisons (*Tractatus de venenis*,
about 1316), mentions the terra sigillata which, he says, causes
vomiting if there is any poison in the stomach. Kings and princes
in the west take it with their meals as a safeguard, and it is called
terra sigillata because stamped with the king's seal. Now, however,
the seals are no longer trustworthy, and Peter cautions the Pope
against what may be offered him as terra sigillata (Thorndike, II,
p. 909).

Earth dug from a grotto in Malta, where St. Paul spent a night,
was formerly used for the cure of many ailments, being esteemed a
cordial, a sudorific, and a certain remedy for the bites and stings of
venomous animals. In the eighteenth century this earth was dis-

tributed from Malta, made up in small round cakes and stamped with the impression of a winged cherub and the words *terra sigillata* (Hill, History of the Materia Medica, p. 206).

In a few wretched villages of Sardinia bread is still prepared from the meal of acorns, which is mixed with a ferruginous argillaceous earth, in order to counteract the tannic acid of the acorns. This earth is called *trokko;* and the bread, *pan' ispeli* (M. L. Wagner, Das ländliche Leben Sardiniens, p. 60, Heidelberg, 1921). This practice corresponds exactly with the acorn bread of the Pomo of California (below, p. 173).

Altheer (p. 93) writes that at Ogliastra in Sardinia a porridge of acorn meal is mixed with a fat clay and that this compound is made into cakes which are sprinkled with ashes or smeared with a little grease and taken as daily food.

The women of Spain and Portugal take pleasure in munching a pottery clay styled bucaro from which vases of a yellow reddish color are made; when dissolved in water or wine, it imparts to these a very agreeable flavor and odor. The bucaro clay is found near Estremoz in the province of Alemtejo, Portugal, and in the province of Estremadura. The almagro, a very fine clay which occurs near Cartagena in the province of Murcia, Spain, is mixed with powdered tobacco in order to render it less volatile and to give it that sweet flavor which is the characteristic of the tobacco of Seville (Camilli, p. 187). Mixed with powdered chili pepper, the same clay is frequently eaten in southern Spain (Altheer, p. 93). The word *almagro* (also *almagra* or *almagre*) is derived from the Arabic *al-maghra* ("red ochre"); this clay is still employed in painting and known in France as *rouge indien* ("Indian red") or *rouge de Perse* ("Persian red").

Deniker states that it is asserted by women that the eating of earth gives a delicate complexion to the face and that the same custom has also been pointed out among women in several countries of Europe, more especially in Spain, where the sandy clay which is used for making the alcarrazas is especially in vogue as an edible earth. The Spanish word *alcarraza*, derived from the Arabic *al-kurrāz* ("earthenware vessel, pitcher"), denotes a porous, unglazed earthenware jar for cooling the water; in the southwestern United States such a jar is commonly called *olla*.

It is said that the ladies of the Spanish aristocracy in the seventeenth century had such a passion for geophagy that the ecclesiastic and secular authorities took steps to combat the evil (Morel-Fatio, Comer Barro. Mélanges de philologie romane dédiés à Carl Wahlund, p. 41, Macon, 1896).

In Macedonia magnesia was sold in the markets and baked in the bread. Another sort of earth was so much in use there that some Ulemas from Anatolia once offered the Grand Vizier various specimens of it as a cheap means of nutrition for the Turkish troops (Altheer, p. 93).

Saint Hildegard of Bingen (1098-1179) describes a complicated cure of leprosy by use of the earth from an ant-hill (L. Thorndike, II, p. 147).

A fossil flour was used in Saxony in times of famine, and its consumption had fatal results. A similar substance was found in Italy, notably in the territory near Magognano in the beginning of the nineteenth century, but it is not on record that this substance styled in Italy "mineral agaric" and "lunar milk" (above, p. 154) was actually consumed (Camilli, p. 188).

The miners in the sandstone mines of the Kyffhäuser ate fine clay placed like butter on bread (known as "stone butter"). The same is reported for miners near Kelbre in Thuringia who used to eat a lithomarge called "stone marrow" (steinmark), a fine clay made liquid or spongy by a small quantity of water (Camilli, p. 187). "Mountain meal" (bergmehl) was resorted to in Germany during the Thirty Years' War for feeding man and cattle (see Hopffe and Zaunick in Bibliography).

"In Finland a kind of earth is occasionally mixed with bread. It consists of empty shells of animalculae, so small and soft that they do not crunch perceptibly between the teeth; it fills the stomach, but gives no real nourishment. In periods of war, chronicles and documents preserved in archives often give intimation of earths containing infusoria having been eaten; speaking of them under the vague and general name of 'mountain meal.' It was thus during the Thirty Years' War in Pomerania (at Kamin or Cammin); in the Lausitz (at Muskau); and in the territory of Dessau (at Klieken); and subsequently in 1719 and 1733 in the fortress of Wittenberg" (A. v. Humboldt, Aspects of Nature, I, p. 196). Ehrenberg (II p. 5) adds Mühlhausen and Oberburgbernheim in Alsace according to the Chronicle of Basle, where earth was baked into bread, and says that the earth-cakes of Klieken served as bread in the fortress Wittenberg, so that the government then found it profitable to sell this treasure of the earth as fiscal property.

During a famine in 1832, the foodstuffs used in the parish Degernä on the frontier of Lapponia contained a meal-like silicious earth mixed with real flour and tree-bark, according to analyses of Bercelius, Retzius, and Ehrenberg. For a long time it has been customary at Umeå, Sweden, to add such earth to wheat flour, and this is said to have no injurious effect on health. Hundreds of car-loads of such earth, especially from Lillhaggsjön Lake in Umeå, mixed with foodstuffs, are said to have served as a nourishment to the Lapps about the same time. Such earth is likewise utilized in Finland; near Laihela, in the region of Vasa in Oesterbotten, Finland, a powder-like white clayish earth (according to Retzius, inorganic) is used as an addition to flour (Ehrenberg II p. 5; and *Bericht über die Verhandlungen der Berliner Akademie*, 1837, pp. 41-43; 1838, p. 7).

NORTH AMERICA

It is commonly believed (and science also has its conventional traditions sometimes half true, sometimes untrue or unproven) that Alexander von Humboldt was the first who drew attention to geophagy among American tribes or even to the subject at all; and when geophagy is spoken of, it is usually Humboldt's illustrious name which is remembered. Humboldt made the subject of geophagy fashionable; his account certainly retains its value, and is still entitled to the interest which it at first aroused, but he was neither the first who discussed the subject (many European writers of the eighteenth century were quite familiar with it as far as Africa, Siberia and Europe are concerned), nor was he the first to point it out with reference to American tribes.

As early as 1527 earth-eating was mentioned by Alvar Nuñez Cabeza de Vaca. Speaking of a tribe called by him Iguaces, who live on wild roots and are much exposed to starvation, he relates that "now and then they kill deer and at times get a fish, but this is so little and their hunger so great that they eat spiders and ant-eggs [the pupas], worms, lizards, salamanders, and serpents, also vipers the bite of which is deadly. They swallow *earth* and wood, and all they can get, the dung of deer and more things I do not mention; and I verily believe, from what I saw, that if there were any stones in the country, they would eat them also." In another passage the same explorer states that the fruit of the mesquite tree (*Prosopis juliflora*) was eaten with *earth*, and then became sweet and very palatable (F. Bandelier, Journey of Cabeza de Vaca, 1905, pp. 89, 127).

Sir Samuel Argoll, in a letter on his voyage to Virginia in 1613, speaks of "the discovery of a strange kind of earth, the virtue of which he did not know; but the Indians eate it for physicke, alleaging that it cureth the sicknesse and paine of the belly" (Purchas, XIX, p. 92).

In reference to clay-eating among the present-day Virginia Indians Dr. Frank G. Speck, professor of anthropology at the University of Pennsylvania, has been good enough to send me the following information: "I recall from my notes that the Pamunkey and the Catawba would confess to eating a little clay at times when they are engaged in making pottery. This they do not as a practice nor, as I recall, for medicinal purposes, but because it tastes agree-

170

ably to them; but they do not make a regular practice of it. They say that it is commenced when they are children, playing about in the clay which has been gathered and cleaned for pot-making by their mothers. Both sexes eat clay. It does not seem to have become a habit among the Indians as among the whites. And I believe there is a connection between it and pot-making, as a theory in its history in the southeast. The Pamunkey mix powdered mussel-shells (Unio) with their pot-clay, and the Catawba sometimes blood, which may be worth considering in the development of the taste." A similar example of women potters enjoying clay while at work is given below (p. 190) for Colombia.

At a later date Dr. Speck communicated to me the following personal observation: "The Catawba women who still make clay pots are given to eating clay in small quantities, because they like the taste of it. This is done when building pots. They say it is good for the health in small measure, acting as a laxative. I found it so too upon trial. Their children also eat it and would be apt to eat too much of it if not controlled by their mothers."

The Zuñi eat the tuber of *Solanum fendleri* (so-called native potato) raw, and after every mouthful a bite of white clay is taken to counteract the unpleasant astringent effect of the potato in the mouth (M. C. Stevenson, Ethnobotany of the Zuñi Indians, Bureau Am. Ethn., 30th Annual Rep., p. 71).

It seems, however, that this procedure was not general among the Zuñi. At least F. H. Cushing (Zuñi Breadstuff, p. 226, repr. in Indian Notes and Monographs of the Museum of the American Indian, VIII), in speaking of the preparation of a diminutive wild potato, which is poisonous in the raw state or whole, but rendered harmless by the removal of the skin, writes that such potatoes were stewed and eaten usually with the addition of wild onions as a relish. He does not refer to clay in this connection, nor to any kind of clay used by the Zuñi in reference to any other food.

The Oraibi of Arizona use a kind of clay which is mixed with potatoes and eaten, hence known as potato-clay (specimen in Field Museum).

J. G. Bourke (The Snake-dance of the Moquis of Arizona, 1884, pp. 70, 252) refers to the Moqui's eating of clay with wild potatoes as a condiment. He adds that the Navaho to a very marked extent and the Apache, Moqui and Zuñi to a smaller degree may be classed among clay-eaters.

Mr. C. Daryll Forde has kindly sent me the following information on the edible clay used by the Hopi: "The edible clay known and used by the Hopi is a white compact material as hard as chalk, but more 'greasy' to the touch and taste. Two sources were known to my informants: (1) the larger supply is obtained from Navaho who bring it in from the Chinlee District, (2) a small local supply also exists in a low hill of sand and shale debris on the west side of Second Mesa near the Mishongnovi spring, Toreva (tojiva). My informants thought that the Navaho themselves did not use it (I was unable to corroborate this with Navaho informants). The Hopi name is *tomöntcöka*. It is always used in association with wild vegetables or berries. The following are two standard recipes:

(1) Kevepsi (berries of a low bush, *keptcoki*, not yet identified) are boiled, the clay is mixed in with them as they cook, and the whole mashed into a paste.

(2) Tumna, the tubers of a wild bush collected in April and May, are boiled and eaten with powdered clay (*au gratin*, so to speak), or the tubers and the clay are mashed together after cooking, or, again, one nibbles at a lump of the clay while eating the main dish."

Mr. E. Simpson and others of the Department of Geology, University of California, have made a physical examination of this clay, with the following report communicated to me by Mr. Forde:—

"The Hopi edible clay is a cream-colored, very fine material with a speckled appearance due to the presence of small, whiter-colored mud ovules. The latter appear to be clay pellets which may have been formed by coagulation in a saline solution, such as would obtain in a saline lake.

"On treating the specimen with water it immediately began to 'dissolve' and rapidly colored all the water in the beaker. In a few hours it had swelled to more than twice its original volume, and had the consistency of soft jelly. The colloidal clay was decanted off and the residue, of which there was extremely little, examined under the microscope. A few very angular grains of quartz, none over a tenth of a millimeter in diameter, were observed together with a few weathered grains of medium plagioclase felspar. Most of the residue, however, was a fibrous chlorite apparently pseudomorphic after biotite.

"The clay was undoubtedly deposited in a lake which was probably saline. It is possibly 'bentonite' or altered volcanic ash, but this cannot be proved by physical examination.

"The property of swelling by taking up considerable quantity of water when immersed suggests that perhaps this clay was of value in giving a sense of repletion to a relatively empty stomach."

W. Hough (in Handbook of American Indians, I, p. 467) writes that "in some localities (among the Pueblos) clay was eaten, either alone or mixed with food or taken in connection with wild potatoes to mitigate the griping effect of this acrid tuber." In this case, accordingly, the clay serves as a soothing medium as among the Ainu in combination with the bulb of a *Corydalis* (above, p. 149).

In acute indigestion the Papago boil for a little while some of the red earth taken from beneath the fire; after being strained a little salt is added, and the mixture is then given to the patient to drink. He has to take this remedy three times, always at mealtime, and he gets nothing or at most very little to eat (A. Hrdlicka, Physiological and Medical Observations among the Indians, 1908, p. 241).

The Pomo of California, in making bread, mix red earth with acorn meal. Dr. S. A. Barrett has been good enough to give me the following information on this point: "The fact of the matter is that they make white bread, as it is called, without a mixture of earth, but this is not esteemed as highly as the black bread, which is as a matter of fact a very dark brown and heavy bread. The two are made, as I recall it, in exactly the same manner, except for this mixture of a very small quantity of a reddish earth, which the Indians say serves as our yeast does. There is nothing, however, in the way of 'raising' of the dough, but the red earth is simply mixed, and the dough is placed in the oven to bake at once. The oven, of course, is nothing more or less than a hole in the ground, which is lined with leaves and filled with layers of this dough and hot stones, the latter being separated from the dough by layers of leaves. It bakes slowly, but is really a very palatable food. I am not sure how this earth actually affects the dough, as I have never had an opportunity to look up the actual chemical composition of this red earth. I cannot state definitely the geographical distribution of this particular custom in California, but as I recall it now, it is not found among the Miwok with whom I have worked to a considerable extent, though not as fully as among the Pomo. The Miwok method of handling acorn meal and bread is quite different from that of the Pomo in several respects. I might add that there is a certain whitish or bluish white clay which was to a certain extent used by the Pomo, though this was not used with anything else and is said by them to be a food of itself. I do not now recall having encountered the use of this whitish or bluish white clay among any of the other Californian tribes with which I came more or less in contact, though it may be that it is such a slight part of their food supply that unless one was specifically hunting for it, it would be very easily overlooked."

In a creation myth of the Cahuilla of California an incidental allusion to earth-eating on the part of the first people is made. Mūkat, in a dispute with Temaīyauit, says, "There will not be enough food for all of them." "They can eat earth," said Temaīyauit. "But they will then eat up all the earth," answered Mūkat, and Temaīyauit replied, "No, for by our power it will be swelling again" (W. D. Strong, Aboriginal Society in Southern California, p. 135, University of California Press, 1929). Of course, it is rather the possibility of eating earth than the fact itself, which is here alluded to; but if eating earth was regarded as possible, actual tests apparently must have been made.

Sir John Richardson (Arctic Searching Expedition, 1852, p. 118) writes, "A pipe-clay is very generally associated with the coal beds, and is frequently found in contact with the lignite. It exists in beds varying in thickness from six inches to a foot, and is generally of a yellowish-white color, but in some places has a light lake-red tint. It is smooth, without grittiness, and when masticated has a flavor somewhat like the kernel of a hazel-nut. The natives eat this earth in times of scarcity and suppose that thereby they prolong their lives."

With reference to this passage, Frank Russell (Explorations in the Far North, p. 133, publ. by University of Iowa) remarks, "I found the bed of edible clay, mentioned by Richardson, near the base of the cliff. It is used for whitewashing at Norman, and is said to have been used as a substitute for soap by the Indians before the introduction of that article by the traders. Norman stands at the mouth of the Bear River near the Bear Rock, a solitary butte over four thousand feet in height." The Indians here in question are Athabascans of northwestern Canada.

V. Stefánsson (Arctic Expedition of the American Museum, Anthr. Papers Am. Mus. Nat. Hist., XIV, 1914, p. 395) has the following entry in his diary under the heading "Edible Earth":—

"Bought to-night a tin full of 'edible clay' from a cutbank on the Kañianirk part of the Colville (S. bank) between the Killirk and Ninñolik branches. The specimen is in flakes and powder. Seller considered the clay a true food, but says it is eaten in large quantities only at times of scarcity or when travelers run out of food. Many eat a little now and then, seller (Kañianirmiut woman) says she puts a little on her tongue almost every day and lets it soak up there till soft. She gets presents every year now of similar stuff up the coast, but the sample sold me has been treasured for years. When clay is to be used in earnest as food, it should be let soak in water over night or longer; it then disintegrates and swells into a thick paste, seems to increase in bulk rather more than rice does in boiling. When about to be eaten, this paste is mixed up in a little more water to make it thinner, and then it is poured into hot water in a pot and cooked 'like flour soup,' i.e., brought to a boil. 'This is good food if one has oil with it; otherwise it constipates you.' The seller, however, considers the clay to be rich in a tasteless and smell-less oil which she says the old men say is old whale-oil that soaked down the cutbank from whales whose bones (lower upper jaws, shoulder-blades, ribs, backbone, etc.) are seen near the top of the cutbank far above."

The Iglulik Eskimo have a tradition relating to the early history of mankind when men had only earth for food. "In earliest times it was very difficult for men to hunt. They were not such skilful hunters as those who live now. They had not so many hunting implements, and did not enjoy an abundance and variety of food such as we have now. In my childhood I heard old people say that once long long ago men ate of the earth. Our forefathers ate of the earth; when halting on a journey and camping, they worked the soil with picks of caribou-horn, breaking up the earth and searching for food. This happened in the days when it was very difficult to kill a caribou, and it is said that they had to make a single animal last all summer and autumn. Therefore they were obliged to seek other food. . . . In those days earth was the principal food of man" (K. Rasmussen, Intellectual Culture of the Iglulik Eskimos, p. 253, Copenhagen, 1929).

A certain outward resemblance of this tradition to the Indic one in the *Vinayavastu* (above, p. 144) is obvious, as is also the diversity of the two stories. In the Indic one earth is considered a superior food of the golden age, in the Eskimoan one it is an inferior food resorted to for lack of better staples in the beginning of life. Of course, such a stage of living, as visualized in the Eskimo tradition, has never existed; it is an afterthought reconstruction, but maybe at the same time a vague reminiscence of earth having formerly been consumed on a larger scale than at the present time.

Lieutenant G. T. Emmons, who has had a thirty-five years' experience with the tribes of the Northwest Coast, assures me that he has never seen or heard of a single case of clay-eating among any of these. Clay or earth is not mentioned either by Harlan I. Smith in his article "Materia Medica of the Bella Coola and Neighboring Tribes of British Columbia" (Annual Report for 1927 of National Museum of Canada, Ottawa, 1929).

It would be erroneous to believe that earth-eating is a privilege of the Indians. It is found among the whites as well, especially in Georgia and Carolina. In 1709 T. Lawson (History of Carolina, p. 206, London, 1714) recorded this observation: "The children [of the Indians] are much addicted to eat dirt, and so are some of the Christians, but roast a bat on a skeiver and make the child that eats dirt eat the roasted rearmouse (bat), and he will never eat dirt again." In 1857 J. R. Cotting published the analysis of a species of clay found in Richmond County, Georgia, which, as announced in the title of his article, "is eagerly sought after and eaten by many people, particularly children." This substance, in its external characters, he writes, resembles lithomarge, or rock marrow; its colors are dark red, yellow, yellowish red, yellowish white, purple and reddish white. He found it associated with other minerals in many parts of the survey, in both the counties of Burke and Richmond, but the purest and most abundant was on land of David F. Dickinson near M'Bean Creek, Richmond County, on the east side of the great road leading from Augusta to Savannah, about fourteen miles from the former place. Here large excavations had been made to obtain this clay, indicating that the demand for it must have been heavy. It has a slight sweetish taste, not unlike calcined magnesia. Its action on the stomach is mechanical, as it contains nothing capable of being decomposed and nothing on which the gastric juice can act. It is composed of silex, oxid of iron, alumina, magnesia, and water.

A boy about fifteen years of age, who was taking his favorite repast at that locality, informed Cotting that he was in the habit of eating daily of that substance, "as much as he could hold in his hand." Cotting asked the boy whether his parents did not inform him better. He replied that he had only a mother and that she ate it too when she was well, but that she was almost always sick. Cotting was informed by people living in the vicinity of the localities where this clay occurs that many deaths have resulted there from no other perceptible cause than from persisting in the use of this clay as a luxury. Cotting adds that this peculiar species of clay is said not to be found north of the Potomac and that a species in some respects similar is found at Bare-hills, Maryland, which, however, is deficient in the proportion of iron and magnesia.

The Redbones (see Handbook of American Indians, I, p. 365) of Carolina are reputed to be clay-eaters.

MEXICO AND CENTRAL AMERICA

The use of earth in ancient Mexico is particularly interesting, especially in its relation to religious ceremonies.

"A peculiar food of the ancient Mexicans seen by the conquerors consisted of cakes made from a sort of ooze which they get out of the great lake, which curdles, and from this they make a bread having a flavor something like cheese" (T. A. Joyce, Mexican Archaeology, p. 155). The source is not quoted by Joyce. Ehrenberg (II p. 3) writes, "Earth-eating was reported in Mexico as early as 1494 and in 1519 by Bernal de Diaz as the relish *tecuitlatl* from the Lake of Mexico, which was confirmed by Hernandez in 1580."

The following documents have reference to this matter. Sahagun (book XI, chap. 3, at end of §5) informs us that "on the Lake of Mexico is found a substance (*urronas*) called *tecuitlatl*, clear blue in color; when it forms a thick layer, it is gathered, spread out on ashes, and formed into cakes which are baked and then eaten."

Bernal Diaz refers to the same matter in two passages (chaps. 92 and 153; translation of A. P. Maudslay, II, p. 73; IV, p. 160). In the former he speaks of fisherwomen and others who sell small loaves made from a sort of ooze gathered on the Lake of Mexico; this ooze curdles or coagulates, and can be cut up into slices the taste of which reminds one a little of our cheese. In the other chapter he writes, "They gathered on the Lake a sort of ooze which when dried had a flavor like cheese." D. Jourdanet (in his French translation of Diaz' work, p. 517) comments that "the Indians of the present time still collect on the banks of the lagoon a mass said to consist of the eggs of gnats mixed with a gelatinous substance which comes from the swarms of these insects (called *agua-utle*); it has indeed a strong flavor like bad cheese."

F. Lopez de Gomara (Historia de Mexico, p. 118, Antwerp, 1554), describing the market, relates, "They eat everything, even earth. At a certain time of the year they sweep up with nets of a fine mesh a fine substance which grows on the water of the lakes of Mexico and coagulates. It is not a plant or earth, but is like mire. There is a lot of it, and they collect much of it, and spreading it out in the way salt is prepared, they empty it out, and there it coagulates and dries. They make of it cakes like bricks. Not only do they sell it in the market, but also they take it to other markets

178

outside the city far away. They eat it as we eat cheese, and it has a slightly saltish taste; with *chilmolli* it is savory. They say, too, that so many birds crave this article on the lake that during the winter they cover many parts of its surface." *Chilmolli* is a ragout or soup in which chili dominates.

F. Juan de Torquemada (Monarchia indiana, 1723, book XIV, chap. 14; II, p. 557), evidently depending on Gomara, has this account: "On the surface of the water of this lake grow some things like finely ground slime, and at a certain time of the year when they are more solidified the Indians gather them with fine meshed nets. They take them out of the water onto the earth or sand of the shore and spread them out till they dry, and then make cakes of them two fingers thick, which subsequently dry out to one finger-breadth when they are ripe. When they are well dry, the people cut them like small bricks, and eat them as though they were of cheese. The Indians think they have a very fine flavor, but they are rather salty. Of this they send a goodly quantity to the markets, and of another food which they call *tecuitlatl*, although at present these two kinds are lost and no longer appear, and I do not know whether the reason is that the Indians have taken to our food and no longer care for their own." The word *tecuitlatl* is listed in the "Diction-naire de la langue nahuatl ou mexicaine" (p. 404) of R. Siméon with the following definition: "A viscous substance (lit. 'excrement of stones') gathered amidst of the plants of Lake Tezcuco; this sub-stance is dried at the sun, and is preserved to be eaten like cheese. The Indians consume it at present and confer upon it the name *cuculito del agua*."

Cf. also the notes of D. Jourdanet in his translation of Sahagun (p. 854).

Midwives give to pregnant women the advice that they must abstain from eating earth and *tiçatl* ("a sort of white earth"), for fear that the infant when born might be sick or disfigured by a bodily defect (Sahagun, book VI, chap. 27). This rule seems to imply that pregnant women in ancient Mexico were in the habit of long-ing for earth.

Joseph de Acosta (Historia natural y moral de las Indias, p. 382, book V, chap. 28, Madrid, 1608) describes a ceremony in ancient Mexico in honor of Tezcatlipoca, god of the night and particularly night winds. During the ceremony the priest blew a pottery flute, and after playing it toward the four points of the compass whereby he meant to indicate that both those present and absent heard him, he placed his finger in the soil, and seizing earth, shoved it into his mouth and ate it as a sign of adoration; the same was done by all who were present (Y aviendo tañido hàzia las quatro partes del mundo, denotando que los presentes y ausentes le oian, ponia el dedo en el suelo, y cogiendo tierra con el metia en la boca, y la comia en señal de adoracion, y lo mismo hazian todos los presentes, etc.). This ceremony was performed ten days before the feast, for the purpose that all might attend this worship in eating earth and demand from the god whatever they pleased. Torquemada (Monarchia indiana, 1723, book X, chap. 14, II, p. 256) describes the same ceremony as follows: "Ten days before the big feast to Tezcatlipoca, in the month Toxcatl, the priest came out of the temple with a flute with a shrill note, and facing in turn all four directions, played it. This was to call all men's attention to the coming feast. Then there was silence, and putting his finger on the ground, he used to take earth, and used to put it in his mouth and eat it as a sign of humility and adoration. Every one did the same, weeping bitterly, throwing himself prostrate on the ground, invoking the obscurity of the night and the wind and asking them with fervor not to leave them shelterless or forget them."

F. Lopez de Gomara (Historia de Mexico, p. 100, Antwerp, 1554) reports that when three thousand nobles came out from Mexico to meet Cortes, "each one as he reached Cortes touched his right hand to the ground, kissed it, bowed down, and passed forward in the order in which they came" (Cada uno, como a Cortes llegaba, tocaba su mano derecha en tierra, besabala, humillabase, y passaba adelante por la orden que venian). The Spanish text is somewhat ambiguous: it is not clear whether they kissed their own hands or the ground, but more probably the latter as a sign of humiliation and adoration as in the ceremony previously described. Gomara (p. 305) relates also that during the ceremonies accompanying the induction of a new ruler in Mexico, nobles as they approached the image of Huitzilopochtli (Vitzilopuchtli), god of war, touched the ground with one of their fingers and then kissed this finger.

Juan Suarez de Cepeda, in 1581, reports that the Tarascans on the west coast of Mexico, on the occasion of an eclipse ("when the mother goddess playing with sun or moon puts her hands over them so that the light is shut off") make noises and eat earth and stones till the eclipse is over ("till the mother goddess returns home"). He further relates, "As soon as the people see the stars which are known as the Pleiades and which in their due course according to the movements of the heavens appear on the horizon, they run to eat, and do eat stones and clods of earth, just as if they were *turrones* [a kind of candy like nougat, very popular in Spain] and honey cakes, and they say they do this so that their teeth may be strengthened, and kept firmly in position so that they do not fall out. And thus they expect will happen to them, feeling like beasts, the opposite effect of what they try for and would wish" (De Cepeda, Relacion de los Indios Colimas de la Nueva Granada. Anales del Museo Nacional, IV, México, 1912, pp. 516, 517). This last sentence would appear to be corrupt; it would seem to suggest that if their teeth do become loosened they are very put out about it.

It was customary among the Aztec that in a certain form of sworn treaty the person rendering the oath put his finger on the soil and then lifted his finger to his mouth as though he was eating earth. In the same manner witnesses also rendered an oath (J. Kohler, Recht der Azteken, pp. 71, 109; and A. H. Post, Grundriss der ethnologischen Jurisprudenz, I, p. 483).

The passages quoted have not been revealed by any previous writer on geophagy, but they are important in showing that the custom is of ancient date in Mexico and roots deeply in religious rites and practices. Lasch (p. 217) states merely that earth-eating is frequent in Mexico, notably among women and children, and that in Guadalajara, San Luis, Puebla, and other places are sold on the markets pastils made of white, lightly baked clay and said to be of good flavor. In regard to the Maya of southern British Honduras Mr. J. Eric Thompson informs me that "they are fond of eating a kind of white chalk which they find in the 'fill' of pyramids. In reply to a question as to why they eat this substance they state that it tastes good and is good for them. Personally I considered it absolutely tasteless. Children in the Maya villages are fond of eating earth. Constant earth-eaters are said to suffer badly from hookworm. Medical authorities with whom I discussed this question differed as to whether hookworm was the cause or the effect of this earth-eating."

The fact that geophagy is still prevalent in Mexico may be gleaned from the following very interesting information kindly sent me by Professor Marshall H. Saville of the Museum of the American Indian, New York:—

"Thirty years ago I visited the town of Etla in the state of Oaxaca, in a valley running west from the Oaxaca valley, and some eighteen miles from the city of Oaxaca. This town is now, and so far as the archaeology is concerned, has always been occupied by the Zapotec Indians.

"I made this visit in order to collect from the various groups of Indians from different parts of the state, who assembled here during the time of the fiestas celebrated annually in honor of the patron saint of Etla.

"The church in which the saint is preserved was built in early colonial times on the pyramidal base of an ancient temple, which, in turn, had been erected on rising ground from which in places the bedrock projected. The ancient Mexicans often took advantages of such eminences, and the Christian priests often razed these old temples to replace them with churches. The fame of the Virgin of Etla is widespread throughout the Indian country of the state of Oaxaca.

"Nearing the town I saw many Indians in family groups returning to their own villages, as this was the last day of the fiesta. Many of them had their faces covered with dust or powder; in fact, they were very dirty. Others were busily engaged in eating powder from a gourd held in their hands. Even the little children were thus engaged. On getting closer to the church I heard the noise of hammering, and saw many Indians industriously hammering off pieces of the rock in the pyramid upon which the church stood. A considerable section of this base looked like a miniature quarry. The rock was obtained by means of stone hammers, and the pieces ground into powder by means of the said stone hammers. I am sorry that I did not get a sample of the rock, nor do I know to what class it belongs. However, it was quite soft and easily reduced to dust.

"I afterwards learned that the Indians not only considered it efficacious for liver troubles, but that coming from this hallowed spot, probably having reference to olden times, taking this powder which was endowed with magical powers, insured their welfare for months to come.

"Father Mayer has just told me that an Indian boy who recently went with him on a collecting trip for us up the Tapajoz River, from Santarem, Amazonia, picked up a clay ball from a site where pottery had been fabricated, and proceeded to eat it, saying that it was 'good to eat.'

"I have seen in the materia medica of native Indian villages in Mexico and Ecuador pieces of soft stone among the herbs, insects, etc., which are sold by the primitive Indian woman for medicine."

According to O. Stoll (Guatemala, 1886, p. 133), the custom of eating certain kinds of earth is generally practised among the Indians of Guatemala, and they do not keep it secret. The earth principally used by them is a light yellowish gray, strongly odorous substance which is a volcanic product weathered away into a powder. It is perfectly insipid and tastes somewhat like chalk. The Indians prize it as a spice of excellent quality and call it "white sweetness" (*sak cab*). Certain it is that this earth is a substitute for tooth-powder and contributes to preserve their white teeth. The quantity eaten at a time is small, as it is merely scattered over the food. Another way of consuming clayish materials is connected with religious ideas. The people who travel to the famous place of pilgrimage, Esquipulas, will take along from there blessed figures of saints made from a powdered earth by the clergy. These figures (*benditos*) are eaten by the devout, or are given away by them to friends and relatives, being credited with the power of relieving existing diseases and preventing sickness.

Stoll affirms that geophagy is a genuine Indian custom which is very ancient; for in the *Popol Vuh* the two magicians, Hunahpu and Xbalanque, rub earth into the roasted birds with which they poison Cabrakan. The fact itself is correct, yet I do not believe that the story of the *Popol Vuh* can be invoked as an example of earth-eating in ancient times. In the text under consideration (translation of Brasseur de Bourbourg, p. 65, Paris, 1861; Villacorta and Rodas, Manuscrito de Chichicastenango, p. 206, Guatemala, 1927), Hunahpu and Xbalanque employ the earth as a ruse to overcome Cabrakan. "This bird," they say, "will be the means of his defeat; in the same manner as white earth will envelop this bird all over through our care, we shall knock him down on the earth, and in the earth we shall bury him." Cabrakan, after eating the bird, staggers and has no more strength on account of the earth rubbed into the bird. Moreover, it was only this one bird which was treated in this manner for the purpose of bringing about Cabrakan's downfall, not, however, the other birds which were plainly roasted at the fire without application of earth. It cannot even be inferred from this passage that birds were generally baked in earth at that time; it was merely a single specific case, a trick devised for the purpose of capturing Cabrakan. The body of the bird was rubbed in with *tizate*, and then white dust was sprinkled around it. The word *tizate* is explained by De Bourbourg as being derived from Nahuatl *tiçatl*, "a whitish earth, very friable, of which they avail themselves to polish metal,

make cement, etc." (see above, p. 179).   The Spanish translation
runs, "Y a uno de ellos (pájaros) le pusieron tizate encima, que es
una tierra blanca, que fué lo que le pusieron."   Nevertheless I am
convinced with Stoll that geophagy is very old in Guatemala and
certainly goes back to pre-Columbian times.

With reference to the treatment of the bird in the preceding
legend it may be called to mind that according to A. Skinner (Material Culture of the Menomini, 1921, p. 194) meat was often roasted
on coals by the Menomini and that small animals were sometimes
rolled up in clay and baked in the hot ashes; this was a favorite
method of dealing with porcupines; when the clay shell was split
open, the quills and hide of the animal adhered to the mold, and the
roast came out clean.

Stoll also mentions the morbid geophagy of children and adults
who devour indiscriminately all kinds of earthy substances.   Popular
opinion ascribes to this habit a number of pathological symptoms,
which is called into doubt by Stoll; he is convinced that many children indulge in this habit without risking disease and that others who
acquire the complex of diseases in question do not really eat earth.

The Guatuso Indians of Costa Rica do not use salt, but are
said by Bishop B. Thiel of San José to enjoy a clayish earth in lieu of
it (K. Sapper, Mittelamerikanische Reisen und Studien, 1902, p. 232).

W. Sheldon (Brief Account of the Caraibs who inhabited the
Antilles, *Transactions Am. Antiquarian Soc.*, I, 1820, p. 412) has
the following note: "The Caraibs as well as the Negroes, when in
a state of melancholy, sometimes hanged themselves; or they would
eat earth and filth until they brought on dropsies or other fatal
disorders, which occasioned their death.   The pernicious habit of
eating earth appears to be endemical in the Westindia islands.
The white Creoles are not free from a propension to this depraved
appetite; and I have heard it much spoken of as prevailing among
the people of Georgia and the Carolinas.   The Caraib slaves would
eat earth whenever they were punished or thwarted."

T. Young (Narrative of a Residence on the Mosquito Shore during the Years 1839–41, p. 76, London, 1842) writes, "The Sambo
girls have a custom of eating charcoal and sand to obtain it fresh
and moist, and they have appeared to enjoy it with great gusto."
The Sambo are descendants of Indians and Negroes who escaped
from a wrecked slave ship, and live on the Mosquito Coast, Nicaragua.

Regarding geophagy of the Negroes in the West Indies, see
above, p. 159.

SOUTH AMERICA

Mention has been made of earth-eating as a means of committing suicide among Negro slaves. The same is reported with reference to the Tupinamba of Brazil by Gabriel Soares de Sousa in his interesting "Noticia do Brazil" (chap. 161, Noticias ultramarinas, III, pt. 1, p. 289), written in 1587. This is one of the earliest accounts of earth-eating in America and certainly the earliest relative to South America; it has thus far been overlooked by every one who has written on the subject. "This people," Soares writes, "has another very great barbarity: when they are seized by disgust or when they are grieved to such a degree that they are determined to die, they begin to eat earth, every day a little, until they emaciate and their face and eyes will swell, and they will finally die; no one can help them or is able to dissuade them from committing suicide, as they affirm that the devil has taught it to them and that he appears to them whenever they are determined to eat earth."

Alexander von Humboldt's observations were made on June 6, 1800, when traveling down the Orinoco he spent a day in the village called La Concepcion de Uruana. His account is as follows:—

"In the midst of this grand and savage nature live many tribes of men, isolated from each other by the extraordinary diversity of their languages: some are nomadic, wholly unacquainted with agriculture, and using ants, gums, and earth as food; these, as the Otomac and Jarure, seem a kind of outcasts from humanity.

"It was a very prevalent report on the coasts of Cumana, New Barcelona, and Caracas, visited by the Franciscan monks of Guiana on their return from the missions, that there were men on the banks of the Orinoco who ate earth. . . . The earth which the Otomac eat is a soft unctuous clay; a true potter's clay, of a yellowish-gray color due to a little oxid of iron. They seek for it in particular spots on the banks of the Orinoco and the Meta, and select it with care. They distinguish the taste of one kind of earth from that of another, and do not consider all clays as equally agreeable to eat. They knead the earth into balls of about five or six inches diameter, which they burn or roast by a weak fire until the outside assumes a reddish tint. The balls are remoistened when about to be eaten. . . . During the periodical swelling of the rivers, which is of two or three months' duration, the Otomac swallow great quantities of earth.

185

We found considerable stores of it in their huts, the clay balls being piled together in pyramidal heaps.   The very intelligent monk, Fray Ramon Bueno, a native of Madrid (who lived twelve years among these Indians), assured us that one of them would eat from three quarters of a pound to a pound and a quarter in a day.   According to the accounts which the Otomac themselves give, this earth forms their principal subsistence during the rainy season, though they eat at the same time occasionally, when they can obtain it, a lizard, a small fish, or a fern root.   They have such a predilection for the clay, that even in the dry season, when they can obtain plenty of fish, they eat a little earth after their meals every day as a kind of dainty. . . .   The Franciscan monk assured me that he could perceive no alteration in their health during the earth-eating season.

"The simple facts are therefore as follows: The Indians eat large quantities of earth without injury to their health; and they themselves regard the earth so eaten as an alimentary substance, i.e., they feel themselves satisfied by eating it, and that for a considerable time; and they attribute this to the earth or clay, and not to the other scanty articles of subsistence which they now and then obtain in addition. . . .   The earth which we brought back with us, and which Vauquelin analyzed, is thoroughly pure and unmixed. . . . That the health of the Otomac should not suffer from eating so much earth appears to me particularly remarkable.   Have they become accustomed to it in the course of several generations?

"In all tropical countries, human beings show an extraordinary and almost irresistible desire to swallow earth; and not alkaline earths, which they might be supposed to crave to neutralize acid, but unctuous and strong-smelling clays. . . .   With the exception of the Otomac, individuals of all other races who indulge for any length of time in the strange desire of earth-eating have their health injured by it.   Why is it that in the temperate and cold zones this morbid craving for earth is so much more rare, and is almost entirely confined, when it is met with, to children and pregnant women; while in the tropics it would appear to be indigenous in all quarters of the globe?"

It must be emphasized that Humboldt himself has not had any personal experience of the effect of geophagy on the Otomac.   As to this point, he has depended entirely on the opinion of the Franciscan friar, Ramon Bueno, and the lay brother, Juan Gonzalez, in whose station he spent the day.   The conclusion that the Otomac are the only people whose health is not impaired by earth-eating (subsequently repeated by many authors) does not seem very plausible; no ill effects are reported, for instance, from Java, Sumatra, Borneo, or Melanesia. Cortambert's observations given below contradict Humboldt's opinion. The conclusion that geophagy is more prevalent in the tropics than in the temperate and cold zones holds good no longer, and is plainly refuted by the facts recorded in this article.

J. Gumilla (Historia del Rio Orinoco, 1791, I, p. 179), said to be credulous and uncritical, denies that the Otomac ever eat pure earth, and states that their clay balls are mixed with maize flour and crocodile's fat; but the two informants of Humboldt affirmed unanimously that the Otomac never added crocodile's fat to their clay balls, and as to maize, they had never heard of it at Uruana.

E. Cortambert (p. 218) gives the following account of the earth-eating habit among the tribes of the upper Orinoco: "This edible earth is a clay blended with iron oxid, reddish yellow in color. It is kneaded into balls or cakes allowed to dry and cooked when to be eaten, rather a ballast for the stomach than a food and commonly used only in times of famine. Although this clay does not contain any nutritive properties, it acts on the principal organ of digestion to such a degree that Indians can subsist on it for several months without any other resources. They sometimes fry it in *seje* oil, and then it offers some really substantial parts. This article of food, in general, does not affect injuriously the health of those who are accustomed to it; but the stomachs unaccustomed to it bear it with difficulty. Obstructions of the viscera and absorption of the chyle are the consequences most to be dreaded by those who want to partake of this strange dish. The Indians who lacking in moderation have a passion for earth considerably fall off in weight, and their reddish color will grow sallow. The taste for clay becomes so intense in some individuals that from houses made of ferruginous clay they will break off pieces and take them into their mouth with avidity. They are discriminating connoisseurs of clay, for not all kinds have the same pleasant taste to their palate; widely varying qualities are distinguished. A few whites in Venezuela have imitated the savages and do not despise cakes of fat earth."

W. E. Roth, in his comprehensive study of the Guiana Indians (Bureau Am. Ethn., 38th Annual Report, p. 225), gives no observations of his own, but quotes J. Gumilla, Humboldt, and J. Crévaux. Among the Otomac, children are given earth to lick and suck by their mothers. Their bread made with alligator fat consists, at least half of it, of chalky earth which, however, does not injure them.

According to J. Crévaux, all the Cayenne Carib are earth-eaters. In each house are found clay balls which the Indians smoke, dry, and eat pulverized. An hour after each meal they will take one of these balls, remove the outer layer that has been blackened, scrape the inside with a knife, and thus obtain a fine powder of which they swallow five or six grams in two doses. In an account of Crévaux's second expedition to South America in 1878-79, given in *Globus* (XL, 1881, p. 262), these observations are made in reference to the Rucuyennes of Guiana. Roth adds that very many children on the upper parts of the Amazon have this strange habit of eating earth, baked clay, pitch wax, and other similar substances; not only Indians, but also Negroes and whites. No conclusion, however, is drawn from this observation, which goes to show that the habit roots in a physiological cause.

In his "Additional Studies of the Arts, Crafts and Customs of the Guiana Indians" (Bureau Am. Ethn., Bull. 91, 1929, p. 18) W. E. Roth adds the following: "In Surinam De Goeje speaks of a hungry Trio widow eating clay."

"Near the Orinoco there is a tribe of savages who feed upon a species of unctuous clay, a practice which, though probably the outgrowth of necessity, is not extremely rare throughout the Amazonian region. This clay, which is said to have a milky and not disagreeable taste, is a species of *marga*, or marl-*subpinguis tenax*, as it is called—which is found in veins of varying color. It is smooth and greasy, dissolving readily in the mouth, and is absorbed into the circulation" (W. G. Mortimer, History of Coca, p. 288, New York, 1901).

T. Whiffen (The North-West Amazons, 1915, p. 124), who stamps clay-eating as a "vice," says that geophagy is very common among all the tribes of the Northwest Amazon, especially with the non-cocainists, the women and children. "As a rule it occurs among the very poorest—the slave clan, those who are least able to obtain such a luxury as salt, and it is found among the female children most of all . . . I never came across any man who ate clay, though I know of a boy who suffered from this neurotic [?] appetite. The clay, if it cannot be otherwise obtained, will be scraped from under the fireplace, and it is always eaten secretly. The Indians look upon geophagy as injurious, but it appears to be ineradicable. I cannot help thinking it must be due to some great 'want' in Indian diet, a physical craving that the ordinary food of the tribes does not satisfy. It is instinctive. In the manufacture of coca they add clay. This suggests that if taken in small quantities it may have a neutralizing and therefore a beneficial effect on some more or less injurious article of daily food. But it rapidly and invariably degenerates into a vice; and the habit appears to have a weakening and wasting effect on the whole body. In some parts of the Amazons, though not with these tribes, the clay is regularly prepared for use, and the vice is shared by other races than the Indian. Children who suffer from this extraordinary craving will swallow anything of a similar character, earth, wax, and Bates even mentions pitch, but they prefer the clay that is scraped from under the spot where the fire has been burning, probably because the chemical processes induced by the heat render it more soluble, easily pulverized, and hence more actually digestive in its action. It has been suggested that this disease was introduced into America by Negro slaves, and is not indigenous. This is a question for the bacteriological expert [?] rather than the traveler to decide, but as it indubitably exists among tribes that have not come in any contact with Negroes or Negro-influenced natives, it would seem to argue on the face of things that the similarity of vicious tastes was due to similarity of causation, rather than to contamination by evil example, unless the ubiquitous microbe is to be held responsible for this ill also.''

Geophagy occurs not rarely, especially among younger individuals on the Amazon (P. Ehrenreich, Beiträge zur Völkerkunde Brasiliens, p. 62).

In regard to the Botocudo P. Ehrenreich (*Zeitschrift für Ethnologie*, XIX, 1887, p. 29) states merely that geophagy is widely diffused among them, and quotes St. Hilaire as saying that saline earths which are not rare in the province of Minas and saline plants serve them for salt the use of which is unknown to them (cf. above, p. 107).

The Bakaïri make dolls of a red loam which is licked by children. This loam, it is said by the natives, was eaten by their forebears before they became acquainted with mandioca. The Bororó drink water mixed with loam as an invigorating beverage, but do not eat loam (K. von den Steinen, Unter den Naturvölkern Zentral-Brasiliens, 1894, pp. 282, 481).

According to T. Koch-Grünberg (Zwei Jahre unter den Indianern, 1910, II, p. 291), edible clay is regarded as quite a delicacy. In his work "Von Roroima zum Orinoco" (III, pp. 298, 311, 337) Koch-

Grünberg mentions balls of dried clay and a fat white clay (probably kaolin, he adds) in form of balls and wrapped up with leaves, used as a relish.

The Juan-Avo or Caripuna who live in the proximity of the cataracts of the Madeira are described by Acunna as devouring earth (C. F. P. von Martius, Beiträge zur Ethnographie und Sprachenkunde Amerika's, I, 1867, p. 415).

The habit is not confined to the Indians; for Negroes and whites have the same propensity. At Pebas, in Peru, Mr. Hauxwell found it impossible to restrain his own children. On the Marañon the half-breeds are mostly addicted to the practice of dirt-eating. Even strangers, English, or the white Peruvians, who have intermarried with Mestizos and have had children by them, find its presence among their little ones the plague of their life. Children commence from the age of four or less, and frequently die from the results in two or three years. Officers there, who have the Indian or half-breed children as servants in their employ, sometimes have to use wire masks to keep them from putting the clay in their mouth; and women, as they lie in bed sleepless and restless, will pull out pieces of mud from the adjoining walls of their room to gratify their strange appetite, or will soothe a squalling brat by tempting it with a lump of the same material (W. L. Distant, *Journal Anthrop. Inst.*, X, 1881, p. 468).

Gilij (Saggio di storia americana, II, p. 311) writes that the Indian women of the village Ranco on the Magdalena River while engaged in making pottery shove large pieces of clay into their mouth.

According to Saffray (*Globus*, XXIII, 1873, p. 8), geophagy occurs rather frequently in some regions on the lower Magdalena River in Colombia, but is not endemic as on the Orinoco. The edible earth consists of a very fatty clay of yellowish or reddish color.

Earth-eating is also reported from southern Brazil, Paraguay, Peru, and Bolivia. In Bolivia a light white clay (called *pasa*) is sold in the markets with victuals, and is also consumed by whites, particularly women. The clay is eaten either in its natural state as it is dug near Oruro, or is purified and fashioned into jars or images of saints. Odoriferous resins are sometimes blended with the clay to improve its taste. J. J. von Tschudi (Reisen durch Südamerika, V, 1869) mentions a lady who daily enjoyed the clay figure of a saint for years.

G. I. Molina (Saggio sulla storia naturale del Chili, p. 50, Bologna, 1810) speaks of a potter's clay, called by him Argilla buccherina and found in the province Santiago, Chile, fine, light in weight, odorous, brown with yellow dots, dissolving in the mouth and sticking to the tongue. The nuns of the capital made delicate pottery from this clay large quantities of which were exported to Peru and Spain under the name "bucchero (bucaro) ware of South America." Water kept in these vessels assumes a pleasant flavor. Peruvian women were in the habit of eating fragments of this pottery (le donne peruane costumano di mangiarne i frammenti como le Mogolesi mangiano il vasellame di Patna); they were presumably attracted to it by its aromatic properties. Compare above, p. 140.

F. Gautier (see Bibliography) found a white clay used in the province of Potosí of Bolivia, but did not hear of any disease accompanied by clay-eating.

Dr. A. Rengger (Reise nach Paraguay in den Jahren 1812 bis 1826, p. 326, Aarau, 1835) writes, "Mr. de St. Hilaire met men who ate earth at Paranagua, Guaratuba, and in other parts of the Province Santa Catharina (in Brazil). He regards it as a degenerate taste. I do not share this opinion, but rather look upon the devouring of earth as a disease, cases of which frequently occurred to me in Paraguay and of which I cured a number of persons. In this country the matter was also looked upon as an evil habit. I have seen several pregnant women addicted to earth, who after delivery lost again this unnatural propensity."

A. N. Schuster (Paraguay, 1929, p. 65), discussing geophagy in Paraguay, regards it as a disease caused by intestinal worms.

BIBLIOGRAPHY

ALTHEER, J. J.—Eetbare aardsoorten en geophagie. Natuurkundig Tijdschrift voor Nederlandsch Indië, Batavia, XIII, 1857, pp. 83-100.

Geophagen in den Indischen Archipel. Beschrijving en onderzoek van eenige aardsoorten, uit de Residentie Kedirie, die door inlanders gegeten worden. Tijdschrift der Vereeniging t. Bevord. der Geneesk. Wetenschapen in Nederlandsch Indië, Batavia, V, 1857, pp. 808-812.

In the Catalogue of the Surgeon General's Library this article is erroneously credited to O. Brummer whose name appears merely in the preface as one who sent several specimens of edible clay to Batavia. These were examined and analyzed by Altheer whose name as that of the author is printed at the end of the article on p. 812.

BIOT (PÈRE).—Note sur des matières pierreuses employées à la Chine dans les temps de famine, sous le nom de Farine de Pierre. Annales de chimie et de physique, LXII, 1839, pp. 215-219.

Translations of Chinese texts by E. Biot (see also Journal asiatique, 1840, p. 290).

BOUCHAL, L.—Geophagie [in Indonesien]. Mitteilungen der anthrop. Ges. Wien, XXIX, 1899, p. [11].

CAMILLI, S.—Observations physiologiques sur le géophagisme. Bulletin des sciences médicales, Paris, XVI, 1829, pp. 185-192.

This is the analysis of an article published in Italian in the *Giornale Arcadio* of 1842 (not accessible to me). The translator, who signs D., antagonizes several of Camilli's theories.

CORTAMBERT, E.—Coup d'oeil sur les productions et sur les peuplades géophages et les autres populations des bords de l'Orénoque. Bull. de la Société de Géographie, 1861, pp. 208-220.

COTTING, J. R.—Analysis of a Specimen of Clay Found in Richmond County, which is eagerly sought after and eaten by many people, particularly by children. Southern Medical and Surgical Journal, Augusta, I, 1837, pp. 288-292.

CRAGIN, F. W.—Observations on Cachexia Africana or Dirt-eating. American Journal of the Medical Sciences, XVII, 1835, pp. 356-364.

DENIKER, J.—The Races of Man, 1906, pp. 145-146.

EHRENBERG, C. G. I—Mikrogeologie. Das Erden und Felsen schaffende Wirken des unsichtbar kleinen selbständigen Lebens auf der Erde. Leipzig, 1854. 2 vols. folio.

II—Über die rothen Erden als Speise der Guinea-Neger. Abhandlungen der Akademie der Wissenschaften zu Berlin, 1868, pp. 1-55.

FERRAND, E.—Terres comestibles de Java. Revue d'ethnographie, V, 1886, pp. 548-549.

GAUTIER, F.—Sur une certaine argile blanche que mangent les Indiens de Bolivie. Actes de la Société scientifique du Chili, Santiago, V, 1895, pp. 85-86.

GOEBEL, A.—Über das Erde-Essen in Persien, und mineralogisch-chemische Untersuchung zweier dergleichen zum Genuss verwendeter Substanzen. Bull. de l'Académie imp. des Sciences de Saint-Pétersbourg, V, 1863, col. 397-407.

HAMY, E. T.—Les géophages du Tonkin. Bull. du Muséum d'histoire naturelle, V, 1899, pp. 64-66.

192

HERINGA, J.—Eetbare aarde van Sumatra. Natuurkundig Tijdschrift voor Nederlandsch Indië, XXXIV, 1874, pp. 185-189.

HEUSINGER. —Die sog. Geophagie oder tropische (besser: Malaria-) Chlorose Krankheit aller Länder und Klimate dargestellt. Cassel, 1852. Non vidi.

HOOPER, D. and MANN, H. H.—Earth-eating and the Earth-eating Habit in India. Memoirs of the Asiatic Society of Bengal, Calcutta, I, 1906, pp. 249-270.

HOPFFE, A.—Über Infusorienerde (Bergmehl). Naturwissenschaftliche Wochenschrift, XVI, 1917, pp. 286-287.

HUMBOLDT, A. VON.—Sur les peuples qui mangent de la terre. Annales des voyages, II, 1809, pp. 248-254.

Personal Narrative of Travels to the Equinoctial Regions of America. London, 1852-53, II, pp. 196, 495.

Ansichten der Natur, third edition, 1849, I, p. 231.

Aspects of Nature. Translated by Sabine. London, 1849, I, pp. 25, 190.

LASCH, R.—Über Geophagie. Mitteilungen der anthropol. Gesellschaft Wien, XXVIII, 1898, pp. 214-222.

MEIGEN, W.—"Essbare Erde" von Deutsch-Neu-Guinea. Monatsberichte der deutschen geologischen Gesellschaft, 1905, pp. 557-564.

MITRA, SARAT CHANDRA.—Note on Clay-eating as a Racial Characteristic. Journal of the Anthropological Society of Bombay, VII, 1904-07, pp. 284-290.

SPENGLER.—Die erdefressenden Menschen. Wochenschrift für die gesammte Heilkunde, Berlin, 1851, pp. 321-327.

THOMPSON, C. J. S.—Terra Sigillata, a Famous Medicament of Ancient Times. Proceedings of the XVIIth Internat. Congress of Medical Sciences, Section XXIII, London, 1913, pp. 433-444.

THORNDIKE, L.—A History of Magic and Experimental Science during the First Thirteen Centuries of our Era. 2 vols., New York, 1923.

THURSTON, E.—Earth-eating. In his Ethnographic Notes in Southern India, pp. 552-554, Madras, 1906.

ZAUNICK, R.—Über "Mehlerde" im Anhaltischen 1617. Naturwissenschaftliche Wochenschrift, XVI, 1917, p. 496.

INDEX

194

201

烟草及其在非洲的使用

Tobacco and Its Use in Africa

BY

BERTHOLD LAUFER, WILFRID D. HAMBLY,

and RALPH LINTON

6 Plates in Photogravure

ANTHROPOLOGY

LEAFLET 29

FIELD MUSEUM OF NATURAL HISTORY

CHICAGO

1930

The Anthropological Leaflets of Field Museum are designed to give brief, non-technical accounts of some of the more interesting beliefs, habits and customs of the races whose life is illustrated in the Museum's exhibits.

LIST OF ANTHROPOLOGICAL LEAFLETS ISSUED TO DATE

STEPHEN C. SIMMS, Director

FIELD MUSEUM OF NATURAL HISTORY
CHICAGO, U. S. A.

Priest of the Hill Angas Smoking a Large Pipe.
From a photograph by C. K. Meek.

FIELD MUSEUM OF NATURAL HISTORY
DEPARTMENT OF ANTHROPOLOGY
CHICAGO, 1930

LEAFLET   NUMBER 29

Tobacco and Its Use in Africa

CONTENTS

[163]

THE INTRODUCTION OF TOBACCO
INTO AFRICA

In 1683 a curious pamphlet in French of thirty pages was published at Cologne by Pierre Marteau. It bears the title "Institution et status de l'ordre des chevaliers de la cajote." Its author was a French officer whose name appears at the end of the preface—De la Motte. The story he has to tell is briefly as follows.

Several French officers met at Hanover and took their meals in the tavern of the widow La Roche (à l'auberge de la Veve la Roche). They were in the habit of passing a tobacco-box around after their meals to prolong conversation through this harmless pleasure. One of the officers, who for a long time had traveled in Africa, proposed to his friends to smoke in African fashion; that is, all together from the same pipe, which had a very large bowl perforated in several places; ten or twelve tubes being inserted into these holes and permitting as many persons to smoke simultaneously. As this manner of using tobacco was considered more entertaining than that then prevailing in Europe, the whole company applauded this proposition. It was therefore resolved to adopt the African custom and to name the society Order of the Cajote, as the tobacco-pipe is thus styled by the Africans. These officers then organized into an order, and "the high and venerable Seigneur brother" De la Motte was instructed to draw up rules and regulations of the order of "the Chevaliers de la Cajote." These are fifty-five in number and occupy the greater part of the booklet, which winds up with a poem in honor of tobacco. Rule 52 provides that ordinary "individual" pipes are rejected by the brethren and are prohibited in their meetings, where the cajote rules supreme. Snuff and chewing were likewise forbidden. It is noteworthy to see this African custom of communal pipe-smoking adopted by French officers as early as in the latter part of the seventeenth century.

In view of the wide diffusion of the tobacco plant over Africa and the many peculiar types of tobacco-pipes evolved by African peoples it has often been asserted that tobacco is a native of the continent. In Asia, however, the cultivation is as widely diffused as in Africa, and is even more intense; and as I have shown in "Tobacco and Its Use in Asia" (Leaflet 18), all civilized nations of Asia have preserved detailed records as to the introduction of the tobacco plant in two species (*Nicotiana tabacum* and *N. rustica*) into their respective countries in the beginning of the seventeenth century.

The question of the origin and development of smoking pipes has nothing to do with the history and distribution of the tobacco plants as botanical species. The fact is now perfectly well established by both botanical and historical evidence that the home of the two species of Nicotiana mentioned is in America. In the same manner as these were introduced from America into Europe and Asia, so they were transmitted to Africa. I have stated that while Asia owes the tobacco plant to America, it owes nothing to America in regard to smoking utensils; for Asiatics have exerted their own ingenuity and produced their smoking apparatus from resources wholly their own. The same statement, with a grain of salt, may be applied to the peoples of Africa who, while at first working after models furnished by Portuguese and Hollanders, display a great deal of acumen and originality in the manufacture of their tobacco-pipes and in their smoking customs.

One of the bulwarks in the arguments of those who have pleaded an African origin of Nicotiana was a casual statement of G. Schweinfurth (The Heart of Africa, I, p. 255) to the effect that *N. rustica* might be indigenous.

[165]

Recently, however, Schweinfurth himself (Festschrift Seler, 1922, p. 532) has conceded that this was an unfounded supposition to which no value should any longer be attached. In the article referred to in which the cultivated plants transmitted from Africa to America and from America to Africa are discussed Schweinfurth champions unequivocally the introduction into Africa of the two Nicotiana species from America, but narrows the routes of transmission to those leading from Europe and Asia to Africa, while there can be no doubt that the Portuguese brought tobacco directly from Brazil and Portugal to West Africa. This is now the consensus of opinion among those most competent to judge, especially those botanists who have studied the cultivated plants of Africa, as the Count De Ficalho (Plantas uteis da Africa portugueza Lisbon, 1884, p. 233), O. Warburg (in A. Engler, Nutzpflanzen Ost-Afrikas, 1895, pp. 255–261), and F. Stuhlmann in his fundamental work "Beiträge zur Kulturgeschichte von Ostafrika" (pp. 367-374).

The fact that the Negroes do not understand how to cultivate tobacco properly and hardly know anything about curing the leaf, and prefer imported tobacco to that of native growth is sufficient proof for the foreign origin of the plant. G. Schweinfurth (The Heart of Africa, II, p. 214) remarks, "In Egypt the Virginian tobacco can be made to grow leaves as large as the palm of one's hand, but in the Negro districts the whole produce is quite diminutive. Negroes always sow tobacco under cover before they plant it out; the midday sun of central Africa is too powerful for the seed, which infallibly perishes in a parched soil."

Reporting a plant otherwise cultivated as "growing wild" was formerly one of the favorite sports and fads of travelers. Numerous plants have the tendency to escape from cultivation and even to become seemingly spontaneous and naturalized in the midst of the native

[166]

flora. American plants like the cashew (*Anacardium occidentale*), the papaya (*Carica papaya*), the guava (*Psidium guayava*), tomato and pineapple, and others all grow seemingly wild in Africa, India, and other tropical regions of the Old World. The guava has escaped from cultivation in such enormous numbers that it forms a characteristic plant of certain floristic regions of Africa. The pineapple is "wild" in Zanzibar in immense quantities, and is encountered in every hedge and thicket; it is likewise "wild" in the forests of Assam and Ceylon, and yet is still known to the natives under the Brazil-Portuguese term *ananas*. The same holds good for Nicotiana: wherever in Africa it was formerly reported "wild" by travelers and botanists, it is but seemingly wild and a fugitive from the hands of man. It produces an enormous number of small seeds which are easily disseminated and propagate themselves spontaneously almost in any soil.

It remains to be considered also that Nicotiana is not the only genus of plants that reached Africa from America, but that there are at least about eighty others which were transmitted from America to Africa alike in the great age of maritime enterprise and colonial expansion and which are listed in the article of Schweinfurth cited above. The most important of these are maize, manihot, batata or sweet potato, three species of bean (*Phaseolus*), tomato, peanut, papaya, guava, pineapple, four species of *Anona*, alligator pear, cashew, passion flower, several species of *Capsicum*, cacao, vanilla, agave.

Following are the arguments in the case to be presented from the standpoint of the historian. The plants cultivated in the soil of Africa in ancient and mediaeval times are perfectly well known to us. Plant remains found in Egyptian tombs and plants depicted on Egyptian monuments have been most carefully investigated and studied, and have familiarized us thoroughly with the flora and agriculture of ancient Egypt. No trace of any Nicotiana

[167]

or of any habit of smoking has ever been discovered there. As to mediaeval times, Arabic travelers and geographers supply us with valuable notes on the cultivated plants of many parts of northern and eastern Africa, and these Arabic records contain no mention of tobacco. There is no European account of Africa written prior to the discovery of America that alludes to anything like tobacco. No reference to tobacco in Africa is made at an earlier date than about a century after the discovery of America. In 1485 the Portuguese discovered the Congo. In 1498 Vasco da Gama rounded the Cape of Good Hope on his way to India. In 1505 the Portuguese laid the foundation of their East-African empire with the occupation of Sofala. In 1506 they discovered the island of Madagascar, and in 1574 they founded the colony of Angola.

None of the early Portuguese navigators touching the west coast of Africa and none of the Portuguese historians relating the expeditions to India lisps a word about tobacco in Africa. Varthema, when he returned from India to Portugal in 1508, sailed around the African continent, touching the new Portuguese possessions, but does not mention tobacco anywhere. Alvares de Almada, who wrote in 1594 a detailed account of Guinea, passes tobacco over with silence; and from this absence of testimony the Count De Ficalho concludes that the introduction of tobacco into Guinea did not take place before the beginning of the seventeenth century.

It is pointed out by several authors that there is no specific account with reference to the introduction of tobacco into Africa. This is not to be wondered at, as Africa is not simply a country, but a continent of vast extent. Moreover, the introduction was not just a single event; the plants or their seeds were deposited at many different localities of the north, west, south, and east coasts at different times by various nations—chiefly the Portuguese, Hollanders, and Arabs. From the coastal

[168]

points the plant spread rapidly over various routes into the interior of the continent, penetrating to its very center.

One of the earliest accounts of tobacco-smoking in Africa is due to William Finch, who visited Sierra Leone in 1607 (Purchas, IV, p. 4): "Tobacco is planted about every man's house, which seemeth half their food: the bowl of their tobacco-pipe is very large, and stands right upward, made of clay well burnt in the fire. In the lower end thereof they thrust in a small hollow cane, a foot and a half long, through which they suck it, both men and women drinking the most part down, each man carrying in his snap-sack a small purse (called *tuffio*) full of tobacco, and his pipe. The women do the like in their wrappers, carrying the pipe in their hands. Unto their tobacco they add nothing but rather take from it: for I have seen them straining forth the juice of the leaves, being green and fresh, before they cut and dry it (making signs that otherwise it would make them drunk), then do they shred it small, and dry it on a sherd upon the coals."

O. F. von der Groeben (Guineische Reise-Beschreibung, 1694, p. 19), who visited the Guinea coast in 1682-83, writes, "The inhabitants of Sierra Leone smoke tobacco—men, women, and children indiscriminately, and are so fond of its fumes that they inhale them not only at daytime, but also at night hang small bags of tobacco around their necks like a precious gem."

O. Dapper, in his "Description of Africa" (1686, pp. 231, 236), writes that in the kingdom of Zenega [Senegal] or country of the Jalofes [Jolofs] tobacco thrived very well, but people did not take pains in cultivating it; if the inhabitants were workmen, he adds, they would harvest tobacco and all sorts of plants and grains in abundance. In the village Gerup, a market was held every fourth day for the sale of clothing, cotton, slaves, tobacco, horses, camels, and cattle. According to the same author,

[169]

tobacco succeeded very well along the river Gambia and above Cassan, and the Portuguese of Juala and Catcheo went there to freight it on sloops. The Negroes of Cap-Verde, Refrisco, Porto d'Ale and Juala traveled to Tinda, Tondeba, and Tankerval to purchase tobacco.

J. Ogilby, in his work "Africa" (1670, pp. 355, 371), informs us, "All along the banks of Gambia and about Cassan, tobacco grows plentifully, which the Portugals fetch with sloops both green and dried, without making up in rolls. The islands Los Idolos, stretching along the coast of Sierra Leone, afford good tobacco."

According to Richard Jobson, who in 1620-21 made a journey on the river Gambia into the interior in quest of gold, the Mandingos received their tobacco from Portuguese slave-traders of Brazil. Stubbs, who traveled in the same region in 1624, saw the Negroes there grow tobacco near their habitations.

Tobacco was extensively grown in Guinea at the end of the seventeenth century, as we learn from W. Bosman's account (Voyage de Guinée, p. 319, London, 1705). He describes the plant as being two feet high, with leaves a hand wide and two or three hands long, and white flowers. "The stench of this villainous herb," he writes, "was so horrible that it was impossible for a sensitive person to be near a smoking Negro. All of them smoked, but those who lived in the Dutch territory and daily communicated with the whites, used Portuguese or rather Brazil tobacco (*Nicotiana tabacum*), which is a bit better and yet smells horribly. Some Negroes had pipes made of reeds more than six feet long, with bowls of stone or clay, in which they placed two or three handfuls of tobacco and had no difficulty in smoking out a pipe thus loaded without stopping. Men and women were so passionately fond of tobacco that they gladly sacrificed their last penny to get it, and would rather hunger than be without it."

It will be noticed that Bosman discriminates between the two species, *Nicotiana rustica* and *N. tabacum*. It appears that the former was introduced at an earlier date, probably by the Portuguese, who cultivated it in Lisbon as early as 1558 (Leaflet 19, p. 49). This rustic species, owing to the extreme strength and intense narcotic qualities which it possesses, has always endeared itself to the Negro. I am inclined to think that the latter was introduced by the Portuguese from Portugal in the latter part of the sixteenth century and that *N. tabacum* followed a little later from Brazil.

The Capuchin missionary Girolamo Merolla, in his "Relatione del viaggio nel regno di Congo nell' Africa meridionale" (Naples, 1692, p. 460), gives as the native Congo name for tobacco the word *fumu* ("smoke," adopted from the Portuguese *fumo*, still used in Brazil). In Plate XIV of his book he figures a Congo "cavalier" (*cavaliere*) and "lady" (*dama*) puffing away clouds of smoke from a long-stemmed pipe. It is interesting to note that Merolla lists four Portuguese words as Congo names of American cultivated plants—*casciu* (Portuguese *caju*), the cashew; *guaiavas*, the guava; *mandioca;* and *mamao* (Portuguese *mamão*), the papaya.

In 1652 the Hollanders took possession of the Cape of Good Hope when Johann van Riebeck founded the first settlement there. Immediately the cultivation of tobacco was taken up there by the Hollanders. The Hottentots soon adopted the habit of smoking and took a great delight in it. W. Ten Rhyne (Churchill's Collection, IV, p. 768), who traveled in Capeland in 1673, saw men and women, children and old men indulge in tobacco. Several eye-witnesses report that the love of a Hottentot woman could be obtained for a pipeful of tobacco (G. Meister, Der orientalische Kunst-Gaertner, 1692, p. 30; F. Leguat, Voyages et avantures, 1708, II, p. 160). La Loubere (Du royaume de Siam, 1691, II, p. 134) hints at the fact that

the passion for tobacco and brandy induced the natives
to admit the Hollanders into their country and made the
Hottentots dance at their will (les fait danser tant qu'on
veut). The Abbé De Choisy (Journal du voyage de Siam,
1687, p. 77), who stopped at the Cape in 1685, says not
unjustly, "The Hollanders gradually advance into the
country which they buy up with tobacco." For the sake
of tobacco the poor and unsophisticated Hottentot was
ready to do anything. For a handful of the leaves he
was then willing to work a whole day (Leguat, p. 157).
Men of the Dutch Company purchased an ox or a sheep
from the natives for tobacco in ropes or coils an inch
thick by measuring with this rope from the front of the
beast to the end of the tail (Leguat, p. 161).

William Dampier (A New Voyage Round the World,
1697, chap. 19), who visited the Cape of Good Hope in
1691, gives this account: "I am told by my Dutch landlord that they kept sheep and bullocks here before the
Dutch settled among them; and that the Inland Hottentots
have still great stocks of cattle and sell them to the Dutch
for rolls of tobacco: and that the price for which they
sell a cow or sheep was as much twisted tobacco as would
reach from the horns or head to the tail; for they are
great lovers of tobacco and will do anything for it."

It is perfectly clear, therefore, that prior to the arrival
of the Hollanders the Hottentot was not acquainted
with tobacco.

The Portuguese may have introduced tobacco into
Madagascar after their discovery of the island in 1506.
Arabic and Persian seafarers and traders also may have
apparently had an equal share in bringing tobacco to the
island and the ports of East Africa, as proved by the early
appearance there of the water-pipe which originated in
Persia. In 1638 Peter Mundy (Travels, III, p. 384) found
tobacco growing in Madagascar. Etienne de Flacourt,
who was French Governor of Fort Dauphin on Madagascar

[172]

from 1648 to 1655, states in his "Histoire de la grande isle de Madagascar" published in 1661 (pp. 30, 101, 143) that "the petun or tabacq or nicotien thrives everywhere and results in the best tobacco of the world; tobacco-pipes were made of bamboo, and a pipe was buried with the dead." He also refers to tobacco cultivation on the islands of Sainte-Marie and Bourbon. On the former fourteen Frenchmen cultivated the herb as early as 1645 (Grandidier, Collection des ouvrages anciens concernant Madagascar, III, p. 203).

The water-pipe was propagated in Egypt and over many other tracts of Africa by the Arabs. As early as 1626 Thomas Herbert found the hooka in use among the inhabitants of Mohilla, one of the four islands forming the Comoro group. In 1638 it was noticed by Peter Mundy, who describes it as being the end of a horn with a short pipe or cane to the end of which they apply a mouthpiece (see Leaflet 18, p. 28).

The countries along the north coast of Africa were supplied with tobacco by European mariners, the Osmans, and the Arabs. One of the earliest references to the use of tobacco in Northwest Africa I have found occurs in the work of Pierre Dan (Histoire de Barbarie et de ses corsaires, 1636, p. 282), who vividly describes the idleness of the men in the bazars and coffee-houses (he describes the preparation and effects of coffee as a novelty), where they spent whole days and nights on drinking coffee and smoking tobacco (souffler le petun). In Algeria oriental types of Nicotiana were grown for a long time before the French occupation. The first French colonists introduced a considerable number of varieties; but only one of these, believed to be derived from Paraguay stock, is now extensively cultivated (Kearney and Means, Agricultural Explorations in Algeria, p. 85, Washington, 1905).

Tobacco became known in Turkey toward the end of the sixteenth century (see Leaflet 19, p. 61) and rapidly

[173]

spread over the whole empire of the Osmans. E. W. Lane, in his classical book "Manners and Customs of the Modern Egyptians" (1871, II, p. 30), writes, "It appears that tobacco was introduced into Turkey, Arabia, and other countries of the East shortly before the beginning of the seventeenth century of the Christian era: that is, not many years after it had begun to be regularly imported into western Europe as an article of commerce from America." He also cites an Arabic author, El Is-hakī, as stating that the custom of smoking tobacco began to be common in Egypt between the years 1601 and 1603. It is curious that in a story of the Arabian Nights (No. 976) is mentioned a powder resembling snuff; but in the opinion of E. Littmann, the eminent orientalist and the latest and most conscientious translator of the Nights, this entire story, in which a coffee-house and coffee-drinking also are referred to, originated in Egypt as late as the sixteenth or seventeenth century.

It may not be amiss to cite here the following judicious observation of Lane: "It may further be remarked, in the way of apology for the pipe, as employed by the Turks and Arabs, that the mild kinds of tobacco generally used by them have a very gentle effect; they calm the nervous system, and, instead of stupefying, sharpen the intellect. The pleasures of Eastern society are certainly much heightened by the pipe, and it affords the peasant a cheap and sober refreshment, and probably often restrains him from less innocent indulgences."

From the marginal countries of the Mediterranean shore tobacco was gradually transplanted into the interior of Africa by the caravan trade. The caravans annually outfitted from Cairo transmitted the product to Nubia, Dongola, Senar, Kordofan, Darfur, and into the countries of the Sudan. The caravans dispatched from Tripolis, Tunis, Algeria, and Morocco freighted it to Biladulgerid and Fezzan as well as into the oases of the Sahara. Over

[174]

the trade routes traversing the desert tobacco penetrated to Timbuktu, Sakatu, Kashna, Bornu, Kanem, and Borgu, where it was highly esteemed and bartered for gold and ivory.

In 1895 the cultivation of tobacco was prohibited in Egypt and in the eastern Sudan, and has been suppressed there ever since.

Schweinfurth has already observed that "it is a great indication of the foreign origin of the tobacco plant that there is not a tribe from the Niger to the Nile which has a native word of their own to denote it. . . . The people ring every kind of change upon the root word and call it *tab, tabba, tabdeet,* or *tom.*" In Hausa it is *taba* (Dalziel, Hausa Botanical Vocabulary, p. 90), likewise so in Tuareg, Sennar, Bornu and Darfur, Uganda and Unyoro. In Senegambia the designation for tobacco is *tamaka,* in Tigre (Abyssinia) *tombak,* in Somali *tumbak,* in Galla *tambo.*

The custom of smoking hemp, either alone or blended with tobacco, is widely diffused over Africa, as may be read in the following chapter. Hemp was introduced into East Africa from India through the medium of the Arabs, as has well been demonstrated by Count De Ficalho (Plantas uteis da Africa portugueza, p. 264), and from the Arabs the Negroes learned the narcotic properties of hemp. João dos Santos (Ethiopia oriental, 1586) testifies that hemp was cultivated throughout Cafraria and that the Kafirs called it by its Indian name *bangue* (*bangh*), as it is still called in the region of Zanzibar (Suaheli *banghi*). The most interesting point in this early Portuguese account is that in the sixteenth century the Kafirs only ate the hemp-leaves, but did not smoke them as they do at present; they could subsist merely on this leaf for several days without eating anything else, our Portuguese author informs us, but when they consumed much of it, they became intoxicated to such a degree as though they had taken a large quantity of wine.

[175]

Peter Kolbe, in his classical "Description of the Cape of Good Hope" (III, p. 290, Amsterdam, 1742) states that hemp was introduced to the Cape by the Hollanders and that it was exclusively grown by them, chiefly for the use of the natives. These, he writes, smoked the seeds and leaves of hemp like tobacco or sometimes mixed tobacco with hemp—a mixture called by them *buspach*. W. Paterson (Narrative of Four Journeys into the Country of the Hottentots and Caffraria, 1790, p. 94) found the native women cultivate tobacco and hemp and classifies these among the plants "which are not indigenous to their country and none of which he found growing spontaneously." It is interesting to note also that Paterson (p. 23) refers to *Mesembryanthemum acinaciforme* L. (family *Ficoideae*), known to us as the Hottentot fig, which he writes "is called *channa* by the natives and is exceedingly esteemed among them, being used both in chewing and in smoking; when mixed with the *dacka* ('hemp'), it is very intoxicating; it appeared to be of that species of hemp which is used in the East Indies by the name of *bang*." This plant Paterson found in a barren stretch of country near the Krome River in the Cape Colony, which he visited in 1777 and which was then called Channa Land (on his accompanying map: Canna Land). This *channa* or *canna* seems to be based on a foreign word of the type of Latin *cannabis*, Spanish *cañamo*, Portuguese *canhamo*, *cannamo* (*canaves*, "hemp-field"), Old Portuguese *alcanave* or *alcaneve* (showing in its connection with the Arabic article *al* its derivation from the Arabic); Persian *kanab*, Arabic *qannab*, Dutch *hennep* and *kennep*.

It should be added that hemp is used in Africa solely as a narcotic, nowhere as a fiber plant. It is at present cultivated down to the land of the Zulus in the south, and is also well acclimatized in the tropical zone. The leaves and immature seeds are simply dried in the sun. As formerly pointed out by me (Leaflet 18, p. 27), there

[176]

1, 2, 7, 8, Clay Pipes; and 3, Pipe with Bowl of Cast Brass. Cameroon, West Africa.
4, Water-pipe; and 5, 6, Wooden Pipes. Ovimbundu, Angola.

is no historical evidence for the opinion that hemp-smoking preceded tobacco-smoking. Neither for ancient India where the use of hemp as a narcotic originated, nor for the Islamic world do we have a single account of hemp-smoking in times anterior to the introduction of tobacco. It is quite certain that the smoking of hemp from a pipe came into vogue only as an imitation of tobacco pipe-smoking, while in earlier times hemp preparations were merely taken internally, either in the form of pills or liquids.

Ibn al-Baitar (1197-1248), an Arabic botanist born at Malaga in Spain and author of a famous treatise on pharmacology, was well familiar with the narcotic effects of hemp, but does not make any mention of hemp-smoking. He encountered the Indian hemp (*qunnab hindi*) only in Egypt, where it was sown in gardens, being called *hashisha*. He observed that it intoxicates those who even take a small quantity of it. He saw the fakirs use it in various manners. Some carefully boiled the leaves and then pressed them into a paste which was made into tabloids. Others dried the leaves, roasted them and triturated them with their hands, mixing them with a little sesame or sugar. This compound was placed in the mouth and slowly masticated, causing excitement and hilarity. It also resulted in intoxication and fits of folly. It appears that hemp was introduced into Egypt by the Arabs from India and Persia.

In Madagascar hemp was smoked in the middle of the seventeenth century, as related by E. de Flacourt (Histoire de la grande isle de Madagascar, p. 145), who gives a vivid description of the pernicious effect of the drug and also refers to the use of the leaf in India, where he says it is called *bangue*. The Malagasy name given by him, *ahets-mangha* (*ahets* means "herbs"), is doubtless based on the Indian *bhanga*.

BERTHOLD LAUFER

[177]

USE OF TOBACCO IN AFRICA

Although there may be differences of opinion as to when and how the tobacco habit spread in Africa, there is not the slightest reason for doubting the enjoyment which is derived by the entire native population from the use of tobacco. Smoking, snuffing, and chewing are widespread practices. In addition to these there is the use of a water-pipe for smoking hemp or possibly a mixture of tobacco and hemp.

The majority of smokers in civilized countries associate the use of a tobacco-pipe with leisurely ease and protracted enjoyment, but in Africa, on the contrary, the aim appears to be rapid intoxication; for even when hemp is not used, the smoker inhales deeply and rapidly in order to produce a comatose condition. Just before this delectable achievement the pipe is handed to the next man in the circle.

Africans regard tobacco as something more than a trivial amusement; in fact, the tobacco habit is often closely connected with public observances, social etiquette, and recognition of difference of rank. The rules regulating the particular way in which tobacco may be used are of a local and arbitrary kind. Thus, the prevalence of smoking, chewing, or snuffing, and the customs determining the employment of tobacco in relation to rank, sex, and age, are of very irregular distribution throughout the continent.

On the contrary, there is a possibility that the use of the water-pipe has, owing to Arab conquest and trading, made an advance along definable routes. Possibly this form of pipe, which is usually associated with hemp-smoking in Africa, made its entry from Asia at one or more points on the coast of east Africa. The pipe is the same in principle in all parts of the continent, though there is great variety of form, and much ingenuity is shown in adapting local materials such as horns, earthenware, gourds, and bamboo for the construction.

The water-pipe has been reported in Abyssinia. It is often found in East Africa from Lake Victoria to Zanzibar. Zulus of South Africa are addicted to its use. In recent years the smoking of hemp in water-pipes has been reported from the southern Congo, Angola, and as far west as Liberia. Along the coast of North Africa and in towns of the Nile valley the nargileh is a form of water-pipe more elaborate and ornamental than any found in other parts of the continent.

The social importance of tobacco is in no way more deserving of attention than are agricultural and commercial aspects of production and distribution. The work of cultivating the domestic tobacco-patch, like all other agricultural operations, falls to the lot of women who generally utilize a piece of waste land where rubbish has been thrown quite near the dwelling. A technique of treatment, for example, planting out young shoots, removal of superfluous foliage, and subsequent drying of the leaves has been evolved. Then, there are local methods of packing the leaves into conical or cylindrical bundles forming a ready currency which has been used in barter over large areas.

In order to understand the importance of tobacco in native life, there is no way better than that of visiting the homes of tribes in many parts of the African continent.

People in Egypt and northern Africa use the hubble-bubble, a water-pipe around which sit the bazar traders complacently smoking in the intervals between the fleecing of one purchaser and another. This is, of course, a modern social use of tobacco without any ritualistic or other import. There has been great divergence of opinion in the Mohammedan world with regard to the permissibility of smoking. Mohammed placed an interdiction on the use of alcohol, but the extension of the prohibition to the use of tobacco is fanciful as the introduction of tobacco into the Old World post-dated the writing of the Koran by ten centuries. Nevertheless there are fanatics, such as the Wahbis of Arabia and the Senussis of Libya, who abhor the use of tobacco and, stranger still, their prejudice extends to the use of coffee.

Tobacco is said to have been brought into Morocco toward the end of the sixteenth century by traders from the town of Timbuktu, a great mediaeval emporium of trade on the northern bend of the Niger. The initial prejudice against the use of tobacco gradually broke down, to be revived again in 1887, when large stocks of the plant were publicly burnt in Morocco. Imprisonments for smoking followed this demonstration, but the interdiction failed through the lack of public support, so that cigarette-smoking is now prevalent again. Native Moorish tobacco of poor quality is grown on the slopes of hills, but the smoker is prone to give flavor to his product by mixing it with Indian hemp (*Cannabis indica*). This compound which is ordinarily called hashish is smoked in a small pipe with a clay bowl and a wooden stem. When used as a sweetmeat ball, a pellet of the mixed herbs is referred to as *m'joon*. The Moorish word for a hemp-smoker is *kiyaf*. Such a one may be readily recognized by the pallor of his face, his half-closed eyes, and listless attitude. Victims of this drug habit are found to suffer moral degeneration, which makes them utterly untrustworthy and unreliable. Opium is eaten by aristocratic Moors of large towns, who would scorn to be seen using the humble pipe with its charge of hemp and tobacco.

Away from the coastal region of Morocco few Moors are to be seen without their snuff-box; the snuffing habit appears to be confined to the men who follow a very general custom of mixing the tobacco dust with powdered

1, Ivory Pipe Bound with Brass Wire. Fangs, Gaboon. 2, Wooden Pipe. Makas, Cameroon.
3, Snuff-box. The clip is placed on the nose after snuff has been taken. When not in
use, it is placed on the ear. From a photograph by Wollaston. 4, Use of the Rifle as
a Pipe in East Africa. From a sketch by Frobenius. 5, Water-pipe. Lower Congo.

shells of walnut-wood ashes. The snuff is laid along the back of the hand from the tip of the index finger, and after this careful preliminary half the length of the train is snuffed by each nostril. Moorish snuff-boxes are often made from young coconuts furnished with ivory probes as stoppers. Decorative designs are added by inlaid silver wire; a high polish is given, and finally a silver chain is added for suspension.

Oscar Lenz, writing in 1884, describes the equipment of smokers in Timbuktu. The pipes had wooden bowls ornamented with inlaid silver, the mouthpiece was of iron, and a string was added for suspension round the neck. The smoker's outfit included a pipe cleaner, pincers for applying a glowing coal to the tobacco, and in some instances flint, steel, and tinder. This traveler quaintly remarks that "he could hardly trust his own eyes," when he saw the smokers rolling their tobacco with butter before charging their pipes. One can well believe the assertion that the taste and smell are so abominable that a stranger cannot smoke the greasy mixture. At that time in the French Sudan snuff was not unknown, for the commodity was used in the form of a pungent yellow powder which men carried in ornamented leather pouches.

Saharan trade routes were used by camel caravans long before historical times; so there has been ample opportunity for the Tuaregs, who are great traders and adventurers, to become acquainted with the use of tobacco. Recently written accounts indicate that the Tuaregs of Air in the southwestern Sahara do not smoke; their use of tobacco is restricted to snuffing and chewing. When the latter custom is followed, green tobacco is reduced to a powder and mixed with saltpetre "to bring out the taste." Snuff placed in the eyes of camels suffering from congestion of blood in the head is said to give relief to the animals.

In the year 1795 Mungo Park penetrated far into the interior of West Africa, where he found the inhabitants of

[181]

Kaarta and Bambarra addicted to the use of tobacco and snuff among all social grades. The pipe-stems were made of wood to which a curious bowl of earthenware was added. At that early date in the opening-up of West African trade, bars of tobacco formed one of the number of articles, such as gin, gun-flints, and gun-powder, each of which had a certain value in relation to native products, the most valuable of which were gold dust and ivory.

The kingdom of Dahomey, now under French administration, has always been of particular interest to students of West Africa. From the fifteenth century onward this coastal province was a striking example of a great despotism. The king, who was absolute, waged incessant warfare with neighboring states. The military system, which included a band of women known as Amazons, was a triumph of military organization. A very reluctant guest of the king was Skertchly, who in 1871 was kept for some months as an honored prisoner and companion to his majesty. Close confinement to the royal compound caused the naturalist to lament that he was unable to pursue his studies in botany and zoology; but this very restriction of liberty resulted in the production of a book which gives detailed accounts of ceremonies, including those in which human sacrifice was made. Skertchly states that on public occasions men of importance were followed each by an attendant, who carried his tobacco-pipe and pouch as insignia of rank. These pipes were invariably of native manufacture. It is said to have been customary for the smoking outfits of officials to be a product of the Amazons' industry.

The bowl of the pipe was generally of reddish yellow clay, though the color might be dark owing to the presence of manganese. Much labor was expended on the carving of bowls which often took the forms of birds, fish, canoes, and human beings. The tube of the pipe was carried in a wooden box having a sliding lid, and the stem appears to

[182]

have been carefully made as was the bowl. Tobacco-pouches of great size were made of goat skin treated with dyes of several colors. In addition to tobacco the pouches held gun-flints, steel, and tinder made from decayed palm-wood. The tobacco is definitely stated to have been of American importation. "Short clay pipes are popular among both sexes, the old women seem to prefer a filthy clay so short that they get as much smoke up their noses as they get into their mouths."

During ceremonial speeches, which were lengthy and tedious, the king's head smoker was prowling about with an immense wooden pipe from which he blew clouds of smoke in the faces of the guests. This may appear rude and aggressive; but, although Skertchly does not say so, the custom was probably a mark of favor. Instances of puffing smoke from the mouth of a headman to the mouth of a guest seated next to him have been recorded in recent times. The official smoker of the Dahomeyan king wore a robe of brown cloth sewn all over with long strips in imitation of tobacco-leaves, while a necklace of pipe-bowls completed his equipment.

Buttikofer states that both men and women of Liberia are passionately addicted to smoking tobacco of foreign importation; but when prices are prohibitive, some solace is obtained by placing a glowing coal in the bowl of the pipe. The fact that "the smoke goes up their noses and intoxicates them" seems to be no deterrent. Very humorously this traveler describes the way in which men, women, and children would gather round him when he smoked. The leader of these uninvited guests came close enough to inhale the smoke as it was ejected. For a few moments he retained it, then blew it into the mouth of a companion, the process being continued until each had received a puff, or the smoke was exhausted. Chewing is also practised in Liberia, though the custom is not so popular as that of smoking.

[183]

Snuff is perfumed with Florida water after it has been pounded with ashes in a mortar made of ivory. In addition to ashes, pounded banana skin and soap are sometimes mixed with snuff, then a pinch of the powder is taken between the thumb and finger, or a silver spoon may be used. Small horns from goats and sheep are the usual snuff-boxes. The water-pipe for smoking hemp is said to have been imported from the Congo to Liberia; the apparatus consists of a clay bowl for the mixture fastened to a gourd which holds water through which the smoke is drawn.

The Kagoros, a head-hunting people of northern Nigeria, have their milder moods when the weapons of war are laid aside in favor of ill-balanced pipes a yard in length, the bowls of which rest on the ground supported by two short legs (Plate I). Every man is expected to carve the wooden bowl for himself, but the manufacture of iron pipe-stems is in the hands of the local blacksmith. Economy is here the rule, so the thrifty head-hunter mixes wood ashes with his tobacco "to make it go farther and to improve the flavor." Women of the Kagoro tribe are not allowed to smoke. Tobacco is snuffed as a remedy for headache, and along near-by trade routes coils of tobacco are a well-known currency. There is undoubtedly a certain Muslim influence at work counteracting the use of tobacco in West Africa. Denial of tobacco to women is perhaps a result of the inferior social standing of them in countries under Mohammedan influence, or there may be just a feeling that the extension of a privilege to women is derogatory to the male dignity. To the Jukuns of northern Nigeria use of tobacco is a repulsive practice.

Nature has been unkind to the would-be snuff-takers of Kivu in the eastern Congo, for many of these people have broad turned-up noses from which the precious powder easily escapes. Native ingenuity has, however, risen to the occasion in the provision of a nose clip of wood

[184]

which is applied when the snuff is taken. When not in use, the native carries the clip behind his ear (Plate III, Fig. 3).

The appreciation of tobacco is so widespread in the Congo region that Bushongo natives of the Southwest have attempted to explain the introduction of the plant by a legend. A man of the Bushongo people, the story runs, astonished his tribesmen by producing a pipe from the trade goods brought from distant places. While smoking in the center of a curious circle, he proceeded to explain the value of tobacco by saying, "When you have had a quarrel with your brother, you may wish to kill him; sit down and smoke a pipe. By the time this is finished, you will think that death is too great a punishment for your brother's offence, and you will decide to let him off with a thrashing. Relight your pipe and smoke on. As the smoke curls upward, you will think that a few harsh words would serve instead of blows. Light your pipe once more and, when the bowl is empty, you will be ready to go to your brother and forgive him."

Torday says that hemp-smoking (Plate III, Fig. 5) is such a widespread and pernicious practice among the Balubas and other peoples of the Congo that in the course of his official duties he decided to burn a large stock of this narcotic, for the loss of which he indemnified the owners. This recompense gave no satisfaction, for all the hemp-smokers were agreed that the indulgence was food, health, and happiness. Life was not worth living if they had no hemp. Hemp-smoking is said to be the curse of the Batetala tribe, according to Hilton-Simpson, who accompanied Torday on his journeys in southwestern Congo Basin. The former observer writes, "I noticed a man squatting on his haunches at the side of my chair. From time to time he made a sweeping motion of his hand toward his face, and I was quite at a loss to understand the movement. It suddenly dawned on me that he

[185]

was directing the smoke into his own mouth, evidently he had left his pipe at home. Three kinds of tobacco-pipe are in use in this region, for in addition to European forms there are gourd water-pipes and bamboo pipes, all of which have pottery bowls. Absence of tobacco is far more serious than the lack of a pipe, for the latter deficiency is easily met by rolling a banana leaf into the form of a cone. Snuff-taking is a somewhat disgusting habit, for a supply is smeared over the top lip and nose to give a more protracted enjoyment."

So long are the pipes of smokers in the central area of Cameroon that a servant is required to apply a light to the bowl. The Museum has a remarkably choice selection of pipes from Cameroon (Case 10, Hall D, Plate II, Figs. 1-3, 7-8), which show to great advantage the skill of the Balis and other tribes in brass-casting, wood-carving, and beadwork; many of the pottery bowls are elaborately formed. Bowls of clay are molded by hand, though pieces of bamboo are often employed to aid the finer modeling. The largest bowls are partially air-dried before the protuberances and appliqué designs are added. The whole product is then fired. Some of the pipes from Cameroon show an unsightly innovation in the wrapping of tinfoil round well-carved wooden stems. Small wooden pipes without ornament are smoked by the Maka of southern Cameroon (Plate III, Fig. 2). There can be little doubt that the large artistic pipes are reserved for the use of chiefs. In fact the ceremonial smoking of such a pipe by a chief, who is officiating as a priest in the ceremony of feeding the ancestral ghosts, has been authentically described. In central Cameroon everybody smokes from the great-grandfather to the child toddling beside him.

The Bali language has several words describing the strength of tobacco, while social conventions center round the smoking habit. The meeting of two men, one of whom is without his pipe, is an amusing incident.

[186]

The more fortunate of the two inhales deeply several times, and at each expiration puffs the clouds into the widely opened mouth of the less fortunate. The two then separate without a word having been spoken. This complimentary greeting is permissible only when the men are social equals. While smoking is a characteristic habit of the grassland area, snuff-taking is more usual in the forest region of Cameroon. The habit of allowing the finger nails to grow to an excessive length as an indication that menial work is not done is known among the Balis; but this custom is distinct from that of the snuff-takers, who allow just one nail to grow indefinitely so that it may serve as a snuff-spoon.

In southwest Cameroon tobacco is said to be next in importance to salt as a medium of exchange. The Jaundes fully appreciate the many brands of imported tobacco, declaring that they could not live without their tobacco-pipes, but for chewing and snuffing of tobacco they have no liking.

Among the Fangs, who live to the north of the Congo estuary, are to be found, according to Tessman, three main types of tobacco-pipe, the most common of which is the short, wooden variety. A second example is provided by the pipe having as its stem a rib of banana leaf more than a yard long, on to which is fixed a tobacco bowl made either of clay or the kernel of a raffia fruit. This perishable stem has to be renewed from time to time. The third variety of pipe has a bowl of clay fixed to a stem of wood or ivory which is neatly bound with brass wire (Plate III, Fig. 1).

An example of the second kind of pipe is to be found on the wall of each communal house. The chief is entitled to the first series of deep inhalations, after which he passes the pipe to the next man. Tobacco is grown either in close proximity to the house, or it may be planted with such field crops as earthnuts. Flowers, fruits, and a

[187]

number of leaves are removed from each plant so that
the remaining leaves may attain a large size and good
flavor, for it is a mistake to allow the plant to mature and
go to seed quickly. In spite of the destruction of flowers
and seed capsules there are always a large number of
seedlings from plants that have escaped these prunings.
Such self-seeded plants are transferred to a new plot when
the tobacco harvest is gathered. Drying is carried out by
suspending the plants from the roof of the hut, and from
the individual harvests the supply of tobacco in the
communal house is replenished.

To the southwest of the Congo region lies Angola, a
Portuguese possession, whose native population has
received less attention than that of any other part of
Africa. One of the few writers on this region is Monteiro,
who published an account of it in 1875. He appears to
have been impressed with the medicinal uses to which
tobacco was adapted. For inflammation of the bowels,
colic, or other violent pains the natives applied to the
abdomen tobacco-leaves which had been dipped in boiling
water. They also chopped the leaves and made them into
a poultice mixed with castor-oil. The Portuguese too are
known to have favored these methods. Tobacco is said
to have been expensive, and the natives have at times to
find solace by placing a piece of charcoal in their pipes.
Some of the inhabitants of Angola mixed their tobacco
with a species of orris root, which they enjoyed because
of its real or fancied resemblance to goats' flesh. After
tobacco-leaves have been dried, they are reduced to
snuff by being beaten on a stone, but the substance is not
ready for use until it has been mixed with the ashes
resulting from the burning of a strongly alkaline bush.
Snuffers who require a stronger stimulant add a quantity
of chili pepper until the desired result is obtained. Snuff-
boxes are of a simple kind formed from slender bamboo or
canes with nodes that serve as bottoms, while the lid is

a wooden plug; they are suspended round the neck by means of a plain string. After a quantity of snuff has been placed in the open palm, the snuffer buries his nose in the mixture, meanwhile giving his nose a rotary motion and snorting loudly. Ingenious porters allow a small stubbly moustache to develop; this serves as an ever ready snuff-box from which a supply may be taken by curling up the lip, hence there is a saving of time when on the march with a load on the head.

The tobacco plant grows near most Angolan villages. Pipes are often well carved. In northern Angola snuffing by the method described in Monteiro's time is still in vogue. It is important to note a present-day reference to the smoking of hemp in Angola. The smoke is drawn through a gourd of water (Hall D, Case 22A; Plate II, Fig. 4), but even so, a few inhalations cause the smoker to cough violently before he passes the pipe. Men say that hemp-smoking makes them warm in the early morning when the highlands are cold.

The Hereros of southwest Africa are a decadent people, who are now glad to beg a pipe of tobacco from any passer-by; but twenty years ago any man of importance among the Hereros and Bergdamaras had his own flourishing tobacco plot cultivated by domestic slaves. As far back as 1870 it was a common sight to see a Herero advancing to a Bergdamara village driving goats that were intended as an exchange for tobacco. At the present time the Hereros are glad to obtain nicotine-soaked dregs which they mix with dried cow-dung. The effect of such a compound is bewildering, the teeth and gums of the smoker become quite black, and in consequence of passing the pipe from mouth to mouth sores are transmitted.

A curious feature connected with the use of tobacco in South Africa is the construction of pipes in or on the ground. This method, which is sometimes described as "earth-smoking," has the advantage of ease and simplic-

[189]

Pondo Girl Smoking. Zulus, Southeast Africa.
From a photograph by Dudley Kidd.

ity.   The smoker needs no apparatus, and while enjoying his weed, he is able to lie at full length on the ground. H. Balfour has, in addition to collecting three varieties of earth-pipe, summarized the various accounts of the ways in which this interesting form of pipe is used (Plate V, Fig. 2). Examples of earth-pipes from the vicinity of Victoria Falls on the Zambezi show that the bowl was formed by scraping together a quantity of moistened red earth to form a mound three inches in diameter and one inch high. The under surface is flat because of its attachment to the ground, and the upper surface is convex. A duct representing the stem of the pipe was formed by withdrawing a hollow grass stem which had been embedded in a wet mass of clay surrounding the bowl. The pipe, said to be the work of a Maklanga native, who was imported into the district as a laborer, would be ready for almost immediate use owing to the quick drying action of the sun. The hollow bowl of such a pipe is formed when the clay is wet, and the shaft of a spear may be used to support the wet earth of the tube until it has hardened. Schulz describes such pipes which are the work of the Bechuanas who smoke hemp in them.

Balfour examined the carbonized remains of the contents of these pipe-bowls, and has come to the conclusion that in several instances hemp had been smoked. Schulz writes that the Bechuana who has filled the bowl of his pipe with hemp places water in his mouth. Then he kneels down and draws in the fumes with deep inspirations; thus the earth-pipe is used as an elementary form of water-pipe or hubble-bubble. Another ingenious form of ground pipe is made by digging a pit to serve as the bowl, from which a duct is made to lead to the surface by boring the soil with a stick. The smoker then extends himself prone on the ground in order to apply his mouth to the surface hole. Sometimes he may use the double pit connected by a tunnel which is made to contain water.

[190]

Bushmen pipes are made from stone, reed, bone, or the horn of an antelope (Plate V, Figs. 1, 5). In the absence of a pipe, pieces of narcotic root are first ignited, then held under the nose. Theal states that among the Bantus of South Africa men drank the leaves of wild hemp which had been pulverized and mixed with water.

One of the most graphic accounts of hemp-smoking has been given by Schulz and Hammar, who were greatly concerned about the life of one of their boys. The youth had deeply inhaled several strong whiffs of hemp in rapid succession from a horn water-pipe. "He fell over in the sand and almost ceased breathing, while his heart beat fainter and fainter and his skin assumed the appearance known as goose-skin. None of the boys evinced any sympathy or endeavored to help him except by hitting him with a stick, meanwhile laughing at him." The initial effects of smoking hemp are different from those described above if the smoker is accustomed to the narcotic. He becomes vaunting and noisy in his narration of stories relating to his prowess in war, then later, after wild gesticulation and rolling of the eyes, he may sink to the ground.

Smoking of hemp is not always followed by such scenes of violence; in fact, spitting out bubbles through a hollow reed may result in a kind of game. Dudley Kidd gives a good illustration of this play in which the bubbles are projected on to the ground as the men try to outflank one another in position.

A collection of snuff-boxes and pipes from South Africa is shown in Cases 25 and 27 of Hall D (Plates IV and V, Figs. 3, 4, 6, 7). The pipes are of simple form, not unlike those used in England, though some of them have a double bowl which makes the plugging of one bowl necessary when the other is in use. A Zulu is seldom seen without his snuff-box which he keeps full of a mixture of ground tobacco and ashes. Snuff-boxes may be small gourds, plain, decorated with incised lines, or ornamented with

[191]

beadwork.   Others are tips of horn which are worn suspended round the neck, on the wrist, in the hair, or even
through the ear-lobe. Some knobkerries have a hole at
the end for the reception of snuff. A few of the objects
in the Museum illustrate a curious and ingenious method
of making snuff-boxes which are characteristic of the
Zulus only. The scrapings of meat and skin resulting
from the dressing of hides are mixed with blood and red
clay until a stiff paste is attained. This paste is plastered
over clay models on which it is allowed to harden as a
surface dressing. A round hole is then made at one end
of the object, and the clay is scooped out, leaving only
the outer covering which forms the snuff-box.

The practice of rolling tobacco-leaves into cigars is
one that does not appear to have recommended itself to
Negroes. Nevertheless, a few of them to the north of the
Zambezi have adopted the Portuguese method of rolling
tobacco in banana leaves which are called "carrottes."
The use of such cigars is local, and the habit has not penetrated south of the Zambezi. Such a cigar is carried
behind the ear whence it is removed many times a day for
a few puffs. Friends express greetings by exchanging whiffs
at their cigars.

In his book "The Zoolu Country" (1834), Gardiner
mentions a Zulu chief who wore an ivory snuff-spoon in
his hair and a cane snuff-box in his ear. At the present
day snuff is ground from crude tobacco which is grown
near almost every kraal, and following the usual African
custom ashes, in this case ashes of aloes, are mixed with
the tobacco. Sneezing is a good omen, provided the words
"may the chief bless me" are spoken. The word "snuff"
has found its way into a proverb, "I sent him for snuff,
and he brought me ashes," used, of course, in reference to
a thoughtless messenger.

Among the Hottentots tobacco is said by Schulze to
have a medicinal value in curing poisoned wounds. When

[192]

PLATE V

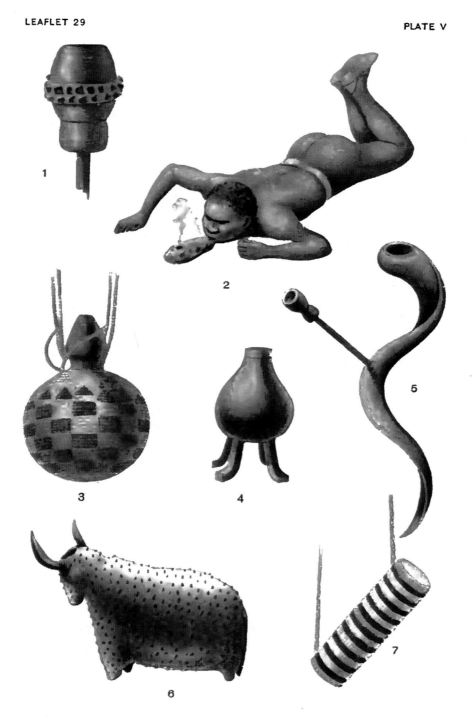

1, Stone Pipe. Bushmen, South Africa. 2, Earth-smoking. South Africa. 3, 4, 6, 7, Snuff-boxes. Zulus. 3 and 7 are covered with beads; 6 is made of clay.

a man is bitten by a poisonous scorpion, search is made for such an animal, which is pulverized and laid on the wound. Meanwhile the patient has to drink water in which tobacco has been steeped.

In British Central Africa the Yaos and Anyanjas have homegrown tobacco for consumption and sale. Considerable attention is given to the plants whose leaf-buds are pinched off to make the remaining leaves attain greater size. After the leaves have been soaked in water and spread in the sun to dry, the Yaos plait them into strands, while the Anyanjas roll them into balls. The Angonis vary the practice by making their tobacco cakes into pyramidal form. Snuffing is a common habit, and chewing is practised; for the latter purpose the tobacco is mixed with powdered shells of snails. Men smoke hemp which they say is as good as food and drink to a tired man, though it is admitted that the hemp "catches their legs." Hemp grows about the villages without any special cultivation.

Although growers of tobacco, the Bambalas appear to be very ignorant of methods of curing the leaves. Toward the end of the rainy season the seed is sown in the shade of the hut. When the seedlings are of a size that can be handled, they are transplanted to some fertile patch, preferably an ant-heap. No effort is made to improve the quality of the leaves by pinching off the suckers and early buds; in fact, all the plants are allowed to run to seed. Of the two varieties of tobacco one has short leaves which are made into flat cakes; this is the stronger kind. The second variety may be distinguished in trade because it is always made into cylindrical packages, each weighing ten pounds. Dried stalks of the water-lily are added to the tobacco in order to make snuff. Both men and women smoke tobacco in long pipes provided with bowls of earthenware and reed stems. The pipe is circulated in the manner described for other peoples. The mixture of

[193]

lime, butter, and other substances to the tobacco has been mentioned, but the climax is reached in the glands of the skunk being added if the tobacco is to be used for snuffing, as stated by Smith and Dale.

The Baronga people living in the south of Portuguese East Africa have a phrase for tobacco, which means "the powder that stimulates the brain." The offering of snuff is a greeting and a necessary preliminary to any conversation. In that part of Africa snuffing is far more common than tobacco-smoking. Junod says, however, that an old custom was that of smoking hens' excrements after the burial of a corpse. At the present day the only smokers are the old women of Lourenço Marques.

The ceremonial use of tobacco is illustrated by the customs prevailing at a Batonga wedding. At this time one of the dancing girls places snuff in the hand of the bride, who gives it to her husband. After he has taken a pinch the girl throws the remainder in his face, then she escapes with the groom in pursuit. The girl has sanctuary by holding a tuft of grass, and refuses to return to the village until money has been given to her. It is considered very bad luck to carry tobacco in the leaf when visiting a lover; at such time snuff only should be carried.

The Negroes of Portuguese East Africa are addicted to the use of leaves of a species of Datura which are smoked through a double-decker gourd whose use gives rise to violent paroxysms of coughing. Cigar-smoking was mentioned as a custom favored by Negroes north of the Zambezi. Maugham adds that in Portuguese East Africa "tobacco is rarely smoked, though it grows with great freedom and luxuriance, in any other form save a cigar of great length." The statement is made that the smoke is inhaled by cupping the hands round the lighted end. This is a local method apparently, and it is difficult to believe that any great satisfaction can be derived from it. The evidence is, however, corroborated by an inde-

[194]

pendent observer, Fulleborne, who also describes the custom of earth-smoking in Konde Land.

The researches of Sir Richard Burton reveal a number of interesting customs relating to the use of tobacco in East Africa about the year 1860. There is no doubt that most of these usages still survive.

Arabs of East Africa, in .the region about Zanzibar, had a prejudice against smoking, but no objection to the chewing of tobacco; consequently the latter custom combined with snuffing is the usual method of using tobacco. Ground coral or pulverized cowrie-shells are added to the snuff by the Suahili, while saltpetre or the ground wood from the middle of a plantain may be added to give flavor to the chewing-tobacco used by the Wanyamwezis.

Near the eastern shore of Lake Tanganyika the inhabitants take liquid snuff. Every man carries a little black earthenware pot from which he pours into his palm a quantity of water in which tobacco leaves have been steeped. This he sniffs into his nostrils, which are then closed with an iron or wooden clip, though the thumb and finger may be used. The liquid is so held for several minutes. Burton describes a women's smoking party at Yombo where the circle included wrinkled old dames and young girls who were provided with pipes of great length. All smoked with intense enjoyment, deeply inhaling, and from time to time cooling their mouths with slices of raw manioc or cobs of green maize roasted over the coals.

Hollis says with reference to the Masais, "Some old men and women chew tobacco mixed with salt, some take snuff, and others smoke pipes." The Masai are a pastoral people who despise agriculture; they do not on this account cultivate tobacco, but obtain it by exchanging butter and lean goats with their neighbors. Everywhere among Africans the habit of snuff-taking is regarded as very refreshing. A Masai story refers to the return of a hunter who called for his stool and snuff-box that he might

[195]

refresh himself before giving an account of his adventures. A Masai youth begins his courtship by most tactfully presenting the father of the girl with a present of tobacco. This is but the forerunner of subsequent gifts of like kind.

In the neighborhood of Lake Victoria Nyanza local custom varies with regard to the particular use of tobacco. Among the Wambugwes men chew, smoke, and snuff. In Ruanda tobacco is chiefly smoked, while in Urundi it is generally snuffed. With the Wanyamwezis and the Washashis the smoking of hemp in a water-pipe (Hall D, Case 24) is a widespread custom. The Nandis are accustomed to observing the custom of hemp-smoking among their neighbors, the Kavirondos, but they themselves have been wise enough to avoid the habit, though they have ready access to wild hemp. The Nandis say that throwing tobacco on a fire during a thunderstorm gives immunity. The Lumbwas (Kipsikis), who are near neighbors of the Nandis, take a form of liquid snuff, but this habit has not extended to the Nandis. Possibly the latter would think it beneath their dignity to borrow a custom from a people whom they regard as their inferiors.

Stanley relates in his book, "How I found Livingstone," that in his time (1870) tobacco played an important part in trade. Tobacco, which was not of very good quality, was made into loaves each of which weighed three pounds. Such a loaf was worth four yards of trade cloth, which was also the exchange value of a black steatite pipe with a stem bound with fine wire. Stanley writes, "The natives (Wanyamwezis) are very fond of using *bhang* with their tobacco. Their nargileh is a very primitive affair made out of a gourd and a hollow stick. One or two inhalations are sufficient to send them into a fit of terrible coughs which seem to rack their frames." See Plate VI, Fig. 5.

The Akikuyus appear to reserve snuff and home-made beer as two of the chief consolations of advancing years, for both these luxuries are generally denied to the young

[196]

men. Snuff is prepared by grinding tobacco-leaves with sheeps' fat, and the offering of the mixture is a courtesy and greeting between strangers.

Roscoe relates a story of the Banyankoles, which is supposed to account for the introduction of tobacco into their territory. A medicine-man from a neighboring tribe, whose king was a friend of the king of Banyankole, arrived with a present of six bags of tobacco which the recipient "found very soothing." Before returning to his own country, the medicine-man showed the Banyankoles how to cultivate and cure tobacco. In this tribe there is a superstition to the effect that presents of tobacco to a prospective bride will make the marriage sterile. Rain-making is an important function of medicine-men in all parts of northeast Africa where the people are dependent on grass and cattle. The rain-maker of the Baris enters the hut of his chief, then fills the bowl of his tobacco-pipe with beads and pebbles. These he shakes out of the bowl several times, and at last he makes a declaration with regard to the rain supply.

The majority of the Bakitaras grow their own tobacco which is used or readily bartered. Both men and women smoke, but boys are not allowed the indulgence until they arrive at puberty, while girls may not use tobacco until they are married. Chewing is a common habit even among the wives of the king, but masticating tobacco in the presence of the king was regarded as a gross insult. When a woman met the king while she was chewing tobacco, she had to get rid of the bolus as quickly as possible because his majesty would in all probability open her mouth, and if any tobacco were seen therein, she would be punished, possibly killed.

On dust heaps near the dwellings of the Bakitaras tobacco plants grow luxuriantly to the height of six feet. The large, broad leaves are dried on mats, after which the coarse parts, such as the ribs, are removed, then the

[197]

finer parts of the lamina are broken up. These are again
dried and tied into packets of a size suitable for barter.
Tobacco for the use of the king was more carefully dried
and made aromatic in a way which the people would not
disclose.

Men who were sent to carry the bride to her husband
dared not enter the kraal at once, but the tedium of their
waiting was relieved with presents of tobacco from the
bride's father. Snuff is not in favor with the Bakitaras.
This is because the act of sneezing, like coughing, spitting,
and blowing the nose, are marks of the greatest disrespect.
Such an action in the presence of the king was equivalent
to an act of gross disobedience.

From early stone-age times it has been customary to
place by a corpse some small offerings of food or stone
implements, which were regarded as indispensable in the
spirit world. The Suks, exceedingly tall people living
near Lake Rudol, make provision of this kind when a
good man dies. They say, "When a good man dies, and
his body is thrown away, we go where his head lies and
bring a little food and tobacco to make his deadness more
endurable. But when a bad man dies, we give him
nothing; we say 'let him die some more.' "

In many parts of Africa there is a belief that the spirit
of a dead man may pass into a snake. When this reptile
enters the house of a living relative, it is entertained among
the Suks with offerings of milk, meat, and tobacco. The
Suks occasionally smoke tobacco from the shank-bone of
a sheep, but more usually snuff is carried in a box suspended from the neck by a chain. The latter indulgence
is enjoyed by all, with the exception of the smallest
children.

Men of a Nilotic tribe named the Langos grow tobacco
plants between their huts, the variety with the yellow
flowers being preferred. Tobacco is smoked only by the
old men, as it is considered harmful to warriors and

[198]

hunters. Young men say that the use of tobacco interferes with their love-making as the girls object to the smell, and women never indulge in the habit. Tobacco is never used by the Langos in the form of snuff.

At the period of Schweinfurth's exploration in the northeast Congo Basin, and among the Dinkas, he found these people growing Virginia tobacco. The plants were carefully raised in the shade as seedlings before being planted out in an exposed situation. The influence of Islam was apparently felt because Schweinfurth states that "the pagan Negroes, so far as they have been uninfluenced by Islam, smoke tobacco. Those who have been so influenced prefer the chewing of the leaf to the enjoyment of the pipe." The Bongos were found to be smoking *Nicotiana rustica,* whose pungent leaves were made up into round cakes by means of a mold, after they had been previously powdered, pressed, and dried. A circumstance related by the explorer suggests that some of the men were in the habit of smoking something stronger than *Nicotiana rustica.* "On one of our marches a Bongo had inhaled to such excess that he fell senseless into the camp fire and was so severely burnt that he had to be carried on a litter for the remainder of the journey."

The Dinkas, like the Bongos, pass the pipe from hand to hand, and with this is sent, for chewing, the wad of bast which has been used to intercept the tobacco juices in the stem of the pipe.

Parkyns, a traveler in Abyssinia in the year 1850, notes the use of the water-pipe as part of the hospitality offered to him at Nassawa.

A detailed account of the tobacco-using habits of the Gallas has been given by Paulitschke. These people smoke tobacco, but their neighbors, the Somalis and Danakils, prefer to chew. So unaccustomed are the Somalis to smoking that they do not know what use to make of a proffered cigar; but in spite of this indifference

[199]

to smoking they actually swallow tobacco, saying that it purifies them. The Gallas, in addition to smoking, are addicted to chewing and snuffing. Snuff is mixed with saltpetre, and the "chew" of tobacco is carried behind the ear. For trade purposes the northwest Gallas make loaves of two pounds weight from tobacco-leaves, and on these the worker presses his finger as a sign of his manufacture.

The evidence collected by a brief regional survey of Africa indicates that a custom, which may at first glance appear trivial, is possibly closely associated with some of the most important aspects of social life. Quite apart from the economic questions connected with the cultivation of tobacco on a large scale by native labor under European supervision, there are several indications that the plant has for perhaps three centuries been affecting the development of trade-routes and systems of barter (Plate VI, Fig. 7). In addition to this, the cultivation of tobacco has stimulated agriculture. What is more important still, the practices of smoking, chewing, and snuffing are centers around which many social usages have been grouped. Incidentally the inquiry has emphasized the strength of customs which appear to be of fortuitous occurrence. Custom is arbitrary, but when formed there are few deviations. In some instances, the evidence makes it clear that Islam has profoundly affected the growth of habits relating to the use of tobacco in Africa.

WILFRID D. HAMBLY

[200]

USE OF TOBACCO IN MADAGASCAR

The natives of Madagascar believe tobacco to be indigenous, but there can be little doubt that it was introduced in early times, probably from Africa. By the latter half of the eighteenth century its use was already universal. The present native tobacco is a large-leaved, white-flowered plant growing from three to four feet high in good soil. It is probably a variety of *Nicotiana tabacum*. The general native name for it is *paraki*, but there are special names for the various prepared forms.

The plant has escaped from cultivation, and is often found growing seemingly wild; throughout most of the island it can hardly be said to be cultivated. Small clumps of it grow in the villages and on the edges of the tilled fields, seeding themselves year after year. There seems to be no attempt to care for it or to improve the quality. On the east coast and in the plateau the leaves are plucked when they begin to turn yellow and impaled side by side on long slivers of bamboo, which are thrust through the stems. When they have been partially dried in the shade, they are hung over the fireplace in the dwelling. Here they become thoroughly dry and smoked. On the west coast the leaves are also dried in the shade, but before they become brittle, they are made up into hard rolls about three inches in diameter and one inch thick. When the tobacco is to be marketed, a number of these rolls are placed side by side, and the whole is enveloped in sheets of banana bark and tied or wrapped with rope.

Three methods of using tobacco are known in the island: making a quid of the green leaf, smoking, and snuff-taking. The use of the quid seems to be limited to the Betsileos in the southern part of the central plateau.

[201]

A fresh leaf is wilted over the fire, rolled in wood ashes and twisted into a cylinder. This is placed between the lower lip and the gum, and is left until the effects become sufficiently strong. It is then removed and carried behind the ear until needed again.

Tobacco is smoked by all tribes of the extreme north, the west, and the south. The Betsimisaraka people, on the northeast coast, and the tribes of the plateau never smoke it, although they all smoke hemp in water-pipes.

Cigars are unknown in Madagascar, and the cigarette is just coming into use, having been introduced by soldiers returning from France. Cigarette-smoking is spreading among the tribes who did not smoke previously, but has so far made little impression on the smoking tribes. The natives employ tobacco-pipes of three types: straight, tubular pipes, pipes with the bowl set at an angle to the stem, and water-pipes. Tubular pipes are used only by the Bara tribe. The simplest form is a plain joint of slender bamboo about six inches long and open at both ends. Powdered tobacco is stuffed into one end with perhaps a wad of leaves or fiber in the middle, to keep it from sifting into the smoker's mouth. At the present time the favorite pipe has a bowl made from the shell of a bottle-necked cartridge, the bottom being cut off and the stem inserted in the neck. There are also a few tubular pipes of native-cast brass.

Angled pipes have the widest distribution of any form, but the shape and material vary with the tribe and with individual fancy. Clay pipes are unknown. The Tsimahety people use small, short pipes with stems of wood or reed and bowls which are sometimes of wood, but more commonly of a soft, red stone resembling catlinite. The northern Sakalava pipes conform to the same general pattern, but the bowls are of wood or nut shells, never stone. The southern Sakalava pipes are larger, and the bowls are of wood, usually lined with metal. The stems

[202]

also are often of metal, or are metal-sheathed. The Mahafaly pipes have plain, wooden stems and small, flaring bowls of graceful form. In the most prized examples the bowl is made from brass or from a green, calcareous stone which takes a high polish. The finest examples look almost like jade.

Water-pipes are used by all the tribes of the southeast coast. The stem is made from a hollowed cornstalk with one of the septa near the bottom left unpierced. A piece of reed three or four inches long is thrust through the side of the stalk just above this septum, forming an acute angle with the stalk. A small bowl, hollowed from a nut or a piece of fresh manioc root, is placed on the end of the reed, and the stalk filled with water. The smoker squats on his haunches, resting the lower end of the pipe on the ground. Most of the tribes who use water-pipes also use small, angled pipes, but the latter are of secondary importance.

All the Malagasy people use tobacco in the form of snuff. Among the tribes who also smoke, men seem to prefer the pipe and the women the snuff, although there is no fixed rule. Snuff is taken in the mouth, never in the nose, a pinch of the powder being thrown on the tongue or placed between the lower lip and gum. Snuff is made by toasting leaf-tobacco in a small pan, pounding it to powder in a mortar and mixing it with sifted wood-ashes. The ashes are prepared from various plants, and in compounding snuff more attention seems to be paid to the ash than to the tobacco, two or three varieties being added in exact proportions. Ashes of various sorts and leaf-tobacco are on sale in all native markets, but the snuff itself is rarely sold, each individual preferring to prepare his own. It is made up a little at a time, as needed, and is said to lose its flavor if kept.

All the tribes carry snuff-boxes. The commonest form consists of a short section of bamboo, slightly less than an inch in diameter, with a bottom and stopper of gourd shell. The surface of these boxes is often etched and then rubbed with soot to bring out the designs in black. The Imerina people sometimes burn naturalistic figures of men, animals and plants on boxes of this type, but this is said to be a recent fashion introduced by an Englishman who had turned native. The same people have discovered how to flatten bamboo without breaking it, giving boxes an oval section. Small bottle-shaped gourds from three to four inches long are also used as snuff-boxes by all the tribes. The natives of the west and south also carry their snuff in boxes made from the tips of cattle-horns. Such boxes are often finely shaped and highly polished. Snuff-boxes made from sections of cattle bone with ends of wood or gourd shell are used sporadically throughout the island.

By far the finest snuff-boxes are made by the Imerinas, who employ a great variety of materials. The most prized are made from the hard, scaly fruit of the raffia palm. The stem end is covered with a large silver plate, while a smaller plate, with a short neck attached, is cemented over the tip. A small piece of etched silver, usually in the form of a leaf, is fastened to the side of the fruit. The stopper, also of silver, is attached to this by a thin chain. Such boxes color and polish with use until they have almost the appearance of tortoise-shell.

Snuff-boxes made from bright-colored univalve shells are also in demand. The opening of the shell is covered with a silver plate and another plate, with neck and stopper, fastened over the tip. Gourd, bamboo and bone boxes are often mounted with silver, and the gourds are cleverly imitated in wood and horn. There are many forms of snuff-boxes, made to suit the individual fancy, one of the most curious being made from a large crab-claw. A fine collection of these boxes is on view in the Madagascar exhibit of the Museum (Hall E).

[204]

Of the three methods of using tobacco in Madagascar, snuff-taking appears to be the oldest. A very similar method is in use around Zanzibar, and it was probably introduced into the island from that region. The simple angled pipes of the western and southern tribes may also be of African origin, for they resemble certain East-African forms more than they do the European ones. Water-pipes, on the other hand, are probably traceable to the Arabs, for their influence is strong in the region where this type is found. The straight pipes of the Baras and the green tobacco quids of the Betsileos appear to be independent local developments.

Smoking in Madagascar is a purely individual matter. A smoker will often pass his pipe to a comrade who has none; but this is an act of kindness, not a necessary courtesy. Snuff, on the other hand, plays a part in the social and religious life of many tribes. Among the non-smoking tribes, the offering of one's snuff-box is a necessary preliminary to polite conversation. Snuff is also one of the commonest offerings to the ancestral spirits, as a small gift accompanying a prayer for a small favor. Among the Betsileos it is not uncommon for a person to promise while still alive that he or she will answer prayers after death, specifying the offering that must accompany the prayer. One old woman said to her family, "If the dead have any power, I will answer prayers for safe return from a journey. This I will do when I have joined the ancestors. When you pray to me, do not forget to put snuff of the sort I now use on my stone, for it is snuff that I love better than anything else."

RALPH LINTON

[205]

BIBLIOGRAPHICAL REFERENCES

BALFOUR, H.—Earth-smoking Pipes from Africa and Central Asia. Man, 1922, No. 45.

BAUMANN, O.—Durch Massailand zur Nilquelle. Berlin, 1894.

BEECH, M. W. H.—The Suk, Their Language and Folklore. Oxford, 1911.

BURTON, R. F.—The Lake Region of Central Africa. London, 1860.

BUTTIKOFER, J.—Reisebilder aus Liberia. Leiden, 1890.

CASALIS, E.—Les Bassoutos. Paris, 1859.

CUNNINGHAM, J. F.—Uganda and Its Peoples. London, 1905.

DE FICALHO.—Plantas uteis da Africa portugueza. Lisbon, 1884.

DRIBERG, J. H.—The Lango. London, 1923.

DUNHILL, A.—The Pipe Book. New York, 1924.

ENGLER, A.—Die Nutzpflanzen Ost-Afrikas. Berlin, 1895.

FITZGERALD, W. W. A.—The Coastlands of British East Africa. London, 1898.

FRITSCH, G.—Die Eingeborenen Süd-Afrikas. Breslau, 1872.

FULLEBORNE, F.—Das Deutsche Njassa und Rowuma Gebiet. Berlin, 1906.

GARDINER, A. F.—Narrative of a Journey to the Zoolu Country. London, 1834.

HARTWICH, C.—Die menschlichen Genussmittel. Leipzig, 1911.

HILTON-SIMPSON, M. W.—Land and Peoples of the Kasai. London, 1911.

HOLLIS, A. C.—The Masai. Oxford, 1905.
    The Nandi. Oxford, 1909.

HOLUB, H.—Seven Years in South Africa. London, 1881, Vol. II.

HUTTER, F.—Nord Hinterland von Kamerun. Braunschweig, 1902.

IRLE, J.—Der Herero. Gütersloh, 1906.

JOHNSTON, H. H.—The Uganda Protectorate. London, 1902.
    Liberia. New York, 1906.

JUNOD, H.—Life of a South African Tribe. Neuchatel, 1912.

KIDD, D.—The Essential Kafir. London, 1904.

KOLLMANN, J.—The Victoria Nyanza. London, 1899.

LENZ, O.—Timbuktu. Leipzig, 1884.

[206]

MALCOLM, F. W. H.—Brass Casting in Cameroon. Man, 1923, No. 1.

MAUGHAM, R. F. C.—Portuguese East Africa. London, 1906.

MEAKIN, B.—The Moors. London, 1902.

MEEK, C. K.—The Northern Tribes of Nigeria. Oxford, 1925.

MONTEIRO, J.—Angola and the River Congo. London, 1875.

PARK, M.—Travels in the Interior of Africa. London, 1799.

PARKYNS, M.—Life in Abyssinia. London, 1853.

PAULITSCHKE, P.—Ethnographie Nordost Afrikas. Berlin, 1893.

RODD, F. R.—People of the Veil. London, 1926.

ROSCOE, J.—The Banyankole. Cambridge, 1923.
     The Bakitara. Cambridge, 1923.

ROUTLEDGE, W. S. and K.—With a Prehistoric People. London, 1910.

SALVIAC, P. M. DE.—Un peuple antique au pays de Menelik. Paris, 1901.

SCHULTZE, L.—Aus Namaland und Kalahari. Jena, 1907.

SCHULZ, A. and HAMMAR, A.—The New Africa. London, 1897.

SCHWEINFURTH, G.—The Heart of Africa. New York, 1874.

SCHULZ, A. and HAMMAR, A.—The New Africa. London, 1897.

SCHWEINFURTH, G.—The Heart of Africa. New York, 1874.
     Was Afrika an Kulturpflanzen Amerika zu verdanken hat und
     was es ihm gab. Festschrift Eduard Seler, 1922. Tobacco:
     p. 532.

SKERTCHLY, J. A.—Dahomey as it is. London, 1875.

SMITH, E. W. and DALE, A. M.—The Ila-Speaking Peoples of
     Northern Rhodesia. London, 1920.

STANLEY, H. M.—How I Found Livingstone. New York, 1891.

STATHAM, J. C. B.—Through Angola. Edinburgh, 1922.

STOWE, G. W.—The Native Races of South Africa. London, 1905.

THEAL, G. M.—History and Ethnography of South Africa. London,
     1901.

TESSMANN, G.—Die Pangwe. Berlin, 1912.

TORDAY, E.—Camp and Tramp in African Wilds. London, 1913.

TREMEARNE, A. J. N.—Tailed Head Hunters of Nigeria. London,
     1912.

WERNER, A.—British Central Africa. London, 1906.

WOLLASTON, A. F. R.—From Ruwenzori to the Congo. London, 1908.

1, Snuff-box from the Fruit of the Raffia-palm. 2, Shell Snuff-box with Silver Mountings.
3, Snuff-box from the Claw of a Crab. 4, Small Snuff-box from a Gourd
Mounted with Silver. 1-4, Imerinas, Madagascar. 5, Water-pipe.
Washashis, East Africa. 6, Pipe with Stone Bowl. Antan-
drays, Madagascar. 7, Package of Home-grown
Tobacco. Wakambas, East Africa.

202

《古代青铜器的保存和铜绿的清除》绪言

FIELD MUSEUM OF NATURAL HISTORY

Founded by Marshall Field, 1893

Museum Technique Series
No. 3

RESTORATION OF ANCIENT BRONZES

AND

CURE OF MALIGNANT PATINA

BY

Henry W. Nichols
ASSOCIATE CURATOR OF GEOLOGY

FOREWORD BY
BERTHOLD LAUFER
CURATOR, DEPARTMENT OF ANTHROPOLOGY

Oliver C. Farrington
CURATOR, DEPARTMENT OF GEOLOGY
EDITOR

CHICAGO, U. S. A.
August 9, 1930

FOREWORD

One of the modern developments of science is the close alliance of chemistry and archaeology for definite practical purposes. The archaeologist cannot dispense with the cooperation of a chemist who enables him to preserve and to restore the material discovered in the soil and entrusted to his care. In Mr. H. W. Nichols, Associate Curator of Geology, Field Museum has a veteran geologist and chemist who combines his technical knowledge and wide experience with an intelligent and sympathetic understanding of archaeological problems, as amply demonstrated by his Report on a Technical Investigation of Ancient Chinese Pottery, inserted in my monograph, The Beginnings of Porcelain in China. Another important contribution made by him to Chinese archaeology—a chemical analysis of one hundred archaic bronzes—still awaits publication.

The condition in which many Egyptian bronzes were found after years of exhibition in Field Museum and in which bronze vessels and implements excavated at Kish arrived here made it desirable to subject them to the Fink electrochemical process which has so successfully been inaugurated by the Metropolitan Museum of Art, New York. Associate Curator Nichols was placed in charge of this work. Familiarity with the process was acquired by treating a number of smaller bronzes to remove disfiguring incrustations. In 1925 a beginning was made with two large and several small bronze figures from the Egyptian collection which exhibited bad cases of malignant patina. By that time the electrochemical process, carefully studied by Mr. Nichols, had been perfected to a high degree, and the treatment of the large bronze figures proved to be entirely successful. Not only was the fatal progressive corrosion checked permanently, but also unsuspected designs with much elaborate detail were revealed. During the four years 1926–29 a total of 360 bronzes, chiefly from Egypt and Mesopotamia, were restored to health by means of the electrochemical process.

The plates accompanying this treatise will teach sufficiently what the results of this process are. Bronze implements which were received here as shapeless and sometimes unrecognizable masses have now been restored to their original forms, and can be classified properly and studied. In many others interesting designs or inlays have been laid bare after the removal of the malignant patina. In each and

5

every case the objects thus treated have obtained a new lease on
their lives and are permanently preserved. The advantages accruing
to archaeology from this method cannot be overvalued and must be
acknowledged with a deep sense of gratitude.

It is hoped that the detailed description of the technique given
by Mr. Nichols on the following pages will benefit other institutions which are obliged to cope with the problems of restoring and
preserving ancient bronzes.

BERTHOLD LAUFER

哥伦布和中国，以及美洲对于东方学的意义

JOURNAL

OF THE

AMERICAN ORIENTAL SOCIETY

EDITED BY

MAX L. MARGOLIS
Dropsie College

W. NORMAN BROWN
University of Pennsylvania

JOHN K. SHRYOCK
University of Pennsylvania

VOLUME 51 · NUMBER 2 · JUNE, 1931

Published June 15, 1931

CONTENTS

PUBLISHED BY THE AMERICAN ORIENTAL SOCIETY

ADDRESS, CARE OF

YALE UNIVERSITY PRESS

NEW HAVEN, CONNECTICUT, U. S. A.

Books for review should be sent to one of the Editors (addresses on the inside of this cover)
Annual subscriptions and orders should be sent to the American Oriental Society,
care of Yale University Press, New Haven, Conn., U. S. A.
(See last page of cover.)

Entered as second-class matter June 1, 1916, at the post office at New Haven, Connecticut,
under the Act of August 24, 1912. Published quarterly.

COLUMBUS AND CATHAY, AND THE MEANING OF AMERICA TO THE ORIENTALIST *

BERTHOLD LAUFER

FIELD MUSEUM OF NATURAL HISTORY

IT IS A PRIVILEGE and a joy to be an orientalist in our day. The Chinese walls that formerly fenced off each branch of oriental learning as a separate mandate are gradually crumbling away. We begin to recognize clearly that all oriental civilizations have been closely connected with one another from very ancient times and that particularistic Monroe doctrines like China for the Chinese, India for the Indians, hold good no longer in scientific research. We are now confronted, for instance, with the spectacle of an early Indian substratum in Western Asia, a wide Sumerian expansion over Iran, northwestern India, central Asia and northern China, and an intimate interaction of Iranian and Chinese civilizations. The ancient history of Asia should be rewritten, not by an individual, but in sympathetic coöperation by the entire brotherhood of orientalists. And more than that—our territorial ambitions may lead us far beyond the natural boundaries of Asia, for Asiatic civilizations could not fail to exert a profound influence over Europe, Africa, the South Sea Islands, and perhaps even Australia. The entire Old World is ours, therefore, but our oriental imperialism is one of peaceful penetration, and, accentuating as it does the unity of mankind and the common origin of human civilization, it tends to work toward the unification and harmony of mankind.

The question I wish to ventilate to-day is: Does America hold out a similar interest to the orientalist? Have we the right to expand our activity into the western hemisphere? Are we privileged to knock at the door of the Americanist, humbly and modestly, of course, and to offer our collaboration in the study of problems in which he is interested? This is a vast and complex subject, also much misunderstood, and on this occasion I can only hope to sketch it in its broadest outlines and to stimulate your interest in this fascinating inquiry.

---

* Presidential Address read at the Annual Meeting of the American Oriental Society, Princeton, April 7, 1931.

87

Our interest in America begins with the history of the discovery of the New World. In fact it was through the medium of Asia that America was discovered. Without the scanty knowledge that ancient Greece and mediaeval Europe possessed of faraway China, America might not have been discovered, or its discovery at any rate would have been long delayed. In 1492 when Columbus set out on his first memorable voyage, he was not actuated by the ambition to discover a new continent, but his principal objective was to find a shorter route to India and the Cathay of Marco Polo, the country of the Great Khan, by sailing in a westerly direction from Europe.

China played an eminent part in all of Columbus's calculations. He was an ardent admirer and a deep student of the memoirs of Marco Polo, his countryman, whose glowing accounts of the Far East left a lasting impression upon his mind. On his first voyage he took along a Latin translation of Marco Polo's travels in which he entered in his own hand numerous notes and observations. This copy of Marco Polo is still preserved in the Colombina Library of Seville.[1]

Another powerful factor that determined Columbus's project was the adoption on his part of the geographical computations of Marinus of Tyre, in which China entered in another subtle way. Marinus, a renowned Greek geographer, who lived in the first century of our era, was a contemporary of the Han Dynasty when the Chinese entertained commercial relations with the Roman Orient and when Chinese silk found a ready market in Syria, Egypt, Greece, and Rome. Domiciled at Tyre in Syria, Marinus had occasion to interview many traveling merchants who returned from the Far East by land or by sea, and gathered from their lips much valuable information regarding the geography of the distant countries of the East. The book he wrote is unfortunately lost, but fragments of it have been preserved by Ptolemy. Marinus was one of the founders of mathematical geography, and computed the extent of the eastern hemisphere from the Isle of Ferro in the Canary Islands to the east coast of China as 225 degrees—a serious error, for this distance is only 130 degrees. This miscalculation, however, was the most fruitful error ever committed by a scholar;

---

[1] Columbus's annotations have been published in the *Raccolta Columbiana*, part I, vol. II, pp. 446 et seq.

for Columbus, in adopting Marinus's computation as correct, quite logically concluded that the ocean stretching west of Europe was rather narrow and that the distance from Europe to the coast of China was quite short, while in reality it was about twice as long as he figured. Here again it was the dawn of a geographical knowledge of China in the West which ultimately led to the discovery of America.

When on October 12, 1492, Columbus reached Guanahani in the Bahamas—now identified with Watling's Island—he was convinced that he had arrived at an island off Cipangu, as Japan was called by Marco Polo. Scars which he noticed on the bodies of several natives elicited the inquiry as to what they meant, and he understood that people from neighboring islands had come to capture them and that they had made defensive war against the invaders. The Admiral was persuaded that their enemies were subjects of the Great Khan of Tartary, who, Marco Polo wrote, were wont to make raids upon the islands off the coast and to enslave the inhabitants. On his first contact the Admiral designated the natives "Indians," and this title was subsequently applied to all the aborigines of the western hemisphere. This name, and like it the name "West Indies," have survived as mementoes of the first discovery, while for a long time the term "The Indies" was used for America in general. On his first and second voyages Columbus took Cuba to be the mainland of Asia, more specifically a part of Mangi, as southern China was called, its eastern end corresponding to the cape of Zaitun. Española (now Hayti) he then identified with Cipangu and the identification was seemingly confirmed by a local Indian name, Cibao. On his third voyage the Admiral sailed farther south and touched the coast of South America near the mouth of the Orinoco River. There, again, he found partial confirmation of his geographical beliefs. The several thousand (7459) islands were there and were inhabited by savages, as had been written by Marco Polo and Sir John Mandeville. On his fourth voyage (1502) Columbus turned southwest of Cuba to make for the India of the Ganges. When the fleet sighted the coast of Honduras, it was recognized as the coast of Ciamba (Champa—in Indo-China). The plan was to follow this coast in a southerly and ultimately westerly direction past Java Major, Pentan, Seilan, the Strait of Malacca, and into the

Indian Ocean to the India of the Ganges. In the letter describing
his first voyage, Columbus writes, " When I reached Juana [i. e.
Cuba] I followed its coast to the westward, and I found it to be
so extensive that I thought that it must be the mainland, the
province of Catayo (la provincia de Catayo)." As late as 1494 he
was convinced that Cuba was tierra firme. " And since there were
neither towns nor villages on the seashore, but only small hamlets,
with the people of which I could not have speech, because they
all fled immediately, I went forward on the same course, thinking
that I should not fail to find great cities and towns." In this idea
he was influenced by Polo's account of large and populous cities
in Cathay, and therefore he despatched from a certain harbor two
men inland to learn whether there were a king or great cities.
They traveled three days' journey and found an infinity of small
hamlets and people without number, but nothing of importance;
for this reason they returned.

Before 1892 it was not doubted that Columbus died in the conviction that he had reached Asia. Since then, however, several
scholars have adopted the view that it had dawned upon Columbus
before his death that he had discovered a new world distinct from
the India and Cathay which had been the original object of his
search. This verdict has been scrutinized at great length and
definitely refuted by George E. Nunn in his book *The Geographical
Conceptions of Columbus: a critical consideration of four problems*
(published by the American Geographical Society, New York,
1924). This author has arrived at the conclusion that no evidence
has as yet been advanced sufficient to disprove the belief that in
1502-03, during his fourth voyage, Columbus believed himself to
be on the coast of Asia and that he died so believing. After him,
Balboa in 1513 so believed. Waldseemüller and the German
cartographers did not reject Columbus's ideas. The writings of
Castañeda, the chronicler of the Coronado expedition, and the
famous Gastaldi map of 1562 are further evidence that many
successors of Columbus continued in the same belief down into the
middle of the sixteenth century.

Not only, however, were Columbus's expeditions and movements
determined by his notions of Asiatic geography, but, what is still
more attractive to us, his mind was imbued with Oriental lore to
such a degree that he projected Asiatic tales into the life of the

aborigines of the New World. Columbus was a man without profound education and learning, and was endowed with a vivid and poetic imagination, which equaled his knowledge of navigation; he was somewhat credulous, deeply religious with a trend toward mysticism, yet a man of extraordinary abilities, keen intelligence, indomitable courage and energy, foresight and sagacity. Whatever fault his critics may have found with him, he remains the man who did the deed.

Of the Oriental lore that exerted a predominant influence on Columbus, the cycle of wondrous peoples was uppermost in his mind. Originating in India, transmitted to the Greeks by Ctesias in his Indica, these stories migrated to China, as well as to Europe, and formed the stock-in-trade of mediaeval mariners who delighted in story-telling.

The outstanding tale is that of the dog-headed people, the Kynokephaloi of Ctesias, usually welded with the fable of the Amazons, in ancient Chinese lore as well as in mediaeval Europe. The frequent close coincidences of Chinese and mediaeval European folklore present an interesting problem that remains to be studied at close range.

At this point I must make a little digression so that you may better appreciate the Columbian tradition which has assumed vast proportions in the history of the Spanish conquest of America.

The Chinese located the wondrous nations in fabulous islands far away in the north-eastern Pacific, and as early as the sixth century A. D. had outlined a complete picture and a fixed scheme of their geography and ethnography. The Annals of the Han Dynasty (*Hou Han shu,* chap. 115, p. 4b) mention a woman's kingdom in an island east of Korea, populated solely by women. In this country there is a marvelous well: when the women gaze into this well, they soon give birth to children. Another Chinese tradition tells of an Amazon kingdom in Ta Ts'in, the Roman or Hellenistic Orient, adding that these women bear children under the influence of water. This, accordingly, was a notion emanating directly from the Near East. Again, we read in another Chinese source (*Ling wai tai ta* by Chou K'ü-fei) of an Amazon island east of Java, where the women disrobe when the south wind blows and conceive from the wind, but give birth only to girls. Qazwini likewise has an account of a women's island in the sea of China, where the women conceive from the wind and bear only girls. The following complex story is contained in the *Nan shi* (chap. 79, p. 4), one of the historical annals which treats the history of China from A. D. 420 to 589. " The women's realm is situated a thousand miles east of Fu-sang. These women have regular bodily forms and a white complexion, but their whole body is covered with hair, and the hair of their head is so long that it trails on the ground. In the second or third month of the year they bathe in a river and thus become pregnant; their children are born in the sixth or seventh month. These women have no nipples, but hair-roots grow on the napes of their necks. Some of these hairs are white, and the white ones contain a sap for suckling their infants. A hundred days after birth the infants are able to walk, and they are fully grown by the fourth year. At the sight of man they are stricken with terror and flee, for they abhor intercourse with males. Like wild animals they subsist on saliferous plants, which have a fragrant odor but are salty of taste. In A. D. 507 a man from Tsin-ngan (in Fu-kien Province) was crossing the sea when he was caught in a storm and cast adrift on a certain island. On going ashore he found it inhabited by men and women who were remarkably differentiated. The women resembled those of China, but their speech was unintelligible. The men, however, had human bodies but heads like those of dogs, and their voices sounded like the barking of dogs. Their food was small pulse, and their garments seemed to be made of cotton. The walls of their houses were of adobe; the houses were circular in shape with an entrance like that to a cave." The most striking parallel to this Chinese story is offered by Adam von Bremen of the eleventh century. In his Gesta Hammaburgensis ecclesiae pontificum (Church History of Hamburg, an important source of mediaeval history) he writes, " It is said that on the east coast of the Baltic there is the Land of the Women (Terra Feminarum) populated by Amazons who, it is asserted by some, conceive merely by drinking water. Others hold that they enter into relations with passing traders, or with their captives, or even with monsters that are said to abound there, and this is the more probable. When they give birth to children, those of the male sex have the heads of dogs (fiunt cynocephali), those of the female sex, however, become most beautiful women. They despise intercourse with men whom they bravely repel when they venture to land. The dog-heads are those that have heads on the breast [this is a confusion with the ἀκέφαλοι of the Alexander Romance, who have no head and who have eyes and mouth on their breasts]. In Russia they may often be seen as prisoners. They bark instead of talking." Here we face the same net distinction between women of human shape and dog-headed men as in the Chinese version.[2]

It is a matter of great interest that this ancient Indian-Hellenistic-Chinese-mediaeval story functioned as godfather to the discovery of America. It is still more interesting to consult Columbus's Diary and to observe how deeply he was affected by this tradition, how it gradually grew and finally assumed definite shape in his mind. In his Diary the Admiral admits, " I do not know the language of the Indians, and these people neither understand me nor any other in my company, while the Indians I have on board often misunderstand." Despite this frank confession, the Diary teems with stories purporting to have emanated from the Indians, while in fact they are offsprings of Euro-Asiatic lore projected into the Indians.

On the 4th of November, 1492, while the Admiral was in Cuba, there is this entry in his Diary, " He understood [from the Indians] that far away there were men with one eye and others with dogs' noses, who were cannibals, and that when they captured an enemy they beheaded him and drank his blood." The dog-heads appear as cannibals in the Romance of Alexander and in Qazwini.

On the 23rd of November, we read in his Journal, " Beyond this cape there stretched out another land or cape, also trending east, which the Indians on board called Bohio (Hayti). They said that it was very large, and that there were people in it who had one eye in their foreheads, and others who were cannibals and of whom they were much afraid."

On the 26th of November, we read, " The Admiral says that all the people whom he has hitherto met have very great fear of those of Caniba or Canima. They affirm that they live in the island of Bohio, which must be very large according to all accounts. The Admiral understood that those of Caniba come to take people from their homes, they being very cowardly, and without knowledge of

---

[2] For more information on the Amazon Traditions see my article in *Festschrift Kuhn*, pp. 204-209.

arms. . . . The natives who were on board declared that the Canibas had only one eye and dogs' faces. The Admiral thought they lied, and was inclined to believe that it was people from the dominions of the Grand Khan who took them into captivity." Caniba is a corruption of Carib, and from Caniba to Cannibal and the Cannibal Islands of ancient Asiatic lore there was but one step; in fact, our word " cannibal " dates from this time, and is derived from the tribal name " Carib." The Spanish change from Caribal to Canibal was made under the influence of *cane, canis* (" dog ") in allusion to the dog-headed people discovered by Columbus. It is a prank of fortune that the term cannibal, while it hails from America contains the germ of an old oriental tradition.

On the 6th of January, 1493, we are informed, " The Admiral also heard of an island farther east, in which there were only women, having been told this by many people." On the 15th of January he decided to repair to Matinino (now Martinique), which was said to be entirely peopled by women, without men.

On the 16th of January the Journal contains the following entry: " The Indians said that by that route they would fall in with the island of Matinino, peopled entirely by women without men, and the Admiral wanted very much to take five or six of them to the Sovereigns of Spain. But he doubted whether the Indians understood the route well, and he could not afford to delay by reason of the leaky condition of the caravels. He, however, believed the story, and that at certain seasons men came to them from the island of Carib,—distant ten or twelve leagues. *If males were born, they were sent to the island of the men; and if females, they remained with their mothers.*"

This is a complete coincidence with the Chinese and European tradition.

The same story is also contained in Columbus's Letter of his First Voyage.[3] Speaking of an island Quaris [either Dominica or Maria Galante], the second at the coming into the Indies, he writes that it is inhabited by a people who are regarded in all the islands as very fierce and who eat human flesh. . . . They are ferocious among these other people who are cowardly to an excessive degree. . . . These are those who have intercourse with the women

---

[3] C. Jane, *Select Documents Illustrating the Four Voyages of Columbus,* I, 1930, p. 16 (Hakluyt Society).

of Matinino, which is the first island met on the way from Spain to the Indies, in which there is not a single man (en la qual no ay hombre ninguno). These women engage in no feminine occupation, save that they use bows and arrows of cane, like those already mentioned, and they arm and protect themselves with plates of copper, of which they have much " (Ellas no usan exercicio femenil, salvo arcos y flechas, como los sobredichos, de cañas, y se arman y cobigan con launas de arambre, de que tienen mucho).

This legend haunted the Spanish conquerors for more than two centuries. The country of the Amazons was discovered in Mexico and South America, chiefly by Francisco de Orellana, a lieutenant of Pizarro, and the Marañon River was finally named the River of the Amazons.[4]

In the letter of his First Voyage Columbus speaks of the Province of Avan in Cuba, where the people are born with tails. Andrés Bernáldez, in his *Historia de los reyes católicos,*[5] comments, " There they told the Admiral that beyond there lay Magón, where all the people had tails, like beasts or small animals, and that for this reason they would find them clothed. This was not so, but it seems that among them it is believed from hearsay, and the foolish among them think that it is so in their simplicity, and I believe that the intelligent did not credit it, since it seems that it was first told as a jest, in mockery of those who went clothed."

Another ancient oriental tradition[6] crops up in the search for the fountain of perpetual youth by the celebrated Ponce de Leon, when with three ships he sailed from Porto Rico in March, 1512. On the 27th he discovered Florida, but the water there proved unsatisfactory, and once more he began his search among the Lucayos or Bahamas, his guide being an old crone who asserted the fountain was to be discovered on Bimini. De Leon returned to Porto Rico and placed in charge of the expedition Juan Perez de Ortubia who reached Bimini and took a bath in the fabled spring. His Indian companion was careful to state that its extraordinary effects would not be apparent for at least twelve months. This

---

[4] For details compare E. Beauvois, " La fable des Amazones chez les indigènes de l'Amérique précolombienne," *Le Muséon,* V, 1904, pp. 287-326.

[5] C. Jane, *op. cit.,* I, p. 138.

[6] See, e. g., A. Wünsche, *Die Sagen vom Lebensbaum und Lebenswasser, altorientalische Mythen* (Ex Oriente Lux, Leipzig, 1905).

fountain of youth is still shown on North Bimini, and its cool and tasteless water is still dispensed to visiting tourists. It is used as a cure for rheumatism, and is said to be valuable in other complaints. Another fountain of rejuvenation, alleged to have been discovered by Ponce de Leon, is pointed out at St. Augustine in northern Florida.[7]

We might expect that the age of new world discoveries and the advance of geography would have dealt the death blow to the ancient fables of wondrous peoples. The reverse is the case, however. Some of them were actually rediscovered. Pigafetta who accompanied Magalhaens on the first voyage round the world records a story told him by an old pilot from Maluco: The inhabitants of an island named Aruchete are not more than a cubit high, and have ears as long as their bodies, so that when they lie down one ear serves them for a mattress, and with the other they cover themselves. This is also an old Indo-Hellenistic creation going back to the days of the Mahābhārata (Karṇaprāvaraṇa, Lambakarṇa, etc.) and reflected in the Ἐνωτοκοῖται of Ctesias and Megasthenes. As early as the first century B. C. the Long-ears (Tan-erh) also appear in Chinese accounts; their ears are so long that they have to pick them up and carry them over their arms.

However paradoxical it may sound, I hope that the day will come when a history of the discovery and conquest of America will be written by an orientalist.

The discovery of America fell like a bombshell into the learned camps of Europe. One must read the writings of the sixteenth century to realize the intense excitement and the tremendous impression produced by this hitherto unknown world, which all of a sudden had emerged from the depths of the ocean, with a new and strange variety of the human species, with numerous novel animals and plants. A new problem had arisen, and the sacred books of the Church and the classics were consulted to solve it, but in vain. The humanists were unable to grasp the idea that high civilizations like those of Mexico and Peru could have developed without the agency of Greeks and Romans, and the fabulous Atlantis was re-

---

[7] Cf. G. J. H. Northcroft, *Sketches of Summerland*, p. 194 (Nassau, 1906); M. Moseley, *The Bahamas Handbook*, p. 62 (Nassau, 1926); E. Beauvois, " La fontaine de jouvence et le Jourdain," *Le Muséon*, III, 1884, pp. 409, 415-417.

vived to land the Greeks in America and to stamp them as ancestors of the Americans. The Jewish descent of the Indians from the lost ten tribes was a favorite theme of many scholars that persisted for a long period. According to Ulloa, Noah's descendants continued their father's shipbuilding activity, set out as navigators, and peopled America.[8] The Phoenicians and Carthaginians likewise appear among the culture heroes of this continent, and "Phoenician" as well as "Hebrew" inscriptions have sometimes been forged and "discovered" on American soil and used in support of such like speculations. The theory that America's cultures originate from Eastern Asia has always had many adherents, and was eagerly defended by Alexander von Humboldt, who expressed his opinion thus: "It seems to me clearly demonstrated that the monuments, the method of time-reckoning, and many myths of America display very striking analogies with the ideas prevailing in eastern Asia, analogies which prove ancient historical connections." In general and to some extent this opinion still holds good, but in the present state of science it must be somewhat modified and can be formulated more satisfactorily and precisely.

I need hardly point out here that the alleged Chinese discovery of Mexico in the fifth century of our era, which kept the public in suspense for more than a hundred years, is pure fiction. The Chinese account of a marvelous island called Fu-sang, on which this deduction was based by De Guignes in the eighteenth century, is devoid of any historical value and presents a fantastic concoction of Buddhist monks. Fu-sang is not a real country, but a product of the imagination, a geographical myth pieced together from many heterogeneous elements emanating from different sources and quarters.[9]

Whatever Chinese influences may be found in America are not due to a migration of individuals or a direct transmission of cultural ideas from China or Japan across the Pacific, but have gradually filtered in over the land route through intertribal communication from northeastern Asia down our northwest coast.

---

[8] R. Andree, "Der Ursprung der amerikanischen Kulturen," *Mitteilungen der anthrop. Gesellschaft Wien*, 1905, pp. 87-98.

[9] Cf. my remarks in *T'oung Pao*, 1915, p. 198; *Memoirs Am. Anthr. Assoc.*, IV, 1917, p. 102; and *Festschrift Kuhn*, p. 207.

Briefly stated, the problem as we now see it is as follows. The appearance of man in the western hemisphere is of comparatively recent date. No remains of an early man comparable to paleolithic man of western Europe have as yet been discovered in America. The Old World is the cradle of mankind, and the American Indian is an immigrant from Asia.

It is justly assumed that at some remote period, which may roughly be estimated at about 25,000-20,000 B. C., North America was populated in many successive waves of immigration from Asia across Bering Strait or the Aleutians. In physical type our Indians are close relatives of the peoples of northern Asia. The stock of culture brought by these first immigrants into this continent was extremely primitive; they were hunters clad in skins, understood the chipping of stone axes and arrowheads, possessed the dog as their sole domestic animal, presumably made crude pottery, cordage, fishing-nets, and baskets, had dugout canoes, and lived in pit-dwellings.

The fact that culture evolved independently in America for many millenniums becomes evident from the great number of diverse linguistic stocks which both in structure and lexicography offer no point of contact or affinity with Old World languages. Many Indian tribes advanced to a highly developed stage of agriculture long before they had the misfortune of being discovered by the Spaniards. All their cultivations are derived from wild plants native to this continent. None of the American cultivated plants occurs in the Old World prior to the age of discovery; on the other hand, not a single cultivated food plant of Asia, Europe, or Africa had found its way to America in pre-Columbian times, not a grain of wheat, barley, or rice. Again, all fundamentals of Asiatic civilizations are strictly absent in ancient aboriginal America, such as agriculture practised by means of the plow and with the ox as draught-animal; all domesticated animals like cattle, horse, camel, reindeer, sheep, goat, swine, and chicken; chariots built on the principle of the wheel; the potter's wheel; stringed musical instruments; roofing tiles; and the art of smelting, forging, and casting iron. All this is sufficient evidence to convince one that in the dim past American and Asiatic cultures must have taken an independent course of development for many millenniums.

But when all this has been said, there is no reason for assuming that America has always marched along in splendid isolation; on the contrary, we recognize more and more that in historical times, at least during the last one or two thousand years, there has been an intimate contact between the two continents and that currents and undercurrents of Asiatic thought have swept over America, especially its northern part. The old hit-and-miss methods formerly employed in such investigations, as forcing conclusions from purely outward resemblances, trivial analogies, or coincidences, have happily been discarded. No one, for instance, will regard any longer the presence of the Swastika in Mexico as an indication of Buddhist propaganda. Intense methods and profound study of particular culture traits, however, are apt to produce results. To cite an example—divination, oracles, riddles, and ordeals are characteristic of most Asiatic civilizations, but are almost completely absent in vast tracts of America, and if such traits occur there sporadically, it is perfectly justifiable to trace them to an outside impetus. Scapulimancy, divination from the cracks produced by scorching over a fire the shoulderblade of an animal, has its centre of gravitation in Central Asia. In America it is restricted to a well-defined area chiefly embracing the Indian tribes of Labrador and Quebec, and Dr. John M. Cooper, of the Catholic University of America, who has made a special study of this subject, admits that it is genetically related to Asiatic scapulimancy. Dr. Cooper has also described scrying in the eastern Algonkian area, a method of divining by gazing into a basin filled with water, corresponding to the crystal-gazing of India and eastern Asia.[10]

We owe a model investigation to Dr. A. I. Hallowell of Philadelphia into the bear ceremonialism in northern Asia and America where the worship of the bear is widely distributed and practically alike in form and content.[11] The same may be said of that primitive type of religion we call shamanism, which also flourished in

---

[10] Frank G. Speck, "Divination by Scapulimancy among the Algonquin of River Desert, Quebec," Indian Notes, *Museum of the American Indian*, V, 1928, pp. 167-173. John M. Cooper, "Northern Algonkian Scrying and Scapulimancy," in *Festschrift Schmidt*, pp. 205-217. In general cf. R. Andree, "Scapulimantia," *Boas Anniversary Volume*, pp. 143-165.

[11] A. I. Hallowell, "Bear Ceremonialism in the Northern Hemisphere," *American Anthropologist*, 1926, pp. 1-175.

ancient China, and is still alive throughout central and northern Asia as well as North America. The association of the shaman with a tambourine recurs throughout this area, even in Lapland.

The game of backgammon—the nard of the Persians, the pachisi of India—appears in ancient Mexico as patolli.[12] Football and other ball games may likewise have penetrated from Asia into America.

The composite bow, i. e., a bow reinforced by layers of sinew, is found in its highest development in Asia. In various forms it covers the highlands of North America well down into Mexico, and is likewise the chief weapon among the Eskimo. As it does not appear in South America, it is justly regarded as an Asiatic intrusion in North America.[13] Even so typical an Indian affair as the moccasin has been shown to be of Asiatic origin.[14]

As I demonstrated in a study of Chinese Armor,[15] the type of plate armor consisting of laminae or plates of metal, bone or ivory lashed together and arranged in parallel horizontal rows was used both in ancient Egypt and Assyria, where it is represented on monuments of King Sargon (722-705 B. C.); through Iranian mediation it spread through Central Asia to China, Korea, and Japan, further into the entire North Pacific area both on the Asiatic and American side, where according to Chinese data it must have been in existence from at least the third century onward. Walrus and narwhal ivory was anciently traded from the arctic shores of North America to China and Japan, and through the medium of the Arabs to India, Persia, Egypt, and Constantinople, long before the discovery of America.[16] I suspect an ancient interrelation of the realistic ivory carvings of the Chukchi and Eskimo and the ivory art of China; likewise an historical contact of ancient Chinese art with the peculiar forms of decorative motives of our Northwest Coast Indians. Systematic excavations in Alaska and northeastern Asia will doubtless have many surprises in store for us.

---

[12] E. B. Tylor, " On the Game of Patolli in Ancient Mexico, and its Probably Asiatic Origin," *Journal Anthrop. Institute*, VIII, pp. 116-129.

[13] C. Wissler, *The American Indian*, p. 133.

[14] R. Dixon, *The Building of Cultures*, p. 128.

[15] *Chinese Clay Figures*, part I, pp. 258-291 (Field Museum Anthrop. Series, XIII, No. 2, 1914).

[16] *Ivory in China*, Field Museum Anthrop. Leaflet, No. 21, 1924.

Still more intimate relations between Asia and America are revealed by a large common fund of beliefs and traditions, and the study of the intercontinental migration of folklore remains the most attractive subject to the Orientalist. The first methodical and critical investigation along this line is presented by the work of Paul Ehrenreich, *Die Mythen und Legenden der südamerikanischen Urvölker und ihre Beziehungen zu denen Nordamerikas und der alten Welt* (Berlin, 1905). The fundamental source-book to be recommended to the investigator is Boas' *Tsimshian Mythology*. The number of correspondences of tales and specific motives on both sides is so overwhelmingly large that the thought of independent origin is virtually excluded. Eastern Asia and northwestern America present a continuous area in folklore; the myths from practically a single group, allied both in form and actual content, and a stream of Asiatic folklore has flowed down the Pacific coast from North to South America.

I have only time to present a few examples.

One of the best examples of intercontinental diffusion is the story of the magic flight or obstacle pursuit: the hero, pursued by his enemies, flings behind him successively a whetstone, a comb, and a vessel of oil or other fluid. The stone turns into a mountain or precipice, the comb into a forest, the fluid into a lake or river. Each of these obstacles detains the pursuer and contributes to the hero's final escape. This story is widely disseminated in the Old World from Morocco to the South Seas and Bering Strait, and it is known in closely related form in America all along the arctic coasts, on the North-Pacific coast, and inland as far as the Mississippi Valley. The Indians of British Columbia have embodied this tale in their most sacred traditions, and for this reason it may be assumed that it has been known there for a long time. The salient point is that the three obstacles occur in the same order in the American versions, and, what is still more significant, the sacred number "three," so prevalent in Eurasian folklore, magic, and ritual, is scarcely ever thus used by the American Indians who replace it either with "four" or "five," but in this particular tradition they have adhered to the number "three." [17]

---

[17] A. L. Kroeber, *Anthropology*, p. 198; F. Boas, "Migrations of Asiatic Races and Cultures to North America," *Scientific Monthly*, 1929, p. 116.

Another Old-World tale still more widely distributed over North and Central America is the test-theme, the trials of a prospective bridegroom by superhuman tasks in order to win his bride.[18]

Boas[19] has found among the Tsimshian the classical story of the war between the Pygmies and the Cranes, three versions of which are also contained in ancient Chinese records. The creation of the earth through an animal diving into water for mud recurs all over North America and northern Asia and even among Finnish tribes.[20]

The concept of the hare in the moon looms up both in ancient India and China. Chavannes assumed Chinese priority and derivation on the part of India. Mayers and Conrady[21] plead for Indian origin and China's indebtedness to India in this respect. The question is more complex, however; for we meet the same notion in Mexican and South-American mythology; the latter, in agreement with the Chinese, also knows of a frog in the moon. In Peru and other parts of South America we encounter many mythological traditions and motives traceable to India, which do not occur in North and Central America and appear to have been imported directly by way of the Malayan or Polynesian islands.[22]

Some years ago I published an article under the title "The Prehistory of Television"[23] in which I treated of mirrors and other magical devices that allow the owner to scan the surrounding country or to see any distant person or object desired. This is a typical Old World motive distributed from India and Iran to western Asia and Europe. Subsequently I was surprised to find two American parallels of this incident—one among the Eskimo and another among the Kekchi of Guatemala,[24] who tell the story

---

[18] R. H. Lowie, "The Test-Theme in North American Mythology," *Journal of American Folklore*, XXI, 1908, pp. 97-148.

[19] *Tsimshian Mythology*, p. 867.

[20] R. H. Lowie, *Primitive Religion*, p. 180.

[21] W. F. Mayers, *Chinese Reader's Manual*, p. 219; A. Conrady, *T'ien-wen*, p. 173.

[22] P. Ehrenreich, *Mythen und Legenden*, pp. 36, 69, 82, 93.

[23] *Scientific Monthly*, 1928, pp. 455-459.

[24] K. Rasmussen, *Intellectual Culture of the Iglulik Eskimo*, p. 205 (Report Fifth Thule Exped., VII, Copenhagen, 1929). J. E. Thompson, *Ethnology of the Mayas*, pp. 127, 173 (Field Museum Anthr. Ser., XVII, No. 2, 1930).

of a man possessing a magical stone in which he could see everything that was happening in the world—the exact counterpart to the cup of King Kai Khosrau in the Shahnameh, that mirrors the world and distant persons. This example goes to show that the Orient may sometimes be nearer to our door than we are inclined to assume.

Where there is smoke there is fire. If numerous tales and myths have found their way from Asia to America over the northern and southern routes, we may expect similar transmissions of other culture traits, such as notions of astronomy, especially the zodiac, calendrical and chronological systems, technical methods and art motives. These investigations are still in the beginning. I have merely touched here superficially one some of the problems confronting us or awaiting further study. There are many others which for lack of time I cannot discuss.[25]

I merely wanted to convey to you the message that America and Asia are closely linked together in a common bond and that the orientalist, if he is so inclined, can contribute his modest share to the elucidation of an early phase o_ American history. With some modification the orientalist also may adopt the timely slogan: See America first!

---

[25] Compare, for instance, A. L. Kroeber, "American Culture and the Northwest Coast," *American Anthropologist*, 1923, pp. 1-20; P. D. Kreichgauer, "Neue Beziehungen zwischen Amerika und der alten Welt," *Festschrift Schmidt*, pp. 366-377; F. Boas, "Migrations of Asiatic Races and Cultures to North America," *Scientific Monthly*, 1929, pp. 110-117; E. M. v. Hornbostel, "Chinesische Ideogramme in Amerika," *Anthropos*, 1930, pp. 953-960 (not convincing); Laufer, *Jade, a Study in Chinese Archaeology and Religion*, 1912, p. 52.

2

204

中国与发现美洲的关系

China and the Discovery of America

BY

BERTHOLD LAUFER
Curator of Anthropology of the Field Museum of Chicago

A Monograph
Published by
China Institute in America
New York City   U. S. A.

China and the Discovery of America

By Berthold Laufer
Curator of Anthropology of the Field Museum of Chicago

IT is now about 150 years since the Chinese first came in contact with Americans when the American ship *Empress of China,* sailing from Boston and rounding the Cape of Good Hope, cast anchor in the harbor of Canton, in the year 1784, under the reign of the great Emperor Chien Lung, who was a contemporary of George Washington. Thus Americans were late arrivals, in fact, the last of foreign peoples to enter into commercial and political relations with China. Europeans, first the Portuguese, then the Spaniards, Hollanders, British and French had preceded them by several centuries. It is no empty saying that from the first days of American-Chinese intercourse the two great countries have been linked by bonds of sympathy which have not existed and do not exist between China and any European power. These bonds of sympathy and friendship have been strengthened from year to year, as witnessed particularly by the ever increasing number of Chinese students and scholars annually flocking to our universities athirst for knowledge.

What, then, have American and Chinese minds in common? I think, a goodly number of very fine traits. First, the spirit of democracy, which has pervaded China for more than 2,000 years, ever since the First Emperor Chin Shi smashed the old feudal system. The principle of government for the benefit of the people certainly is American, but it is equally Chinese and goes back to the fourth century B. C. when Mêng-tse (Mencius), the most gifted of Confucius' disciples, proclaimed the doctrine, "The people are the most important element in a nation, and the sovereign is the least important of all." Second, the spirit of religious tolerance. I know of no more tolerant nation than the Chinese. Third, the lack of a caste system and lack of an hereditary nobility. China was always guided and governed by an aristocracy of intellect, not of birth; the old system of free competition by civil service examinations recruited the best talent from all ranks of society. Fourth, Americans and Chinese do not suffer from the obsession of that great evil, the race superiority complex, they are averse to armed force, they are friends of peace, and are animated by

3

a deep sense of justice and fair play toward all, regardless of race, color, or creed. Fifth, and this is the greatest asset that the two nations have in common, they have an unbounded, almost religiously fanatic, faith in the power of education and knowledge as the best guarantor of progress, as the best possible safeguard of the permanence of our social structure and institutions. With this capital of a common historical tradition and mentality—democracy, tolerance, equality, justice, and education—we are well prepared to stand the test and storms of the time.

I have just said that Americans came to China as late as 1784. But two hundred years earlier the Chinese received their first knowledge of the existence of America. The name "America" first appears in Chinese records at the end of the 16th century, in the official historical Annals of the Ming Dynasty. Let us consult the *Ming Shi*. In chap. 326, which deals with foreign countries, we are amazed to find a notice entitled I-ta-li-ya (Italy). "Italy is a country situated in the great western ocean. In times of old it had no intercourse with China. A man of this country, Li Ma-tou by name (Matteo Ricci), arrived at the capital in the period Wan-li (1573-1620, more exactly in 1601) and displayed a map of the world (*wan kuo chuan tu*, lit. a complete map of the ten thousand countries; a novel experience to the Chinese!), explaining that there are in the world five great continents. The first of these is called Asia with more than a hundred countries, of which China is the first. The second is Ou-lo-pa (Europe), the third Li-wei-ya (Libya, the old name for Africa). The fourth is A-mo-li-kia (America), vast in extent and divided into a northern and southern continent which, however, are connected with each other." This is the first record of America in Chinese literature. Ricci was an Italian Jesuit, the first missionary who came to China, a very learned and cultured man, well versed in mathematics and astronomy; he translated Euclid's Geometry into Chinese, and became very proficient at writing Chinese. Ricci's world map caused a sensation and made a profound impression on the Chinese scholars of his time. His map was subsequently printed in Chinese with Chinese prefaces and a number of good stories. A few copies of this map have survived, and reprints were recently published in England (*Geographical Journal*, 1917) and China.

The first messenger from America to China, however, arrived long before Ricci landed at Canton in 1582. This messenger was Indian corn or maize, the staple cereal of our American Indians, which spread to China between 1550 and 1570 and soon acquired a great economic importance. Our Indians were highly accomplished agriculturists and had cultivated for many centuries numerous useful plants peculiar to

4

America and unknown to the Old World. Most of these were introduced into China, many as early as the Wan-li period, such as the peanut, tobacco, the potato, the batata or sweet potato, the pineapple, guava, papaya, anona, chili pepper, tomato, and many others. All these American plants as ambassadors of good will migrated to China long before any American set his foot on Chinese soil.

While it took the Chinese several centuries to obtain a correct knowledge of the geography of America and Europe, European ignorance with reference to China was much more appalling. Curiously enough, it happened twice in history that China became known in the West under a double set of names and that it was not recognized that this duplicity of names referred to the same country. The Chinese never had a proper appellation to designate their own nation. Their favorite custom was to style themselves after a favorite dynasty, for instance, the men of Han or the men of Tang, in allusion to these great dynasties. China had a front porch and a back porch; the former facing the Pacific ocean, the latter connecting the country with Central and Western Asia. The Greeks had only a very vague knowledge of China and knew her under two different names: Serica and Thinai or Sinai. The name "Serica" is derived from a Greek word *ser* which denotes the silkworm and which probably is based on a Chinese word (this is too difficult a problem to discuss here). The Chinese were styled by the Greeks Seres, "the silk people," silk being the product that achieved fame for China throughout the world. The other name "Sinai" reached the West from Southern China over the maritime route, and is connected with the Ch'in dynasty, the first powerful dynasty that unified and welded China into a great empire. Our word "China" is the direct offspring of the name of this dynasty. The term Chinese properly means "people of the Ch'in dynasty," a perfectly honorable name. The Greeks never realized that Serica and Sinai were one and the same people, but believed that they represented two different countries.

During the middle ages China became known in Europe under a new name, Cathay or Kitai. This word is traceable to the Kitan, a Tungusic tribe from the Sungari River that conquered China and ruled as the Liao dynasty from A. D. 907 to 1125. The Mongols and Turks still call China Kitat; and the Russians, Kitai. In English, Cathay is now restricted to poets and highbrows. Marco Polo and many other travelers who visited China under the Mongol dynasty in the 13th century never speak of China, but only of Cathay, and describe Peking under the Mongol-Turkish name Cambaluc or Khanbalik, which means the capital of the Great Khan, the title of the Mongol sovereigns. Europe was puzzled and kept in suspense for

5

several centuries over the question as to whether China and Cathay were one and the same or two different countries. The common belief was that Cathay was a country far north of China and was so represented in many maps. A Portuguese Jesuit, Benedict Goës, who was stationed at Agra in India, undertook in 1603 a long and arduous journey from India overland to Khotan and from there through Turkestan to Kansu in northwest China, for the specific purpose of settling this problem definitely. Only by the middle of the 17th century did it dawn on Europe that China and Cathay, Peking and Cambaluc were one and the same.

Perhaps it will startle you when I now tell you that America was discovered through the medium of China. Historical events are determined by the law of cause and effect, and seemingly insignificant causes may result in effects of tremendous importance. Without the scanty knowledge that the ancient Greeks and mediaeval Europe possessed of faraway China, America would not have been discovered, or its discovery at any rate would have been long delayed.

In 1492 when Columbus set out on his first memorable voyage of discovery, he was seeking no new continent, but a shorter route to India and the Cathay of Marco Polo by sailing in a westerly direction from Europe. Columbus died with the conviction that the new countries which he had discovered were the eastern shores of Asia, the land of the Great Khan,—a belief which was alive for some twenty years even after his death. China played a prominent role in all of Columbus' calculations. He was an ardent admirer and a deep student of Marco Polo who was his countryman, and whose glowing descriptions of the Far East left a lasting impression upon his mind. On his first voyage Columbus was accompanied by a Latin translation of Marco Polo's travels which was constantly in his hands and in which he entered 45 notes and observations. This copy is still preserved in the Colombina Library of Seville in Spain. Columbus, when he discovered the Bahamas, Cuba and Hispaniola (now Haiti) in the West Indies, believed that he had reached the Cipangu of Marco Polo (identical with Chinese Ji-pen kuo, "land of the sunrise"), Japan; and when on his later voyages he discovered the mainland, the continent of America, he imagined that he was on the outskirts of China. China was a link in the providential chain which at last dragged the New World to light.

Again, in another subtle way, China played a significant role in the discovery of America. In the first century of our era there lived at Tyre in Syria a renowned Greek geographer, Marinus. He was a contemporary of the Han dynasty when the Chinese had active commercial relations with the Roman Orient and when Chinese silk by way

6

of Central Asia found a ready market in Syria and Egypt, in Greece and Rome. Marinus interviewed many traveling merchants who returned from the Far East by land or by sea, and gathered from their lips much valuable information regarding the geography of the distant countries of the East. The book he wrote is unfortunately lost, and only fragments of it are preserved in the work of Ptolemy. Marinus was one of the founders of mathematical geography and made many calculations expressed in longitudes and latitudes. The extent of the eastern hemisphere from the isle of Ferro in the Canary Islands on the west coast of Africa to the east coast of China he determined as 225 degrees—a serious error; for this distance is only 130 degrees. Marinus, accordingly, overestimated the size of the Asiatic continent. This miscalculation, however, was the most fruitful and far-reaching error ever committed by a scholar, for it gave Columbus the impetus to his daring plan. Convinced that Marinus' computation was correct, Columbus believed (and quite logically) that the ocean stretching west of Europe was quite narrow and that the distance from Europe to China was quite short, while in reality it was about twice as long as he figured. Here again it was the dawn of a geographical knowledge of China in the West which ultimately led to the discovery of America. We see that errors may lead to truth and that there is an inward coherency of all historical events. History is more romantic than fiction. It is a fascinating picture: China and Greece, Marco Polo and the Great Khan, Columbus and America, all closely interwoven, and finally the United States arising and, with us, Chinese friends whose ancestors indirectly helped to discover this country! *(From address delivered before Chinese students at a reception tendered to them by the Friends of China of Chicago.)*

7

205

鸬鹚在中国和日本的驯养

FIELD MUSEUM OF NATURAL HISTORY

FOUNDED BY MARSHALL FIELD, 1893

PUBLICATION 300

ANTHROPOLOGICAL SERIES VOLUME XVIII, No. 3

THE DOMESTICATION OF
THE CORMORANT IN CHINA AND JAPAN

BY

BERTHOLD LAUFER

CURATOR, DEPARTMENT OF ANTHROPOLOGY

4 Plates in Photogravure

CHICAGO, U. S. A.

1931

CONTENTS

201

LIST OF ILLUSTRATIONS

203

THE DOMESTICATION OF THE CORMORANT IN CHINA AND JAPAN

INTRODUCTION

However much has been written on cormorant fishing in China and Japan, beginning with Friar Odoric of Pordenone, no one has ever made a serious study of the subject, nor has any one ever consulted the Chinese sources relating to it. The present study attempts to fill this gap. For more than twenty-five years I have been interested in the domestications of animals, especially in problems as to when, where, how, and why domestications originated and developed. In the course of these studies when I perused all general books on domestications and a great many monographs on specific subjects, I was struck by the fact that China and Japan are hardly mentioned in this literature and, if so occasionally, data and conclusions are usually wrong. This state of affairs should not be allowed to continue. The Chinese have preserved a vast amount of interesting material, both in literary records and works of art, bearing upon domesticated animals, which if properly used and correctly interpreted, is bound to be of great service to our science.

The problem of the domestication of the cormorant is the more interesting, as it is a typically and characteristically Chinese domestication. Of all nations of the world, the Chinese is the only one that has brought the cormorant into a complete and perfect state of domestication, the birds propagating and being bred in captivity. The fact is the more striking, as the cormorant is a cosmopolitan diffused over nearly the entire world, so that other nations had the same opportunity, but they did not seize it, nay, probably did not even see it. In Japan, the cormorant is semi-domesticated at present, but there is a possibility that there also it was truly domesticated in former ages. Here the interesting problem arises as to the relationship of Chinese and Japanese cormorant fishing, and it will be seen that it is a complex question which if studied at close range is widely different from what at the outset might be expected.

Dr. Gudger (I and II, consult Bibliography at end) has in recent years published two interesting articles on cormorant fishing in China and Japan. Being an ichthyologist and primarily interested in

205

methods of fishing, he has treated the subject from this point of view and in a bibliographical manner. He passes in review, usually quoting the text completely, the more important accounts of cormorant fishing extant in European literature from Odoric down to recent times. Few of these accounts are of importance, or add much to our knowledge of the subject; most of them give surface observations of the fishing method, but say little or nothing about the cormorant itself. Fortune and Fauvel are praiseworthy exceptions and have given us data of scientific value. The more recent, the more stereotyped and duller the accounts of travelers become, and it is difficult to judge what is due to their own observations and what they copied from their predecessors. In the eighteenth and nineteenth centuries when China was merely a cabinet of curiosities in the eyes of European readers, cormorant fishing was not allowed to be wanting in any book on "China and the Chinese" and held its place alongside with birds' nests, dog and cat flesh, crippled feet, eunuchs, punishments and tortures.

Cormorant fishing was practised in Europe as a transient sport toward the end of the sixteenth or beginning of the seventeenth century when it appeared almost simultaneously at both the English and French courts.

James I took great delight in fishing with trained cormorants (as he did also in watching his tame otters, which were trained for a similar purpose), and John Wood was appointed "master of the royal cormorants." In 1618 the king decided to build a house and make ponds for his cormorants, ospreys, and otters at Westminster. In 1609 fishing cormorants were demonstrated at Fontainebleau before Louis XIII when he was dauphin. In the nineteenth century cormorant fishing was revived in England by Captain F. H. Salvin and in France by P. A. Pichot.

Harting (p. 427) holds that it is not unlikely that the sport was first made known in Europe by the Hollanders, who besides being enterprising navigators and traders in the East, have in all ages been known as skilful falconers and great bird fanciers. Likewise Pichot (p. 27) observes that cormorant fishing has come to us from the Far East and that it appears to have been introduced into Europe by the Hollanders in the beginning of the sixteenth century. Freeman and Salvin (p. 328) report two instances of cormorants having been brought to England from Holland, where they had been trained.

Yule (Cathay, new ed. by Cordier, II, p. 189) is mistaken in saying that the English bird was formerly used for fishing both in

England and in Holland quite in the Chinese way. This, a priori, is utterly impossible, as the Chinese bird is thoroughly domesticated, while his European cousin was never domesticated, but merely trained for hunting fish.

In England the training of the cormorant was practised in adaptation of that of the falcon. The birds were hoodwinked when carried out of their enclosures to the fish-ponds so that they might not be frightened. The hoods were taken off on arrival at the fishing ground. When returning from fishing, the keepers called them to their fist, and the birds were carried on the gloved hand like a falcon. Nearly all the sportsmen of England and France interested in the cormorant were originally falconers. Thus Pichot (p. 27) admits that he learned the use of the cormorant from John Barr, a falconer from Scotland. This falconry method of cormorant fishing was never practised in China, and is peculiar to Europe. While it has but little interest to the student of cormorant domestication, the notes of European cormorant trainers like Harting, Salvin, and Pichot on the behavior of the bird are apt to offer him valuable suggestions.

Hunting with cormorants remained restricted to Holland, England, and France. In Germany where cormorants were occasionally hunted (J. Wimmer, Geschichte des deutschen Bodens, 1905, p. 363) no attempts at training them were made; neither in Scandinavia. Olaus Magnus (Compendious History of the Goths, Swedes, and Vandals, 1558, p. 199) briefly describes the cormorant under the name "water-crow or eel-rook," but does not allude to fishing with cormorants.

The cormorant of China does not constitute, as was formerly assumed, a species of its own, being paraded under such hard names as *Hydrocorax sinensis* Vieillot, or *Pelecarus sinensis* Latham, or *Phalacrocorax sinensis*. According to Armand David (Les oiseaux de la Chine, p. 532) and other ornithologists, the Chinese cormorant is identical with that of Europe, and must simply be termed *Phalacrocorax carbo* Schr. Swinhoe. The species is diffused as far north as Kamchatka, and is very common along the entire coast of China and on lakes and rivers in the interior of the country, as well as in Mongolia. Moreover, this species is widely distributed along the Atlantic coast of North and South America, in South Greenland, Iceland, the Faroe Islands, Europe, Asia, Africa, Australia, and New Zealand.

CHINESE TERMINOLOGY

鷀 *i*, 鸕鷀 *lu-ts'e* (written language), also pronounced *lu-se*, *Phalacrocorax carbo*. The *Er ya* 爾 雅 gives the name of the cormorant as *i* or *ts'e i* 鷀 鷀, explained as *lu ts'e* and defined as a bird "with a beak curved like a hook and subsisting on fish" 觜 角 曲 如 鉤 食 魚.

Li Shi-chen 李 時 珍, in his *Pen ts'ao kang mu* 本 草 綱 目, cites the *Yün shu* 韻 書 (cf. Watters, Essays on the Chinese Language, p. 40) to the effect that both *lu* 盧 and *tse* 茲 mean "black," and that these names are conferred upon the bird with reference to its deep-black color (cf. 黲 *tse*, "black"). This explanation goes back to the *Ts'ang kie p'ien* 蒼 頡 篇, on which see Watters, Essays, p. 26, and Pelliot, Le Chou King, *Mém. concernant l'Asie Orientale*, II, 1916, p. 137). According to the *Pen ts'ao*, the word *i* (in some editions written 鷀) is the cry or call of the bird itself (鷀 者 其 聲 自 呼 也), and was hence adopted as the name for the bird.

S. Wells Williams (*Chinese Repository*, VII, 1839, p. 54) asserts that the etymology of *lu ts'e* is "the black [bird] in the reeds." Evidently he thought in this connection of 蘆 *lu* ("a kind of reed"), but I am not aware of the fact that this derivation is given by any Chinese author. The supposition given above seems quite plausible, and the bird's name would simply mean "the black one."

In Lo-lo-p'o, one of the Lo-lo dialects, the cormorant, according to Liétard, is called *vɪ-dzö-mo*. The element *dzö* obviously corresponds to Chinese 鷀, anciently *dzi* (Shanghai *ze*).

The *Yi ts'ie king yin i* 一 切 經 音 義 (chap. 19, p. 1b), compiled by Hüan Ying 玄 應 toward the middle of the seventh century, defines the cormorant after the *Tse lin* 字 林 as "resembling the *i* 鷀 (Giles No. 5490, 'the fish-hawk'), but being black, an aquatic bird with a beak curved like a hook and subsisting on fish, also called *shwi ya* 水 鴉 ('water crow')." The latter term corresponds in meaning to our cormorant, derived from Med. Lat. *corvus marinus* ("sea-crow." Cf. German *wasserrabe, seerabe*; Dutch *waterraaf*).

The *Mong liang lu* 夢 梁 錄, written by Wu Tse-mu 吳 自 牧 (chap. 18, p. 15) of the Sung in A.D. 1274, classes the cormorant among the birds of Hang-chou under the name *lu-ho* 鸕 鶘, adding that it is also called *lu-ts'e*.

A synonym for the cormorant given in the *Pen ts'ao kang mu* and in several gazetteers of Fu-kien Province (e.g. *Hing-hua fu chi*

208

與 化 府 志, chap. 14, p. 9) is 蜀 水 花 *Shu shwi hua*, i.e. "water flower of Shu" (Se-ch'wan). In the *Ch'ang-t'ing hien chi* 長 汀 縣 志 (chap. 30, p. 58) this name is explained as referring to the cormorant's dung, which is used as medicine; it is ground to a powder with water, and when administered to a man, has the effect on him that he will give up wine.

慈 老 *ts'e lao*, in the Gazetteer of the District of Ch'ang-t'ing (chap. 30, p. 58).

青 鶿 *ts'ing lu*, 鴂 鶿 *kiao lu*, in Palladius' Chinese-Russian Dictionary (also Giles, No. 1316).

Designations of the bird in the colloquial language are:

烏 頭 網 *wu t'ou wang*, "black-headed net" (*Ts'ing i lu*, tenth century, see below, p. 221).

水 老 鴨 *shwi lao ya*, "aquatic old duck" (*Pen ts'ao yen i*, A.D. 1116).

水 老 鴉 *shwi lao ya*, "aquatic old crow." Especially used in Che-kiang.

魚 鴉 *yü ya*, "fish crow."

釣 魚 郎 *kou yü lang*, "fish-catching gentleman." Giles (Glossary of Reference, p. 96) remarks that this name is borrowed from that of the kingfisher.

摸 魚 公 *mo yü kung*, "Mr. Fish-diver."

鶿 賊 *lu tsei*, "cormorant, the robber" (in *Tan-t'u hien chi* 丹 徒 縣 志).

烏 鬼 *wu kwei*, "black devil." This local Se-ch'wan term is discussed in detail below (p. 214). While in the famous passage of Tu Fu this term in all probability does not denote the cormorant, it has been made to denote the bird from the Sung period onward. This term reminds us of Milton's (*Paradise Lost*, IV, 196) comparison of Satan with a cormorant ("and on the tree of life, . . . sat like a cormorant"). A. Newton (Enc. Brit., VII, p. 162) thinks that this similitude is prompted by the bird's habit of sitting on an elevated perch, often with extended wings, and in this attitude remaining motionless for a considerable time as though hanging itself out to dry.

魚 鷹 *yü ying*, "fish," or rather, "fishing falcon" (in northern China, particularly Shan-tung and Chi-li). This term properly denotes the fishhawk or osprey, and at Peking and Tientsin is especially applied to the common tern (O. F. von Möllendorff, Vertebrata, pp. 77, 102).

It is an interesting fact that the cormorant appears in several place-names. According to the Geography of the Ming (*Ta Ming i t'ung chi*), as quoted in *Pien tse lei pien* (chap. 210, p. 5), there is a Cormorant Cliff (Lu-ts'e Yen 巖, a hundred *li* west of King-ning hien 景 寧 縣 in the prefecture of Ch'u-chou 處 州 府, Che-kiang; there is a spring there a drink from which would cure disease; subsequently it was struck by a bolt of lightning and formed a pool; there were fishermen who dared not cast their nets in it.   A Cormorant Lake (Lu-ts'e Hu 湖) is located forty *li* southwest of Hai-yen hien 海 鹽 縣 in the prefecture of Kia-hing 嘉 興, Che-kiang, and measures over forty *li* in circumference.   A Cormorant Embankment (Lu-ts'e Pei 陂) existed on the River Yüan 洹 水 in the district Nei-hwang 內 黃, which formerly belonged to Ta-ming fu 大 名 府, Chi-li (now to Chang-te fu 彰 德 府, Ho-nan), five *li* southwest from the old city of Nei-hwang, measuring eighty *li* in circumference; it was an advantageous place for fishing, protected by the natives.   A Cormorant Islet (Lu-ts'e Chou 洲) is located in the district Shang-kao 上 高 in the prefecture Jui-chou 瑞 州 府, Kiang-si, east of the Lo-han Rock 羅 漢 石.

The earliest European illustration of a Chinese cormorant and fishing-boat is contained in Johan Nieuhoff's Dutch Embassy (1669, p. 134), and is titled "the bird Louwa."   As this is Dutch spelling, the diphthong *ou* is the equivalent of *au*.   The author does not give any European name.   A. A. Fauvel (La Province chinoise du Chan-toung, p. 293, Bruxelles, 1892) identifies this *louwa* with *lao-wa* which he says is still current in China.   I have never heard this word, and have in vain looked for it in dictionaries.   I imagine that it should be written 老 窪 *lao wa*, *wa* being a name for the heron with which the cormorant is sometimes confounded.   E. H. Parker (Up the Yangtse, p. 270, Shanghai, 1899) states that he heard the cormorant call *lao wa* in the upper Yangtse region.   In Mesny's *Chinese Miscellany* (IV, 1905, p. 228) the cormorant is called *shwi-lao-wa*.

In Yen-chou, Che-kiang, according to an observation of E. H. Parker (*J. China Br. R. A. S.*, XIX, 1885, p. 40), the heron (*lu-se* 鷺 鶿) is called the cormorant, and the cormorant (*gang ngo* 衚 鵝) the heron.   I note from R. S. Maclay's Dictionary in the Foochow Dialect (p. 505, Foochow, 1870) that in Fu-chou colloquial the cormorant is *lo li* or *lo si*, *lo* being 鷺 *lu* ("heron").

JAPANESE TERMINOLOGY

u (a native Japanese word) 鵜, *Phalacrocorax carbo*. This character, read *t'i* in Chinese, refers in China to the pelican. It is not exactly clear why this character was adopted in Japan for the designation of the cormorant, but it is intelligible since pelican and cormorant are closely allied birds, both belonging to the Palmipedes. R. Hemeling, in his English-Chinese Dictionary (1916), assigns to *t'i* the meaning "cormorant." I do not know whether or in how far this is correct.

shimatsu 島 津 (Manyōshu), 鵁 慈, *P. capillatus*.

umi-u, *P. capillatus*.

hime-u 姬 鵜, *P. pelagicus*.

chishima-u 千 島 鵜, *P. bicristatus*.

kawatsu, *P. carbo hanedae* Kuroda, a smaller species caught mainly on the coast of Tōkyo Bay and used for fishing in the streams near Tōkyo.

u-bune 鵜 船, a boat used in fishing with cormorants.

u-kai 鵜 飼, cormorant fisher.

u-tsukai 鵜 遣 飼, do.

u-nawa 鵜 繩, a line or straw rope used in cormorant fishing.

eboshi 烏 帽 子, head-dress of the cormorant-trainer.

u-jo, chief cormorant-trainer.

211

HISTORICAL DATA

The earliest mention of the use of trained cormorants for fishing occurs in the Chinese Annals of the Sui Dynasty (A.D. 590–617), but with reference to Japan.

In the *Sui shu* (chap. 81, p. 7) it is on record that "in Japan they suspend small rings from the necks of cormorants, and have them dive into the water to catch fish, and that they can catch over a hundred a day" (倭國以小環挂鸕鷀項令入水捕魚日得百餘頭).

This brief information presumably given by the Japanese envoy who visited the Chinese court in A.D. 607 (cf. O. Nachod, Geschichte von Japan, I, pp. 207, 270) conveys the impression that this method of fishing was a novel affair to the Chinese chronicler. It is curious that neither at the time of the Sui nor under the T'ang do we have a single account relating to this matter, so far as China is concerned. The only historical text cited in the cyclopaedia *T'ai p'ing yü lan* 太平御覽 (chap. 925, pp. 8b–9a), published by Li Fang 李昉 in A.D. 983, is the above passage of the *Sui shu*, while three other texts quoted there have merely reference to superstitions or medical prescriptions in reference to the bird, but not a word is said about its being trained in China. The same holds good for the *T'u shu tsi ch'eng*, where the historical notices of the bird 紀事 open with the text of the Sui Annals, while the *Ts'ing i lu* 清異錄 of the tenth century is given as the first record of trained cormorants in China.

The information given in the *Sui shu* is amply confirmed by Japanese sources. The cormorant (*u* 鵜) was utilized for fishing in ancient Japan. In the *Kojiki* 古事記, completed in A.D. 712, the Emperor Jimmu 神武天皇 addresses in a poem "the keepers of cormorants, the birds of the island" (translation of B. H. Chamberlain, p. 144). The same poem, which in Chamberlain's opinion probably dates from a far earlier age, is found in the *Nihongi* 日本記 of A.D. 720 (translation of W. G. Aston, I, p. 126). In the same work we meet the son of Nihe-motsu 包苴擔, who was the first ancestor of the cormorant-keepers (*u-kahi* or *u-kai*) of Ata (*ibid.*, p. 119) 阿太養鵜部始祖. Again, an allusion to cormorants "diving into the water to catch fish" is made in the same chronicle under the year A.D. 459 (*ibid.*, p. 341).

Cormorants must have been abundant and popular in ancient Japan, and also played a role in mythology (K. Florenz, Quellen

212

der Shinto Religion, p. 68). The father of the Emperor Jimmu bore the name Ugayafuki-aezu-no-Mikoto 鸕 慈 草 葺 不 合 尊 ("Cormorant-rush-thatch unfinished."—Aston, Nihongi, I, pp. 95, 98). The lying-in hut (ubuya 産 屋) was thatched with cormorant feathers. In the record of a census made in A.D. 702 appears the name U-kai-be ("clan of cormorant-keepers") no Mezurame 鵜 養 部 月 都 良 賣 of the province of Mino. According to the Record of the Customs of the Province of Mino 美 濃 國 新 續 風 土 記, there were during the period Engi 延 喜 (A.D. 901–922) seven houses of cormorant fishermen on the Nagara River 長 良 川 in Mino; these prepared special dried ayu 鮎 (sweet-fish) for the use of the emperor and annually presented the fishes to the imperial household. The ayu is a salmonid about a foot in length and found only in the clear upland and mountain streams of central Japan. The Manyōshu 萬 葉 集 (chap. 19) contains a poem 潜 鸕 歌 by Ōtomo no Yakamochi, written in the province of Etchū.

Among the Hundred Laws of Ieyasu there is one (No. 24) in which it is said, "Formerly there were people who asserted that the hunt with cormorants and falcons should be abolished, and yet this is not an idle pleasure or a useless destruction of life. It is an old custom with the princes of China and Japan that they offer their hunting spoils to the emperor," etc. (T. Kempermann, *Mitteilungen der deutschen Ges. Ostasiens*, I, p. II).

Two Japanese scholars, Ikenoya (1917) and Kuroda (1926), have written treatises on the history of cormorant fishing in Japan, and excerpts from their data are given in E. W. Gudger's article "Fishing with the Cormorant in Japan" (see Bibliography at end).

Gudger (II p. 9) states, "So far as I have searched, unlike similar works for China, none of the early European voyagers to Japan, not even Kaempfer, figures or even refers to cormorant fishing in Japan." I have found, however, the following interesting note in the Diary of Richard Cocks (ed. of Hakluyt Society, I, p. 285) referring to the year 1617:

"Soyemon Dono made a fishing over against English howse with cormorants made fast to long cordes behind their winges, and bridles from thence before their neckes to keepe the fish from entring their bodies, so that when they took it they could take yt out of their throates again." This is exactly the method still followed in Japan at the present time.

While there is good evidence for cormorant fishing in ancient Japan extending from the fifth and sixth to the eighth century, contemporaneous evidence for China is entirely lacking. To be sure, ancient references to the cormorant in dictionaries, poetry, and the *Pen ts'ao* literature are plentiful (see below, pp. 221 and 224), but these passages are reticent as to the training of the bird. The first and earliest document that contains a notice of trained cormorants used by man is the *Ts'ing i lu*, a work of the tenth century (below, p. 221). It is supposed that a passage in a poem of Tu Fu alludes to domesticated cormorants; unfortunately, however, the bird does not appear there under its real name, but the phrase "black devil" is interpreted as such. This passage has been the subject of a lively controversy among Chinese authors, and I shall briefly review their opinions. At the same time we have occasion to touch upon several texts of the Sung period which refer to cormorant fishing.

In a poem of Tu Fu 杜 甫 (A.D. 712–770) the following verse occurs:

家 家 養 烏 鬼
頓 頓 食 黃 魚

This means literally:

"All families (or: in the houses they) raise the black devil
And at every meal feed on the yellow fish."

S. Wells Williams (*Chinese Repository*, VII, p. 542) translates, "Every family trains the black devil, which, often diving, seizes the yellow fish." The text, however, contains nothing about "diving" or "seizing." If *wu kwei* really be the cormorant, it may be admitted that the cormorant may occasionally feed on yellow fish; but if *wu kwei*, as interpreted by other scholars, is something quite different, this conception of the passage is absurd. It is quite clear that *kia kia* is the subject of the second clause and that the families eat the yellow fish. The latter is explained in the commentary to the *Er ya* as the name given east of the river 江 東 (i.e. in the lower Yangtse region) to the sturgeon, *chan* 鱣, which is from twenty to thirty feet long (*Pien tse lei pien*, chap. 135, p. 17b, where also the above passage from Tu Fu is quoted, the title of the poem being given as 戲 作 俳 諧 體 詩. Cf. also *Sü po wu chi* 續 博 物 志, chap. 2, p. 6, ed. of *Pai hai*). It is obvious that a cormorant can not catch a fish of the dimensions and weight of a sturgeon. The weight of a full-grown bird is about seven pounds, and a one-and-a-half pound fish is about the maximum weight a bird can carry. Long after I wrote this, I found in the *T'ung ya* 通 雅 (chap. 45, p. 21b) that Ma Yung-k'ing 馬 永 鄉, who lived in the first part of the twelfth century, has made the same criticism ("Can a cormorant catch a yellow fish?" And Fang I-chi 方 以 智, author of the *T'ung ya*, comments, "Why is it necessary that he catches yellow fish? Why should it not be a yellow-cheeked fish 黃 頰 魚 *hwang kia yü* [Möllendorff, *Vertebrata*, p. 107]?"). The *Pien tse lei pien* gives a quotation from another poem of Tu Fu, entitled "The Yellow Fish" (cf. also *Neng kai chai man lu*, chap. 6, p. 23) and cites the *Yu yang tsa tsu* as saying that in Shu (Se-ch'wan), whenever a yellow fish is killed, it will invariably rain.

The phrase "black devil" (*wu kwei*) has aroused much comment from authors of the Sung period; and, it will be seen, is credited with several different meanings. The general situation is well summed up by Wang Mou.

Wang Mou 王 懋, in his *Ye ko ts'ung shu* 野 客 叢 書 (chap. 26, pp. 5b–6, ed. of *Pai hai*), the preface of which is dated A.D. 1202, has the following discourse on the subject under the heading *Wu kwei* ("Black Devil"):

"As to the line of old Tu 'in the houses they raise the black devil,' there are several explanations. The *Lan chen tse* 嬾 眞 子 [by Ma Yung-k'ing 馬 永 卿: Wylie, *Notes*, p. 164] interprets *wu kwei* as 'pig.' Ts'ai Kwan-fu 蔡 寬 夫 [author of *Ts'ai shi shi hwa* 蔡 氏 詩 話, biography in *Sung shi*, chap. 356] explains it as 'the seven gods of the dark wilderness' 烏 野 七 神. The *Leng chai ye hwa* 冷 齋 夜 話 [by Hui Hung 惠 洪, end of eleventh century: Wylie, *Notes*, p. 164] regards it as 'the devils of the Black Man,' *Wu Man kwei* 烏 蠻 鬼. Shen Ts'un-chung [i.e. Shen Kwa 沈 括, A.D. 1029–93], in his *Mong k'i pi t'an* 夢 溪 筆 談, the *Siang su tsa ki* 緗 素 雜 記, the *Yü yin ts'ung hwa* 漁 隱 叢 話 [by Hu Tse 胡 仔 of the Sung], the *Lu nung shi* 陸 農 師, and the *P'i ya* 埤 雅 interpret it as the cormorant (*lu-ts'e*). These four explanations differ widely. However, solely the explanation of the *Leng chai* is correct. Referring to the chapter 'Record of the Southern Man' in the T'ang Annals 唐 書 南 蠻 傳, we read that 'they commonly esteem shamans and devils and in the great tribes they have Great Devil Chiefs 大 鬼 主, while in the families they establish a Small Devil Chief 小 鬼 主; the White Man (Pai Man) form a single clan, the Black Man (Wu Man) five clans; the latter are so called, because the dresses of their women are made of black silk, while those of the White Man are made of white silk.' Further, in examining the comment of the *Leng chai*, Liu Yü-si 劉 禹 錫 (A.D. 772–842), in his poem Nan-chung 南 中, says, 'They sacrifice in excess to numerous dark devils 青 鬼, for white-haired men are scarce among the inhabitants.' There is another saying with reference to the so-called dark devils: The cause of the prestige of the Man tribes of the rivers and gorges of Kwang-nan rests on the fact that they have such names as dark devils and black devils.

"In the lines of Tu the 'yellow fish' is antithetical to the 'black devil,' from which it follows that the devils of the Black Man are meant. Let us further examine a poem of Yüan Wei-chi 元微之 [i.e. Yüan Chen 元稹, A.D. 779–831], who says, 'In their rustic taste they prize frogs or clams; in their domestic worship, all serve the crow or raven' (鄉味尤珍蛤家神悉事烏). Yüan Chen further says, 'In case of disease they carry a raven around in a procession, praising it as a demon; the sorcerers predict (the outcome of the disease) by means of a tile which with them takes the place of a tortoise (病賽烏稱鬼巫占瓦代龜).' According to the commentary, the people of the south, when infected by a disease, carry a black devil around in a procession. These explanations are similar to, but not identical with that given above. As regards the Record of the Southern Man (in the T'ang Annals), the word *wu* 烏 has the meaning 'black,' but in the poem of Yüan Chen, as the word *ko* 蛤 is antithetical to *wu* 烏, it must have the meaning 'raven'."

This discourse is instructive in many respects. We note above all what difficulties Sung authors had in correctly understanding and interpreting a passage in a T'ang poem and how widely divergent their opinions were. I do not see that we poor epigones can do much better. There is no doubt that *wu kwei*, apparently restricted to Se-ch'wan, had several different meanings during the Sung period. I shall give additional information from the sources cited by Wang Mou and likewise from others.

Wu kwei, as Ma Yung-k'ing informs us, was applied to pigs. This is confirmed by the *Man sou shi hwa* 漫叟詩話 (cited in *Pien tse lei pien*, chap. 207, p. 3b): "The people of Se-ch'wan are fond of pork, and all families there are engaged in the rearing of swine. Whenever they call their pigs, they emit the sound *wu kwei* 烏鬼. Hence they call a pig *wu kwei* ('black devil')."

Hia-hou Tsie 夏侯節 (in the *T'ung ya*, chap. 45, p. 21b, he is called 夏太夫), a scholar living in the gorges of Se-ch'wan 峽中士人, is quoted as saying, "The black devil is a pig. Many families inhabiting the gorges serve the devils and raise a particular pig, not, however, to be sacrificed to the devils, as this would be useless, but they put this pig among the common herd, and from time to time call the devils and the pig, keeping them apart, and this is the proper thing to do" (*Ning-hua hien chi*, 甯化縣志, chap. 2, p. 128). In the *T'ung ya*, however, the same scholar is quoted as saying, "The black devil is a pig. The families raise a particular pig for the purpose of sacrificing it to the devils." If this be correct, the designation *wu kwei* for the sacrificial pig may have arisen from the fact that it was intended as a sacrifice to the *wu kwei*. Yang Shen 楊 愼 (1488–1559) writes that the people in the gorges raise young chicks which, provided with copper or tin rings, are offered to the spirits, and these are called *wu kwei* (*T'ung ya*, chap. 45, p. 21b).

Wu kwei was also applied to the raven. *Corvus torquatus* is still designated "devil bird" 鬼 鳥 (Möllendorff, Vertebrata, pp. 88, 89).

Lo Yüan 羅 願, author of the *Er ya i* (twelfth century), writes, "In the opinion of some people, the inhabitants of the defiles of Se-ch'wan term the cormorant *wu kwei*, but these folks, on the other hand, serve the raven and call it devil; hence *wu kwei* is not the cormorant" (或 云 峽 人 謂 爲 烏 鬼 然 峽 人 乃 事 烏 爲 鬼 非 此 物 也). Regarding another explanation as "raven" see below, p. 219.

On the other hand, it cannot be doubted that *wu kwei* began under the Sung to denote the cormorant, and Wang Mou cites four author-ities to this effect. This, of course, does not mean that the cormorant can be interpreted into the passage of Tu Fu. Wang Mou, however, errs in listing Shen Kwa among the pro-cormorantists; on the con-trary, as will be seen presently, Shen Kwa, is averse to this explana-tion. Wang Mou's misunderstanding is copied from the *Leng chai ye hwa* (chap. 4, p. 7b, ed. of *Pai hai*), which he quotes, and this error was probably caused by the fact that Shen Kwa cites the *K'wei chou t'u king* which seems to be the first Sung work to explain *wu kwei* as cormorant and the text of which is given alike in the *P'i ya* and the *Mong k'i pi t'an* written between A.D. 1086 and 1093.

Shen Kwa, in his *Mong k'i pi t'an* (chap. 16, p. 1, ed. of *Pai hai*) writes as follows: "The scholar Liu K'o 士 人 劉 克 extensively inspected strange writings. One day he lighted on a poem of Tu Fu [as quoted above]. The general explanation of this passage now given is that all say that in the gorges of K'wei-chou there are up to the present time 'devil families' (*kwei hu* 鬼 戶), and these are simply savages 乃 夷 人 也; their chieftain is called a devil's chief (*kwei chu* 鬼 主). However, I have never heard of a phrase like *wu kwei* 然 不 聞 有 烏 鬼 之 說. Moreover, as regards the devil families, these are the savages who are so called, but not something raised by man (as the cormorant is).

"The *K'wei-chou t'u king* 夔 州 圖 經 says that 'the inhabitants of the gorges of Se-ch'wan call the cormorant "black devil" (*wu kwei*) and that the people of Shu 蜀 人 dwelling along the water-courses raise this bird and fasten a cord to its neck whenever they send it into the water to dive for fish; when the cormorant has caught a fish, they pull him out of the water by means of this cord.' In this manner it is still practised up to this time, and I can testify to the truth of it, for I was in Se-ch'wan myself and saw people there raise cormorants (*lu-ts'e*) and employ them for the purpose of catching fish. Only I do not know that the people of Se-ch'wan call the cormorant black devil (但 不 知 謂 之 烏 鬼 耳)."

Here we have good and trustworthy testimony for domesticated cormorants in Se-ch'wan during the eleventh century. As to the term *wu kwei*, one man's word is as good as another's. Shen Kwa asserts merely that he did not hear the term used in Se-ch'wan for the cormorant, while others affirm they did. There is no reason to doubt that in the eleventh century *wu kwei* was a popular designation of the cormorant locally in certain parts of Se-ch'wan. We have no evidence, however, that this term was so used three centuries earlier during the lifetime of Tu Fu. No text of the T'ang period informs us that *wu kwei* then had the significance "cormorant." This is not a T'ang tradition, but a Sung tradition, and what happened was that Sung scholars who heard of that local usage interpreted this meaning back into the line of Tu Fu.

Fan Chen 范 鎮, author of the *Tung chai ki shi* 東 齋 記 事, who lived in the latter part of the eleventh century, deserves special consideration as he was a native of Hwa-yang 華 陽 in Se-ch'wan and must have been well informed on the affairs of his country. This work is reprinted in the *Shou shan ko ts'ung shu*, vol. 84, where his notice of cormorants is not given; it is cited, however, alike in the *Tsing k'ang siang su tsa ki* to be noted presently and in the *Se-ch'wan t'ung chi* 四 川 通 志 (chap. 74, p. 21), as follows:

"The fishermen of Shu raise cormorants to the number of ten and daily catch fish to the weight of ten catties. They tie a cord around the neck of the bird, so that a small fish may just pass through his throat, while he cannot swallow a big fish. From time to time the fishermen take the birds out of the water and let them dive into the water again. The birds are very tame, and will attend to everything as if they had a human heart. Whether they have caught a fish or not, they return to the boat, and those familiar with the flock will feed them and then cause them to return to their work. This method is comparable to falconry, but saves the trouble of riding on horseback and running. The profit obtained from this business is very large."

The *Siang su tsa ki* cited by Wang Mou is called with its complete title *Tsing k'ang siang su tsa ki* 靖 康 緗 素 雜 記. This work, as implied by the title, was written in the Tsing-k'ang period (A.D. 1126-27), and is from the hand of Hwang Ch'ao-ying 黃 朝 英 of the Sung (cf. Wylie, Notes, p. 159: early twelfth century). It is reprinted in the *Shou shan ko ts'ung shu*, vol. 69, and the text in question, entitled *Wu kwei*, is in chap. 5, pp. 2b–3a (the *T'ang Sung ts'ung shu* contains only a greatly abridged form of this work of only thirteen pages, which does not contain this text).   The author refers to the *Mong k'i pi t'an*, saying that solely Liu K'o 劉 克 has explained the meaning of *wu kwei* correctly; then he cites the *Tung chai ki shi*, translated above, noting that its author, Fan, does not know either that the cormorant should be intended by the "black devil" of Tu Fu. Finally he cites the cormorant account of the *Sui shu;* and, I believe, he is the only Sung author who has done so and seen the identity of cormorant fishing in Japan and China.   Judging from this text, Wang Mou is wrong in classifying Hwang Ch'ao-ying among writers who interpret Tu Fu's *wu kwei* as the cormorant; but in the *Se-ch'wan t'ung chi* 四 川 通 志 (chap. 74, p. 21) Hwang is quoted thus: "The people in the gorges (of Se-ch'wan) call the cormorant 'black devil' and raise it for the purpose of catching fish, making cormorant-fishing their business."   If Hwang Ch'ao-ying should really have made this statement, which is not contained in the edition of the *Shou shan ko ts'ung shu,* and which contradicts the text of the latter, Wang Mou would be right.

Exactly the same information as in the *Se-ch'wan t'ung chi* is given in the *K'wei-chou fu chi* 夔 州 府 志, 1891, chap. 14, p. 4.

Wu Tseng 吳 曾, author of the *Neng kai chai man lu* 能 改 齋 漫 錄 (chap. 6, p. 21b; middle of twelfth century), cites the passage from Yüan Chen's poem with the commentary in the same manner as Wang Mou and decides that the *wu kwei* of Tu Fu has the same meaning.   He also attributes to Shen Kwa the notion that the *wu kwei* of Tu Fu is the cormorant, adding, "I do not know on what evidence this is based" (不 知 又 何 所 據 也).

The opinion expressed in the *Leng chai ye hwa* (chap. 4, p. 7, ed. of *Pai hai*) is shared by Kwo T'wan 郭 彖 in his *K'wei kü chi* 睽 車 志 (Wylie, Notes, p. 197; after *T'u shu tsi ch'eng;* I have not been able to trace this passage in the edition of *Pai hai*), who denies that in Tu Fu's verse the term *wu kwei* refers to the cormorant, but who asserts that it alludes to the Black Man devils 烏 蠻 鬼 and that this is the only correct explanation.   The *Leng chai ye hwa* says that "the people along the roads in the gorges of Se-ch'wan sacrifice to the devils of the Black Man."   Fang I-chi adds this comment: "On the roads of Pa-tung 巴 東 there are ravens.   People traveling by boat must throw away their meat and give it to the ravens as provision; otherwise they will have no luck.   Should these ravens be the black devils?"

In the *Wen kien lu* 聞見錄 by Shao Po-wen 邵伯温 of the Sung dynasty it is said (*Pien tse lei pien*, chap. 207, p. 3b), "On the eleventh day of the first month of every year the people in the gorges of K'wei 夔峽之人 offer sacrificial animals and wine in the fields on behalf of Ts'ao Ts'ao 為曹; on this occasion there is a drill of soldiers with a great deal of noise, and this they call 'raising the black devil'" 養烏鬼. The *Wen kien lu* is reprinted in the *Tsin tai pi shu* and *Hio tsin t'ao yüan* to which I have not access at present.

It is certain that the term *wu kwei* hails from Se-ch'wan.

It cannot be said that the Sung writers who plead for *wu kwei* as the cormorant have made out a strong case in their favor. They are all obsessed with the sole passage of Tu Fu, and do not fall back on any other text. Above all, they fail to explain why *wu kwei*, a local Se-ch'wan colloquial term, was transferred to the cormorant. From their explanations it is perfectly clear why pig and raven came to be called "black devil." Here the relation of these creatures to devils or spirits is conspicuous; not so, however, in the case of the cormorant who lacks any association with the supernatural world. Why this useful helpmate of man who contributes so much to human economy should be dubbed a black devil is simply unintelligible.

Palladius (Chinese-Russian Dictionary, I, p. 127) explains *wu kwei*, "raven, as a demon. In the south of China, in case of any disease, the female shamans make an offering and divine by means of the tortoise. Swine-mother. Cormorant." Giles, in his Dictionary, gives merely the meaning "cormorant."

I shall not decide how the above verse of Tu Fu is to be interpreted. I have merely reproduced the various opinions of Chinese authors from which it follows that the definition of *wu kwei* as "cormorant" in Tu Fu's verse is at least exceedingly doubtful. This study is not a contribution to Chinese poetry, but one to the domestication of the cormorant. As I have studied the latter subject for many years, I feel obliged to say that it seems to me most improbable that the cormorant can be understood in the passage of Tu Fu; for to say that "the families raise the cormorant" would imply that the domestication of the bird was quite general (at least in one part of China or the other) during the eighth century. Now, if this had been the case, we should assuredly expect to find other contemporaneous texts corroborating this matter, but no such text of the T'ang period, as far as I know, has as yet come to the fore. It seems fairly clear that the interpretation of *wu kwei* as "cormorant" in Tu Fu's verse arose only in the Sung period when the subject of cormorant fishing became known to scholars, and personally I am convinced that the cormorant is not visualized by Tu Fu. But even if this were the case, it would simply be an isolated scrap of evidence to which no great historical significance can be attached. In my opinion the line of Tu Fu means, "The families when they perform the ceremony known as 'caring for the black devil,' feed (on this festive occasion) on sturgeon." To be sure, Tu Fu, like Li Po, Wang Wei, and other poets of the T'ang period, was familiar with the cormorant; he mentions it in a verse that begins 門外鸕鷀久不來, but here the bird appears under its regular name *lu-ts'e,* and so it is designated by other writers, e.g. Yüan Chen (E. von Zach, Ein Briefwechsel in Versen, pp. 216, 220). The salient point is that the cormorant is mentioned and discussed by Sui and T'ang authors, but there is dead silence on their part as to the domestication of the bird.

The *I wu chi* 異物志 of Yang Fu 楊孚, which is possibly a work of the Sui dynasty, for instance, says that the cormorant dives in deep water in order to catch fish on which it subsists; it is a water-bird and nests in high trees. A superstition dealt with below under the heading "folk-lore" is added to the text, but nothing is said about the employment of the bird in the service of man.

In so-called classical literature the cormorant appears not to be mentioned. It is not referred to in the *K'in king* 禽經 ascribed to Chang Hua 張華 (third century A.D.), nor in the *Mao shi ts'ao mu niao shou ch'ung yü su* 毛詩草木鳥獸蟲魚疏 of Lu Ki 陸璣 (A.D. 260–303).

The earliest text that mentions the cormorant as trained for fishing is contained in the *Ts'ing i lu* 清異錄 (chap. 上, pp. 57b–58), written by T'ao Ku 陶穀, who lived in the tenth century (A.D. 902–970, according to Giles, Biogr. Dict., No. 1898):

"For the purpose of catching fish they use cormorants (*lu-ts'e*) which are quick and alert to a high degree. In the district of Tang-t'u 當塗 there are ponds covered with aquatic plants and rocky hills, where people live scattered on farms and in cottages and raise cormorants in their houses 畜鸕鷀於家. They make a small boat fast on the bank and daily send a man 一丁 to catch fish for the supply of the families. A municipal official (*yi yü* 邑尉), when he passed this place, noticed it and said to the hill-people, 'The small boat is the arena for receiving the meat 小舟即納膾場; the cormorant is the small official (*siao yü* 小尉).' It is said also that the fishermen of the rivers and lakes use the cormorant and call it 'the small official.' Other fishermen bestow on the cormorant the epithet 'black-headed net' (*wu t'ou wang* 烏頭網)."

Tang-t'u forms the prefectural city of T'ai-p'ing 太 平 府, An-hui Province, on the lower Yangtse. How little dependence can be placed on the *T'u shu tsi ch'eng* is shown by the fact that this important document is quoted therein in the following abbreviated form: "For the purpose of catching fish they use cormorants which are quick and alert to a high degree. The people therefore speak of the small boat as the arena for receiving the meat and of the cormorant as the small official." The locality and all the significant points of the story are coldly deleted. I allow this to stand as I wrote it, but afterwards when I consulted the edition of the *Ts'ing i lu* in the *T'ang Sung ts'ung shu* (chap. 2, p. 13b), I found there to my surprise the text as given in the *T'u shu*. There must be, accordingly, different manuscripts of the *Ts'ing i lu*.

The document in question is characteristic of the attitude of Chinese scholarship. Here we are confronted with the first mention of the domesticated cormorant on Chinese soil and might justly expect an expression of surprise or astonishment at so unusual an achievement, or an inquiry on the author's part into the origin of this extraordinary practice. It is simply taken for granted, however, without further comment. What the author is interested in is not the novel fact, but a literary *bon mot* for which his story serves as an explanation. In fact, his notice is entitled 納 臉 場 (腸 in the edition before me, printed in 1875 by 陳 氏 庸 閒 齋, is a misprint) 小 尉, and it is this new, learned term that he is intent on introducing to his scholarly readers. This term has not proved a success, for it has not been adopted, so far as I know, by any subsequent author. The case is analogous to that of the Sung writers who got excited over the term *wu kwei*. The story itself is perfectly clear in demonstrating that during the tenth century the cormorant was domesticated and reared for catching fish in certain places of Tang-t'u, apparently out of the way and not easily accessible to officials, who noted merely the *fait accompli* without bothering about questions of origin and development. Simple folks, fishermen by avocation, had accomplished what an official would never have thought of; the process of taming and training the bird, ultimately resulting in its domestication, had remained unnoticed by the learned, and no record of it is preserved.

This philological attitude is characteristic of the Chinese scholarly mind. Words, phrases, characters, inscriptions, etc., have always found attention and were made the subject of profound studies, while the subject-matter itself was neglected. It is hardly conceivable that a matter so characteristically Chinese as the domesti-

cation of the cormorant, which is an interesting scientific problem
to us, has left Chinese scholars completely indifferent. Ch'en Hao-tse
陳 淏 子, in his *Hwa king* 花 鋭, published in 1688, devotes chapter VI of his work to animals kept by man. Of birds he deals, for
example, with crane, peacock, egret, parrot, falcon, eagle, pheasant,
pigeon, etc., but the cormorant is not even mentioned. The *Ts'e
yüan* 辭 源, now revered and quoted by European sinologists as a
sort of Bible, exhibits the same defect, and in giving a superficial
definition of the cormorant does not even mention the fact of its
domestication or its employment in the human household; the term
wu kwei is given as a synonym of the cormorant, but not even the
passage of Tu Fu is cited, nor is there any reference to the controversy which it has aroused!

In one respect T'ao Ku's story does not enlighten us. He does
not describe the method of fishing itself, a subject which did not
greatly interest him; and while he says that cormorants were reared
in the habitations of fishermen in the Tang-t'u District, which
indicates that the birds to some extent must have been domesticated
in that locality at that time (about the middle of the tenth century),
we are unable to judge to what degree the domestication of the birds
had then progressed. Was it still in the primary stage? Or had it
far advanced? The question is important, for we are anxious to
know to what time the beginnings of this domestication go back.
If it was in a perfect state in T'ao Ku's days, we have to concede
that a considerable span of time must have elapsed before this state
was reached; or, if it then was in its initial stages (and this is more
probable), this concession becomes superfluous. T'ao Ku's succinct
note does not give us a direct clue to the solution of this problem.
While the domestication of the cormorant requires a great deal of
patience and endurance, it is not excessively difficult, and in my
estimation it is not necessary to date the preliminary steps leading
to the domestication in China farther back than about the beginning
of the tenth century. More will be said about this point in the
chapter on the Process of Domestication.

The texts of the *K'wei chou t'u king, Mong k'i pi t'an,* and *Tung
chai ki shi* have been quoted above. These refer alike to the province of Se-ch'wan, which has played an important role in the domestication of the cormorant, and agree in emphasizing the cord tied to
the bird's neck—a sure sign of its domestication. There can be no
doubt that during the Sung period it reached the state of perfection.

In the *Er ya i* 爾 雅 翼, written by Lo Yüan 羅 願 in the twelfth
century, it is stated, "At present there are in Shu (Se-ch'wan) many

people inhabiting the water-courses who keep and raise cormorants by tying a cord around their necks. In this manner small fishes may pass the bird's throat, while it is unable to gulp down large fishes. From time to time the fishermen call the birds and relieve them of their fishes; then they are sent out again. They are so docile and familiar with man that signs are sufficient for them to grasp their masters' intentions. When they finally return to the boat, whether with fishes or without, they are detained and fed, and then are allowed to return home. This method is comparable to hunting with kites and sparrow-hawks without the trouble of hustling around. The profit from this business is rather large, for the fishermen raise several tens of birds and daily obtain several tens of catties of fish. The fishes coming out of the birds' throats have a strong odor, being affected by the unpleasant saliva of the birds [the cormorant pockets its prey in its oesophagus]. After they have come out of the water, the fishes are spread out on rocks and dried in the sun."

The attitude of the T'ang and Sung *Pen ts'ao* literature toward the cormorant, as far as its domestication is concerned, is negative. The *Sin siu pen ts'ao* 新修本草 (chap. 15, p. 25b of the facsimile edition published in Japan) does not give a definition or description of the bird; in fact, the article in question is not entitled "The Cormorant," but "The Cormorant's Ordure" 鸕鷀矢, with the synonym *Shu shwi hwa*; the ordure, it is said, removes black spots and pimples from the face; and this is followed by a quotation from T'ao Hung-king, who gives a bit of folk-lore concerning the propagation of the bird (below, p. 255). Ch'en Ts'ang-k'i 陳藏器, author of the *Pen ts'ao shi i* 本草拾遺, does not go beyond this; and the *Pen ts'ao yen i* strings its harp on the same note (text given below, p. 255). T'ang Shen-wei 唐慎微, in his *Cheng lei pen ts'ao* 證類本草 of A.D. 1108 (chap. 19, p. 19b), is not either interested in the cormorant itself, although he pictures it in a naively crude drawing. He is contented to reiterate the data of the *Sin siu pen ts'ao*. The T'ang and Sung herbalists, accordingly, restricted themselves to pharmacological and folkloristic notes without manifesting any real interest in the bird itself. The fact that the T'ang authors of *Pen ts'aos* are silent as to the bird's employment for fishing is, of course, inconclusive; the Sung Pentsaoists do not mention it either, although at their time this was an accomplished fact.

Li Shi-chen 李時珍, in his *Pen ts'ao kang mu* 本草綱目, is the first herbalist who discusses the cormorant with some degree of intelligence:

"Cormorants occur everywhere in districts where water is found. The bird resembles the fish-hawk (*yi* 鶚), but is smaller than the latter. In color it is black like a crow. It has a long and slightly curved bill. It is expert in diving into water and catching fish. During the daytime the birds gather on islands; at night they roost in the trees of forests. The ordure of the birds is poisonous, and the trees on which for a long time they have perched will decay. The fishermen in the southern parts of the country keep them by tens, tying them together, and thus they catch fish for them. The passage in Tu Fu's poem that 'families raise black devils and at every meal feed on yellow fish' is referred by some to this species. There is another kind resembling the cormorant, but with a head like that of a snake, a long neck, and moulting during the winter; it roosts on the banks of mountain-streams, and at the sight of men is unable to walk and dives into the water. This is identical with what the *Er ya* calls *yao t'ou* 鴉 頭 or *yü kiao* 魚 鮫; it is not used in the pharmacopoeia."

This species, under the name *yü kiao* or simply *kiao* ("shark"), is also mentioned by Ch'en Ts'ang-k'i of the T'ang, who describes it as having a slender head and a long body and being white in the upper part of the neck (cf. *Cheng lei pen ts'ao*, chap. 19, p. 19b). The description of this species, as given by Li Shi-chen, is almost identical with that in the *Er ya i*, save that there the reference to the two names is wanting.

During the Ming period cormorant fishing appears to have been flourishing. Sü Fang 徐 芳 of the Ming writes, without specifying localities, that "cormorants were reared by many people along the rivers, being carried on small rafts. Fishing was done in stagnant water or in places where the river formed eddies and where fishes congregated. The birds dived deeply into the water and swiftly brought up small fishes; when their strength failed in carrying big ones, they broke them up. A small ring was tied around their necks, so that they could not swallow fishes of large size; when they caught such, the fishermen took them away at once. Small fishes entered their throats as far as the spot where the ring was placed, but the birds could not swallow them on account of the bones. When the fishes had piled up and the birds were hungry, they were fed with a couple of fishes. The birds were greedy and insatiable, but the fishermen were satisfied and reaped a large profit" (text in *T'u shu tsi ch'eng*, XIX, chap. 45, *i wen*).

To the Buddhists, naturally, cormorant breeding and what is associated with it has been a thorn in the flesh. In a Buddhistic tract written in the colloquial language and published by L. Wieger (Moral Tenets and Customs in China, p. 203, Ho-kien-fu, 1913), the officials are urged to prohibit the keeping of cormorants as well as fishing or catching crabs. Here the term *yü ying* 魚 鷹 is used.

GEOGRAPHICAL DISTRIBUTION

The center of cormorant fishing is the lower Yangtse Basin including the provinces of An-hui, Kiang-su, and Che-kiang, the present T'ai-p'ing fu on the Yangtse being pointed out as early as the tenth century. The region around Lake T'ai (T'ai hu) and the entire country intersected by a net of canals around Wu-si, Su-chou, Hang-chou, Chu chou, Shao-hing, Fung-hua, and Ning-po swarm with cormorants kept by fishermen. The easternmost point to which the trained cormorant advanced is Ting-hai on the Island of Chu-san (*Ting-hai t'ing chi* 定 海 廳 志, chap. 34, p. 42). The most celebrated of the localities of Che-kiang is T'ang-si-chen, a small town situated 50 *li* northwest of Hang-chou, whose inhabitants are reputed to possess a secret which insures to them a decided success in the rearing of cormorants (Fauvel, p. 230). Nearly all district and prefectural gazetteers of the province allude to the industry; for example, *Ts'ing-t'ien hien chi* 青 田 縣 志 (1875, chap. 4, p. 29); *Shao-hing fu chi* 紹 興 府 志, chap. 11, p. 31; *Ts'e-k'i hien chi* 慈 谿 縣 志, chap. 54, p. 9. In the region of Shao-hing and Ning-po I observed cormorant fishing many times in 1901.

From Che-kiang the practice spread to the province southward, Fu-kien. A careful examination of the prefectural and district gazetteers of this province in the Library of Congress has led me to the following result. Fishing with cormorants occurs all along the seacoast of Fu-kien from Fu-ning fu in the north down to Fu-chou, Hing-hua, Ts'üan-chou, and Chang-chou in the south, all along the Min River from Yen-p'ing to Fu-chou; further, in Lung-yen chou, Yung-ch'un chou, as far west as T'ing-chou fu (汀 州 府 志, chap. 8, p. 19) and as far north as Kien-ning fu. It may be said, therefore, that cormorant fishing is generally practised over the entire province.

As to Fu-chou, cormorant fishing was observed by R. Fortune (I p. 88) in 1843 and by G. Smith in 1845 in the suburb of Nan-tai (*Chinese Repository*, XV, 1846, p. 207). Freeman and Salvin (p. 328) also state that they met with people who saw it about Fu-chou fu, at the mouth of the river. Cormorant fishing is still actively pursued there.

The two coterminous provinces Che-kiang and Fu-kien have always exchanged cultural products, and most probably cormorant fishing spread from Che-kiang to Fu-kien. The Gazetteer of Hing-hua fu 興 化 府 志 (chap. 14, p. 6) in Fu-kien stresses the

227

fact that "the people of Che-kiang are in the habit of raising cormorants who dive to catch fish, swallowing the small ones, but bringing the big ones to their master."

In Fu-chou, Fu-kien, it seems to be customary that when a bird has brought a fish to the surface, the boatman paddles his raft to the spot and casts a net into the river, hauling bird and fish on board (G. Smith, *Chinese Repository*, XV, 1846, p. 207). This unnecessary procedure goes to show that the fishermen in question had passed to the use of the cormorant from a former method of fishing with nets. J. Doolittle (Social Life of the Chinese, I, 1865, p. 55) intimates that the fishermen of Fu-chou take bird and fish out of the water with a dip-net only in case the fish is a large one and a struggle ensues between bird and fish.

In An-hui Province, to which our earliest Chinese account relates, the district of Wu-ho 五 河 縣 enjoys a reputation for breeding cormorants (Korrigan, p. 39). This method of fishing is practised all along the Huai River 淮 河, which traverses Ho-nan and the northern part of An-hui.

As to the province of Kiang-si, we have observations of cormorant fishing made in the prefecture of Nan-an 南 安 府 by Father Ripa in 1710 (F. Prandi, Memoirs of Father Ripa, 1844, p. 40). The P'o-yang Lake in this province, as well as the Tung-t'ing Lake in the adjoining province of Hu-nan, swarms with fishermen who avail themselves of the cormorant. The birds coming from Hu-nan, as well as from Ho-nan, enjoy a special reputation. The district of Siang-yin 湘 陰 in the prefecture of Ch'ang-sha, and Li chou 澧 州, as well, are emphasized as cormorant-breeding localities in the *Hu-nan fang wu chi* 湖 南 方 物 志 (chap. 2, p. 17; chap. 7, p. 14. Regarding this work see below, p. 237).

In Kwang-tung Province cormorant fishing is practised at Ts'ung-hua 從 化, in the prefecture of Kwang-chou (witnessed by J. H. Gray, China, II, 1878, p. 297). De Guignes (Voyages à Peking, Manille etc. faits dans l'intervalle des années 1784 à 1801, I, 1808, pp. 271, 289, 293) observed in 1794 cormorant fishing in the prefecture of Shao-chou 韶 州 府. Dyer Ball mentions the North River above Canton and the river above Ch'ao-chou fu 潮 州 府.

Dr. Gudger (I pp. 37, 38) has reproduced two photographs taken by Dr. C. K. Edmunds in 1907 on the Fu River, a tributary of the Si Kiang or West River. Dr. Edmunds states that he found fishing with the cormorant everywhere practised on the lower sections of the Grand Canal and the connecting canals in the Yangtse Delta and throughout South China generally.

That there is cormorant breeding in Kwei-chou Province may be inferred from a note of E. H. Parker (Up the Yangtse, p. 233, Shanghai, 1899), who observed fishing cormorants in the gorges of the upper Yangtse and remarks, "They are said to come from the wild lands of Yün-nan and Kwei-chou, notably from the Wu-chiang River in the latter province. There is a well-known place on the Se-ch'wan and Yün-nan frontier, called Lao-wa Tan, which perhaps may have some connection with the catching of these birds." E. G. Kemp (The Face of China, 1909, p. 201) gives a colored picture of the Lao-wa Tan river and village, saying that the name means "cormorant rapid."

Cormorant fishing in Yün-nan Province is attested by T'an Ts'ui 檀 萃 in his *Tien hai yü heng chi* 滇 海 虞 衡 志 (chap. 6, p. 3b, ed. of *Wen ying lou yü ti ts'ung shu*), published in 1799. "In the southern part of Tien (Yün-nan) the inhabitants of many mountains and rivers rear cormorants by catching them. Though it cannot be said that 'all families raise black devils' [quoting Tu Fu], they occur in certain places. In the same manner as they rear falcons to seize pheasants and hares, they raise cormorants for the purpose of catching fish. These birds perfectly understand the commands of men and exert themselves for men's benefit. They are also styled 'aquatic old crow.' It happens that several birds unite forces in catching a large fish, some pecking the eyes of the fish, others its fins, others its tail and dorsal fin. When the fish is thus exhausted, they lift it together out of the water, and their master takes the fish. Truly, a clever performance!" See also *Yün-nan t'ung chi k'ao* (chap. 68, p. 23), which in the main cites T'an Ts'ui.

His observation as to several birds seizing a large fish is quite correct and has also been made by several European writers. "And, what is more wonderful still, if one of the cormorants gets hold of a fish of large size, so large that he would have some difficulty in taking it to the boat, some of the others, seeing his dilemma, hasten to his assistance, and with their efforts united capture the animal and haul him off to the boat" (R. Fortune). Harting once saw three cormorants trying to tackle a large eel which one of them had brought up, but from its size and weight could not hold; the other two came to his assistance, and the three worried it like hounds with a fox.

F. Garnier (Voyage d'exploration en Indo-Chine, I, 1873, p. 517) noticed the fishing with cormorants on the lake of Ta-li fu, Yün-nan, saying that the fishermen cast rice on the water as a bait to the fish. Prince Henri d'Orléans (From Tonkin to India, 1898, p. 141) observed cormorant fishing on the lake of Ta-li fu, giving a sketch of a boat with eight birds.

The province of Se-ch'wan has been an active center of cormorant breeding and fishing ever since the days of the Sung dynasty. We have seen in the section of Historical Data that the prefecture of K'wei-chou 夔州府 played an eminent role in this respect during the middle ages. The *Se-ch'wan t'ung chi* 四川通志 (chap. 74, pp. 21, 28a) gives the same prefecture, as well as Mei chou 眉州, among localities where cormorants are kept. They are likewise used in the districts of Ch'eng-tu and Hwa-yang (成都縣志, chap. 3, p. 10; 華陽縣志, chap. 42, p. 49b).

On the whole, cormorant fishing occurs intensely in central, western, and southern China. The foregoing citations of localities should merely be taken as examples. It is impossible and also unnecessary to give a complete list of such localities. In the north of China keeping of cormorants is comparatively rare, but sporadically it does occur wherever conditions are suitable.

As to Shan-tung, F. von Richthofen (Schantung, 1898, p. 97) limits cormorant fishing to the western part of the province, which is quite natural, as the employment of the bird followed the Grand Canal. John Barrow's (Travels in China, 1804, p. 506) brief remarks refer to Tsi-ning chou.

J. Nieuhoff's (or Neuhof's) account (Gesantschaft der ost-indischen Geselschaft, 1669, p. 134) of cormorant fishing refers to Ning-yang 寧陽 in Yen-chou fu, western Shan-tung. In another passage (p. 353) Nieuhof writes that the bird *louwa*, as he calls it (the name cormorant was unknown to him), occurs throughout Sina.

Dr. Gudger (I p. 39) has reproduced a photograph taken in 1921 by Mrs. Mary G. Lucas of cormorants ready for fishing on a stream outside of Peking. I have never seen cormorant fishing or heard of it in the environment of Peking or in other places of Chi-li, but Sowerby (p. 72) refers domesticated cormorants to the vicinity of the lakes in Pao-ting fu and to the stretches of water to the west of Tientsin.

That cormorant fishing occurs on the Wei River in Shen-si is known to me only from a photograph published by Clark and Sowerby, "Through Shen-kan," plate XX, which is thus captioned, but no description of it is given.

As has been shown, cormorant fishing is distributed in China over a vast stretch of territory. The cause of this wide distribution lies in the fact that the bird has been truly domesticated and is bred in captivity, with the result that hundreds of birds thus bred can be easily transported from one locality to another where there are prospective fishing grounds. In opposition to China, cormorant fishing is restricted in Japan to certain localities and practically to a single fish, the *ayu* aforementioned. In Japan, the trained cormorants are recruited and always replenished from wild stock, so that no active trade in the birds could develop. The principal and famous old center of cormorant fishing in Japan is the Nagara River 長良川, the town of Gifu 岐阜 being the metropolis of the industry.

Cormorant fishing is further practised on Lake Biwa and in the northern part of Kyūshu in the Naka and Sawara Rivers, in the department of Fukuoka 福岡縣, province of Chikugo; and in the Sagami 相摸, Tama 多摩 or 玉川 and Ara 荒川 Rivers near Tōkyo, which enter into the Sea of Japan.

A study of the geographical distribution, unfortunately, does not allow us to recognize exactly in what territory of China the cormorant was first domesticated and how it was diffused from this center to other localities. The documents fail us. Certain it is that northern China must be excluded from the places where the domestication might have originated. The lower Yangtse Basin would seem to have been the logical center. I would not lay too much stress on the fact that the majority of Sung authors refer to cormorant fishing in Se-ch'wan; the fact remains that they were not interested in the subject itself, but that they were exercised over the significance of the term *wu kwei;* and since the latter hailed from Se-ch'wan, the existence of trained cormorants there had to be emphasized. While in this manner we get good evidence for Se-ch'wan, there is no reason to believe that the trained cormorant was monopolized by this province during the Sung period; it must certainly have persisted in the prefecture of T'ai-p'ing on the lower Yangtse, where it is reported in the tenth century. Assuming that there the primeval domestication took its origin, it is not difficult to realize how from An-hui and Che-kiang it spread to Fu-kien, Kiang-si, Hu-nan, and Kwang-tung, how another movement sent the domestication up the Yangtse to Se-ch'wan and Yün-nan, and finally how the Imperial Canal promoted its northward migration through Kiang-su, Shan-tung, and Chi-li.

RELATION OF JAPANESE TO CHINESE
CORMORANT FISHING

The first problem that confronts us is, What is the relation of Japanese to Chinese cormorant fishing? Did one nation acquire the domestication from the other? Or what was the historical development? The only author who has ever ventilated this question is Dr. Gudger (II p. 8), who argues as follows: "If now the commonly held belief be accepted that Chinese culture and civilization (including the use of the cormorant in fishing) antedated that of Japan, then, since we have dates for the sending of Chinese embassies to Japan, we need not find it difficult to believe that the Japanese learned this method of taking fish from the Chinese, and indeed possibly got their first birds from these embassies." This conclusion, first of all, is based on wrong premises; and, secondly, is a rather sweeping generalization unsupported by any evidence. Gudger did not have the historical side of the question straight. He quotes the notice of the *Sui shu* indirectly by translating it from Hervey-Saint-Denys' rendering of Ma Twan-lin's *Wen hien t'ung k'ao*, and is thus led to the belief that it refers to the thirteenth (instead of the sixth or seventh) century. Ma Twan-lin's work, of course, has no independent value and presents merely a compilation of older sources; the quotation given by Dr. Gudger is merely copied from the account of Japan contained in the Sui Annals. As shown above, this is the earliest extant reference to cormorant fishing in the world, and is much earlier than any Chinese references to the practice in China. The first unmistakable notice of cormorant training in China is not older than the tenth century, so that according to our Chinese documentary evidence the use of the bird in Japan antedates that in China by at least three centuries, while according to Japanese sources it is even much older. For this reason no serious historian will rush at the conclusion that the Japanese have simply adopted the domestication from the Chinese, or will indulge in such comfortable speculations as the one that the Japanese possibly got their first birds from Chinese embassies. If this had been the case, Japanese writers would have been sincere enough to admit it. Whenever the Japanese received cultural elements from China or Korea, they placed such indebtedness on record. We have no right to conclude that because much of ancient Japanese culture is derived from China,

232

everything Japanese must have radiated from the same center; this has to be proved for each and every case, and we have to be mindful of the fact that there are numerous Japanese cultural traits which cannot be laid at the threshold of China.   L. Reinhardt (Kultur-geschichte der Nutztiere, 1912, p. 403), without giving any reason, alleges that the Japanese learned the fishing with cormorants from the Chinese.

Let us consider the opposite possibility that the Chinese might be indebted to Japan for the trained cormorant.   E. H. Parker, in translating Ma Twan-lin's account of Japan (*Transactions As. Soc. of Japan*, XXII, 1894, p. 44), comments on the cormorant passage, "It would thus seem that the Chinese owe at least one idea to the Japanese."   The Chinese knew of this Japanese fishing method in the beginning of the seventh century, and the fact was recorded in the official dynastic annals.   The records of foreign countries contained in the *Sui shu*, as well as the biographical portion, were completed in A.D. 636.   It would not be impossible that a Chinese official who read this notice might have conceived the idea of inducing his compatriots to try the same experiment; but, if an official had taken the initiative in this matter, some official record of this event would surely have been preserved.   And then we should expect to see this experiment carried out soon afterwards, say, in the beginning of the T'ang period; yet we face this long gap of nearly three centuries between the Sui and the Wu-tai periods which cannot be bridged over.   The possibility that the Japanese account of the *Sui shu* or *Pei shi* should have struck the eyes of a fisherman of An-hui, Che-kiang, or Se-ch'wan appears to me so remote that it does not merit a discussion.

The plain fact remains that Chinese sources do not admit any indebtedness to Japan in matters of cormorant fishing, in the same manner as the Japanese on their part do not credit it to the Chinese. And this fact, in my opinion, carries weight.   I have already insisted on Japanese honesty, and I plead the same degree of honesty for the Chinese.   Every one familiar with the history of Chinese civilization knows only too well that the Chinese have always been frank and upright in acknowledging foreign loans.   To us trained in scientific thought the domestication of the cormorant appears a significant affair; and, as pointed out, it is a unique phenomenon in the history of the world.   To the Japanese and Chinese, however, who alone of all peoples accomplished the permanent training of the bird, it is something insignificant over which they never made any fuss.   These two nations were always distinguished for modesty, reserve, lack of conceit and ego worship. Neither preserved the name of the first fisherman or men who did the deed. It was too trifling a matter. But if the Greeks or Romans had accomplished it, what pride would swell the chests of our classicists! Of course, the Greeks would have handed down the name of the "inventor" with a romantic story of how his genius was inspired by the gods, and every school boy in our midst would be obliged to learn it by heart. As matters stand now, he is spared this thrilling or sad experience, and things peculiarly Japanese and Chinese do not bother our public. The thirteenth edition of the Encyclopaedia Britannica devotes an entire column to the cormorant with a lengthy story of its training in Europe, but maintains cold silence about its domestication in Japan and China! In the fourteenth and last edition the article in question is curtailed to a half column, and it is said, "The practice is nearly obsolete in Europe, though still common in China." And this is all.

As the evidence stands, there is but one conclusion admissible, and this is that the domestication of the cormorant and everything connected therewith was independently achieved in China and Japan. This conclusion is corroborated by many facts which lie in the domestication itself. The principal facts are as follows:

The method of using and treating the cormorant in Japan is fundamentally and radically different from that of China. Here the Latin saying "si duo idem faciunt, idem non est" holds good. What the Japanese practise may be briefly defined under the name of the harness or team method. A cord or rein of spruce fiber, about twelve feet long, is attached to the body of each bird, and the master lowers the birds one by one, altogether a team of twelve, into the stream and gathers all reins in his left hand, manipulating the various lines thereafter with his right hand, as occasion requires, to keep them free of tangles. This method is absolutely unknown anywhere in China. In China, the cormorant has reached a perfect stage of domestication, is reared in captivity, and is the born slave of his master. Nothing like this is at present done in Japan, where the cormorants pressed into service are all caught from wild stock on the coast of the Owari Gulf and immediately receive their training. The Japanese method in all its various details is as different from the Chinese as both Japanese and Chinese methods are at variance with that formerly adopted in Europe. What these methods are will be more fully discussed in the following chapter. The point to be made here is that in view of these principal differences Chinese and Japanese utilization of the cormorant cannot have a common basis of origin. The two are entirely distinct. The only point in common to the two civilizations is that the cormorant is used by man for the purpose of catching fish; but this is all, and here the coincidence comes to an end. The discrepancies outbalance in weight the outward similarity. Different causes, different methods and technique have merely yielded a similar result.

Another observation remains to be made. It is a significant fact that cormorant fishing was never practised in Korea and that it is unknown there at present (I speak advisedly, as I had occasion to interrogate Korean students on this point). Accordingly, cormorant fishing does not belong to that series of culture elements which spread from China to Korea and were further transmitted from Korea to Japan. In fact, the Chinese cormorant domestication did not spread at all; it was not communicated to any of those nations which came under the spell of Chinese civilization. The aboriginal tribes inhabiting western and southern China did not adopt it; neither did the Annamese or any other peoples of Indo-China or the Siamese, Burmese, or Malayans. The domestication of the cormorant and the peculiar method of raising, training, and using the bird have remained a purely and typically Chinese affair. And this method did not spread to Japan. Japan has evolved a method of her own and peculiarly herself without the aid of outward influences.

The question of a possible mutual stimulus, that may be raised, dwindles into insignificance and is wholly immaterial, as compared with the basic processes. If it is a question of priority and precedence in originality, the balance of the evidence certainly favors the Japanese.

THE PROCESS OF DOMESTICATION

The only scholar who has treated of the cormorant as a domestication, unfortunately too briefly, is Eduard Hahn (Die Haustiere,
1896, pp. 347–350). No other book dealing seriously with domesticated animals—and there are many of these—mentions the cormorant. Hahn points out that the cormorant as a domesticated animal
is an achievement of Chinese civilization, which shows the patience
and intelligence of the Chinese at their best. He refers to Friar
Odoric, Armand David, and R. Fortune. It was vaguely known to
him that fishing with cormorants is practised in Japan, and he offers
a curious misunderstanding of a passage in *Journal asiatique* (1871,
p. 403), "If it be correct that the mythical Matwan-lin is Japan,
this method must go back to very ancient times, as far back as the
sixth century A.D." L. Reinhardt, in his popular book, "Kultur-
geschichte der Nutztiere" (1912, pp. 400–403), gives a few notes on
the life of the wild bird, copies the historical information of Hahn
(without acknowledgment and adding some errors of his own), and
says very little about the domestication itself.

The wild cormorant is not difficult to tame. The only statement
to the contrary I have found is made by L. Reinhardt (*op. cit.*, p. 403),
who writes, "As the excellent ornithologist Naumann with good
reason designates the cormorant as difficult to tame and fond of
biting, the great patience and endurance of the Chinese in making
it a domestic animal must be the more acknowledged." Naumann
may have been a good ornithologist, but has never attempted to
train a cormorant. Those who did venture to dissent from him.

Harting (p. 438), who had much experience in training cormorants,
states that "cormorants are by no means difficult to train, and do
not require half the care and attention which has to be bestowed
upon hawks for example." Pichot (p. 27) writes in the same spirit.
"This web-footed marine bird is very easily tamed. His heart is very
near to his stomach, and one may be reached by way of the other. As
I was engaged in falconry, I was naturally led to practising cormorant
fishing. This is a delightful sport and the easier, as the cormorant
becomes rapidly familiar; if you feed him out of your hand, you will
have trouble to prevent him from following you everywhere, ascending the stairs behind you, perching on your furniture, and leaving
on all pieces incontestable traces of his rapid and abundant digestion."

236

In China, the cormorant has attained the stage of a complete and perfect domestication, the birds propagating and being reared in captivity. It is said by Sowerby (p. 73) that the domestic stocks are replenished from time to time by the capture of wild birds or by robbing the latter's nests; it is very sensible, of course, to refresh the blood of a domestic species occasionally by interbreeding with a wild congener, but I believe this is but seldom done. The early travelers to China were merely content to record their surface observations of cormorant fishing without inquiring into the life and habits of the bird. Nieuhoff was the only one who possessed enough intelligence to ask the question whether the birds also propagated and bred many young, and was given the reply that "this happens, but very slowly and little." (Dr. Gudger [I p. 12] cites Nieuhoff in Ogilby's English translation of 1669, which in the cormorant section is inaccurate and deficient.)

A statement concerning the use of domesticated birds is also found in a Chinese source. The *Hu-nan fang wu chi* 湖 南 方 物 志 ("Record of the Local Products of Hu-nan Province"), written in 1818 by Ki Chung-liang 驥 仲 良 and Tsiang Siang-wei 蔣 琅 維 (chap. 7, p. 14; edition before me printed in 1864) cites from the *San ch'ang wu chai Ch'ang shwo* 三 長 物 齋 長 說 the following: "In the south the cormorants are all attached to the fishermen's houses, and are fed and reared by them. I have never heard that birds born in freedom (or wild birds) are used for fishing" (南 中 鸕 鷀 係 漁 家 粲 養 未 聞 有 野 生 者). The date of the book *San ch'ang*, etc., is not directly known to me, but as farther on Li Shi-chen's *Pen ts'ao kang mu* is quoted in it, it must be a product of the seventeenth or eighteenth century; the title of the book is *Ch'ang shwo*, and it was published by the *San ch'ang wu* Studio which issued also a *ts'ung shu*.

The effect of the domestication is shown in the complete submission and subordination of the birds, who become as docile and obedient as dogs, knowing their master and his boat and understanding his commands perfectly well, and in their outward appearance they display variation in color, a marked characteristic of all domestic breeds. Sowerby (p. 72) observed in Chi-li many pied or even pure white individuals.

The females lay yearly from three to nine eggs in the first and eighth month. The eggs are green in color and the size of a duck's egg; what is called the white of the egg is light greenish; the eggs are never consumed. The eggs of the first month are the only ones retained for hatching, for the reason that the young birds will grow up in the spring; if those of the eighth month were hatched, the young, who are extremely sensitive to cold weather, it is feared, would not live through the winter. The eggs are always hatched by hens, not by the cormorant mother. The only author who gives a different account is Dabry de Thiersant, who writes that "the cormorants prepare in a spot retired and dark a nest of straw, on which the female lays her eggs which she herself covers all the time." I have but little confidence in this statement, in view of the fact that the author shows himself rather uncritical and credulous in his notes on the cormorant; for instance, when he asserts that on the tenth day after birth the fledglings are taken by the trainer out on his boat and seek their places on the common perch; while the training, in fact, begins only two months from the date of birth.

The fact that the eggs are given to hens to hatch is attributed by several authors to the female cormorants being bad mothers (Fauvel, p. 230). "Curiously enough the mothers are so careless that they cannot be trusted to rear their own young" (Gordon-Cumming in Dyer Ball, p. 182). This comment savors of Chinese mentality, being made in response to a question, and is either really believed by the breeder or is just elicited to satisfy the curiosity of an importune inquirer. This explanation, of course, is absurd, for "the young of the wild species are assiduously fed and cared for by the parents" (Sowerby); and how with this alleged lack of maternal affection could the species have spread over the entire world? The female kept in captivity lacks her nest in high trees and the natural conditions of her life, and this rather is the reason why she declines to incubate the eggs. Moreover, experience has taught the breeders that safer and faster results are attained from hens than from female cormorants.

The fledglings come out of the egg after a month's incubation. They cannot stand on their legs, and are very sensitive to cold. They are transferred to cotton-stuffed baskets which are kept in a warm room. The young birds are enveloped in cotton wool and fed with small morsels of bean-curd (tou-fu) and pellets of raw fish, preferably eel, if procurable. Cormorants are inordinately fond of eels and prefer them to all other fish; where eels are plentiful they will even catch nothing else. Fortune was given the information that for five days the young are fed with eels' blood and that after five days they can be fed with eels' flesh chopped fine. Dabry de Thiersant denies that this is done in the central provinces, and states that during seven days the young receive three times a day finely minced meat which they prefer to any other food; afterwards small fish are added to the meat. There is no doubt that in matters of feeding and training a good many local variations exist. Lean and weak birds are fed on dogs' flesh (*Ning-hua hien chi* 寧 化 縣 志, chap. 2, p. 128). "After the tenth month the cormorants are given dog's flesh which will keep their bodies warm and protect them from the cold; even when breaking through the ice and plunging into the water, they will not die from a chill" (*Pen ts'ao kang mu shi i,* chap. 9, p. 7b).

Under the heading "Method of rearing cormorants," Fang I-chi 方 以 智, in his *Wu li siao shi* 物 理 小 識 (chap. 10, p. 5), gives the following brief notes: "For a period of six months the (young) cormorants are susceptible to cold. Their keepers wrap them in refuse cotton and feed them with pepper. In autumn and winter the birds are turned loose into the water to catch fish. Those incubated in the summer are so strong that they can peck up the eyeballs of large fish." If pepper is given the young birds, it must be mixed, of course, with some food, which Fang I-chi forgets to mention, and may act as a sort of tonic.

At the end of a month feathers begin to cover the down, and the quantity of fish is increased for their diet, while the proportion of bean-curd is reduced. After the second month the birds have doubled their size, and are fit for the market, a female being half as much in price as a male. This difference in price is due to the superior strength of the male who is able to capture larger fish.

When the nursery days are over, the schooling of the birds begins at once, and they are turned over to a professional trainer. Their wings are clipped to prevent their flying away. The first lesson is as follows. A string is tied to one of the bird's legs, the other end of the string being fastened to a stake set in the bank of the pond or canal. The bird is driven into the water, the trainer whistling a peculiar call and, if necessary, enforcing obedience by the persuasive strokes of a bamboo, the great educational means of China. Small live fish are thrown to the bird who will pounce upon them greedily, as he was previously kept on a reduced diet. He, is now called back by a different whistle signal and, to make him understand, is first pulled back by means of the string or line until he has learned to comprehend the call and to obey it spontaneously. This procedure is repeated daily for about a month till the bird is accustomed to his master's voice and commands.

The second lesson is given in a boat along the same line as previously, and lasts four or five weeks according to the bird's intelligence.

When the young birds are accompanied by older well-trained birds, which is usually done, the time of instruction is shortened considerably, as they will quickly learn from their elders. At the end of this period they are relieved of the leash or line. Birds not properly trained by that time are regarded as stupid and hopeless. A period of seven or eight months may be considered sufficient to turn him into an expert fisher.

There is a good deal of individual character in the birds. Some are intelligent, alive, and alert; others are dull, lazy, and sulky. Some are more expert at diving and catching than others. European writers have described the cormorant as "solemn, weird, uncanny"; but such words are elicited by impressions of our mind, and are not objective characteristics of the bird. What is more important is that each bird possesses enough individual characteristics to enable his master to recognize him among a flock of other birds working in the water; and *vice versa*, the birds are endowed with sufficient intelligence to know their master and their boat; they always occupy the same place assigned to them in the boat. Many fishermen name their birds, and will always call them by their names.

"These birds differ much in their tempers, some being most spiteful and savage, as well as shy and disobedient in the field, whilst others are just the reverse. They are very sagacious, knowing the places where they have caught fish and are likely to meet with them again, etc. They are also capable of great affection. Honest 'Isaac Walton' was particularly fond of his master and, singularly enough, he would not allow any one else to approach without showing fight. When angry, the cry somewhat resembles the gobble of a turkey, and when pleased, it is a loud guttural sound, like 'haw, haw!'" (Freeman and Salvin, Falconry, p. 331).

The domesticated cormorants of China live in basket-like cages, or in the summer are also left in the open where they are tied to their perches by means of a leather thong fastened to one of their legs.

The fisherman who uses the birds is not necessarily the breeder. There are special establishments which make it their business to breed the birds and to sell them or lease them to the fishermen. Fortune mentions such a large establishment, which he visited in 1848, thirty or forty miles from Shanghai and between that town and Chapu, where a pair was then sold at from six to eight dollars. The tenants usually repay the owner of the birds with a certain quantity of fish. Under this system the birds are worked as hard as possible, for the fisherman's sole interest is in catching as large a booty as possible in order to gain a surplus for himself in excess to what he owes to his landlord. Under these circumstances the friendly and sympathetic relations that exist between man and his

domestic animals must be completely lost. The wholesale breeding
of birds, however, has one advantage that they can be distributed
and sold in the entire country.

The earlier writers on China represented cormorant fishing as a
sort of royal monopoly. R. Willes (see p. 245) has a fantastic
account of the king of China possessing a good store of barges full
of sea-crows allowed a monthly provision of rice; these barges the
king bestows upon his greatest magistrates. Juan Gonzalez de
Mendoza has a similar story, and Nieuhoff has the fishermen pay
an annual tax to the emperor for the use of the birds. With reference
to Lord Macartney, who stated that exorbitant taxes had to be paid
to the emperor for the permission to use cormorants, Fauvel remarks
that he never heard of such in Che-kiang and that they probably
existed nowhere in his time. Of course, the cormorant fishermen,
like every one else, paid an annual tax, but these taxes never were
excessive, nor has a government monopoly on cormorants ever
prevailed, nor has the government to all appearances ever taken an
interest in the whole business. It is known to every one who was
in the China of the Manchu dynasty that the people were in the
habit of complaining and sobbing to foreign visitors about high
taxation, oppression, and extortion, while in fact they were the most
lightly taxed people in the world, comparatively speaking. In
Chinese sources I have found nothing alluding to a cormorant monopoly or special taxation. The only item I have found is as follows:

In the Gazetteer of Ning-hua 寧 化 縣 志 (chap. 2, p. 128), in
the prefecture of T'ing-chou, Fu-kien, it is stated, "At present the
inhabitants of Yen and Kien 延 建 人 (Yen-p'ing fu and Kien-ning
fu), who raise these birds, pay their taxes to the officials in rice 輸
米 于 官.''

Authors like Fortune and Fauvel have also given price quotations
for the birds, which can hardly claim any validity at present. I
mention only that the *Ning-hua hien chi* (chap. 2, p. 128) gives the
price of a single bird as ten taels (ounces of silver), and this may
be regarded as an average value.

Most writers who have described cormorant fishing, although
their accounts are incomplete and deficient, do not fail to mention
the ring or strap around the bird's neck, and do not get tired of
repeating the worn-out statement that this is done to prevent him
from swallowing the fish which he catches. Friar Odoric (Yule,
Cathay, new ed., II, p. 190) is the first of European travelers who
noted that the fisherman tied a cord round the birds' necks that
they might not be able to swallow the fish which they caught. True

it is Chinese authors make the same statement. Several have been
quoted to this effect (p. 224), and another may be added here. "Those
who raise the cormorant tie a cord around its neck, so that small
fishes can pass its throat, while big fishes can not; from time to time
the fishermen call the bird and take the fish away from it; and then
they send it out again" (*Yung-ch'un chou chi* 永 春 州 志, chap. 7,
p. 22; the same in *Ning-hua hien chi*, chap. 2, p. 127; both localities
are in Fu-kien).

The fact itself is correct, but is not logically expressed. That
the cormorant is prevented from swallowing a large fish is the effect
of the ring or strap, but not the cause of it, which is quite different.
The ring is the symbol of the cormorant's bondage, and was the
original expedient that brought its domestication about. It takes
the place of the dog's neck-collar, the horse's bit, the water-buffalo's
nose-ring, the falcon's leash and hood. In order to govern and keep
a cormorant, man required some means of grasping and holding him,
and a cord of hemp slung around his neck and terminating in a line
which man could seize was the means he devised. Man's first thought
in training the bird was to hold him in check, to bridle him, to direct
him; he did not think, at first, of preventing the bird from swallowing
large fish; this resulted as a secondary effect from the use of the strap
or ring.

The ring is called *chüan* 圈 (*Ma-kia-hiang t'ing chi* 馬 家 巷 廳 志,
chap. 12, p. 11); in the *Sui shu,* as pointed out above (p. 212), *hwan.*

Great care must be exercised in placing the cord or strap around
the bird's neck; it must be fastened in such a way that it will not
slip farther down upon his neck, as this would be apt to choke him.

Different materials are used for the ring in various localities;
straw, hemp, bast, tow, bamboo (Kwang-tung), rattan, and even
iron (Maffei and Nieuhoff) are reported. The use of iron is unnecessary, even foolish, as its weight will hamper the bird's free movements.

That the neck-collar is merely a stepping-stone in the gradual
development of the domestication becomes clear from the fact that
in the last and highest stage of development the neck-collar is simply
discarded, especially in Che-kiang. At this stage of the game the
birds are disciplined to such a degree of perfection that they fish
in unrestrained and absolute freedom. A well-trained cormorant,
while on duty, will not swallow any fish whether large or small;
he knows his business and his lord.

G. Staunton (Account of an Embassy to the Emperor of China,
1797, II, p. 388), for instance, reports that "the birds appeared to
be so well trained, that it did not require either ring or cord about

their throats to prevent them from swallowing any portion of their prey." His observations were made on the Imperial Canal near Hang-chou. Milne (Life in China, 1857, p. 307) writes that he could find neither ring nor cord about the necks of any of them to prevent the swallowing of fish.

De Guignes (Voyages à Peking, I, pp. 271, 289) states expressly that in Kwang-tung Province, where he noticed cormorant fishing in three different places (in 1794), the birds were free and appeared well tamed; "they did not have, as P. du Halde says, a collar around their necks."

What Friar Odoric describes (Yule, Cathay, II, p. 190) is also a free method of fishing which indicates a highly developed stage in the evolution of domestication, though of a somewhat lower degree than the preceding one. The water-fowl were let loose without being driven, and straightway began to dive, catching great numbers of fish and putting these of their own accord into the baskets, so that before long all the three baskets were full. Of course, as mentioned previously, Odoric's birds were equipped with the neck-strap; and when their task was finished, they were tied to perches in the boat.

In some parts of the country the fishermen haul the bird with his catch out of the water in a dip-net, and methods of managing the bird vary according to locality. For instance, at Tsi-ning chou and probably in other places too, the fisherman who commands a flock of ten or twelve birds will not rush all of them into the water at once, but will allow only one or two of them to dive at a time; and when he perceives that they are fatigued, he will take them in and feed them and then dispatch another pair into the water. This humane method has the advantage that it will give the poor workers a good chance for rest. On the Min River in Se-ch'wan cormorants, after diving, are brought up to the surface in baskets of much the same shape as the birds (E. G. Kemp, The Face of China, 1909, p. 180).

Some of these varying practices, particularly those bearing on minor details, may simply be due to local variations of custom; but in the main, the basic differences are an index of the various stages through which the development of the domestication has run. It is clear that the more the bird's movements are restricted, the more restraint is imposed upon him, the older this stage of development must be. It was gradually recognized that the laws deemed necessary for his enslavement need not be too rigidly enforced, that the creature was attached to his master and would not forsake him, that barriers could be slowly removed and a greater amount of liberty be restored to him. In this point the Chinese have manifested admirable wisdom, and have advanced far beyond the Japanese. This indeed is the goal for which all cormorant domestication must strive— granting the bird a maximum degree of freedom. This makes a happy cormorant and a more successful and therefore happier fisherman.

Another implement indispensable to the cormorant trainer and fisherman is a long bamboo pole, which serves a twofold purpose— propelling the raft or boat tenanted by the fisherman with the heavier end of the pole and directing and controlling the movements of the birds with the other. He uses the pole as the conductor of an orches- tra does his baton. When the theatre of action is reached, he signals to the birds to dive by beating the water with the pole. When a bird fails to attend to his business, a blow of the bamboo on the water near where the bird floats accompanied by an angry shouting, is sufficient to remind him of his duty. Whenever a bird gets tired, or when his gullet is filled with rich booty, the pole is stretched forth so that he may jump and perch on it, and be lifted into the boat. On the same bamboo pole the birds are carried from their home to the water's edge and back home when their task is done (compare Plate XIII).

The methods of fishing with cormorants have frequently been described and are beyond the scope of this study which is concerned with the cormorant as a domestication; but I wish to mention one point, as it has not been brought out by previous authors. There are two principal methods of fishing—the solitary one and the method of group fishing. A man single-handed, especially on a raft, is able to manage three or four cormorants and to attend to the whole business (see Plates XV–XVI); a variation of this is duet fishing when two men join in dividing labors, one steering and propelling the boat, the other tending the birds and the fish caught by them. Rafts are chiefly used in the southern provinces. Boats are either single, or two of them are placed side by side and connected by a plank; the latter are more stable (illustration in Korrigan, p. 42). The raft owners usually operate with from two to four birds; the boat masters, with ten or twelve. In group fishing a fishermen's gild or association gets together and makes common cause. A fleet of small boats moves into action and spreads out in a line or crescent formation, setting the frightened fish moving and driving the birds in front. As each fisherman knows his own birds and as each bird knows his boat and his place on it, everything proceeds in orderly fashion. The concerted action in this manoeuvring naturally insures a larger haul of fish. The solitary method assuredly is older than the group or community method, and is the one pointed out in our earliest Chinese source, the *Ts'ing i lu*, which advisedly refers to "a single man" (above, p. 221). The first mention of community fishing occurs in R. Willes' Reports of the Province of China (about 1565), based on the data of Portuguese, chiefly Galeotto Pereira (Hakluyt, Glasgow ed., II, p. 327). Here it is said (I modernize the old English spelling), "At the hour appointed to fish, all the barges are brought together in a circle, where the river is shallow, and the crows tied together under the wings are let leap down into the water, some under, some above, worth looking upon; each one as he has filled his bag, goes to his own barge and empties it, which done he returns to fish again. . . . There were in that city where I was, twenty barges at the least of these aforesaid crows."

From what has been said about the present-day training of cormorants, it is not difficult to imagine what the steps in the primeval process of the domestication have been during the tenth century. A wild young cormorant was ensnared, and the Chinese have always been skilful bird-catchers. A cage with a perch, a cord and a leash or line, a bamboo pole were all the paraphernalia required. A noose was tied around the bird's neck, and along a line he was immediately dispatched into the water. The first man who did it merits greater admiration for the originality of his idea than for what he accomplished, while the bird as the natural fish-hunter is deserving of greater credit for the achievement. At the outset it is difficult to realize what keeps the enslaved cormorant in bondage, or why he continues hard labor for an employer who has so little to offer him in return. The service of domesticated animals is based on a silent pact which gives them advantages not enjoyed by their wild congeners: proper shelter, protection from rapacious enemies, adequate food, and assurance of a constant and regular food supply. The cormorant to some extent suffers from what the modern Chinese would call an "unequal treaty." He is hardly fed by man, but looks out for his own meals, catching his own fish. In some parts of China he is given a morsel to eat after every catch, and some fishermen even feed their birds with their own hands, stroking their necks to facilitate the downward movement of the food, which the birds are said to like very much. They are also fed with morsels of bean-curd, but this alone can hardly be a sufficient attraction for the bird to remain in his state of socage. He has no natural enemies, as chickens and pigeons have, from whom he would need protection. His quarters are by no means palatial or sumptuous, and there is but little sentimentality in a fisherman's heart. Even granted that he treats his birds well in his own interest, the cormorant's psychology is not perfectly clear and requires further elucidation. I have never heard of a cormorant attempting to break loose in order to gain his liberty, and with his wings clipped and his spirit broken it is questionable how far he would get; probably he would be pursued and soon captured by his owner. Born and raised in captivity, he is ignorant of the sweetness of liberty and looks upon slavery as his natural lot.

While the domestication of the cormorant has passed through several successive stages and has improved by degrees, it is not necessary to imagine that the primeval or initiatory process was a superhuman task which required a long span of time. I have therefore suggested that in accordance with the present state of our knowledge the beginnings of the domestication in China should not be dated farther back than the tenth century A.D.

In its natural state the cormorant is said to live twenty to twenty-five years. It is an interesting fact that in Japan the captured and trained birds reach a much higher age than the domesticated ones of China. According to Ikenoya, the Japanese birds will live to the age of twelve. Palmer, as quoted by Chamberlain, estimates that they work well up to fifteen, often up to nineteen or twenty years of age. According to Kuroda, birds from four to eight years old are the best; beyond this age they begin to slow down in their work, but they can be employed up to about fifteen years of age.

According to Jametel, the cormorants of China begin to lose their plumage from their fourth year, and they usually die before reaching the age of six years. Fauvel (p. 233) states that a good cormorant can serve during five years, but at the lapse of this period begins to lose its feathers and will die soon. Dabry de Thiersant, however, was informed that the birds are serviceable up to the age of ten years. Whatever the exact facts on either side may be, there is no doubt that the average is much higher in Japan than in China. And the reason for this is not far to seek. The Japanese birds are kept busy only a five months' season and rest during the winter, while the Chinese birds are worked and overworked the whole year round, except during extreme frost in the winter. The Chinese have not yet learned that domestic animals also need a vacation and time for rest and play and that this concession will prolong their lives and intensify their ability to work. There is no doubt also that the Japanese treat and nurse their birds better than the Chinese. In the summer, the cormorant quarters in Japan are even surrounded with mosquito nets. Lack of cleanliness in their cages must necessarily breed disease and doom many birds in China to a premature death.

But aside from this, the Chinese method of domestication is infinitely superior and preferable to the Japanese method of catching young birds and training them. They are caught on their roosting places around Shinoshima in the province of Owari. Their wings are clipped, and they are sent blindfolded to Gifu. At first they are very vicious, and are kept tied up. They are daily taken out at noon, under the leash, on the river and allowed to dive, catch, and swallow from one to two pounds of fish. After about fifteen days they are taught to catch and disgorge fish.

At Gifu a cord is attached to the bird's foot and passes under its belly up to its neck where it connects with the ring. Twelve birds form a team, and their cords are fastened to a single line directed by one master. In this manner the cormorants are kept close to the boat and hampered in their free movements. Pichot (p. 32), in criticizing this method, remarks that the true sport consists in having the cormorant work with liberty without any other means of restraint than the leather ring around his neck. I would go a step farther and say that, as demonstrated by the example of Chinese fishermen, even the ring can be dispensed with and that a well-trained cormorant may be given the "freedom of the seas." It might be advisable for the Japanese to send a commission of experts to China for a thorough study of the Chinese system of cormorant breeding with a view to apply it at home, or at least to improve the domestic system. On the other hand, I am sure, Japanese cormorant experts could teach the Chinese a great deal in the proper care of their birds.

The complex and cumbrous apparatus set in motion by the Japanese is unconsciously inspired by the fear lest the captured bird might escape, and he is fettered and closely watched every minute. Too much unnecessary fuss is made about the whole business, and too many fads and frills are connected with it. In the fishing method followed by the Japanese the birds are unnecessarily excited, and their agitation when in the water is increased by the burning sparks which fall into the water from the braziers or cressets on the boat intended to illuminate the nightly scene. The weird light of lanterns, the noise from music and songs on the boats of the pleasure-seekers make the birds still more nervous.

The Japanese procedure in fishing with harnessed cormorant teams has the one advantage that the keeper has absolute control over each member of his team and can pull him out of the water from his particular line instantaneously he has caught some fish. The Chinaman may lose some time in giving commands or reprimanding or punishing lazy or recalcitrant birds, but he is more humane and more sportsmanlike in allowing his birds some freedom of action; and freedom, after all, is what makes good sport. The Japanese method, although in its outward appearance a sport, is, in fact, not a sport, at least so at the present time. I am inclined to think that it was different in ancient and mediaeval Japan when the cormorant was still regarded as a sacred bird. It was this sacred, mythological character of the bird which prompted the ancient Japanese to keep it in captivity, and, I am disposed to believe, to domesticate it, although I cannot produce any documentary evidence to this effect. In the modern Japanese and European sources I have been able to consult nothing is said about the birds propagating in captivity or the breeding of birds born in captivity; I should be very grateful for any information on this point, as it is an important matter for the history of domestications.

In opposition to Japan cormorant fishing in China is carried on during the day. This is reported alike by all observers. Sowerby (p. 73), however, asserts that "sometimes and in some parts of China the fishing is done at night, when great flares are carried on the boats, which serve to attract the fish and also to help the birds to see them." It would be interesting to have more specific information on this point, especially as to the localities where it is done and as to the methods applied.

In Japan, in opposition to China, cormorant fishing is usually carried on during the night. In China it is an industry from which fishermen gain a livelihood, while in Japan it has rather the character of a sport connected with pleasure parties and spectacular festivals for the entertainment of illustrious visitors and ordinary sightseers. This, however, to all appearances, is a modern development which does not hold good for ancient and mediaeval Japan. This recent sporting tendency may be largely due to two imperial visits at Gifu in 1878 and 1880.

Where cormorant fishing is a commercial enterprise in Japan, it is also carried on in daytime only, in one locality both night and day.

Another difference between China and Japan is that Japanese cormorant fishing is practically restricted to the *ayu* (above, p. 213), while the Chinese without discrimination take any fish the cormorant is able to hunt.

RELATION OF CORMORANT TO OTTER FISHING AND EGRET TAMING

A question that remains to be answered is whether there is any relation between otter fishing, as still practised in the upper Yangtse Basin, and cormorant fishing. It would seem so at the surface, judging from a remark of Sung K'i 宋 祁 (A.D. 998–1061), who in his *Pi ki* 筆 記 (known as 宋 景 文 公 筆 記) makes a certain Wang Tse-huan 王 子 幻 say that he saw with his own eyes at Yung-chou (in Hu-nan) tame otters kept for fishing in the place of (or, as substitutes for) cormorants 永 州 養 馴 獺 以 代 鸕 鷀 沒 水 捕 魚 and that these daily caught about ten catties of fish, enough to supply the want of a family. Wang Tse-huan was apparently familiar with cormorant fishing, while the sight of otter fishing was a novel experience to him. It cannot be said, however, with any regard to historical truth that the otter replaced the cormorant in certain parts of the country; for otter fishing was practised at an earlier date than cormorant fishing and was known under the T'ang. Translations of two passages to this effect from T'ang authors were transmitted by me to Dr. Gudger, who published them in his article Fishing with the Otter (pp. 198–199). As cormorant fishing was in all probability inaugurated in the Lower Yangtse Valley in the beginning of the tenth century, the two events are distinct as to space and time, and it is hardly necessary to assume an interrelation of the two. It may be, of course, that news of otter fishing on the upper Yangtse reached fishermen in the lower course of the river and might have suggested to them a similar idea which may have set them to thinking about the cormorant. For the rest, the two events are entirely different. The cormorant was gradually brought into a state of domestication, while the otter could merely be tamed, and has always remained in the feral state.

The egret (*lu* 鷺, *Ardea egretta*) was also kept in captivity, although it is not stated for which purpose. The earliest notice to this effect I have been able to find occurs in the *Mao shi ts'ao mu niao shou ch'ung yü su* 毛 詩 草 木 烏 獸 蟲 魚 疏 (chap. B, p. 4b, ed. of *T'ang Sung ts'ung shu*), where it is said, "At present the people of Wu also raise egrets" (今 吳 人 亦 養 焉). The authorship of this work is ascribed to Lu Ki 陸 璣 (A.D. 260 or 261–303), although my edition of the *T'ang Sung ts'ung shu* makes him "T'ang"; but as Legge

(Classics, IV, p. 178) says that "the original work was lost and that now current was compiled, it is not known when or by whom, mainly from K'ung Ying-ta's constant quotation from it," it is difficult to date the above passage. The *Wu lei siang kan chi* 物 類 相 感 志, ascribed to Su Shi (A.D. 1036–1101), says that "the heron is kept by men in ponds and becomes as tame as domestic fowl; whenever the day Pai-lu 白 露 (8th of September) appears, the herons fly away and are gone." This evidently refers to the southward migration of the birds. The same information is given in the *P'i ya* 埤 雅 by Lu Tien 陸 佃 (A.D. 1042–1102), who writes that "the people of the present time raise white herons intensely and that there are birds quite tame and docile; when they have left on the day Pai-lu, they cannot be kept again." The *Hwa king* 花 鏡, written by Ch'en Hao-tse 陳 淏 子 in 1688 (chap. 6, p. 4b), says that many people keep these birds in ponds and pools.

Li Fang 李 昉 (A.D. 924–995), compiler of the *T'ai p'ing yü lan,* is said to have raised five herons whom he called "cloud guests" (*yün ko* 雲 客), according to the *Ts'e lin hai ts'o* 詞 林 海 錯.

The Gazetteer of Shao-hing 邵 興 府 志 (chap. 11, p. 31) gives the following information: "The egret is snow-white in color, and on its crest has a silk-like bunch over a foot long; when the bird desires to catch fish, it droops this feather-bunch. Many people living on the banks of rivers north of the mountains keep the egret in their houses, and the birds become so tame that they do not fly away; only during the day Pai Lu 白 露 (8th of September) is it necessary to cage the birds for the entire day so that they do not escape."

If this information be correct (and it should be verified in the locality), egret taming may bear some relation to cormorant training. It is a curious coincidence that the employment of the heron in the service of man begins about the same time as, or a little earlier than, cormorant training, but it seems never to have reached a great practical importance.

ICONOGRAPHY

The oldest representation of a cormorant known to me is a carving in jade of the Chou period in Field Museum, Chicago. Illustrated in Laufer, Archaic Chinese Jades, 1927, plate XXVI, fig. 7, and as a vignette on page 205 of this article.

It is singular that cormorant fishing has not inspired any great Chinese artist. To be sure, there are many drawings and pictures of the subject of Ming and Ts'ing periods, but all these are mechanical productions of small or no artistic value. I feel almost confident in saying that no T'ang or Sung artist has ever painted a cormorant. The *Süan ho hwa p'u* 宣 和 畫 譜 enumerates many pictures of herons, even herons engaged in fishing, for instance, by Hwang Kü-ts'ai, but not a single cormorant painting.

Fishing with cormorants is depicted in a Chinese painting attributed to the Ming period (weak as a painting and teaching but little about the method of fishing), reproduced in O. Sirén, Chinese Paintings in American Collections, plate 138.

A finger-painting by Kao K'i-p'ei, representing a cormorant fisher, is in the possession of Mr. Benjamin March, Detroit, to whom I am indebted for kindly placing at my disposal a photograph of it here reproduced in Plate XIV. The painting (46 by 22 inches) is on paper, in black ink on a blue background wash and some tan-orange in the foreground. It illustrates well the method of solitary fishing described above (p. 244). The bare-legged fisherman is cautiously propelling his five-bamboo raft with a long pole, and one of his two cormorants is spying the water for fish. The picture is inscribed in grass characters as follows: 康 熙 乙 巳 元 宵 前 一 日 鐵 嶺 高 其 佩 指 戲, i.e. "Finger-play (finger-painting) of Kao K'i-p'ei of T'ie-ling (in Fung-t'ien fu, Sheng-king, Manchuria, place of his birth), done on the day before the first full moon of the year (the Feast of Lanterns) of the year *i-se* of the K'ang-hi period" (1665). As the artist died in 1734, he must have been very young when he sketched this picture; the date of his birth seems to be unknown. The date 1665, at any rate, is apt to rouse suspicion, and there may be something wrong about it and the entire legend; 1725, an *i-se* year, would be more probable, but this falls within the Yung-cheng period. The year 1665 is separated from 1734 by 69 years; presuming that in 1665 Kao was about 20 years old, he must have lived to the age of 89; this is not impossible, but it is harder to

251

believe that in his youth he should have done this picture which bears the ear-marks of a work of maturity. Hirth (Scraps, p. 30) remarks that his best period seems to fall in the years 1700–15. Field Museum possesses a finger ink-sketch by him, representing two hawks fluttering around a bare tree-trunk with a date corresponding to 1685, and another, undated, representing a carp swimming upstream and stretching its head out of the water. Both pictures are reproduced in Laufer, History of the Finger-print System, *Smithsonian Report for 1912*, plates 5 and 6. Kao K'i-p'ei, as is well known, was a great exponent of the art of finger-painting, and was a really good artist.

The woodcut inserted in the *T'u shu tsi ch'eng* (XIX, chap. 45) shows a single bird perching on a rocky platform and overlooking the water; the figure is fairly exact, except the beak, the upper mandible of which is but slightly curved instead of terminating in a hook. The *Pen ts'ao kang mu* contains an engraving of a cormorant floating downstream with a small fish in its beak.

The *San ts'ai t'u hui* (1607, sect. Birds and Animals, p. 18) illustrates a cormorant on the bank of a river, a rather sorry specimen. The figure and scene are very similar to the illustration in the *Cheng lei pen ts'ao* (above, p. 224).

Good examples of Chinese ink-drawings of cormorant fishing are reproduced in the book of P. Korrigan (p. 38) and in Gray's China, II, opp. p. 297.

G. E. Freeman and F. H. Salvin (p. 328) entertained the idea that "this ingenious method of catching fish was most likely invented by the Chinese, and must be of very great antiquity, if we may judge by the representation in old China ware and other Chinese illustrations," to which is added in a foot-note, "We have seen cormorant fishing represented upon some ancient china cups at Leagram Hall, Lancashire, the seat of J. Weld, Esq."

On a white porcelain bowl of the Yung-cheng period (1723–35), brought from China by Mr. C. T. Yao of New York and presented to Field Museum by the American Friends of China, Chicago, various scenes in the occupations of fishermen are represented in enamel colors, among others a man standing in a boat and carrying on his shoulders a bamboo pole on which two cormorants perch (Plate XIII).

The *Ku yü t'u p'u* (chap. 71, p. 13) illustrates a jade spoon or ladle terminating in a cormorant's head 碧 玉 鸬 鹚 杓, placed in a water receptacle in the shape of a tazza; evidently made in allusion to the cormorant wine-vessel of Li Po (see *Pien tse lei pien*, chap. 210, p. 5, or *Ts'e yüan: lu-ts'e*). A curious coincidence is represented by a spoon of the Eastern Dakota Indians, used in the feasts of the Medicine Lodge, which is provided with a handle carved to represent a cormorant's head (in Field Museum, Cat. No. 60411).

I do not enumerate the illustrations of cormorant fishing in China contained in the older European books, as these are reproduced by Gudger (I) with critical annotations and as in this age of photography they have but little scientific value.

The *National Geographic Magazine* of June, 1927 (p. 704), contains a good reproduction of an excellent photograph of a cormorant fisher taken by Dr. Camillo Schneider in western Yün-nan.

A photograph of cormorant fishing on the Wei-ho in Shen-si is reproduced by Clark and Sowerby (plate XX), without description; it shows a single fisherman standing astride on two small boats joined together, operating with three birds.

H. Kraemer's "Der Mensch und die Erde" (X, opp. p. 288) contains a colored plate entitled "Fishing with trained cormorants in China" after a painting by F. de Haenen. A fisherman is shown in the act of removing a fish from the bill of a cormorant which has just reached the edge of the boat; three birds are swimming in the water, and a confused mass of ropes is visible. The picture is rather fantastic than instructive. In the caption accompanying the plate it is said that the birds dive to a depth of 50 meters (!) and swim under water for two or three minutes with immense velocity. In the text which purports to trace the development of fishing all over the world nothing is said about cormorant fishing.

Yukihide Tosa (fifteenth century) has painted in colors an excellent scene representing cormorant fishing; a man in the boat governs two birds with strings held in his left hand, closely watching them; one bird is shown in the act of diving. A color reproduction of this picture is in *Kokka*, IV, No. 47, plate 1.

Kōrin Ogata, who died in 1716, is the creator of a masterly picture in ink on silk, showing an old fisherman in a boat with torchlight, eagerly watching his two cormorants in the water; one of the birds holds an *ayu* in its beak. The picture belongs to Baron Iwasaki of Tōkyo, and is reproduced in *Kokka*, XXX, No. 352, 1919, plate VI. It is said there that the theme was favorite with Kōrin and that this work belongs to his best.

Cormorant fishing at night in the Nagara River is illustrated in a colored print by Yeisen said to be "very rare" (reproduced in

Japanese Color Prints of Lindsay Russell, New York, 1920, p. 22).
It is evidently identical with No. 55 of the Sixty-nine Stations of the
Kisokaido by Hiroshige and Yeisen.

A *tsuba* by Hironaga in Field Museum shows a fisherman with a
lighted torch in the water, holding the ropes attached to a cormorant
whose bill reaches up toward a fish (H. C. Gunsaulus, Japanese
Sword-mounts, plate L, fig. 1).

Pichot's book illustrates seven boats fishing with cormorants in
Lake Gifu, Japan; a lantern made in the same place and painted
with cormorant fishing-boats; the count R. de Najac holding his
cormorant "Pole Nord" or "Carême"; and a French engraving of the
eighteenth century showing falconers and pêcheurs au cormoran.

"Fishing with cormorants on the Nagara River, Gifu" is the
subject of an illustration in an article on The Fisheries of Japan by
Jihei Hashiguchi in *Far Eastern Review*, XIV, 1918, p. 319.

Dr. Gudger's article (II) contains many good reproductions of
fine photographs representing cormorant fishing in Japan, many of
these having been supplied by the Municipal Office of Gifu.

FOLK-LORE OF THE CORMORANT

The Chinese folk-lore of the cormorant is not particularly interesting, but some notions entertained regarding the bird are worthy
of mention. It is an old popular conception first pointed out by the
calligrapher Wang Hi-chi 王羲之 (A.D. 321–379) and by T'ao Hung-
king 陶宏景 (A.D. 452–536) that "this bird is not born from eggs,
but spits its fledglings out of its mouth." Such an absurd idea could,
of course, obtain only at a time when the bird's life was unknown,
and no attempt at training it had been made. Ch'en Ts'ang-k'i,
the physician of the K'ai-yüan period (A.D. 713–741), writes that
"this bird is viviparous 胎生 and brings its young forth from its
mouth like the hare vomits its offspring; hence women at the time
of childbirth, when holding this bird, will have an easy delivery."
In the *Yu yang tsa tsu* (chap. 16, p. 2) it is said, "The hare spits its
young out, the cormorant spits its fledglings out."

The *I wu chi* of Yang Fu (*T'ai p'ing yü lan*, chap. 925, p. 8b),
quoted on p. 221, adds to this superstition that the number of young
born from the mouth is large, at least seven or eight, and that five
or six are connected with one another and come out like a silk thread.
In the Buddhistic dictionary *Yi ts'ie king yin i* the number of young
ones brought forth from the mouth at one birth is given as eight
or nine.

A parallel to the notion that holding a cormorant will bring
about easy delivery occurs in ancient Japan, where for the same
purpose a cormorant feather was grasped in the hand of a parturient
woman (Aston, Nihongi, I, p. 98).

K'ou Tsung-shi 寇宗奭, in his *Pen ts'ao yen i* 本草衍義 (chap. 16,
p. 9, ed. of Lu Sin-yüan) written in A.D. 1116, gives the following
account of the cormorant: "T'ao Yin-kü [T'ao Hung-king 陶宏景]
asserts that this bird is not born from eggs, but vomits the fledglings
out of its mouth. The people of the present time call it 'old water-
duck.' The birds nest in large trees where they flock in large numbers. The trees in which they lodge for a long time will decay.
Their droppings are poisonous. Pregnant women do not dare
eat this bird on account of the fledglings being vomited out of its
mouth. Ch'en Ts'ang-k'i, on the other hand, states that, in order
to insure an easy childbirth, one should let a woman, when her
hour approaches, hold a bird. While T'ao Siang-hi served as an
official at Li-chou [in Hu-nan], there was a large tree behind the

255

house of this gentleman. In the crown of this tree there were thirty or forty cormorant nests, where at evening the birds could be observed in the act of mating. Egg-shells of green color were found spread over the ground. How should this bird then obtain its young by vomiting them forth from its mouth! Such a thing has never been verified, and is nothing but baseless talk of the people."

This is one of the rare instances where a superstition is refuted by actual observation. S. Wells Williams (*Chinese Repository*, VII, 1839, p. 542), referring to this belief, asserts that "Li Shi-chen very wisely puts such accounts among errata." Li Shi-chen, however, does not make any comment on this point; the criticism in question is solely due to K'ou Tsung-shi.

According to Wang Hi-chi, the ordure of the cormorant is white and dispels black spots on the face (apparently a skin-disease). According to the *Fang shu* 方書, evidently a book of medical prescriptions, cormorant's ordure is called "water-flower of Shu" 蜀 水 花, it is rubbed into a powder and administered in water; it has the effect of causing men to renounce wine; the bird's head is a good remedy for fish-bones sticking in the throat (*Ko chi king yüan*, chap. 80, p. 3b). According to the *Pen ts'ao kang mu*, the "water-flower of Shu" is even mentioned in the *Pie lu*, and T'ao Hung-king comments, "It is plentiful in the valleys with streams; it is necessary only to get hold of it oneself, as what is offered in the markets cannot be trusted."

As the cormorant is able to swallow a fish, bone and all, it is easily understood that in the pharmacopoeia parts of the bird are recommended as relieving one from fish-bones sticking in the throat. T'ao Hung-king prescribes for this purpose the bird's bones to be burnt and mixed with lime and water; this medicine will force fish-bones down the throat. Fan Wang 范 汪, a physician, at the time of the Eastern Tsin dynasty (A.D. 317–419) recommends to swallow a cormorant's beak or to burn a cormorant's wing (prepared in the same manner as the bones previously) as a remedy against choking from fish-bones; even an inch square of a cormorant administered will bring the bone down, if only the bird's name is called out (*T'ai p'ing yü lan*, chap. 925, p. 9). Li Shi-chen extols the bird's crop which must be swallowed, as very efficient for the same purpose.

Finally the cormorant appears in one story as a rain bird. In A.D. 797, at the time of a drought, prayers for rain were offered in the Dragon Hall of the Hing-k'ing Palace 祈 雨 于 興 慶 宮 龍 堂, when a flock of white cormorants appeared above a pond, grouped as though conducting the imperial barge; on the following morning it rained (*Nan pu sin shu* 南 部 新 書, written by Ts'ien Yi 錢 易 about A.D. 975, chap. 丙, p. 2b, ed. of *Yüe ya t'ang ts'ung shu*. The story is told with some greater detail in *Kiu T'ang shu;* see *T'ai p'ing yü lan*, chap. 925, p. 8b, or *Yüan kien lei han*, chap. 427, p. 8b).

The description of the cormorants as "white" in the above text seems somewhat anomalous; perhaps there is confusion with herons. In England it was regarded as a sign of rain or wind when cormorants and gulls bathed themselves much, pruned their feathers, flickered or flapped their wings (J. Brand, Observations on the Popular Antiquities of Great Britain, III, 1888, p. 218).

Two popular sayings in the Amoy dialect are noted by Francken and De Grijs (Chineesch-Hollandsch Woordenboek van het Emoi Dialekt, p. 365, Batavia, 1882):

鸕 鶿 箍 項 *lo tsi k'o am*, to have a ring around the neck like a cormorant; i.e., not to be wholly one's own master.

鸕 鶿 未 知 尾 復 臭 *lo tsi bu tsai bu ao ts'ao*, the cormorant is not conscious of the odor penetrating from under its tail; i.e., not to see one's defects.

In England the voracity of the bird was proverbial, and Shakespeare likens to it a man of large appetite, as "the cormorant belly" (*Coriolanus*, I, 1), "cormorant devouring Time" (*Love's Labour's Lost*, I, 1), "this cormorant war" (*Troilus and Cressida*, II, 2). Compare T. F. Thiselton Dyer, Folk-lore of Shakespeare, 1884, p.108. Harting (Ornithology of Shakespeare, p. 260) writes, "Although Shakespeare mentions the cormorant in several of his plays, he has nowhere alluded to the sport of using these birds, when trained, for fishing; a fact which is singular, since he often speaks of the then popular pastime of hawking, and he did not die until some years after James I had made fishing with cormorants a fashionable amusement." E. Phipson (Animal-lore of Shakespeare's Time, 1883, p. 285) also writes that Shakespeare's references to the cormorant are only as an emblem of insatiable appetite.

The scanty information known to the ancients about the cormorant (if indeed it refers to this genus) has been collected by O. Keller (Die antike Tierwelt, II, p. 239).

A mythology of the cormorant exists only in ancient Japan (see above, p. 212) and among the Tlingit and some other Indian tribes along the northwest coast of America (for references see O. Dähnhardt, Natursagen, 1910, III, pt. 1, pp. 28, 29, 77, 105, 147, 232).

BIBLIOGRAPHY

The earlier works on China which make reference to cormorant fishing have not been included here, as Dr. Gudger (I) has canvassed this ground, nor is mention made of modern works on China which have a casual reference to the subject without contributing anything new or worth while.

ANON.—Cormorant Fishing. East of Asia Magazine, Shanghai, II, 1903, pp. 95–97.
Brief description inaccurate in several points. 4 ill.

BALL, J. DYER.—Things Chinese, 4th ed., Shanghai, 1903, pp. 181–183.

BELVALLETTE, ALFRED.—Traité de fauconnerie et d'autourserie. Suivi d'une étude sur la pêche au cormoran. Paris, 1903.

BROWN, LUCY FLETCHER.—Fishing with the Birds of Gifu. Japan, XIV, 1925, pp. 23–24, 31. 3 ill.

CHAMBERLAIN, BASIL H.—Things Japanese, 5th ed. London, 1905, pp. 105–108.

COCHRANE, MAY L.—Harnessed Birds of Gifu. Asia, XXV, 1925, pp. 301–305.

DABRY DE THIERSANT, P.—La pisciculture et la pêche en Chine. Paris, 1872, pp. 171–172, plate XIX, fig. 1.

DAVID, ARMAND, and OUSTALET, EMILE.—Les oiseaux de la Chine. Paris, 1877, pp. 532–533.
Brief description of the species.

DOOLITTLE, JUSTUS.—Social Life of the Chinese. London, 1868, pp 36–38. 1 ill.

FAUVEL, ALBERT-AUGUSTE.—Promenades d'un naturaliste dans l'archipel des Chusan et sur les cotes du Chekiang. Cherbourg, 1880.
Cormorant fishing: pp. 230–233 (valuable observations).

FLOERICKE, K.—Kormoranfischerei. Kosmos, XI, Stuttgart, 1914, pp. 30–33.

FORTUNE, ROBERT.—I. Ten Years' Wanderings in the Northern Provinces of China. 2d ed., London, 1847, I, pp. 98–103. 3d ed., London, 1853, I, pp. 86–90.
II. Two Visits to the Tea Countries of China, 3d ed., London, 1853, I, pp. 86–90.

FREEMAN, G. E., and SALVIN, F. H.—Falconry. Its Claims, History, and Practice. To which are added Remarks on Training the Otter and Cormorant. London, 1859.
Fishing with Cormorants, pp. 327–349.
Fishing with Otters, pp. 350–352.

GRAY, JOHN HENRY, ARCHDEACON.—China, II, pp. 297–298. ill. London, 1878.

GUDGER, E. W.—I. Fishing with the Cormorant in China. The American Naturalist, LX, 1926, pp. 5–41. 16 ill.
II. Fishing with the Cormorant in Japan. The Scientific Monthly, XXIX, 1929, pp. 5–38. 31 ill.
III. Fishing with the Otter. The American Naturalist, LXI, 1927, pp. 193–225. 6 ill.

HAHN, EDUARD.—Die Haustiere. Leipzig, 1896, pp. 347–350.

HARTING, JAMES E.—Essays on Sport and Natural History. London, 1883, pp. 423–440: Fishing with Cormorants.

IKENOYA, S.—Cormorant Fishing. Japan Magazine, 1917, pp. 31–32.

JAMETEL, MAURICE.—La Chine inconnue. Paris (Rouam), 1886. Chap. XII: Le faucon à poisson, son éducation, pp. 207–213.
Information on the training of the cormorant copied from Fauvel.

258

JORDAN, DAVID STARR.—Fishing for Japanese Samlets on the Jewel River. Outing, XL, 1902, pp. 23–25. 1 ill.

Republished in Jordan's Guide to the Study of Fishes, II, New York, 1905, pp. 116–118; and Fishes, New York, 1925, pp. 142–144. 2 ill.

JOUY, P. L.—On Cormorant Fishing in Japan. American Naturalist, XXII, 1888, pp. 1–3.

KORRIGAN, POL.—Causerie sur la pêche fluviale en Chine. Chang-hai, Imprimerie de la Mission Catholique, 1909.

An excellent booklet. Cormorant, pp. 39–43.

KURODA, NAGAMICHI.—Cormorant Fishing on the Nagara River. Japanese Magazine, Tokyo, XVI, 1926, pp. 303–320. 16 ill.

(LE COMTE) LE COUTEULX DE CANTELEU.—La pêche au cormoran. Paris, Revue britannique, 1870.

He was the owner of a flock of cormorants at his castle Saint-Martin.

LEONHARDT, E.—Aus China. Deutsche Fischerei Correspondenz, V, 1901, June, p. 7; July, p. 3.

A few data on the training of the cormorant in China.

PICHOT, PIERRE-AMÉDÉE.—Les oiseaux de sport. Paris (A. Legoupy), 1903.

Cormorant: pp. 27–35; chiefly with reference to France and Japan.

RUPPRECHT PRINZ VON BAYERN.—Reise-Erinnerungen aus Ost-Asien. München, 1906.

Cormorant fishing in Tamagawa west of Tokyo, pp. 322–323.

SCHMIDT, M.—Fortpflanzung des gemeinen Cormorans in Gefangenschaft. Der Zoologische Garten, XI, 1870, pp. 12–18.

Interesting data on the nesting habits and breeding of the cormorants.

SEEBOHM, HENRY.—On the Cormorants of Japan and China. Ibis, 5th series, V, 1885, pp. 270–271.

Ornithological classification and description.

SMITH, HUGH M.—I. Japan, the Paramount Fishing Nation. Transactions American Fisheries Society, 33rd meeting, 1904, pp. 129–132. 4 ill.

II. The Fisheries of Japan. National Geographic Magazine, Washington, XVI, 1905, pp. 213–214. 3 ill.

SOKOLOWSKY, ALEXANDER.—Der Kormoran in seinen Beziehungen zur menschlichen Wirtschaft. Weltwirtschaftliches Archiv, Zeitschrift für allgemeine und spezielle Weltwirtschaftslehre, Jena, XII, 1918, pp. 315–320.

With a short bibliography.

SOURBETS, G., and SAINT-MARC, C. DE.—Précis de fauconnerie, suivi de l'éducation du cormoran. Niort, Clouzot, 1887.

SOWERBY, ALFRED DE C.—The Cormorant in China. China Journal of Science and Arts, IV, 1926, pp. 72–74. 4 ill.

SPECIAL CATALOGUE of the Ningpo Collection of Exhibits. International Fishery Exhibition, Berlin, 1880. Also in Ibis, IV, 1880, pp. 375–376; and Special Catalogue of the Chinese Collection in Great International Fisheries Exhibition, London, 1883.

Information on cormorant by A. A. Fauvel.

STONE, JABEZ K.—Cormorant Fishing at Gifu. Japan, VIII, 1919, pp. 5–7, 44. 5 ill.

WILLIAMS, S. WELLS.--Notices in Natural History: the Loo-sze or Fishing Cormorant. Chinese Repository, VII, 1839, Canton, pp. 541–543.

Very incomplete translation of the text of the Pen ts'ao.

INDEX

An-hui, cormorant fishing in, 221, 222, 228
Aston, 212, 213, 255
ayu, 213, 231, 248, 253

Buddhists, attitude of toward cormorant fishing, 225

Chamberlain, 221
Ch'ang-t'ing hien chi, 209
Che-kiang, cormorant fishing in, 227, 242
Ch'en Hao-tse, 223, 250
Ch'en Ts'ang-k'i, 224, 225, 255
Cheng lei pen ts'ao, 224, 225, 252
Chi-li, cormorant fishing in, 230, 237
Cocks, 213
cormorant, Chinese terminology of, 208; folk-lore of, 255; geographical distribution of, 207, 227; iconography of, 251; Japanese terminology of, 211; process of domestication of, 236
Cormorant Cliff, 210
Cormorant Embankment, 210
Cormorant Islet, 210
Cormorant Lake, 210

Dabry de Thiersant, 238, 246, 258
De Guignes, 228, 243
Doolittle, 228
duet fishing, 244

eels, preferred by cormorants to other fish, 238
eggs, of cormorant, 237
egrets, tamed and kept in captivity, 249–250
England, cormorant fishing in, 206, 207
Er ya, 208, 214, 225
Er ya i, 217, 223, 225
Europe, cormorant fishing in, 206

falconry, cormorant fishing compared with, 218, 229
Fan Chen, 218
Fan Wang, 256
Fang I-chi, 214, 219, 239
Fang shu, 256
Fauvel, 210, 227, 238, 241, 246, 258
fishermen's gild, 244
Florenz, 212
Fortune, 206, 227, 229, 238, 240, 241, 258
France, cormorant fishing in, 206
Fu-kien, cormorant fishing in, 227–228

Giles, 208, 209, 220, 221
group fishing, 244, 245
Gudger, 205, 213, 230, 232, 237, 249, 253, 254, 258

Hahn, 236, 258
Hai-hou Tsie, 216
harness method of cormorant fishing in Japan, 234, 247
Harting, 206, 207, 229, 236, 257, 258
Hemeling, 211
Hing-hua fu chi, 208, 227
Hironaga, 254
Hiroshige, 254
Holland, cormorant fishing in, 206, 207
Ho-nan, cormorant fishing in, 228
Hu-nan, cormorant fishing in, 228; other fishing in, 249
Hu-nan fang wu chi, 228, 235, 237
Hu Tse, 215
Hüan Ying, 208
Hui Hung, 215
Hwa king, 223, 250
Hwang Ch'ao-ying, 218, 219
Hwang Kü-ts'ai, 251

I wu chi, 221, 255
Ikenoya, 213, 246, 258
Indo-China, cormorant fishing absent in, 235

James I, fishing with cormorants, 206, 257
Jametel, 246, 258
Japan, cormorant training and fishing in, 212, 213, 231, 246–248

Kao K'i-p'ei, 251, 252
Kiang-si, cormorant fishing in, 228
K'in king, 221
Kojiki, 212
Korea, cormorant fishing absent in, 235
Kōrin Ogata, 253
Korrigan, 228, 244, 252, 259
K'ou Tsung-shi, 255, 256
Ku yü t'u p'u, 252
Kuroda, 213, 246, 259
Kwang-tung, cormorant fishing in, 228, 243
Kwei-chou, cormorant fishing in, 229
K'wei-chou fu chi, 219
K'wei-chou t'u king, 217
K'wei kü chi, 219
Kwo T'wan, 219

Lan chen tse, 215

260

FISHERMEN CARRYING TWO CORMORANTS ON A BAMBOO POLE

Painting in enamel colors on a white porcelain jar

Yung-cheng Period (1723–35)

CORMORANT FISHER, FINGER-PAINTING
BY KAO K'I-P'EI

FISHERMAN ON BAMBOO RAFT WITH CORMORANTS

FISHING WITH CORMORANTS   ON BAMBOO RAFT

CORMORANTS PERCHING ON BAMBOO RAFT

Photographs taken by Floyd Tangier Smith on Min River near Fuchow,
Fu-kien Province

FISHERMAN HOLDING CORMORANT

PLACING THE CATCH IN A BASKET

FEEDING THE CORMORANT AFTER THE CATCH

Photographs taken by Floyd Tangier Smith on Min River near Fuchow,
Fu-kien Province

206

东亚的感应梦

THE

JOURNAL OF
AMERICAN FOLK-LORE.

VOLUME 44

EDITED BY
RUTH BENEDICT.

Associate Editors.

GEORGE LYMAN KITTREDGE.　　　C.-MARIUS BARBEAU.
GLADYS A. REICHARD.　　　　　　ELSIE CLEWS PARSONS.
AURELIO M. ESPINOSA.　　　　　　FRANZ BOAS.

NEW YORK
PUBLISHED BY THE AMERICAN FOLK-LORE SOCIETY.

KRAUS REPRINT CO.
New York

INSPIRATIONAL DREAMS IN EASTERN ASIA.[1]

By Berthold Laufer.

In discussing briefly the subject of inspirational dreams I must disclaim at the outset any technical knowledge of dream psychology. There is a vast literature on this subject chiefly produced by psychiatrists like Freud and Kraepelin and psychologists, most of which I confess I have not read and may never hope to have the time to read. The best introduction to the subject for our purposes is that of an orientalist, Georg Jacob, whose book is entitled "Märchen und Traum" (Hannover, 1923), written with special reference to the Orient and provided with a good bibliography.

My main object in presenting these remarks is to call the attention of our ethnologists interested in the inspirational dreams of the North American Indians to some striking parallels in eastern Asia and to stimulate them to a more profound investigation of this interesting problem, which has been but little studied and which I believe will be better comprehended with due consideration of corresponding phenomena in other culture areas.

Although the scientific study of dreams is still in its initial stages and the interpretation of cases and symptoms varies to a great extent, it is quite safe to assert now that dreams have exerted an enormous influence on the formation of human behavior and culture. Many motives of legends and fairy tales have justly been traced to dreams; many mythical concepts and motives of art and even entire works of art have been inspired by them. It is infinitely more probable that the majority of fabulous monsters and chimaeras which so abundantly pervade all oriental arts owe their origin to visions and dreams than, as has been suggested, to the discovery and imaginative reproduction of real fossil monsters. As may be expected, the exaggerations of the specialist have also been at work in this field; and as we have pan-sexualists, who reduce everything to sex, so there are also "pan-dreamists" (or pan-oneiromantists) who exert themselves to trace all happenings to dreams.

Both India and China possess an ancient literature on dreams, and from earliest times have had a special class of dream interpreters to whose verdicts great attention was paid. Many hundreds, more probably even thousands, of dreams are recorded in the Chinese annals and in the biographies of individuals, and have had a sometimes far-reaching effect on the course of historical events; but despite this abundance of material no one has ever made a special study of Indian or Chinese dreams. Of all

---

[1] Read at the Meeting of the Central Section of the American Anthropological Association, Milwaukee, May 9, 1930.

categories of dreams the inspirational dream is the most interesting, because it has proved a creative force in literature, science, and art, or stimulated ambition or provoked activity of one sort or another.

An inspirational dream opens the history of the conversion to Buddhism of the Mongol emperors. This conversion was accomplished by the Grand Lama of the famous Sa-skya monastery in Tibet, 50 miles north of Mount Everest. This was Blo-gros rgyal-mts'an, usually called by his title P'ags-pa ("Reverend"). He was summoned to Peking by the emperor Kubilai in A. D. 1261, who received from him the Hevajra consecration. Hevajra is the name of a mystic Buddha.

The Mongol chronicler Sanang Setsen narrates that on the first day when the Tibetan church dignitary had an audience with the Grand Khan he was unable to answer any questions put to him by his majesty and could not even comprehend a single word of what he said. Deeply alarmed, he begged the emperor to be excused and to be allowed to continue the conversation the following day. The cause of his incompetence was that the text of the Tantras (magic spells) of Hevajra, which formerly was the property of the Sa-skya family, was now in the hands of the Grand Khan, and that P'ags-pa had not seen the book. In consequence the saint passed a night full of sorrow and anguish. During that night he had a dream of an old man, who had the appearance of a Brahman, with snow-white hair tied into a knot on the crown of his head, holding in his hand a flute made from a human thighbone. This old man ordered P'ags-pa to light a lamp, and then produced a box from which he took a book that contained the text of the Hevajratantra. P'ags-pa perused the book in his dream and memorized its contents. On the following day he was able to answer the emperor's questions, gave him his benediction, and received the title "king of the doctrine in the three countries [i. e. China, Tibet, and Mongolia], the holy Lama." The old man who had appeared to him in his dream was the tutelary god Mahākāla ("the Great Black One"), a form of Çiva. In fact, a brass or bronze image of Mahākāla in the form of a Brahman, holding a bone flute, such as P'ags-pa saw in his vision, still exists.[1] This role of Çiva-Mahākāla as an inspirationalist is of ancient date in India. The most perfect grammar produced in India, that of Pāṇini, is ascribed to an inspiration of this god.

Inspirations obtained by dreams play an extensive role in the latest, mediaeval phase of Buddhism, the Yoga or Tantra school, which combined the teachings of Mahāyāna Buddhism with the mystic doctrines of the Yoga system of Patañjali. This system consisted chiefly in the practice of magic and sorcery aiming at endowment with supernatural, miracle-working power. This was attained by recitation of mystic formulas *(tantra)*, litanies, or spells accompanied by music and certain positions or distortions of the fingers *(mudrā)*. Mental concentration on a single

---

[1] A. Grünwedel, **Mythologie des Buddhismus**, pp. 54, 63—64, 176.

point was required with a view to annihilate thought, and resulted in a mental and physical coma, auto-suggestions, visions, and dreams. The spells and the magical powers obtained from them were in the hands of certain deities, who had to be exorcised, so that it is no wonder that these appeared to the visionaries in their dreams and inspired them to write certain books or to visit a certain country for the purpose of propagating religion. Thus, Çānti is instructed by the goddess Tārā in a dream to go to Ceylon to preach the Mahāyāna; Abhayākaragupta is visited twice in a dream by the goddess Vajravārāhi, who exhorted him to write manuals and obtain salvation through her.[1] Other saints, by beholding the face of a god in a dream, obtained the highest miraculous powers, such as lightening their bodies, flying through the air, assuming any shape, reaching any place, etc.

The Chinese theory of dreams is as follows: A double soul is distinguished in every individual — a material or animal soul called *p'o*, which regulates the functions of life, is indissolubly attached to the body and goes down to earth with it after death; and a spiritual soul, called *hun*, which governs the functions of reason, is able to leave the body and at death goes to heaven, carrying with it an appearance of physical form. A dream arises when the connection of the body with the spiritual soul is interrupted. The body lives as long as the material soul dwells in it, but is doomed to die as soon as it escapes. The spiritual soul, however, may leave the body without endangering its life. This is the case in swoons, trances, and dreams. The soul separates from the body and enters into communication with spirits; it may freely interview the souls of the departed or have speech with the gods. At the end of the dream the soul returns to the body.

The Chinese are perhaps the only people who have conceived a way of representing dreams pictorially. From the head of the sleeper radiates a fluttering band or the dream-path in form of a lane on which are drawn or painted the figures appearing in the dreamer's vision. A beautiful example of this kind may be seen in Field Museum, carved on a rhinoceros-horn cup where a handsome maiden asleep dreams of her Prince Charming.

The form of dream known as incubation or temple sleep formerly played a prominent role in China. Every town has a tutelary god styled "the father of the walls and moats" (Ch'eng huang ye) and worshiped in a special temple. These city gods are defunct and deified officials who, during their tenure of office, had merited appreciation from the city under their administration. The city god cares for the welfare of the inhabitants under his jurisdiction, and is the mediator between this and the other life, keeping an account of all good and evil deeds of his protégés and reporting these to the gods of Heaven and Hell. The worship of city

---

[1] Grünwedel, Taranatha's Edelsteinmine, pp. 110, 112, 113.

gods reached its climax toward the end of the fourteenth century under the first emperor of the Ming dynasty. At that time it was obligatory for all officials of higher ranks when entering a walled city to pass the first night in the temple of the city god, in order to receive his instructions in a dream. In case of a difficult point in law judges will spend the night in the city god's temple, in the hope that the god will appear to them in a dream and enlighten them on the case in question. Sometimes a district magistrate or judge will adjourn his court to the temple hall in order to give solemnity to the trial, to intimidate the witnesses or to encourage them to speak the truth. The temple sleep is always preceded by bathing and fasting, and the official betakes himself to the temple in a solemn procession. Sometimes a written petition is burnt before the altar of the god with proper ritual, and then the official retires to rest in an adjoining room. There are stories to the effect that ashes of the burnt document had, traced in them, the name of the guilty person and the place where he lived, whereupon he was duly brought to justice.[1]

Not only officials, but also plain people, in case of a difficult decision, resort to the expedient of seeking a dream by visiting a temple; burning incense, they invoke the deity to favor them with a dream that will shed light on the subject of their perplexity. They frequently go to sleep before the image of the god. Should they have a dream, they rise and inquire by means of a certain process of divination whether the dream was really sent by the god in answer to their prayer. When an affirmative reply is received, they proceed to study the character of the dream or consult a dream interpreter to decide on a course of action.[2] In the province of Fu-kien it is customary for people to sleep on a grave for the purpose of provoking a revelation in a dream.[3] This custom reminds one of what Herodotus (IV, 172) reports anent the Nasamones who predict the future by visiting the graves of their ancestors, where they pray and fall asleep and act in accordance with their dreams.

Examples of inspirational dreams are known from China prior to the introduction of Buddhism. Thus, Confucius in his earlier years had frequent dreams of Chou Kung, his ideal in political wisdom, whose principles and institutions he endeavored to put into practice; and when he grew old and disillusioned, he complained that "for a long time he had not dreamt, as he was wont to do, of the Duke of Chou."[4] Confucius is also said to have received a premonition of his coming death through a dream, in which he saw himself seated between the two pillars of the platform in front of his house, receiving offerings due to the dead. Seven days later he died.[5]

---

[1] cf. G. Willoughby-Meade, Chinese Ghouls and Goblins, p. 85.
[2] cf. J. Doolittle, Social Life of the Chinese, 1868, p. 449.
[3] De Groot, Fêtes, p. 593.
[4] *Lun yü*, VII, 52.
[5] Legge, Li ki, I, p. 138; Chavannes, Se-ma Ts'ien, V, p. 424.

Another inspirational dream of ancient date, in which a portrait is in question, is of interest. When Wu Ting (1324—1266 B. C.), a ruler of the Yin or Shang dynasty, lost the services of his aged teacher Kan-p'an, who advised him on government affairs, he was in quest of a new counselor. Therefore he addressed a prayer to Shang-ti, the supreme god, requesting that he should reveal to him in a dream the man capable of acting as his prime minister. In his dream he really saw the likeness of the man selected by God, but he could not find him among the high dignitaries of the empire, though he searched all over the country. A portrait was then made of the man, as he had appeared to the emperor in his dream, and this was circulated throughout the empire. Finally this led to the discovery of the man in the person of a common workman, Fu Yüe by name, who was raised to the post of prime minister.[1]

Tao K'an, a celebrated Chinese statesman (A. D. 259—334), once had a dream which led to his advancement. He dreamed that he scaled the heights of heaven with the aid of eight wings, and passed through eight of the celestial doors, but was driven back from the ninth by the warder, who cast him down to earth. When he landed there, the wings on his left side were broken. Subsequently he entered public life, and was appointed governor of eight provinces, which was interpreted as a realization of his dream.

T'ang Li-yüan, when he crossed the Yang-tse River, noted in it the body of a woman, pulled it out of the water, and buried it. During the night he dreamed that he found himself in a place like a recess deep in a mountain; the bright moon had just risen above the horizon, a gentle breeze played with his garments, and in the distance was audible a tune produced by a reed organ, the tones of which melted sweetly away. Suddenly a beautiful woman appeared at the edge of the woods and sang this tune: 'The melodies of the Purple Mansion (the heavenly spheres) can be clearly heard in succession in a pure clear night such as this.' Subsequently T'ang Li-yüan presented himself in the capital at the examinations for the highest literary degree. The theme assigned to the candidates for writing an essay was as follows: 'In the Hou Mountains, Wang Tse-tsin is heard playing on a reed organ during moonlit nights.' Li-yüan now used the lines he had heard in his dream as the third and fourth lines in his composition. In consequence he was successful and won the degree of *tsin-shi*. The people considered this as a reward bestowed upon him by the spirit of the woman for whom he had provided burial.

A well-defined artistic composition connected with dreams is presented by the Arhats or, as the Chinese say, Lo-han. The Arhats are the most advanced disciples of the Buddha, who have reached the highest degree of saintship, and are considered the most powerful protectors of the Buddhist religion. They usually appear in a group of sixteen to which

---

[1] *Shu king*, IV, VIII, 2—3.

at a later time two were added in China. They play the same role in Chinese religion and art as the apostles in Christendom, and their portraits and statues rank among the foremost productions of Chinese art. The creator of the Arhat types was a Buddhistic monk, Kwan Hiu, who lived from A. D. 832 to 912, up to the age of 81. He became a novice in the monastic order when he had reached the age of seven, and belonged to the sect known as Dhyāna ("contemplation"), the chief aim of which was to obtain self-perfection and salvation through inward concentration and meditation. From meditation to visionary dreams there is but one step. Kwan Hiu was a precocious youth famed for his poetry among his contemporaries, but he soon developed into a greater artist with the brush, specializing in portraits of the Arhats which designated a novel departure from the established routine. In his biography it is said, "Every time he desired to paint one of the venerable saints, he first recited a prayer, and then in his dreams obtained the respective figure of the Arhat. Awakening, he fixed this dream picture in his mind and painted it accordingly, so that his portraits did not conform to the customary standard."[1] Like a Crow Indian, who sought directions for the decoration of his shield in a dream, Kwan Hiu was in deliberate quest of a vision. In another source *(Yi chou ming hua lu)* it is on record, "Kwan Hiu painted the pictures of the sixteen Arhats with long bushy eyebrows, with drooping cheeks and high noses, leaning against a pine-tree or a rock, or seated in a landscape, men of a strangely foreign appearance or a Hindu face. When people marveled at his pictures and interrogated him, he replied, 'I paint what I see in my dreams.'" In honor of this artist, a temple was erected at She (in Hui-chou fu, An-hui-Province), called the "Hall of the Arhats Corresponding to Dreams."

This account of Kwan Hiu's Arhat portraits is by no means fanciful, but thoroughly authentic and in consonance with the facts. Hundreds and thousands of Arhats have been painted in China in the course of many centuries, and whatever differences of style and composition there may be, there are only two fundamental types of Arhat — a naturalistic type and a dream type traceable to Kwan Hiu as his father. The naturalistic type is based on close observation of Indians who came from India to China, and the Chinese artists reveled in portraying exactly their prominent racial features. What Kwan Hiu did in consequence of his dreamy visions was to exaggerate with grotesque humor the Indian racial characteristics, equipping them with strangely formed, hill-shaped heads, peculiarly curved noses, protruding eyeballs, excessively long eyebrows, wild curly beards, crooked and bent bodies in almost impossible positions; although seemingly grotesque caricatures, yet personifications of sublime sainthood. Since they do not conform to any reality, we are

---

[1] S. Lévi and E. Chavannes, Les seize Arhat protecteurs de la loi, *Journal asiatique*, 1916, II, p. 300.

compelled to admit that they must be the result of dream revelations and that the tradition of Kan Hiu's dream pictures is correct.[1]

Tang Hou, author of a treatise on painting *(Hua kien)*, has this story: "Hui Tsung painted with his own hand a picture entitled 'A Dream Journey to the Other World.' The inhabitants, several thousands in number, were about half the size of one's little finger. All things in heaven and earth, most beautifully executed, were to be found therein — cities with their suburbs, palaces, houses, banners, pennants, bells, drums, beautiful girls, souls of men, clouds, red glows, mists, the Milky Way, birds, cattle, dragons, and horses. Gazing at this picture makes one feel a longing to travel away into space and forget the world of men. Verily it is a marvelous work."[2] Such dream pictures describing scenes in the beyond were numerous, or rather the things and spirits encountered in the other world were revealed in dreams. Another famous subject of this kind is the emperor Ming Huang's aerial journey into the palace of the moon. In general it may be asserted that all notions of a supernatural world and the beliefs in the immortality of the soul have largely been inspired and influenced by visions and dreams.

Both in India and China we must discriminate between two types of inspirational dream — the real one or subjectively true one and the dream as a purely literary motive or pattern. The latter, naturally, is a later development which has grown out of the former. The development is the same as in art where conventionalized motives go back to spontaneous or realistic ones or where copies and copies of copies are traceable to one original prototype. India and China offer hundreds of examples of the literary dream pattern. In the 550 Jātakas or birth-stories of the Buddha, when he still was a Bodhisatva and appeared in all sorts of animal incarnations, there are at least a dozen stories in which a queen or princess dreams of a golden peacock, a golden stag, a golden deer, or a six-tusked elephant able to preach the law of salvation; she falls ill from yearning for this wonder of nature until the animal is captured by order of the king after many strenuous efforts and queen and king are ultimately converted to Buddhism by the Bodhisatva. In the legendary biography of Buddha nearly every important event is accompanied by a prophetic dream. In a similar manner the introduction of Buddhism into China is ascribed by tradition to a dream of the emperor Ming of the Han dynasty (A.D. 61) in which a golden image appeared to him soaring in the air above his palace. This statue was interpreted by the emperor's brother as that of Buddha, and the dream resulted in an embassy being sent to India for teachers and scriptures. This is merely a legend invented at a later time by the clergy for the edification of the pious;

---

[1] For reproductions of Kwan Hiu's paintings see Laufer, *T'ang, Sung and Yüan Paintings,* plates VII—VIII.
[2] Giles, Introd. to History of Chinese Pictorial Art, 2d ed., p. 136.

Buddhism, in fact, was known in China long before this alleged dream episode.

The distinction here made, *mutatis mutandis*, is also traceable among the Plains Indians, save that the historical connections are lost here. According to Wissler, the putative dream designs of the Dakota do not differ in principle from other designs, and a dream design is not so much a distinct type of design as an illustration of the manner in which Dakota philosophy accounts for the origin of the present styles of decorative art.[1] In my opinion there must have been in the past a group of Dakota designs inspired by dream notions which were handed down from generation to generation; the decorative motives may have changed in the course of time, while the original dream story persisted or evolved into a mere pattern.

Inspirational dreams occasionally occur in European folklore also; for instance, in the cycle known as "the dream of the treasure on the bridge." A fishmonger dreams that he will find his fortune under a certain bridge, but when he betakes himself to the spot, he finds there only a beggar in tatters. Disgusted, he is about to get away when the beggar reveals to him a spot where he will find a hidden treasure that will make him a rich man.[2] This treasure motive also occurs in China, and I will conclude with a Chinese dream story that at the same time contains an amusingly correct explanation of dream phenomena. It is a monologue recited on the stage in a fast tempo to the accompaniment of castanettes.[3] "I want to tell you a fine story. It is regrettable how unjust Heaven is; he sends rain and snow down upon us, but no lumps of silver. Last night I lay on my bed, tossing around sleeplessly. I was awake from the first till the second nightwatch, and again from the second till the third was beaten. Then I saw a dream in my sleep. I dreamt of a trasure buried to the south of the village. I therefore seized spade and hoe and went out into the field to dig for the treasure. I really was in luck; after a few blows with spade and hoe I hit upon the treasure. It was a whole cellar full of silver ingots, wrapped in a large rush-mat. I lifted the mat and looked beneath it. Ah, I burst into laughter: there were a coral tree thirty feet high, genuine red carnelian and white agate. I picked up seven or eight sacks full of diamond-points, six large baskets full of cat's-eyes, thirty-three striking clocks, sixty-four ladies' watches, fine boots and caps, handsome garments, fine fashionable hand-bags, seventy-two large gold ingots, and in addition thirty-three thousand, three hundred and thirty-three silver shoes inlaid with enamel work. Now I had so much gold and silver that I didn't know what to do with it. Should I buy land for it and till the

---

[1] Cf. R. Lowie, Primitive Religion, p. 289.

[2] F. Liebrecht, Zur Volkskunde, p. 93.

[3] W. Grube, Chinesische Schattenspiele, p. 440.

ground? I was afraid of drought and inundations. Or should I go into the grain business? But the rats might eat up my whole stock. Or should I lend money on interest? But I lacked bail. Or should I open a pawnshop? I feared that I might lose money; for if the manager would run away with the capital, where should I look for him? Pondering over these thousand difficulties I got so agitated that I awoke with excitement, and lo and behold, it was merely a dream! Both my hands had fumbled with the bed and had caught the tinder-box with flint and steel: this was the silver shoes of my dream. Then I had seized the brass tobacco-pipe: this was the ingots of gold. After groping around in the dark for a while, I hit upon a green-headed large scorpion which stung me so that I screamed and yelled with all my might."

207

飞行的历史背景

The

OPEN COURT

Devoted to the Science of Religion,
the Religion of Science, and the Extension of the Religious Parliament Idea

FOUNDED BY EDWARD C. HEGELER

AUGUST, 1931

VOLUME XLV   NUMBER 903

Price 20 Cents

The Open Court Publishing Company

Wieboldt Hall, 339 East Chicago Avenue
Chicago, Illinois

THE PREHISTORY OF AVIATION
BY BERTHOLD LAUFER

THE desire to fly is as old as mankind; in all ages man's imagination has been stirred by the sight of flying birds and seized
by the ambition to sail upon the wind like one of them. There is a
long record of ventures, experiments and failures, and the romance
of flying still remains one of the most fascinating in the history of
mankind.

It is to man's ingrained love for the fabulous, for the wondrous
and extraordinary, to which we are indebted for the preservation of
ancient records of flight. The prehistory of mechanical science is
shrouded in mystery because primitive man was unable to render
an intelligent account of it. Just as natural phenomena were regarded by him as wonders wrought by supernatural agencies, so
any mechanical devices were interpreted as witchcraft. Every investigator and skilled artificer of prehistoric and early historic days
has gone down in history as an enchanter or wizard who had made
a pact with demoniacal powers. Many of the so-called magicians
were simply clever mechanics whose work was beyond the com-
prehension of their contemporaries and whose achievements were
so singular and awe-inspiring that they were believed to have been
inspired by supernatural forces. This is the reason that those who
made attempts at aerial flights were so often associated with magic
and necromantic art and why in our middle ages solely witches and
devils were endowed with the faculty of flying.

Ancient traditions regarding mechanical wonders must, therefore, be divested of their legendary garb and exposed in their historical nucleus, but we owe to them the preservation of many re-

Extract from Publication 253, Volume XVIII, No. 1, Anthropological
Series, Field Museum of Natural History, Chicago.

cords, for the dry and bare bones of historical events are apt to be relegated to the waste basket.

The imaginative faculty of the human mind does not conceive things that have no reality in existence. The product of our imagination is always elicited by something that we have at least reason to believe exists. The question is : if these myths exist, how did they arise, and what germ of fact lies behind them.

In the same manner that astrology was the precursor of astronomy and alchemy evolved into the science of chemistry, so there is an abundance of lore which godfathers the history of aviation. To distinguish that primeval stage from the present accomplished fact we will simply speak of it as the prehistory of aviation and we will show that our modern progress is not due solely to the efforts of the present generation.

The Romance of Flying in China

At the threshold of the earliest recorded history of China an imperial flyer[1] appears, the emperor Shun who lived in the third millennium before our era : and he is not only the first flyer recorded in history but also the very first who made a successful descent in a parachute.

Shun's early life teemed with thrilling adventures. His mother died when he was quite young, and his father, Ku Sou, took a second wife by whom he had a son. He grew very fond of this son and gradually conceived a dislike for Shun which resulted in several conspiracies against the poor youngster's life. In spite of this, however, Shun continued in exemplary conduct towards his father and step-mother. His filial piety attracted the attention of the wise and worthy emperor, Yao. Yao had two daughters who instructed Shun in the art of flying like a bird. In the commentary to the annals of the Bamboo Books, Shun is described as a flyer. Se-ma Ts'ien has preserved the following tradition. "Ku Sou bade his son, Shun, build a granary and ascend it, and thereupon set the structure on fire. Shun who stood on top of the tower, spread out two large reed hats which he used as a parachute in making his descent and landed on the ground unscathed." Considering the fact that Chinese reed hats are umbrella-shaped, circular

[1] Bladud, the legendary tenth king of Britain, is said to have made wings of feathers by means of which he attempted an aerial flight which resulted in his death in 852 B. C.

and very large (two to three feet in diameter) this feat would not seem impossible.[2] Shun later married the two sisters, and their father gave him a share in the government.

Winged flight, however, seldom appears as a real attempt. The emperor Shun is practically the sole example and seems to have found few imitators.

Chinese writers fable about a country of flying folk, Yü Min, located on an island in the southeastern ocean, a people with long jaws, bird-beaks, red eyes and white heads, covered with hair and feathers resembling human beings, but born from eggs.

The conception of bird-men is quite familiar to Chinese mythology and is often represented in Chinese art. Lei Kung, the god of thunder and lightning, has wings attached to his shoulders (usually those of a bat) by means of which he flies to wherever he wishes to produce a thunderstorm.

The first description of an air journey is found in a poem by Kü Yüan, who, having lost his position as statesman by the intrigues of his rivals, found solace from his disgrace by writing. In his poem, he surveys the earth to its four extreme points, travels all over the sky, then descends again in a flying chariot drawn by dragons.

This idea is not alien to Chinese art. An aerial contest between a dragon chariot and winged beings astride scaly and horned dragons is represented on a gravestone of the Han period (second century A.D.)

Huang Ti, one of the ancient legendary emperors, attained immortality by mounting a long-bearded dragon, strong enough to transport his wives also and ministers—more than seventy persons. The officials of lower rank who were not able to find a seat on the dragon's back clung to the hairs of the dragon's beard, like strap hangers in the street cars. These, however, gave way, and the passengers were plunged to the ground, and also dropped the emperor's bow. The multitude of spectators reverentially watched the apotheosis and when Huang Ti had reached his destination, they picked up the hairs and his bow.

When the imagination of a nation is filled with the romance of air, when the very air is populated with winged genii and flying chariots, and when such subjects are glorified by art, it is the logi-

[2]Leonardo da Vinci was the first in our midst to conceive the idea of a parachute.

AERIAL CONTEST OF DRAGON-CHARIOT AND DRAGON-RIDERS
Stone Bas-relief of Han Period, A.D. 147. Shan-tung, China

cal step that imagination leads one or another to attempt the construction of some kind of an airship.

The history of the ancient Emperors, the *Ti wang shi ki*, contains the following notice: "Ki-kung-shi was able to make a flying
chariot which, driven by a fair wind, traveled a great distance. At
the time of the emperor Ch'eng T'ang the west wind blew Ki-kung-
shi's chariot as far as Yü-chou. The emperor ordered this chariot
to be destroyed so that it should not become known to the people.
Ten years later, when the east wind blew, the emperor caused another chariot to be built by Ki-kung and sent him back in it."

The term "flying chariot" (*fei ch'o*) used in this passage is
now current in China to designate an aeroplane.

Another account ascribes this invention to the Ki-kung nation,
who are one-armed, three-eyed hermaphrodites. Most likely two
distinct legends have here become contaminated.

A wood engraving of Ki-kung's chariot of comparatively recent origin reconstructed from the slender fabric of the ancient
tradition is reproduced here. The Chinese draughtsman is decidedly wrong about producing a two-wheeled chariot as the sole indication of motive power given in the account itself is the wind.
In ancient China only two devices were known to set a vehicle in
motion, namely, a sail and a kite. A sail alone cannot lift a vehicle
into the air, but this can be accomplished by several powerful kites.
Therefore, Ki-kung's chariot was probably built on the aerostatic
principle, being driven by a combination of sails and kites.

Possibly the chariot was similar to the aerial boat designed by
Francesco Lana which was to be lifted by four copper globes from

AERIAL CONTEST OF DRAGON-CHARIOT AND DRAGON-RIDERS

which all the air had been extracted. The boat is then propelled by oars and sails.

Kung-shu Tse, a contemporary of Confucius, also called Lu Pan, is said to have carved a magpie from bamboo and wood; when completed he caused it to fly, and only after three days did it come down to earth. According to another tradition, Kung-shu made an ascent on a wooden kite in order to spy on a city which he desired to capture. This invention is sometimes ascribed to Mo Ti, and a great deal of confusion surrounds the accounts. As early as the first century of our era, real knowledge of this contrivance was lost.

This wooden bird and its affinity, the dove of Archytas, meet with a curious parallel in the west. The astronomer, Regiomontanus, who lived in Nurenberg in the 15th century is said to have constructed an eagle which he sent out high in the air to meet the emperor and accompanied him to the city gates. Considering the fact that such similar contrivances are reported from different parts of the world at widely varying times we cannot help concluding that a grain of truth must underlie these accounts, even though we grant that they are exaggerated. Perhaps Kung-shu's bird was a glider, or perhaps it was attached to or raised by a kite.

Starting from realistic means of flight, Chinese efforts developed into mysticism and magic. In the second century B. C. alchemistic lore began to infiltrate from the west. The notion of flight was a link of paramount importance in the chain of mystic dreams which held the people enthralled for many centuries. Alchemists sought the elixir of life, people ascended to heaven upon drinking concoctions, or upon the back of cranes, of ducks or tigers.

In this later history two singular ideas come to the fore: levitation by means of starvation and by means of remedies taken internally—live on air to conquer the air. These doctrines and practices of Taoism are partially traceable to India.

T'ao Hung-king, a distinguished physician and adept in the mysteries of Taoism, compounded a "flying elixir" of gold, cinnabar, azurite and sulphur. It was said to have the color of hoar frost and snow and to have a bitter taste. When swallowed it produced levitation of the body. It is the only example in the history of the world of teaching to fly by means of medicine taken internally.

KITES AS PRECURSORS TO AEROPLANES

Kites were first invented and put to a practical test in ancient China. The toy we used to fly in our boyhood days is but a poor degenerate orphan compared to the Chinese kites which are wonders of technique and art. The ordinary Chinese kites are made of a light framework of bamboo over which is spread a sheet of strong paper, painted in brilliant hues with human or animal figures. The figures are designed for a distant vista and may seem, at a close proximity, distorted, but from a distance appear most beautiful, and waving and soaring as the kite moves on like a real bird. They are maintained by a long tough cord wound over a reel which is held in the hand and is continually turned as the paper plane rises or falls. The most complicated one of these is the centipede kite. One in the American Natural History Museum in New York measures 40 ft. in length. Mechanically kites are constructed on the principle underlying the behavior of a soaring bird, which performs its movements with peculiar warped and curved surfaces.

The ninth day of the ninth month in the autumn is devoted to the festival called Ch'ung-yang. Friends join for a picnic in the hills and set kites adrift. This also is the day for holding kite contests. The cord near the kite is stiffened with cut glass. The kite-flyer manoeuvres to get his kite to windward of that of his rival, allows his cord to drift against that of his rival, and by a sudden jerk cuts it through, so that the hostile kite is brought down.

A musical kite was invented in the tenth century by Li Ye who fastened a bamboo flute to the kite's head. Sometimes two or three flutes are attached one above the other, more frequently, however, a musical bow made of light willow-wood or bamboo, and strung with a silken cord is attached to the kites.

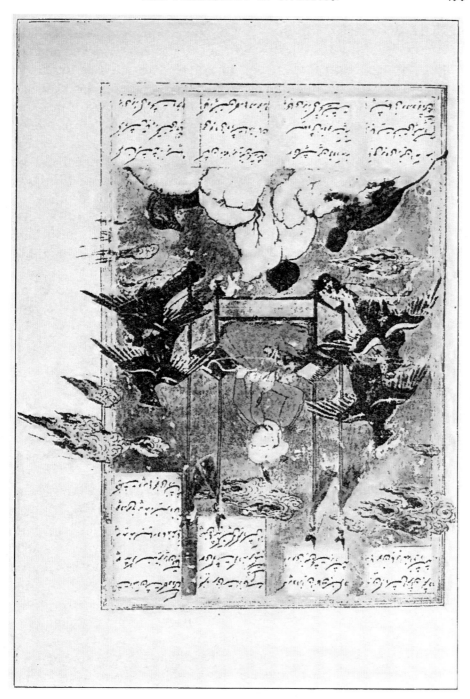

KAI KAWUS FLIGHT TO HEAVEN
From a Persian Illustrated Manuscript of the Shahnameh, Dated 1587-88
Courtesy of Metropolitan Museum of Art, New York
Courtesy of Field Museum

Kites were originally used for military signaling. The idea that a kite drives away bad spirits is of local and recent development— found more in Korea than in China and bears no relation to the origin of kites, and cannot be clearly traced. They seem not to have existed in times of early antiquity, and therefore they are not mentioned in the treatise on the art of war by Sun Wu in the 6th century B. C.

Kung-shu's wooden bird was not a flying kite. The earliest notion of this looms up in the life of Han Sin who died in 196 B. C., one of the three heroes who assisted Lu Pang in ascending the throne as the first emperor of the Han dynasty. He wanted to dig a tunnel to the palace and in order to measure the distance he is said to have flown a kite. Some say he measured the cord, others say that he ascended the kite, but it is most probable that he introduced kites into warfare using them in trigonometrical calculations of the distance from the hostile army. The story is however not well authenticated for it appears only in comparatively late sources and Han Sin's kite is said to have been made of paper, while paper was invented only 300 years later.

Chinese authors are wont to speak of paper kites. Paper was invented in A. D. 105. Ever since paper has come into use, kites have been made of this material, and no other material has been used for them. But the framework might have been covered by some other light material, silk or hemp. Chinese records, however, are reticent on this point.

From China knowledge of kites was diffused to all other nations of Eastern Asia, that experienced the influence of Chinese civilization such as Korea, Japan, and nearby countries. In some parts of Indonesia, kites are put to the practical purpose of catching fish. Kites were introduced into India through Malay or Chinese immigrants. Kite-flying is a popular amusement in the spring and contests are held for high stakes.

In Siam, kite flying is a state ceremony, as well as a public festivity connected with agriculture and the northeast monsoon.

All data at our disposal goes to prove that the kite spread from the far east westward, to the near east and finally to Europe and makes its debut there as a Chinese contrivance.

In European literature kites are first described by the Italian Giovanni Batista in his book on natural magic, and the Jesuit Athan-

asius Kircher who also wrote a book on China which is based on information received from members of his order working in China.
Kites were flown in England as a pastime. In the middle of the seventeenth century they were employed for the purpose of letting off
fireworks. They were finally used in Europe by Alexander Wilson and in the United States by Benjamin Franklin for scientific
purposes in making temperature and electrical experiments. The
classical experiment of Benjamin Franklin which identified lightning
with electricity is, of course, well known.

Both in China and Japan there are stories current about men
riding on kites. Athanasius Kircher mentions that in his time kites
were made of such dimensions that they were capable of lifting a
man.

About the year 1826, the principle of the kite was turned to a
practical purpose by George Pocock a schoolmaster of Bristol, who
found that by attaching several kites one beneath another they
could be elevated above the clouds. In January the following year
he claimed to have covered several miles between Bristol and Mal-
borough at twenty miles an hour. He proposed to use kites for
shipwrecking and to tow boats, and for military purposes to elevate
a man for reconnaissances and signaling.

In 1876, Joseph Simmons claims that he was drawn into the
air to a height of 600 feet by means of two superimposed kites and
adjusting his weight by guy lines to the earth. Others have also
reported such success.

Laurence Hargrave, an Australian, introduced a new principle,
the cellular construction of kites. This type of box kite formed
the starting point of Alexander Graham Bell's researches and construction of tetrahedral and triangular kites. The wings of the
modern biplane are closely modeled after the Hargrave box kite.
The man-lifting kite has developed into an aeroplane. The speed
plane of our day is but a first cousin to the kite.

Another Chinese apparatus deserves mention here as it served
as a source of inspiration to Sir George Cayley one of the great
pioneers of modern aviation. He says that his first experiments
were made with a Chinese aerial top which served at once to illustrate the principle of the helicopter and air-screw. Though
but a toy a few inches long, its capacity to demonstrate certain
principles in aeronautics made a lasting impression on his youthful mind.

THE DAWN OF AIRSHIPS IN ANCIENT INDIA

Although the Aryan Indians of the Vedic period had numerous aerial deities, such as the Gandharvas, elves "haunting the fathomless spaces of air," no allusion is made in the Rigveda to their manner of locomotion. The Vedic gods did not fly, but preferred driving in luminous cars drawn by fleet horses, cows, goats or spotted deer. Indra, the favorite national god, primarily a storm and thunder god, is borne in a golden chariot drawn by tawny chargers as an eagle is borne on its wings, faster than thought.

A myth of post-Vedic times tells of quaking mountains with wings gifted with the power of flight. They flew around like birds, alighted wherever they pleased and with their incessant motion made the earth unsteady. With his thunderbolt, Indra clipped their wings and settled them permanently in their places; their wings were transformed into thunder clouds.

The Açvine (horsemen), the twin dieties, probably representing the dawn and the morning star, traverse heaven and earth in a single day, drawn in a sun-like chariot by horses or birds or swans or eagles. OtherVedic gods, Surya, the sun god, Agni, the personification of the sacrificial fire, drive in chariots or are represented as birds. Pushan, who is closely connected with the sun, moves in golden ships sailing over the aerial ocean. The sun on one hand appears as a boat in which Varuna, the god of the sky, navigated the aerial sea, and on the other hand as a chariot with Varuna as the charioteer. This conception arose from the experience of seeing the sun set in the sea.

The Maruts, the gods of the winds, are described as having yolked the winds as steeds to their pole: that is, their chariot is driven by winds.

In post-Vedic literature the Indians profess to have had two distinct types of flying machines, the Garuda airship of native manufacture constructed on the principle of bird-flight, and the Yavana airship ascribed to the Greeks whose manufacture was scrupulously guarded as a secret. Whether the ancient Indians ever really navigated the air or whether their dirigibles are fiction is irrelevant. The main thing is they had the idea, and their ideas about aeronautics were not worse or more defective than those of Europe from the 16th century to the first part of the 19th century. They saw two points clearly—that aircraft must operate on the principle of the flight of birds and that a mechanism is required to start the machine, to keep it in midair, and to make it descend. They devoted considerable thought to problems of the air and efforts were made to construct aircraft of various types.

The Greek records are silent as to aircraft, so that we do not know whether the Greeks really, as asserted, did supply them with flying machines. Certainly Greek mechanists and artisans enjoyed a high reputation in India, and marvelous inventions were ascribed to them, such as marvelous automata, movable figures of beautiful women.

The vehicle of the god Vishnu is Garuda, a celestial bird originally a solar bird. This mythological conception proved very fertile in stimulating imagination and according to Indian stories led to construction of airships, and attempts at flying.

The most popular collection of Indian folk-lore contains the story of the weaver as Vishnu. A weaver became infatuated with the king's daughter. His friend, a carpenter, made a wooden airship for him in the shape of a Garuda, which was set in motion by a switch or spring. Equipped with all the paraphernalia of the god, he flew to the seventh story of the palace where the princess had her apartment. She took him for Vishnu, and he married her according to the rites of the Gandharvas (by mutual consent). To her father's questions, she said she was the consort of a god. The king thereupon became overbearing to his neighbors who made war upon him. He implored the pseudo-Vishnu for help through his daughter. He accordingly appeared above the battlefield with bow and arrow, ready to die. But Vishnu, not wishing his authority to suffer among men, as it would if he allowed the weaver to die, entered his body and scattered the enemy. After the victory the weaver told the whole story to the king, who rewarded him and married him to his daughter. The most interesting point to this story is that a garuda is used to rout an enemy.

Another garuda is described in a collection of old Indian stories. The wife of a rich man's son is stolen, and a carpenter's son, to rescue her, builds a wooden Garuda. It is supplied with three springs, one in front to make it go upward, on the side to make it float smoothly along, and one beneath to make it descend.

Again reference to airships is found in a collection of stories written during the 11th century. Vasavadatta desired to mount an aerial chariot and visit the earth. The carpenters who were summoned said that flying machines were known only to the Greeks. Later in the same story, Viçvila makes an aerial journey on a mechanical cock, but says that the secret should be revealed to no one but a Greek. Pukvasaka, his father-in-law, commanded him to build a flying machine, but Viçvila who had learned the secret from the Greek artisan not daring to reveal it, fled with his wife during the night on the cock, to the country whence he had come. The artisans were flogged; meanwhile a stranger appeared who said, "Do not flog the artisans; I will build a flying machine." In the nick of time he produced a flying machine in the form of a Garuda. When the Queen refused to mount it alone, the stranger said it could carry the entire city. So the king and his personnel, his wives and officials set out and circumnavigated the earth. On his return, he did honors to the craftsman.

In a Sanskrit romance of the 7th century, a king, desirous of marvels, was carried away, no one knows whither, on an aerial car made by a Greek who had been taken prisoner. The term used in this passage means "a mechanical vehicle which travels on the surface of the air."

As regards winged flight, only one example is known to me from Indian literature. The Katha Sarit Sagara contains the following tale: A young Brahman, having seen the prince of the Siddhas flying through the air, wished to rival him, and fastening wings of grass to his side and continually leaping up he tried to learn to fly. The prince took pity on the boy who was making such an earnest effort, and by means of his magic power took the boy on his shoulder and made him one of his followers.

In Indian art, particularly in the sculpture of the Buddhists, winged beings in the act of flying are frequently represented.

Among the marvelous abilities promised as a reward for yoga practice was "traversing the air." What has been observed as flying by modern yogins proved to be hopping close to the surface of the ground without seemingly touching it.

More interesting, however, are two charming motifs of folklore presented by India to the world, magic boots and the enchanted flying horse.

FROM BABYLON AND PERSIA TO THE GREEKS AND THE ARABS

From the Euphrates Valley large fragments have been recovered of a legend of the sovereign Etana who, as a reward for having helped a wounded eagle, is carried on his back to the dwelling of the gods. They reach the heaven of Anu and halt at the gate of the ecliptic. The eagle is next urging Etana on to the dwelling place of Ishtar, six hours distant, but either his strength is exhausted or the goddess intervenes, for a precipitous descent begins. They fall through space three double hours and finally reach the ground. The close of the story is wanting, but the purpose of the flight has failed.

This is the only record of flight recorded in cuneiform literature. Although it is found in Babylon and several different cylinder seals illustrate the legend, it is thought to be of Iranian or possibly Aryan origin.

There is an ancient Persian tradition of especial interest which was transmitted to Europe at an early date. In the semi-legendary history of Iran, there was a king, Kai Kawus, who was easily led astray by passion. He built seven palaces on Mt. Alburz, then he tried to restrain the demons of Mazandaran, one of which retaliated and sowed the seeds of discontent in his heart, so that he set his mind on attaining supremacy in the celestial abode. He built a throne, supported and raised by four starving eagles. As an incentive for the birds to fly, four pieces of flesh were fastened to the top of four spears planted on the sides of the throne. The flight was of short duration; the strange vehicle soon came down in a crash and the grandees found the king unconscious in a forest.

The Iranian motif of an aerial vehicle lifted by starved eagles was adopted by the Greek Romances of Alexander the Great which became widely known throughout the middle ages.

Of all the flying stories of classical antiquity the one that has left the most lasting impression and inspired the greatest number of imitators is that of Daedalus (Cunning Worker). He incurred the wrath of king Minus and, in order to escape imprisonment, fashioned a pair of artificial wings coated with wax for himself and his son. They mounted and flew westward over the sea. Icarus, however, disregarded his father's advice and flew too near the sun; the wax on his wings softened and melted, and he fell headlong into the sea.

It does not matter if the story is or is not true. It is the flight of human imagination, the impulses and visions of a genius, very often his errors, which have stimulated inventions and progress.

The Daedalus story finds an echo in the Germanic saga of Wayland the Smith, the artificer of marvelous weapons. King Nidung endeavored to keep him in his service by cutting the sinews of his feet, thus laming him forever. Wayland forged a feather robe and revealed his purpose to the king from the tower of the castle and flew home to Seeland.

The most notable of the Greek gods and goddesses who flew through space were Perseus and Hermes with winged helmet and shoes. Fantastic conveyances were used on the Greek stage to give the illusion of persons being lifted upward or descending from the air.

Archytas, a Greek philosopher, mathematician and statesman, who lived in Italy in 428 B.C. attained great skill as a practical mehanician. His flying dove of wood was one of the wonders of antiquity. From the accounts we have, it is not clear just what it was. It is described as being a wooden figure balanced by a weight that was suspended from a pulley. It is said to have soared in the air and been put into motion by a current of air "hidden and enclosed" in its interior. Some scholars incline to the opinion that it was an anticipation of the hot air balloon, others that it was an aerostat or glider, for it is said that it could fly but not rise again after falling. It may also have been on the order of Lu Pan's wooden kite.

Lucian, the delightful satirist and divine liar of the second century of our era, tells of an air voyage where the flyer, Menippus, goes Daedalus one better by refraining from the use of wax. He fastened an eagle's and a vulture's wing to each side by straps with handles for grips. Thus he essayed to fly, at first leaping and flapping, keeping close to the ground as geese do, later becoming bold enough to fly to Olympus, and to the moon. This story gave the impetus to the class of fiction known as "voyages imaginaires."

Such a voyage is described by Francis Godwin in his romanc "The Man in the Moone." His hero, Gonzales was abandoned on an uninhabited island, St. Helena. He trained a flock of birds to fly together bearing a burden. Then he devised a mechanism whereby he could distribute his weight at the start of the flight. At first

THE AERIAL VOYAGE OF DOMINGO GONSALES
From F. Godwin's Man in the Moone, 1638

Courtesy of Field Museum

he experimented with a lamb, then he was himself carried aloft. "For I hold it far more honor to have been the first flying man than to be another Neptune that first adventured to sail upon the sea." This sentiment, "to be the flying man," finds its earliest expression here.

The Arabs, heirs to Greek philosophy, and science, made considerable progress in mechanical devices. About the year 875, an Arabian, known as the Sage of Spain, who was the first to manufacture glass, invented a contrivance to make his body rise into the air. He made wings, clothed himself with feathers and flew quite a distance, but as he had not considered what would happen during his descent, he fell and injured his buttocks. He was ignorant, the Arabic chronicler adds, that a bird falls only on its rump, and had forgotten to make a tail for himself.

There is another story of a flying architect from the tenth century, who erected a huge tower for King Shapur I. The king, not wanting anyone else to profit by his genius, left him on the top of the tower. The architect built a pair of wooden wings, fastened them to his body, and driven by the wind, flew to a place of safety. This story bears a remarkable resemblance to the Daedalus story.

In Constantinople, at the festivities held in honor of a visiting Sultan, a Saracen wanted to show his skill in flying. He announced he would fly from the tower of the hippodrome across the racecourse. He was clad in white garments, large and wide, braced with rods of willow-wood laid over a framework. He delayed for a long time and the crowd became impatient; but finally, when the wind was favorable, he soared like a bird and seemed to fly in the air.

Oliver of Malmesbury, an English astrologer of the eleventh century, is said to have attached wings to his hands and feet and attempted to fly off from a tower. He attributed his fall to the lack of a tail. This bears a striking resemblance to the Arabic story above mentioned.

John Damion, an Italian by birth and a physician at the court of King James, claimed he could overtake an embassy to France. He fastened wings of bird feathers on himself and hopped off the top of Stirling Castle, but he fell and broke his legs. He blamed his misfortune on the fact that there were some chicken feathers in his wing which showed a natural affinity to return to the barnyard.

Giovanni Battista Danti, a mathematician of Perugia, is said to have attempted winged flights over the lake Trasimeno.

Roger Bacon was to some extent under the influence of Arabic science. He had all the superstitions of his contemporaries in regard to flying. He suggested that flying machines could be made so that a man "seated in the midst of the machine, revolving some sort of device by means of which wings artificially composed may beat the air after the manner of a flying bird." Bacon's place is at the end of the line in the prehistory of aviation. His ideas of flying are the echo of the ancient idea that we have traced from China and India, Persia and Arabia.

The modern history of aviation begins with Leonardo da Vinci.

THE AIR MAIL OF ANCIENT TIMES

Air-mail service was first established in the United States in 1918 when the New York-Washington mail route (218 miles) was established. While our air-mail is an epoch-making innovation and an achievement of modern times, there was also a prehistoric air-mail which is no less admirable, carried on the wings of pigeons. This institution we owe also to the Orient.

The first Chinese who made use of carrier pigeons is Chang Kiu-ling (A.D. 673-740) a statesman and poet, who corresponded with his relatives by means of a flock of carrier pigeons, which he called his flying slaves (fei nu). The messages were attached to the feet of the birds who were taught how to deliver them. The government of China never employed pigeons for carrying important messages, but their use remained restricted to private correspondence chiefly for bringing news of the arrival of cargoes and the ruling prices of markets.

In India the use of carrier pigeons goes back to great antiquity and may with certainty be assumed to have been in full swing in the beginning of our era. Kings of India received news about the movement of hostile troops by domesticated pigeons. In Indian stories various kinds of birds appear as harbingers of messages, the white wild goose, for instance, the crow, and frequently parrots.

As regards Persia, many pigeons were kept on their sea going vessels, capable of flying several thousand li (Chinese miles). These were released and they returned home bearing tidings as it were that everything on board was well. In medieval times Persian authors repeatedly refer to the conveyance of letters by pigeon mail. The pigeon also appears in love songs as messenger and bearer of love letters.

The use among the Greeks and Romans of carrier pigeons is restricted to isolated instances were news is carried of victory in the Olympian games or to a besieged city. Since there is no mention made of their being trained for message bearing, it was probably of no great significance among the ancients and probably died out during the days of the decline of the Roman Empire.

Mesopotamia appears to be the home of the domesticated pigeon, and the domestication of the bird was accomplished as early as pre-Semitic times by the Sumerians. Among the Semites, pigeons are closely connected with religious practices. They are sacred to the goddess Ishtar. It is unknown when and where pigeons were first trained for conveying messages. Both in Egypt and Mesopotamia the practice was unknown, but it is improbable that the practice could have developed where clay tablets were the common writing material.

The dove which Noah sent from the ark three times represents an entirely distinct class in the category of land-spying birds which navigators released when they had lost their bearings and were in quest of land. These birds never returned to their ships.

In the 9th century when the Vikings sailed from Norway, they kept birds on board which were set free from time to time amid sea, and with their aid, succeeded in discovering Iceland. Land expeditions would also be accompanied by land-spying birds and would settle in a territory where the birds would descend.

In the present state of our knowledge we can only assert with safety that the highest development in the use of pigeon messengers was reached in the empire of the Caliphs and under the Mohammedan dynasties of Egypt, where the whole business was organized and systematized on a scientific basis, while of course, isolated cases occured many centuries earlier. Indo-Iranian peoples may very well have given the first impetus to the training of carrier pigeons. Under the Caliph, Nūr-ed-din a regular air mail was established. Pigeons were kept in all castles and fortresses of his empire. Under the Caliph Ahmed Naser-lidin-allah air mail developed into a regular institution. Although many were engaged in the business of raising pigeons, their prices reached amazing figures. A well trained pair sold as high as 1000 gold pieces. Bagdad was the central station for air-mail until it was conquered by the Mongols in 1258.

One of the most curious incidents in the history of airmail was when the Caliph Aziz (975-996) had a great desire for a dish of cherries from Balbek. His Vezir caused 600 pigeons to be despatched from Balbek to Cairo each of which carried attached to either leg a small silk bag containing a cherry. This is the first record of parcel post by air-mail.

Stanley Lane Poole, in his "History of Egypt in the Middle-Ages," writes of Beybars (1266-77) "the most famous and energetic of all the Bahri Mamluks, that he established a well-organized system of posts, including the pigeon post. . . . The pigeons were kept in cots in the citadel and at various stages which were farther apart than those of the horses. The bird would stop at the first post-cot where its letter would be attached to the wing of another pigeon for the next stage. The royal pigeons had a distinguishing mark and when one of these arrived at the citadel with a dispatch, none was permitted to detach the parchment save the Sultan himself ; and so stringent were the rules that were he dining or sleeping or in the bath, he would nevertheless be informed at once of the arrival and would immediately proceed to disencumber the bird of its message." The letters were written on a fine tissue paper and were fastened beneath the wings and later to the tail feather.

During the middle ages the European nations became acquainted with pigeon air-mail when the cross and the crescent clashed during the Crusades. There are stories on record which depict the wonder and amazement of the Christian soldiers at witnessing this novel experience. They brought carrier pigeons back from the Orient. Medieval knights used them in sending communications from castle to castle. The convents did so also. All pigeons used in medieval Europe for air-mail purposes were of Oriental origin.

The first employment of pigeons for military purposes in Europe was during the siege of Harlem by the Spaniards in 1573. The garrison received advices by pigeon mail announcing the approach of a relief army under the Prince of Orania.

It is said that Rotchschild of London had his agents join Napoleon's army and send him first hand information by air-mail.

whereby he managed his financial speculations. Reuter started his career by organizing a pigeon post from Aix-la-Chapelle to Brussels. A newspaper reporter equipped with a small pigeon cage was not a rare sight. During the siege of Paris in 1870 the only news from the outside world that reached the city was conveyed by the wings of pigeons. In the world war pigeons were extensively utilized and achieved brilliant records of flight under great difficulties.

Pigeons are still bred and kept in large numbers for messenger service and racing. In good weather young birds will fly about 300 miles in seven to nine hours and flights of 600 miles in one day have been accomplished by older birds.

KI-KUNG'S FLYING CHARIOT
Chinese Woodcut from T'u shu tsi ch'eng

Courtesy of Field Museum

208

新几内亚的烟草

American Anthropologist

NEW SERIES

ORGAN OF THE AMERICAN ANTHROPOLOGICAL ASSOCIA-
TION, THE ANTHROPOLOGICAL SOCIETY OF
WASHINGTON, AND THE AMERICAN
ETHNOLOGICAL SOCIETY OF
NEW YORK

ROBERT H. LOWIE, *Editor*, Berkeley, California
FRANK G. SPECK and E. W. GIFFORD, *Associate Editors*

VOLUME 33

MENASHA, WISCONSIN, U. S. A.

PUBLISHED FOR

THE AMERICAN ANTHROPOLOGICAL ASSOCIATION

1931

The method of using the two plants is the same. Speaking of pitchuri, he says,

Die Zubereitung der Prieme gleicht ganz der aus Tabak: zunächst erwärmt man die zerkauten Blätter und vermengt sie dann mit Akazienasche.

Although Dr. Hartwich[15] regards the New Guinea tobacco as of American origin, of Australia he says:

Australien muss uns dabei von grösstem Interesse sein, weil es derjenige Erdteil ist, in welchem die Menschen vor Aukunft der Europäer zweifellos Tabak benutzten und zwar einheimischen, der also nicht auf irgendeine Weise von America gekommen war.

And yet Dr. Merrill, apparently basing his statement solely on an opinion of Professor A. R. Radcliffe Brown, says categorically that "the aborigines in Australia made no use of the Australian native *Nicotiana*," and he uses this "fact," as he calls it, to help prove how highly improbable a local discovery of smoking in New Guinea would be!

The somewhat polemic character of the discussion so far is because it seems to me that Dr. Merrill has not correctly represented either my own paper or the problem as a whole. Any new evidence is welcomed by no one more than by myself. The question is not yet decided, however. There are too many unknown regions and unidentified tobacco plants in New Guinea, New Britain, and Bougainville, and the difficulties of cultural diffusion too great, to admit of guesswork. It is the calm assumption by nearly every one, including the ethnologists, that all cultivated tobacco must have come from America, that prevents the acquiring of real evidence. Dr. Merrill's paper is an illustration of this.

If the cultivation and use of tobacco does owe its origin to America, its spread, as shown in New Guinea and the other islands, is a remarkable case of cultural diffusion, and more interesting and important to the ethnologist than the other supposition. If one cultural element can pass independently from tribe to tribe, from culture area to culture area, through or over sharp cultural boundaries,and across many almost isolated regions, and all in two or three hundred years, what mixtures may not have arisen in the past centuries, and why worry about the migration of peoples when cultural elements are so independently migratory?

ALBERT B. LEWIS

TOBACCO IN NEW GUINEA: AN EPILOGUE

As Dr. Lewis' contribution to the use of tobacco in New Guinea is one of six leaflets prepared by various members of the Museum staff under my direction and edited by me in the Field Museum series of Anthropology Leaflets (15–19 and 29), I may be allowed to add a few remarks to my colleague's rejoinder to Dr. Merrill. No one who has read these six leaflets will accuse me or even suspect me of being an anti-American heretic, for I have strictly upheld and, I venture to hope, have also proved the introduction of both *Nicotiana tabacum* and *N. rustica* from America into Asia, Europe, and Africa. In regard to Melanesia and Australia, however, I

---

[15] C. Hartwich, Die menschlichen Genussmittel (Leipzig), p. 117, 1911.

believe the situation is somewhat different, and this despite Dr. Merrill's criticism, which is not convincing to me. Dr. Merrill is well known as an eminent botanist from whose writings I have learned a great deal, especially with reference to the cultivated plants of the Philippines, a subject in which he is an undisputed authority. The "ethnological myth" invented by Dr. Merrill has been refuted above by Lewis. The botanical data of Engler-Prantl to which he refers, combined with a statement of A. de Candolle (Origin of Cultivated Plants, p. 142), who records two native species, *Nicotiana suaveolens* for New Holland and *N. fragrans* for the Isle of Pines near New Caledonia and who states that these two are foreign to America, were sufficient botanical evidence to an ethnologist for assuming a native *Nicotiana* species in Melanesia and Australia.

Dr. Merrill states from hearsay that *N. tabacum* occurs in New Guinea. I should be the last to doubt it; its non-occurrence there would be next to a miracle. This fact, however, does not disprove that a native species might not occur there. Dr. Merrill's informants have assuredly not explored the entire length and breadth of New Guinea.

Tobacco was known in New Guinea at least in the beginning of the seventeenth century, for it is mentioned by Jacob Le Maire, who sailed along the coasts ot the island in the year 1616 (Australian Navigations, ed. by De Villiers, pp. 223, 226). On the 23rd of July of that year Le Maire, when he was a short distance from the land, reports that he was followed by six big canoes of natives bringing dried fish, coconuts, bananas, tobacco, and small fruit-like plums. The date in question is rather early and almost coincides with the first introduction of the tobacco plant or plants into Japan, China, Java, India, and Persia. Le Maire's notice, of course, is inconclusive as to whether the tobacco to which he alludes was imported or indigenous. Dr. Merrill's supposition that tobacco was introduced into New Guinea from Amboina is purely subjective and does not constitute historical evidence. I have written a very detailed history of the American cultivated plants in their distribution over the Old World mainly for the purpose of doing away with all the wild guesswork and speculation from which this subject has hitherto suffered, and replacing it with facts based upon documentary evidence; it will be seen that many of the guesses made, especially with reference to Spanish and Portuguese agency, are wrong and that the transmission of American plants was frequently effected through quite different channels.

The worst of Dr. Merrill's guesses is his suggestion that "the type of pipe and method of smoking tobacco in New Guinea is merely an adaptation of the pipe and opium smoking." In the first place, there is not a ghost of resemblance or affinity between the New Guinea method of tobacco smoking and opium smoking. In the second place, this alleged interaction is historically impossible, for opium smoking sprang up as a sequel of tobacco smoking only in the beginning of the eighteenth century, when the opium-pipe was first invented by Chinese in Formosa (see our Leaflet 18, pp. 23–24). In other words, opium-pipe and opium smoking are rather recent events, hardly two hundred years old, while the peculiar method of smoking tobacco in New Guinea from tubes in which the smoke is stored must be many centuries old. This method does not occur anywhere in America, Asia, Africa, or Europe; it is unique and peculiar to and characteristic of New Guinea, and must therefore have originated in New Guinea.

Chinese contacts with New Guinea are not of ancient date, as supposed by Dr. Merrill. New Guinea is not even mentioned in any Chinese records, nor do early Chinese contacts in the Philippines, as Dr. Merrill boldly asserts, antedate or at least approximate the beginnings of the Christian era. Chinese trade with a few of the Philippine islands was established from the tenth to the thirteenth century, but closer relations and foundations of Chinese trading posts in the islands are not earlier than the end of the fourteenth and the fifteenth century.

BERTHOLD LAUFER

RECENT DISCOVERIES OF FOSSIL HUMAN REMAINS

Through the courtesy of Mr. Henry Field, we are able to publish the following note, quoted from a letter written by Professor Wilhelm Freudenberg regarding his recent discoveries near Heidelberg.

"I may tell you that Professor Boule in Paris suggested that part of my fossils from Bammenthal belong to *Homo heidelbergensis*, while the other fossils, which I named *Hemianthropus osborni*, suggested to him some resemblances to *Homo sapiens* and not to the anthropoids. The jugal bone resembles cynocephalid forms in the direction of the squamose processes, while the cheek part is truly human but very convex. This may be a transitional form in some respects, like *Eoanthropus dawsoni*, which is also suggested by the very short diaphysis of the femur, which resembles that of a chimpanzee. A new genus occurs with *Elephas trogontheri*, the steppe elephant. *Homo heidelbergensis*, however, is found with *Elephas antiquus*, the woodland elephant. The former may, on the other hand, be a relation to *Pithecanthropus* and *Sinanthropus*, which is proved apparently by the very large brain case, of which a fragment has been found, together with an isolated part of the orbits, which are very large and round, and in this respect similar to the African anthropoids. The jugal bone combined with the orbits is of a different type compared to that of *Hemianthropus*.

"While with Professor Sir Arthur Keith, F. R. S., it was stated in the Royal College of Surgeons that the gorilla is not present at Bammenthal or Mauer, as the cancellae on the side part of the supraorbital ridge are present in the fossil *Homo heidelbergensis* or related form but not in the gorilla. They are present in *Sinanthropus*, as I saw in photographs sent to Professor G. Elliot Smith of London by Professor Davidson Black of Peking. This information throws a new light on the earliest history of mankind."

209

中国古代的造纸和印刷术

Paper and Printing
IN ANCIENT CHINA

BERTHOLD LAUFER

Printed for THE CAXTON CLUB Chicago
·M·C·M·X·X·X·I·

Paper and Printing in Ancient China

Paper and Printing in Ancient China

THE ANCIENT SUMERIANS, Babylonians, Egyptians, and Greeks may have reached a flourishing civilization long before the Chinese, but all their achievements, however great, do not equal in importance the invention of paper which we owe to the Chinese and the art of printing that was born from it. Printing has been, and still is, the supreme factor in the progress of civilization. The Chinese as the inventors of paper were the first who printed books, many centuries before Gutenberg, and not only that—they have also made typography a fine art and produced books which belong to the finest examples of the craft. They have been a book-loving people for ages. The primary conditions of printing are paper, writing-brush, ink, and ink-pallet or ink-stone, which are still regarded by the Chinese as the four great emblems of scholarship and culture that form the fundamentals of their civilization. These four constituents the Chinese may justly claim as their own, inventing them and perfecting them entirely from their own resources, unaided by any other nation; and this arsenal has largely contributed to make them a nation of learned, studious, well-bred and cultured men.

The turning point in the history of printing is the invention of paper by Ts'ai Lun in A.D. 105. In order to understand and appreciate this event correctly, it is necessary to have some idea of what writing-materials were prior to that date and in what condition the early documents and books were before the art of printing came into being. I shall therefore discuss, first, Chinese books before the invention of paper and, second, Chinese books after the invention.

The earliest means of communication in ancient China of which we have any knowledge reminds us of the quippus of the ancient Peruvians.

In a prehistoric age we find in China knotted cords in use for the conveyance of messages, chiefly in the transaction of government business. Lao-tse, the famed philosopher, in a sentimental yearning for the past, expressed the desire that he might bring his people back to the ancient usage of knotted cords; that is, the simple life of old. The Tibetans have a

6

tradition to the same effect, and several aboriginal tribes in the south of China availed themselves of this method as late as the twelfth century of our era.

In early historic times, calendars, calculations, and contracts were made by means of wooden tallies in which notches were carved with a knife. Even when writing had long been in use, contracts made by means of notches in a wooden stick were continued for simple business transactions — such as deeds, bonds, or obligations. The wooden stick was notched on either side and then split and equally divided between the two contracting parties, the creditor receiving the left half, the debtor the right half of the tally. When the time arrived for fulfilling the contract, the two halves were joined together to make sure that the notches of the one tallied with those of the other. When the debt was liquidated, all the creditor had to do was to break up his portion of the wooden contract. This was called "breaking the contract," which meant as much as "fulfilling one's obligation." Credit systems were always highly developed in China, and there is an old story on record that a tailor even made garments on credit for a duke and that whenever the duke was in a position to render a payment on the instalment plan, the wooden tally was smashed by the happy tailor. Even when contracts were subsequently drawn up in writing, the

7

notches were retained as a means of checking the two halves or verifying the twin documents.

A survival of this practice still characterizes the modern banking system. Our banking methods are based on the signature and identification of the individual. Neither is required in China. A Chinese draft is made out in duplicate on a single long sheet of paper, containing the same matter on the right and left sides, one column of writing running exactly down the center. The document is evenly cut into halves across this line, the right half being given the bearer of the draft, the other half being mailed to the bank on which the draft is issued. When the bearer presents his half of the draft at this bank, it is carefully checked off and tallied with the other half which meanwhile arrived at the bank by mail; and if the two halves are found to fit perfectly together, payment is made, no receipt and signature being required.

These drafts bear a striking resemblance to the indentures used in old England. Hamlet, in musing over a lawyer's skull, exclaims, "Will his vouchers vouch him no more of his purchases, and double ones too, than the length and breadth of a pair of indentures?" They were referred to as a pair, as both copies of a deed were written on one piece of parchment or paper and then cut asunder in a serrated or sinuous line (a remi-

8

niscence of the notches in tallies), so that when brought together again the two edges exactly tallied and proved that they formed part of the same document.

A fundamental of culture in eastern Asia is divination. Divination was based on the bones of certain animals. In central Asia divination was practised from the shoulder-blade of a sheep which was scorched over a fire, and from the cracks thus arising in the bone the future was predicted. In ancient China the carapace of a tortoise was utilized in fortune-telling, and this magical procedure probably gave the impetus to the origin of writing. The tortoise was regarded as a sacred animal imbued with a knowledge of the future. In 1899 a deposit of several thousand fragments of bones, chiefly tortoise-shell, was discovered at Chang-te fu, Honan Province. These bones forming a sort of archives are engraved with inscriptions of a very archaic style, representing the earliest form of Chinese script we now possess, and were used for purposes of divination. They date in general to about 1500 B.C. The oracles and in some cases the answers were incised into these bones. We meet, for instance, inscriptions such as these: "We consulted the oracle to ascertain whether the harvest will be abundant," or "The oracle was consulted, as we wish to know whether God will order a sufficient rainfall so that we may obtain an adequate food-

9

supply," or "If we go a-hunting to-morrow, shall we capture any game?" Divination has always dominated the whole life of the Chinese from the cradle to the coffin, and no business was transacted, no marriage concluded, no burial undertaken, without consulting a fortune-teller. These ancient augurs formed a special profession, in their social position comparable to the lawyer of our society. In the same manner as the modern financier and captain of industry consults his lawyer on all important questions, so the Chinese did not make a move in the most trivial matters without asking a diviner's advice.

Further, we have from the early dynasties inscriptions on objects of bronze such as vases, bowls, bells and weapons, cast by means of the lost-wax process, the characters being traced in the wax mould, and being either incised or raised in the bronze. Tablets of jade were used for writing by the emperors; tablets of ivory, by the nobles and higher officials. The most common writing-material, however, particularly under the Chou dynasty (1122-247 B.C.), consisted of bamboo slips or square wooden splints which were perforated at their upper ends and fastened together by means of a silk cord or fine leather strip. The main difference between the utilization of bamboo and wood was this, that a message containing upwards of a hundred words was written on bamboo slips; when it contained less than

10

a hundred words, on wooden boards. The bamboo tablets were naturally narrow, and could be piled up in any required number, formed into a pack. The wooden documents, being too heavy to allow of a combination of many, served only for brief texts, as official acts and regulations, statistics of the population, and prayers, but they could not be united into books.

The early canonical or classical literature was handed down on bamboo slips of different lengths, each slip as a rule containing a single line of writing varying from eight to twenty-five or thirty words, and inscribed on one side only. A great number of such tablets was naturally required to make a book. Such books, of course, were exposed to many causes of destruction, chiefly from humidity and pernicious insects, so that bamboo books of early antiquity have long since disappeared, but a large number of wooden documents of the Han period have come to light in Chinese Turkestan. Another inconvenience of these books was their heavy weight. A curious incident in allusion to this fact is recorded anent the emperor Ts'in Shi who was compelled to examine daily state documents to the weight of a hundred and twenty pounds. Neither writing-brush nor ink was invented in those early days, and the bamboo and wooden memoranda were inscribed by means of a pointed bamboo or wooden stylus (pi) dipped in a black varnish (ts'i).

II

my history of ink embodied in the book of Frank B. Wiborg, Printing Ink, New York, Harper Bros., 1926, pp. 1-76.)

Paper was invented in China by Ts'ai Lun in A.D. 105 when he conceived the idea of manufacturing with refuse material of vegetable origin a substance light and economical at the same time, which would replace advantageously the writing-materials used up to his time. A record of this memorable event is contained in the biography of Ts'ai Lun, which is embodied in the Annals of the Later Han dynasty (chap. 108). Ts'ai Lun was born at Kwei-yang, a city in Kwei-chou Province in southern China. In A.D. 75 he entered the service of the Emperor Ho, and in A.D. 89 was appointed director of the imperial arsenals. He was deeply given to study, and whenever he was off duty, he would shut himself up for that purpose. The passage relating his memorable discovery runs as follows: "From times of old, documents had been written on bamboo boards fastened together. There was also paper made of silk refuse (chi). But silk was too expensive, and the bamboo boards were too heavy; both were inconvenient. Therefore Ts'ai Lun conceived the idea of utilizing tree-bark or bast-fiber, hemp, and also old rags and fishing-nets for making paper. In A.D. 105 he submitted his invention to the emperor, who lauded his skill. From this moment there was no one who did not use his paper, and throughout the

14

empire, all people called it the 'paper of the honorable Ts'ai.'"

This brief and sober account reveals what the writing-materials were in the times before Ts'ai Lun, what his innovation consisted of, and what impression it made on his contemporaries. It should not be understood that the ingredients enumerated in this passage were mixed together and resulted in paper; but each substance was the principal constituent to make a particular kind of paper. Paper may be obtained from many and various plant-fibers by a process of cleaning, maceration, and drying. The paper of Ts'ai Lun was in fact distinguished, according to the material used, as "hemp paper," "bark paper," etc. He substituted vegetal fibers for the fibers of animal origin used previously; in principle he utilized two kinds of materials—the raw fibers of bark and hemp and the worked-up fibers of rags such as ropes and fishing-nets. He survived his invention for thirteen years, and was ennobled in A.D. 114 as marquis by the empress dowager. He was no favorite, however, with the empress; and when his patroness, the empress dowager, died, the empress began to intrigue against him. Driven to despair, he died a suicide by swallowing a dose of poison.

Two different places were pointed out in later times as the seat of Ts'ai's operations. According to one report, he had lived near Lei-yang in Hunan Province, where near his resi-

15

principal center for paper-making, and this region continues the manufacture to the present day. From Italy the art spread to France and Germany, somewhat later to England, where it began to flourish when the Revocation of the Edict of Nantes in 1685 sent many French paper-makers into exile to England and America. In 1690 (a millennium and a half after the first invention) the first paper-mill was organized in this country by William Rittenhouse at Roxborough near Philadelphia.

During the early centuries of our era, paper of a great variety, paper sized and loaded to improve its quality for writing, paper of various colors, writing paper, wrapping paper, even paper napkins and toilet paper, all were in general use in China. The method of sizing paper with starch is also an improvement initiated by the Chinese. Their ancient practice of extracting the fiber from the bark and other parts of plants by means of maceration is in principle identical with our modern method of extracting cellulose by means of chemical processes.

In the tenth century the Chinese conceived the idea of issuing printed paper money, which reached its climax under the Mongol emperors. It aroused the greatest admiration of Marco Polo, who devotes to it one of the most interesting chapters of his visit to the Grand Khan. The Mongol rulers introduced paper-bills into Persia under the Chinese name *chao*,

18

and in 1293 established a printing-office at Tabriz, where paper-notes were turned out by the Chinese method of block-printing.

We owe to China in particular also our paper-hangings or wall-paper. The walls and ceilings of rooms are invariably decorated in China with paper, on which different patterns are printed from wooden blocks. The paper is confined in size to foot-square sheets.

During the middle ages, Europe had only linen, silk, and leather tapestry. French missionaries in China sent to France some specimens of colored Chinese wall-paper, which stimulated a Frenchman, Le François by name, to establish a factory at Rouen in 1630 for the purpose of imitating the Chinese papers. This Rouen paper was exported to England where it became known as "flock-paper." The English claim a previous invention by Jeremy Lanyer who, in 1634, had used Chinese and Japanese processes. It was, however, as late as the middle of the eighteenth century that real colored papers were made in France and England. The actual importers of Chinese wall-papers painted or printed with pictures were Dutch merchants, who traded them also to France, England, and Germany, where they were used to decorate screens, desks, chimney-pieces, etc., toward the end of the seventeenth century. They were called pagoda-papers. The prices paid for these papers were exorbitant.

19

In 1770 there was advertised for sale in Paris 24 sheets of China paper with figures and gilt ornaments 10 feet high and 3½ feet wide, at 24 livres apiece, to be sold altogether, or in lots of 8 sheets each. By that time entire rooms were papered. In 1779 an apartment in Paris was advertised to let, having a pretty boudoir with China paper of 13 sheets in small figures representing arts and crafts.

Wall-paper was brought to America in 1735. Its manufacture was introduced into the United States in 1790 by two Frenchmen, Bouler and Charden, and only three or four firms engaged in the business before 1844. In that year the first machine for printing wall-paper was put up in the Howell factory at Philadelphia. Up to that date, wall-paper had been made in small sheets of 22 x 32 inches according to the Chinese fashion. After the establishment of machinery continuous rolls or webs of paper came into general vogue. The dependence on Chinese models is illustrated also in the two processes hitherto applied to wall-paper. The patterns were either put in with stencils and the background with a brush, or by means of block-printing, the design being engraved on a wooden block.

In old colonial mansions of Massachusetts, specimens of Chinese wall-paper are still to be found, some even imported in A.D. 1750 and still in a good state of preservation. Many of

20

the older American papers exhibit their relationship to the Chinese in that the decoration is not repeated, but runs continuous about the entire room or contains a scenic representation. An interesting book on this subject was written by Kate Sanborn, "Old Time Wall Papers, an account of the pictorial papers on our forefathers' walls" (Greenwich, Conn., 1905).

When the invention of rag-paper was made, the Chinese were in the possession of all technical materials that make the primary conditions for printing: an extensive literature; a suitable and economic medium, easily manufactured in large quantity, for taking print; and ink as the medium to fix permanently to the paper the written thought. And yet, it is astounding, centuries elapsed before any steps were taken in the direction of printing. This is the more amazing, as printing of an embryonic type was practised long ago by means of seals made of clay or metal in which the script stands out in the negative in the same manner as in the later block-prints. The case is of psychological interest inasmuch as it shows that new inventions depend not merely on the existence of mechanical appliances, but to a much higher degree on the mental attitude of society. Some dynamic force is required to set the slumbering spark afire, in order to create the demand for printing. The art of printing is the first step toward democracy, the education of the people, and na-

21

tional awakening. To this the intellectual minority in all countries of Asia and Europe was at first bitterly opposed. In ancient Egypt, if the idea of printing had ever been proposed, it would at once have been nipped in the bud by the caste of jealous priests. A similar situation prevailed among the Brahmans of ancient India, where the sacred hymns of the Veda were memorized and transmitted for ages from generation to generation merely by memory. Even at a time when an alphabet was introduced, the Brahmans first refused steadfastly to commit their sacred texts to writing, and but slowly and reluctantly yielded to this innovation which threatened to break down their monopoly and the prerogatives of their caste. In India, in opposition to China, it was the spoken word which was looked upon as a fetish. In China, it was the written word that was regarded with fervent reverence and treated as a fetish. This worship of the written word ultimately led to its permanent preservation in print, while in India this idea was always detested. Despite her close contact with China, India did not adopt from her paper and printing. Paper was introduced into India only in the Mohammedan period by the Arabs after the twelfth century, and only as late as the sixteenth century was the first printing press set up by the Portuguese at Goa. The first book printed there was Garcia da Orta's "Colloquies on the Drugs of India," 1563.

22

We might expect that printing should have arisen in the circle of Confucian scholars who certainly had an interest in preserving and diffusing their philosophical writings; but this expectation is not fulfilled, perhaps for the reason that the Confucian doctrines appealed largely to the intellectual minority, not to the masses. The people found more satisfaction in the tangible teachings of Buddhism which promised them speedy salvation in the paradise of Amitābha. The earliest attempts at printing from wooden blocks were therefore made in the camp of Buddhistic priests, and consisted of charms, especially for the healing of diseases, prayer formulas, and engravings of religious images, made with the avowed object of appealing to the sentiments of the people. The earliest of these charms extant were printed in Japan about A.D. 770 by order of the empress Shōtoku in fulfilment of a vow: one million charms were printed, placed in a million of small wooden pagoda models and distributed among the populace, these charms being believed to be efficient in expelling the demons of disease. Several of these are still preserved. The first object of printing, therefore, was not, as we might imagine, a desire for the diffusion of knowledge, but a desire on the part of an empress to acquire religious merit and to safeguard her people from the ravages of epidemics. The meaning of the charms was even unintelligible

23

to the people, for they were written in Sanskrit transcribed in Chinese characters.

It is uncertain to what date block-printing in China goes back. While no accurate date can be fixed and while there is no record of an inventor of block-printing, which was achieved by a gradual process, there are indications that the initial stages are traceable to the sixth and seventh centuries. During the ninth century printing from wooden blocks was practised in the farthest west of China, the province of Sech'uan, which seems to be the home of the art. Again, the books printed there were not intended for scholars, but for the people. They consisted chiefly of works on divination, dreams, geomancy, and elementary school books, but, as our Chinese informant writes, they were so spotty and blotted that they were difficult to read.

The earliest printed book in existence was discovered in the library of the cave temples of Tun-huang by Sir Aurel Stein in 1907, and is now in the British Museum. It is the Chinese version of a Buddhist Sūtra, the Vajracchedikā. It bears the date A.D. 868, contained in the colophon at the end. It reads, "Printed on May 11, 868, by Wang Chie, for free general distribution, that the memory of his parents be reverently perpetuated." In this case the printer performed an act of filial piety. The text is printed on a roll of paper 16 feet long and 1 foot

24

wide; it consists of seven sheets pasted together, and was printed from seven blocks. The frontispiece is the earliest datable wood-cut. Printed books and single sheets, however, were exceedingly rare in the temple library of Tun-huang; among thousands of manuscript rolls only four printed rolls were found, and a not inconsiderable number of charms from single-page block-prints. The latter presumably were of local manufacture, the former imported from Sech'uan. Printing was practised under the T'ang (A.D. 618-906) to a limited extent only, and did not supersede the manuscripts which were evidently regarded as more meritorious.

A certain official, Feng Tao (A.D. 881-954), is credited by some with the invention of block-printing. He was a versatile politician who served under no less than ten emperors of four different houses. Presenting himself at the court of the second emperor of the Liao dynasty, he asked for a post. He said he had no home, no money, and very little brains—a statement which appears to have recommended him strongly to the sovereign, who at once appointed him grand tutor to the heir-apparent. Block-printing was certainly known long before Feng Tao's time, but he was the first who applied it to the printing of the Confucian classics, and this is the reason why Confucian scholars have stamped him as the inventor of block-printing.

25

Under Feng Tao's direction the ancient canonical literature with its host of commentaries was printed for the first time in 130 volumes. The work of editing and printing lasted for twenty-one years and was completed in A.D. 953. Up to the year 1064 the private printing of the classics was forbidden. All printing was the prerogative of the government and had to give the ortho-dox accepted text. Of Feng Tao's edition nothing unfortunately has come down to us. The great renaissance of culture that took place under the Sung dynasty (A.D. 960-1279) resulted in an enormous output of literature and a corresponding advance in the art of printing.

As to the technique of the block-print process, it is simplicity itself: the book undergoes only three principal stages. What is composition among us is performed by a professional calligra-phist, who receives the manuscript from the author's hands and writes it out in clear and uniform style on thin sheets of paper. Prefaces contributed by friends (and most books are introduced by a number of prefaces) are usually facsimiled from their own hand-writings, and may even be written in an archaic or highly ornamental style. Calligraphy, like drawing and painting, is an art, and the three are closely interrelated. A point worthy of note is that the Chinese scribe, as well as the draftsman and painter, is deprived of the privilege of making corrections or

26

alterations. Chinese paper and silk are highly absorbent materials, and a stroke of the brush, once made, will stand forever, and cannot be erased. The artist therefore must be sure of a firm hand and a scrupulously thoughtful and precise technique.

When the manuscript is completed, it is sent to the block-engraver. The single sheets are pasted over the finely planed and smooth wooden blocks, usually of pear-tree wood, the writing turned face downward. As the paper is thin and transparent, the writing is perfectly displayed through the back. Then commences the task of the engraver who with a set of gouges, picks, and chisels pares the surface of the block around the characters, so that the script in negative stands out in a flat relief.

In this state the blocks are finally transmitted to the printer whose requirements are limited to just two brushes. He uses a round, bell-shaped brush of coir-palm fibers for rubbing ink over the block. Then he places a sheet of paper over it and takes the impression by means of a flat, handled brush, which takes the place of our press. The printed sheet, of course, represents an exact facsimile of the original manuscript, and the printer cannot make any mistakes. A single sheet, as a rule, consists of two pages with a margin in the center, that contains the title of the book on top, chapter and folio number in the middle and

27

usually title of the particular chapter at the foot. These sheets are folded and then stitched at the ends.

The Field Museum has an exhibit illustrating the whole process of wood-engraving and printing in China, Japan, and Tibet, also an exhibit of writing-materials, paper, brushes, inks, pigments, and ink-pallets. The oldest printing blocks in existence are likewise preserved in the Museum. They are engraved with floral designs and must have been made before the year A.D. 1108. They were found in the ancient city of Chü-lu in southern Chili, which was submerged by a flood in that year.

The results obtained by the economic process of block-printing are stupendous. It is best adapted to the genius of Chinese writing which employs many thousands of characters, and has many advantages over movable types which are expensive, difficult to store and to arrange and hard to find when needed in setting type. Block-printing could easily be established anywhere and made literature accessible to every one at a moderate cost; it is a democratic art. Above all, it has always satisfied the æsthetic sense of the Chinese in that the block-prints preserve accurately the beauty of form of the characters and the hand-writing of the individual. No two Chinese hand-writings are alike, and hardly two block-prints can be found in the same style of writing if based on the hands

28

of different individuals. In typography, of course, the type is standardized and conventional.

One of the disadvantages of block-printing was the storing, arrangement, and preservation of blocks which were easily destroyed by humidity or fire. For the printing of the Tripitaka, the sacred scriptures of the Buddhists, a copy of which is in the Newberry Library (printed in 1736-38), 28,411 blocks were required. It was customary, especially under the Ming, in the government printing office at Nanking, if single blocks were lost, to re-engrave these and to add the date on the margin. Thus, the Newberry Library has several editions of the dynastic histories made up from blocks of different dates such as 1368, 1530, 1531, 1533, 1572, etc.

As to the form of the Chinese book, it was originally in the form of a roll. The manuscripts written on silk under the Han were kept in rolls, likewise all manuscripts and xylographs from the second down to the tenth century. The Buddhists introduced the folded book, somewhat on the order of our railroad time-tables, and still retain it for their sacred literature. The stitched and paged book, as we have it now, is not older than the Sung dynasty and goes back to the eleventh century. How it originated is not yet ascertained.

The earliest printed book in existence in America is pre-

29

served in the Newberry Library, Chicago, and is dated A.D. 1167. It is entitled *T'ang Liu sien-sheng wen tsi* (No. 1174) and contains, in twelve volumes, the collected poems and essays of Liu Tsung-yüan (A.D. 773-819), one of the celebrated poets and essayists of the T'ang dynasty. The work next in date secured by me for the Newberry Library is a general history of China known as *Tung kien kang mu* by the philosopher Chu Hi, published in 1172; and it is a complete copy of this first edition in a hundred volumes, which is in the possession of the Library. It is a rare and fine specimen of Sung printing and perhaps the most extensive work of that period now known. Several Sung editions are also in the Library of Congress and in the Gest Chinese Research Library at McGill University, Montreal. One of the recent acquisitions of the latter is an edition of the Tripitaka of the Sung and Yüan periods, of different dates, the earliest actual print going back to A.D. 1232. This collection was first printed in A.D. 972.

The Chinese were also the first who conceived the idea of the printed newspaper. The Peking Gazette, "the News of the Capital" *(Ching Pao)*, is the oldest daily paper in existence. It began to appear in A.D. 713 under the T'ang dynasty, and has since been issued daily until the collapse of the Manchu dynasty in 1911. It contained the imperial rescripts, decrees, and all official

30

news relating to interior and foreign affairs. It was printed in two editions, in a Government edition sent to officials throughout the empire, and in a popular edition sent to regular subscribers in the capital and the provinces; also a manuscript edition could be obtained.

One of the forerunners of printing is represented by inscriptions carved in stone tablets. China is a land of inscriptions many thousands of which are still in existence, and epigraphy is a favorite occupation of her scholars. As early as A.D. 175 the texts of the Confucian classics were inscribed on stone for their permanent preservation. Subsequently these texts were deeply incised in stones, and paper rubbings were taken from them for distribution among scholars. Stone tablets were the recognized method of preserving exact copies of fine calligraphy and drawings of great masters. Taking squeezes of all sorts of inscriptions has developed into a regular trade since the days of the Tang dynasty, and in Chinese collections are still preserved rubbings taken in the Sung and Ming periods, chiefly from inscriptions now lost, that are highly prized.

To obtain facsimile rubbings of inscriptions on stone or bronze, the Chinese use sheets of thin, but tough paper, which is folded, slightly soaked in water, and then evenly applied to the surface of the inscription. The paper is pressed in with a

31

wooden mallet and forced into every depression by means of a soft brush. When the paper becomes sufficiently dry, they go over it with a stuffed pad of cotton lightly dipped into liquid ink. When taken off, the paper shows a perfect impression of the inscription coming out in white on a black background. The men doing this work form a special profession, and as ancient inscriptions are numerous, there is a lively trade carried on in such rubbings to supply the demands of scholars.

At an early date the Chinese experimented with movable type. In the period A.D. 1041-49 a commoner, Pi Sheng by name, is said to have made a set of clay types which were locked in an iron frame for printing, but no print made from these types has survived. Under the Mongol dynasty, in the fourteenth century, type was cast of tin, and subsequently made of wood. Wooden types were made by Wang Cheng, a geographer and agriculturist, who likewise devised a revolving table upon which the types were arranged, and from these wooden types he printed his *Nung shu*, a work on agriculture. In a chapter of this work published in A.D. 1313 he records a history of his set of movable type, stating that the characters were first engraved in wooden blocks which were then sawed apart into individual types.

Printing from movable metal type on a large scale was first

32

practised in Korea, and reached there its highest development. A set of 100,000 copper types was cast in A.D. 1403 by order of the king, and actually used in the publication of many books up to the year 1544. The Japanese Government General of Chosen reported in 1916 that it has taken care of old types of metal, clay, and wood, formerly in the possession of the imperial household of Korea, to the extent of about half a million pieces.

A revival of type-printing took place in China at the end of the seventeenth century. At the suggestion of Jesuit missionaries the emperor K'ang-hi had a font of 250,000 movable copper types cast which were used for the printing of an extensive cyclopædia the *T'u shu tsi ch'eng*, in six thousand volumes, completed in A.D. 1726. In A.D. 1736 there was a shortage of currency, so that this font was sent to the melting pot for the minting of copper coins. It was replaced in A.D. 1773 by a font of wooden type which was used for printing the catalogue of the emperor K'ien-lung's library and other books. Printing from movable type was an expensive undertaking requiring large capital and was entirely carried on by the government, ceasing when government support was withdrawn. Block-printing was found more practical and reasonable for private and commercial purposes. By the nineteenth century the use of type had come to an end in China, Korea, and Japan, and was reintroduced

33

from the West as an entirely new art. At present European typography and even paper and printer's ink dominate China, but one cannot say that the productions of these modern presses are as elegant, graceful, and artistic as the time-honored block-print.

210

人类学部的目标和研究对象

DEPARTMENT OF ANTHROPOLOGY —ITS AIMS AND OBJECTS

Departments of anthropology have been founded at many of our leading universities and in the larger natural history museums. The scientist who desires to make anthropology his lifetime vocation has therefore the choice between an academic and a museum career. The teaching of anthropology is, of course, an important task, as investigators must be trained to shoulder the burden of their predecessors, and our museums must look to the universities for a supply of competent men.

The anthropologist devoting his energies to museum work enjoys a wide sphere of activities and an unlimited range of opportunity; he may be explorer, research-worker, author, lecturer and educator at the same time. A hall in the museum covering the ethnology or archaeology of a certain country or group of tribes, properly arranged and labeled, has the same value and offers the same advantages as a university lecture course on the same subject—with two notable differences, however: the university course is given for the benefit of a limited number of students, while the silent course offered by a museum hall will reach many hundreds and thousands of people daily. Moreover, it is a permanent institution, a visual demonstration of facts and data accompanied by lectures printed on labels, while the class room instruction naturally is transient and evanescent and lacks the actual demonstration of culture objects, models, and groups.

At present eighteen large halls are installed with labeled exhibits in the Department of Anthropology of Field Museum. These cover all parts of the world and represent the equivalent of eighteen lecture courses, which is far more than all university departments of anthropology combined are able to offer. Any visitor to the Museum who is determined to study these collections carefully case by case and to digest the information given on the labels will receive a liberal education in anthropology and a thorough knowledge of the cultural achievements of mankind.

The label is the bond that links the Museum with the public. A label may be very concise, consisting of only a line or two, and yet it will embody the results of long and painstaking research and considerable thought.

It is hoped to publish a guide for each of the halls. Three such guides have already been issued, and a fourth is in press now. The object of this series of handbooks is to furnish the synthesis to the analytic collections, to present a survey of the region or culture in question and to depict in particular its geographical, historical, social and religious background. These booklets are amply provided with illustrations, maps, and bibliographies, and are gotten up in an attractive style.

While all resources are thus supplied by the Museum for an intensive study and appreciation of all phases of human cultures, the Department is not content with the mere role of disseminating knowledge of its science, but it is also eager to perform a distinct service to the public. The practical value of the art of primitive and oriental nations to our own art and industries is now generally recognized, and the creators of new and better ideas have always discovered in the Museum's collections many suggestions and inspirations. Art students, artists, craftsmen, designers, and manufacturers have made liberal use of decorative forms and designs such as those shown in the American Indian, ancient Egyptian, Chinese, South Pacific and other collections.

A new study room has just been opened in the quarters of the Department of Anthropology on the third floor of the building. It is spacious, well lighted, attractively furnished and equipped with study material from all parts of the world, arranged in wall cases. This room is open to all who desire to pursue specific studies in any anthropological subject or to apply material to any legitimate purpose in art or industry.

—BERTHOLD LAUFER

(An article on the purposes and functions of the Department of Botany will appear next month, and similar articles on the Departments of Geology and Zoology in succeeding months.)

MODEL OF FAMOUS HORSE

"Man o' War," one-time race track favorite, has been immortalized by the placing of a model, showing his sleek lines, on permanent exhibition in Field Museum.

The famous race horse was selected to represent the highest development of the modern horse in the Museum's series of models illustrating the evolution of the horse from a four-toed animal about the size of a cat, through the various stages of development to the present day.

The model of Man o' War is the work of Frederick Blaschke, sculptor of Cold Spring-on-Hudson, New York. It is one-fifth actual size, and was made from life by Mr. Blaschke shortly after Man o' War's retirement from the turf.

With the addition of the model of Man o' War, the Museum's horse evolution exhibit shows six stages of development. The display begins with the *Eohippus* or "dawn horse," which had four toes on the fore feet and three on the hind feet. It grew no

Man o' War

Model of famous race horse presented to Field Museum by the sculptor, Frederick Blaschke, and added to series illustrating evolution of horse.

larger than a cat, and lived about 55,000,000 years ago, according to Dr. Oliver C. Farrington, Curator of Geology. Next is shown the *Mesohippus*, a three-toed horse about the size of a collie dog, which lived about 35,000,000 years ago. Following this are a slender-limbed small desert horse, of 19,000,000 years ago; a larger one-toed horse of some 7,000,000 years back; and finally the modern horse as represented by Man o' War.

In addition to the models, fossil skulls and feet of each of these are on exhibition. Although the horse appears to have originated in North America, soon spreading to South America, and appearing later in Asia and Europe, it was completely exterminated on the American continents in prehistoric times, and modern horses here are descended chiefly from European and Asiatic stock.

211

中国玉器新展厅开放并展出 1200 件藏品

Field Museum News

Published Monthly by Field Museum of Natural History, Chicago

| Vol. 2 | DECEMBER, 1931 | No. 12 |

NEW HALL OF CHINESE JADES IS OPENED; COLLECTION OF 1,200 DISPLAYED

One of the world's finest and most comprehensive collections of Chinese jades, valued at several hundred thousand dollars, and comprising more than 1,200 objects carved in a myriad variety of forms, was placed on exhibition at Field Museum with the opening last month of a new hall (Hall 30 on the second floor) devoted entirely to jades. The jades range from ancient pieces of the archaic period which began at an unknown time roughly estimated at 2000 B.C., down to the end of the eighteenth century—a span of nearly 4,000 years in the development of one of the most important of the fine arts of China.

In connection with the wealth and treasures of the Orient one naturally thinks first of all of King Solomon, and remembers Christ's saying, "Consider the lilies of the field, how they grow; they toil not, neither do they spin; yet I say unto you, that even Solomon in all his glory was not arrayed like one of these." It may now be assumed also that Solomon never beheld and never owned a single piece of jade, although he was a contemporary of the Chou dynasty when the carving of jade was a highly developed art in China, and when the Chinese sovereigns, as high priests of the nation, performed a function strikingly similar to that of the High Priest of Jerusalem. Like the latter, the ruler of the old Chinese empire had received his sacred mandate from Heaven, the supreme deity of the universe, and by his command ruled as the Son of Heaven.

The emperor was responsible to Heaven for his conduct and actions, being the mediator between Heaven and his nation. His virtues resulted in prosperity, his evil manners caused misery and calamities in the empire. The sovereign was believed to be able to commune and consult with Heaven through the medium of a perforated disk of jade; for this stone was endowed with supernatural qualities, supposed to be engendered by solar light and capable of transmitting messages to transcendental powers. When the great Emperor K'ang-hi in 1688 conferred a posthumous honor on his deceased grandmother and had a document to this effect carved on slabs of jade (shown in the Museum's jade collection), he was actuated by the belief that his ancestress in heaven would actually take notice of this encomium.

Jade was to the Chinese the most precious substance produced by nature, and the favorite material for placing in graves. It was believed to preserve the body and to aid in its resurrection. Many of the archaic pieces (which are well represented in the Museum collection) are carved from a kind of jade no longer obtainable, as the supply was scarce and soon became exhausted. Owing to long burial and chemical action of the soil, most ancient jades have undergone alterations in composition and color. In many instances these color changes have enhanced the beauty of the objects.

The popular saying that dead men tell no tales is a fallacy. Dead men do tell tales. Every detective knows it, and every

White Jade Incense Burner
Carved all over in open work comparable with most exquisite lace. Ming period (fifteenth–sixteenth century), China. About one-third actual size.

archaeologist who has learned to profit from the detective's methods knows it as well. The dead man tells us a vivid tale through the testimony of the objects interred with him in his grave. The jades unearthed from Chinese tombs are not dead and dumb stones, but speak an eloquent language to him who is eager to listen with sympathy to their voices. They reveal to us amazing stories, the earliest mythological concepts, man's intimate associations with the great cosmic powers, his love of nature, the content and meaning of his worship, his family bonds, his joys and sorrows, his yearning for immortal life, his constant solicitude about the hereafter. They are hymns to nature and the creator. The interpretation of the significance of all the manifold symbolism connected with these jades, their peculiar forms, and colors, is the result of many years of hard study and research, and the 2,000 labels of the eight cases in the Museum's jade hall offer a liberal education in Chinese art, religious thought, and symbolism.

Jade implements were fashioned as early as the neolithic age of China, and at first were on a par with common stone implements. When the belief gained ground that jade was a material of particular and superior virtues, however, it was set apart in a category of its own and was used exclusively for ceremonial and religious purposes. A stone chisel served for daily use, while a jade chisel was endowed with magical properties that would bring luck to its owner, who carefully kept it during his lifetime and had it buried with him. Axes, hammers, knives, daggers, and swords were likewise reproduced in the precious material and functioned in the grave as dispensers of light, demon-killers and dispellers of nefarious influences. Large swords and knives were emblems of sovereign power and also played a part in religious rites. Examples of all these types of jade objects are included in the collection at Field Museum.

In ancient times it was customary to send to the funeral of a deceased relative or friend an ornament of jade which was placed on the tongue of the corpse. This was the last tribute paid by the mourner to his departed friend. These ornaments were usually carved in the shape of a cicada. In the same manner as the larva creeps into the ground and rises again in the state of the pupa, until finally the cicada emerges, so the dead were believed to awaken to a new life. The cicada amulet therefore was an emblem of resurrection, an expression of faith and hope. The mourner's last gift signified that he desired to hear again some day the voice of his dear one. Many such amulets are exhibited in the jade hall.

Various novel uses of jade are illustrated in the Museum collection. The ancient Chinese notion of the shape of the earth, flat and square outside, and rounded in the interior, is illustrated by many emblems of the earth deity carved from jade in that shape. Of interest is a pair of sandals made of jade, and worn by ancient sovereigns during the imperial sacrifice to the deity of heaven. Jade handles for walking sticks, in the shape of pigeons, are included in the collection. The pigeon was believed to have special powers for digesting food, and gifts of these sticks to old men implied wishes of continued good health.

Many objects have historic interest. There is an imperial seal of jade, weighing six pounds, which was conferred upon the Empress Jui, consort of Emperor Kia-k'ing of the Manchu dynasty on February 12, 1796, when she received her first official appointment as empress of China.

Among pieces outstanding in novelty are jade chopsticks to please the vanity of an

(Continued on page 4)

TWO PROGRAMS FOR CHILDREN —RAYMOND FOUNDATION

A special additional program, as well as the final entertainment of the regular autumn series for children, will be given at the Museum during December. Both programs are provided by the James Nelson and Anna Louise Raymond Foundation for Public School and Children's Lectures.

The final program in the autumn series will be given on Saturday morning, December 5. Four films will be shown: "Winter Birds," "Snowflakes," "Mr. Groundhog Wakes Up," and "Skating in the Spreewald."

The special program will be given on Saturday morning, December 19. Two films chosen for their extraordinary interest and appeal have been chosen: "I Am from Siam," and "The Beaver People."

Both programs will be given twice, at 10 A.M. and 11 A.M., in the James Simpson Theatre of the Museum. Children from all parts of Chicago and suburbs are invited to attend.

17,000 PLANTS PHOTOGRAPHED

The joint project of the Rockefeller Foundation and Field Museum of Natural History to provide for botanists of the United States a complete reference collection of photographs of historic specimens of tropical plants of the western hemisphere has resulted to date in an assemblage of more than 17,000 such photographs.

J. Francis Macbride, Assistant Curator of Taxonomy, is still in Europe, where he has been since 1929, supervising the work of making these pictures. The original type specimens of famous botanists sent from Europe in America's early days, whose collections are now in European museums and herbaria, are being photographed. These include the first collections of plants made in America, chiefly by botanists sent by Spanish kings to investigate the resources of their then new territories. This work reached its climax about 1785 when Charles III of Spain ordered a scientific survey of all Spanish dominions in America.

The specimens being photographed include those from which scientists obtained their earliest accurate knowledge of the important plants which yield quinine, cocaine, rubber and other valuable products of commerce. Many of the plants photographed have never before been represented in botanical collections in the United States. The present project will give American botanists and students access to these without the former necessity of a trip to Europe. Copies of the photographs made by Field Museum and the Rockefeller Foundation will be available at cost to institutions and individuals all over the world.

HALL OF JADES OPENED

(Continued from page 1)

epicure, several sets of chimes made from jade, a pair of jade flutes of full size carved in imitation of bamboo, and intricately designed jade trees of chrysanthemums and pomegranates. There is a "longevity mountain," a landscape carved from a solid block of jade, with clusters of fungi representing immortality, and two cranes which were symbols of longevity. Large pieces in the collection include a jade incense burner delicately carved in an open work floral design as intricate and exquisite as fine lace; a bell of jade; a square green jade box used by officials of the Manchu dynasty for keeping seals; and "scepters of good

augury" which were considered to be magical wands.

Scores of figures of animals and birds carved from jade are shown, some in conventionalized and some in naturalistic art forms. Many kinds of jewelry, and many charms are included. Two lizards carved on a loving cup are emblematic of marital love.

In addition to jade, one case in the new hall contains Chinese art objects of rock-crystal, quartz, agate, tourmaline, turquois, amber and ancient glass.

Green Jade Monster
Used as an offering in a grave. Han period (about first century A.D.), China. About one-third actual size.

The foundation of the collection displayed in this hall was laid by the Blackstone Expedition to China, 1908–10, under the leadership of the Curator of Anthropology. Many additions were made during a subsequent expedition in 1923, known as the Marshall Field Expedition to China, also led by the Curator. In 1927 the Bahr collection of Chinese jades was acquired by the Museum with a fund contributed jointly by Mrs. George T. Smith, Mrs. John J. Borland, Miss Kate S. Buckingham, Martin A. Ryerson, Julius Rosenwald, Otto C. Doering, and Martin C. Schwab. Other objects were presented by individuals, chiefly John J. Abbott, American Friends of China, R. Bensabott, Inc., the late Richard T. Crane, Jr., Dr. I. W. Drummond, Fritz von Frantzius (deceased), Charles B. Goodspeed, H. N. Higinbotham (deceased), Linus Long, J. A. L. Moeller, Mrs. William H. Moore and Mrs. George T. Smith.

—BERTHOLD LAUFER

Many metals known to few people, with collections of objects illustrating their uses, are on exhibition in the Department of Geology.

Gifts to the Museum

Following is a list of some of the principal gifts received during the last month:

From Abbé Henri Breuil—41 prehistoric flint implements, France; from Stanley Field—23 figures, busts and heads of types of various races; from Harper Kelley—parts of a Magdalenian skeleton, France; from Dr. G. von Bonin—an ink stone, China; from Edmond I. Woodbury—10 wooden articles, Peru Indians; from Professor L. H. Bailey—250 herbarium specimens, Canal Zone; from C. H. Lankester—81 herbarium specimens, Costa Rica; from T. R. Williams—8 mahogany panels, Africa, Cuba, India and Mexico; from James Zetek—317 herbarium specimens, Barro Colorado Island and Canal Zone; from John Bigane and Sons—3 specimens fossil plants, Pennsylvania; from Walter Anthony Ragozzei—4 photographs of pillars produced by erosion, California; from S. R. Sweet—7 specimens skulls and jaws of fossil vertebrates, Nebraska; from E. A. Mueller—127 specimens fulgurites, Michigan; from Frank von Drasek—13 specimens acicular apatite and brookite, Arkansas; from Mrs. William H. Hess—weaver-bird's nest, India; from D. C. Lowrie—345 salamanders, Tennessee; from Count Degenhard Wurmbrand—a mounted birdskin, Austria; from C. Irving Wright—a large tarpon, Florida; from Thomas Abbott—35 crickets, China.

212

关于人类种族的放映大厅

THE PROJECTED HALL OF THE RACES OF MANKIND (CHAUNCEY KEEP MEMORIAL HALL)

By Berthold Laufer

Curator, Department of Anthropology

In 1935 or thereabout a convention of impressive magnitude is to take place in Field Museum. On this occasion the most perfect representatives of all living races will be assembled here. In order to facilitate study of their characteristic features and preserve them permanently, they will have been transformed from life into bronze, and will thus be presented to the public as durable monuments.

The hall selected for this unique convention is named Chauncey Keep Memorial Hall in honor of the late Chauncey Keep, a highly esteemed member of the Museum's Board of Trustees from 1915 until his death on August 12, 1929. A legacy of $50,000 left to the Museum by Mr. Keep will be applied to the cost of the exhibits in this ball. Added to this is a gift of $18,000 from Mrs. Charles Schweppe for the creation of a large central group in the hall. The balance of the cost of tbis hall, exceeding $100,000, is generously contributed by Marshall Field, whose continued interest in the work of the institution has been manifested in so many ways. Mr. Field's gift for this project is made in token of his affection and esteem for his friend, Mr. Keep.

The center of Chauncey Keep Memorial Hall will be occupied by a monumental bronze group—a triad representing in life size a white, a yellow, and a black man grouped in a circle. The group is surmounted by a globe, upon which are outlined the five continents as the habitat of the human species. The object of this monument is to emphasize the unity of mankind—man as a well-defined, fundamentally uniform species, which has spread all over the surface of the earth and conquered almost every habitable spot. While to some degree this triumvirate is symbolic, each figure in it is an outstanding type embodying the highest qualities of his race and worthy of minute study. This is the group presented by Mrs. Schweppe.

Radiating from this imposing central monument will be an avenue of primitive man, lined with twenty-seven life-size bronze figures of American Indians, Eskimos, Malayans, Africans, and Asiatics. These will not be standing at attention, but each will appear in lively action befitting the behavior of his particular group. To cite a few examples: the primitive Vedda of Ceylon is to be equipped with a bow, the native of Australia will be shown in the act of throwing a spear, the Bushman of the Kalahari steppe will display his prowess in archery while his spouse and offspring admiringly look on. A Solomon Islander will be seen about to climb a coconut palm, while natives of Java will be setting cocks to fight. Daboa of the African Sara tribe, in graceful movements of her slender body, will perform a coquettish dance, while an old Negro pounds an accompaniment on a drum. All these figures and groups, modeled from live subjects after years of painstaking study, will be absolutely correct in every detail of their anatomical structure and their accoutrements. Besides the life-size figures there will be numerous bronze busts and heads to illustrate the numerous variations of human types within the principal races.

The creator of all these bronzes is Miss Malvina Hoffman, an artist and sculptor of extraordinary ability and international reputation. Miss Hoffman studied painting under John Alexander, and sculpture under Herbert Adams and Gutzon Borglum of New York, as well as under the great master, Auguste Rodin of Paris. She has received numerous prizes and gold medals at exhibitions in Paris, New York, Philadelphia, and San Francisco, and many of her sculptures are on permanent exhibition in the Metropolitan Museum of Art in New York, American Museum of Natural History in New York, Academy of Rome, Art Museum of Stockholm, and Luxembourg Musée of Paris. Field Museum, however, will be the repository not merely of the largest number of her works, but of her finest and maturest creations. All her statuary is dramatically conceived and intense with life and motion. It is far removed from the ordinary plaster busts of racial types. Miss Hoffman is at present journeying in the Far East, stopping in Hawaii, Japan, China, Indo-China, Java, and India, to complete her task for the Museum.

The contents of Chauncey Keep Memorial Hall will include other material in addition to the work of Miss Hoffman. While her sculptures will dominate the hall, giving a clear and vivid impression of the appearance of man, special exhibits are required to illustrate many physical characteristics of mankind in greater detail. Exhibits of this class will include complete normal human skeletons, both male and female; a comparative series of skeletons of the principal races; and a human skeleton in comparison with the anthropoid apes, man's closest relatives in the animal world. Another exhibit will illustrate the capacity of the cranium, the size and characteristics of the brain, and its variations in apes and humans. Instructive charts will give information on the extensive variation of skin and eye color, and hair samples will demonstrate the structure, color, and differentiation of hair in the various races. Bodily proportions, as exemplified by the two extremes of giants and dwarfs, will receive due attention, as will bodily disfigurations such as artificial deformation and molding of the head.

Another section of this hall will be devoted to demography—charts and tables of vital statistics conveying information on birth and death rates, frequency of plural births, infant mortality, relative fertility of races, effects of disease and epidemics on the population, growth of population, longevity, effects of intermarriage and heredity, and other problems of general interest. A special feature will be made of the racial problems of the United States, with particular reference to our Negro population.

EXPEDITION AT KISH RESUMES OPERATIONS

The ninth season of excavations on the site of the ancient city of Kish by the Field Museum-Oxford University Joint Expedition to Mesopotamia has begun. Professor Stephen Langdon of Oxford continues as director of the expedition, but he will remain in England where he will conduct research upon the antiquities unearthed at Kish, as they are shipped to him. L. C. Watelin, in charge of operations in the field for several years past, will again head the party at work on the excavations.

Kish is believed to be the seat of the world's earliest civilization. To date the expedition has uncovered temples and palaces identified with Sargon I and Nebuchadnezzar; has found traces of the great flood recorded in the Bible; and has collected a vast amount of pottery, inscribed tablets, gold, silver and jewelry, remains of ancient chariots, and skeletal remains of human beings and domestic animals. As a result of studies of these things made by Professor Langdon much has been learned of the history and cultures of Babylonia back to about 5,500 years ago. Further revelations, as well as additional treasures for the Museum, are expected to result from the continuance of this work.

The expedition is financed on behalf of Field Museum by Marshall Field, and on behalf of Oxford by Herbert Weld and other British philanthropists.

Museum Luncheon for 600 Children

Six hundred children, members of the Four-H Clubs, an organization for farm youth, will attend a luncheon in the children's dining room at Field Museum on December 3, following a tour of the Museum's exhibition halls. The tour and luncheon have been arranged by G. H. Noble, Chairman of the National Committee for Boys' and Girls' Club Work. The children will be conducted on the tour by guide-lecturers of the staff of the James Nelson and Anna Louise Raymond Foundation for Public School and Children's Lectures. They are coming to Chicago to attend the International Live Stock Exposition (November 28 –December 5), and several other groups of children are also expected at the Museum during the exposition week.

New Exhibit of Geese

An exhibit of representative North American geese and swans has been installed in one of the bird halls at Field Museum. Fourteen species of geese and two of swans are shown. Those which have at any time been recorded in Illinois are marked with red stars, and of these there are nine.

Among the species shown are Canada goose, Richardson's goose, brant, black brant, Ross's goose, greater snow goose, blue goose, white fronted goose, pink footed goose, emperor goose, trumpeter swan, and whistling swan. The birds were mounted by Taxidermist Ashley Hine of the Museum staff.

CAREY–RYAN EXPEDITION SENDS SPECIMENS

Excellent specimens of the seladang (gaur ox or Indian bison) and of Indian water buffalo have been received at Field Museum, as a result of the Carey-Ryan Expedition to Indo-China, which recently returned. This expedition was financed by G. F. Ryan of Lutherville, Maryland, and was led by George E. Carey, Jr., of Baltimore, jointly with Mr. Ryan.

The Museum has received also collections of tree trunks, bark, leaves and other such materials from the forests in which these animals live, which will be used to construct scenic reproductions of natural backgrounds for the groups of animals where they are mounted. The exhibits will form part of the series of Asiatic mammal habitat groups in William V. Kelley Hall.

Messrs. Ryan and Carey had many adventures, the most thrilling of which was when a man-eating tiger attacked their hunting camp one night. The tiger dragged a coolie who belonged to the hunters' caravan from the camp, and later the unfortunate native's dead body was found. During the night the tiger revisited the camp several times, and, although the hunters opened fire with their rifles each time, the animal escaped.

A 400-pound lodestone, with unusually strong magnetism, is exhibited in the Department of Geology.

213

马球游戏

THE GAME OF POLO

By BERTHOLD LAUFER
Curator, Department of Anthropology

Many Museum visitors viewing the exhibit of four Chinese clay figures of women on horseback engaged in a polo match, exclaim in surprise, "We never knew polo was played in China, and that Chinese women indulged in athletic sports." Yet polo has had a long and honorable history in China, and has been a favorite subject of many illustrious painters and sculptors.

The clay figures in question, which were buried with sport-loving noblemen in the eighth century of our era, are the earliest monuments to polo now extant. The first great polo match on record was played in A.D. 709, at the imperial court of Ch'ang-an, between Chinese princes and Tibetan ambassadors who had arrived from Lhasa to receive a daughter of the Chinese emperor who was to marry the king of Tibet.

Polo was first played about the beginning of our era by Iranian tribes of nomadic horsemen inhabiting Central Asia, and from this center both the polo horses and the game were transmitted to Persia and China. In its origin it was not a game, but rather an exercise in preparation for war, and a trial of skill and endurance, on a par with archery.

In China polo was vigorously cultivated by several emperors of the T'ang dynasty, and also under the Sung dynasty, during which it was adopted as an exercise in the army. Under the Manchu dynasty the game became extinct.

There is a story of an old general, who used to place a pile of ten coins in the polo court, and galloping his horse strike one off with his club each time he passed, knocking the coin up seventy to eighty feet in the air.

The polo sticks are described as terminating in a point like the crescent moon, and are therefore styled "moon sticks." In Chinese paintings they appear provided with a scoop or ladle, exactly as in Persia. The balls were of an elastic vermilion painted wood, but leather balls are also mentioned. The players formed two teams and contended for the same ball. The goal was set up at the mouth of the course and consisted of two stakes connected by a board on top, making an open gate, in which was suspended a net to receive the ball. The side able to strike the ball into the net was the winner. The horses were gorgeously adorned with pheasant feathers, tassels, bells,

and metal mirrors. Once tossed into the air, the ball was not allowed to fall to the ground, and the highest ambition was to keep it spinning in the air, so that it never became detached from the stick.

It is a singular fact that in China donkeys and mules as well as horses were trained for polo. From ancient times Shantung Province has been celebrated for its enormous

Polo Player
Chinese mortuary clay figure of woman polo player.
One of a pair presented by David Weber.

donkeys, and it was there that the initiative was taken to train them for the game. In the year 826 an official of Shantung sent a present of polo donkeys to the imperial court and four renowned players who performed before the emperor. The prince of Ting-siang under the T'ang taught his ladies to play polo on donkey-back. The Museum owns several Chinese paintings representing women on donkeys playing polo.

Bird Collecting Expedition

Staff Taxidermist Ashley Hine was dispatched to California toward the end of last month to conduct an expedition which will make collections of important birds needed for addition to the Museum's North American ornithological series. A special effort will be made to obtain specimens of many small birds which are to be found in the middle and southern parts of the state during the next few months.

An Important Plant Collection

Field Museum has received in exchange from the Royal Museum of Stockholm, through Dr. Gunnar Samuelsson, a valuable collection of 1,336 specimens of plants for the Herbarium. The sending consists in part of 450 specimens collected in the State of Paraná, Brazil, by the late Per Dusén. These include many rare species not represented previously in the Museum Herbarium, and they are the more desirable because of the extreme care used in their preparation.

An equally desirable portion of the sending consists of 640 plants collected in Cuba by Dr. Erik L. Ekman.

Museum hours in March: Daily, 9 A.M. to 5 P.M.

Chancellor Collection Arrives

A collection of some 400 fishes, and numerous corals and other marine invertebrates, collected by the recently returned Chancellor–Stuart–Field Museum Expedition to Aitutaki, Cook Islands, was received at the Museum last month. Among the fishes are many remarkable for their curious forms and their beautiful coloration, and these will make excellent subjects for exhibits which are to be prepared in the near future. Material for addition to the study collections was also received.

Philip M. Chancellor, who sponsored and led this expedition and the previous Chancellor–Stuart–Field Museum Expedition to the South Pacific in 1929-30, is now engaged in supervising the making of a motion picture film, "The Dragon Lizard of Komodo." Part of this film, which will have sound effects, was made on the first expedition, and some scenes were taken at Field Museum as a result of the exhibition here of the Komodo lizard reproduction made from one of the specimens Mr. Chancellor collected.

214

与中国成为朋友的美国人向博物馆捐赠礼物

Field Museum is open every day of the year during the hours indicated below:

November, December, January   9 A.M. to 4:30 P.M.
February, March, April, October   9 A.M. to 5:00 P.M.
May, June, July, August, September 9 A.M. to 6:00 P.M.

Admission is free to Members on all days. Other adults are admitted free on Thursdays, Saturdays and Sundays; non-members pay 25 cents on other days. Children are admitted free on all days. Students and faculty members of educational institutions are admitted free any day upon presentation of credentials.

The Library of the Museum, containing some 92,000 volumes on natural history subjects, is open for reference daily except Sunday.

Traveling exhibits are circulated in the schools of Chicago by the Museum's Department of the N. W. Harris Public School Extension.

Lectures for school classrooms and assemblies, and special entertainments and lecture tours for children at the Museum, are provided by the James Nelson and Anna Louise Raymond Foundation for Public School and Children's Lectures.

Announcements of courses of free illustrated lectures on science and travel for the public, and special lectures for Members of the Museum, will appear in FIELD MUSEUM NEWS.

There is a cafeteria in the Museum where luncheon is served for visitors. Other rooms are provided for those bringing their lunches.

Busses of the Chicago Motor Coach Company (Jackson Boulevard Line, No. 26) provide service direct to the Museum. Free transfers are available to and from other lines of the company.

Members are requested to inform the Museum promptly of changes of address.

DEATH OF MARTIN A. RYERSON GREAT LOSS TO MUSEUM

Martin A. Ryerson, one of the original Incorporators of Field Museum, and actively associated with the operation of the institution as a Trustee and as First Vice-President from the Museum's earliest days, died on August 11 at his summer home in Lake Geneva, Wisconsin. He was 75 years old.

Mr. Ryerson was one of the Museum's staunchest friends and most active workers. He participated in the preliminary steps which led to the establishment of the institution, and was one of the members of the original Board of Trustees, being elected in 1893 and remaining on the Board until his death. In 1894 he was elected First Vice-President, and remained in that office also until his death. He served as a member of the Executive Committee from 1894 to 1914, and as a member of the Finance Committee from 1901 to 1932. His many generous gifts to the Museum, both in

money and in additions to the collections, placed his name high on the institution's list of Contributors, while the deep interest he displayed in every detail of the Museum's operations, and the eminent services he rendered to it, resulted in his election in 1922 as an Honorary Member. He was also a Corporate Member from the beginning of the Museum's existence, and became a Life Member about 1896.

As an Officer and Trustee of the Museum Mr. Ryerson found time, despite his heavy burden of widespread business interests, to devote much thought and effort to the building up of a natural history institution of which Chicago could be proud. His sage advice and many suggestions were of great value in the deliberations of the Board of Trustees. His affection for and interest in the Museum are evidenced in countless instances

Chicago Daily News Photo

Martin A. Ryerson

by the work he performed for it, and the many valuable gifts he made to it.

Mr. Ryerson as a young man attended Harvard University where he was graduated with a degree in law, and in later years three honorary degrees were conferred upon him—Master of Arts by Yale, Doctor of Laws by the University of Chicago and Doctor of Laws by Kenyon College. He became a trustee of the University of Chicago at the time of its founding in 1890, and for thirty years served as president of its board of trustees. He was active also in the affairs of the Art Institute of Chicago, where he became a trustee in 1890, served as president in 1925-26, and was honorary president from 1926 until his death. He was associated also in the work of many other civic institutions, among them the Rockefeller Foundation of Washington, the Chicago Old People's Home, the Sprague Memorial Institute, and the Chicago Orphan Asylum. In Grand Rapids, Michigan, city of his birth, he founded an excellent library.

Notably successful in business, Mr. Ryerson was a leader in the management of various important industrial organizations and leading banking institutions.

MAJOR JOHN COATS DEAD

Word of the death in August of Major John Coats, co-leader of two important Field Museum expeditions, and Patron, Contributor and Corporate Member of the Museum, has just been received in a letter from George G. Carey, Jr., of Baltimore. Major Coats died at his home in Ayrshire, Scotland.

With Captain Harold A. White of New York, Major Coats jointly financed and actively participated in the Harold White-John Coats Abyssinian Expedition of Field Museum in 1928-29, and the Harold White-John Coats Central African Expedition in 1930-31. The first of these obtained material for a large group of various animals at a water hole, now in preparation at the Museum, and also specimens for a habitat group of lions and aardvarks. The second expedition collected various mammals, including a number of excellent specimens of the rare bongo.

Major Coats's skill in the field commanded the respect and admiration of his comrades, while his personality inspired their deep

affection. Since the return of the expeditions, he had evinced the greatest interest in the work of preparing the groups at the Museum, and had hoped and expected to be present at the time of the opening of the water hole group when completed.

BUSY SUMMER IS REPORTED BY RAYMOND FOUNDATION

During the summer just ending, much educational work both for Chicago children, and for visiting groups of children from outside the city, has been carried on by the James Nelson and Anna Louise Raymond Foundation for Public School and Children's Lectures. In addition to the summer series of free entertainments during July and August, the Foundation staff lecturers have conducted many parties of children on tours of the museum exhibits. Among the out-of-town groups have been boys who had earned summer trips by securing subscriptions for magazines, groups sponsored by various civic organizations, and railroad and motor bus excursion parties. There were also groups of children who had been taking courses in various subjects over the summer radio school conducted by the *Chicago Daily News*, who were brought to the Museum to study material correlative with the broadcast lessons.

AMERICAN FRIENDS OF CHINA MAKE GIFTS TO MUSEUM

Two important objects were acquired this month from a fund presented to the Museum last year by the American Friends of China, Chicago. One is a unique figurine, six inches high, of celadon porcelain of the Sung period (A.D. 960-1279), representing the God of the North, who is conceived as a powerful warrior clad in a suit of armor. Seated erect on a rock, he sets his right foot on a tortoise held in the grasp of a snake wriggling around the tortoise's body. Both animals, in a previous incarnation, were mighty demons who were subdued by the God of the North and now accompany him as faithful attendants. This figurine felicitously supplements two other images of the same god in the Museum's collections, one of wood lacquered and gilded, and another of soft porcelain glazed in two shades of blue, both of the Ming period.

The other gift due to the Friends of China is a beautiful cover of cut velvet, fifty inches square, made for the palace under the reign of the emperor K'ien-lung (1736-95). The designs, an elaborate symphony of peonies and foliage woven in orange red, purple, violet blue, yellowish green and gold, stand out vigorously under the nap.   —B.L.

215

现代时尚的古代溯源

Field Museum News

Published Monthly by Field Museum of Natural History, Chicago

Vol. 3   JULY, 1932   No.

MODERN FASHIONS HAD ORIGIN IN ANCIENT TIMES

By Berthold Laufer
Curator, Department of Anthropology

The former kaiser is not the inventor of the mustache with turned-up tips; the Medicis did not invent the collar named for them; the Paris fashion dictators did not originate the décolleté and the high-waisted gown. There is little if any originality in modern fashions. Most of these were anticipated by peoples of the Far East many centuries ago. Spend a half-hour or so studying the exhibit of ancient Chinese clay figures just reinstalled at Field Museum, and you will be convinced.

First, with regard to the kaiser—the upright mustache was anciently worn in central Asia by equestrian tribes of Iranian and Turkish extraction. It was a privilege of the military aristocracy, the outcome of a superiority complex—an ornament regarded as accentuating manliness and martial prowess, and intended to strike terror into the hearts of the enemy. The Chinese were deeply impressed by these warriors, and modeled statues of them in clay, some of which are now in the Museum. They buried these statues with their dead as guardians of the grave. Armed to the teeth and enveloped in heavy suits of sheet armor, these valiant knights were supposed to be ready to fight for their dead master. Many of them are bedecked with the mustache à la kaiser, which was foreign to the Chinese, who cultivated a mustache only after reaching the age of forty, and wore it with long, drooping whiskers at the end.

The Museum has three unique clay statuettes noteworthy for their skillful modeling and delicate painted designs. They represent a princess of the T'ang dynasty (A.D. 618–906), seated, with her two ladies-in-waiting standing on either side. These may be regarded as actual portraits. They show silk dresses exquisitely painted in dainty colors, with designs on the borders indicating embroidery. The princess wears a high chignon in the form of a snail, held by a golden hoop. Her wrists are adorned with golden bracelets. Her décolleté, combined with short sleeves cut off above the elbows, is a noteworthy feature accentuating the modernistic note. She wears a rose-colored silk jacket with collar, the borders of which are embroidered. The skirt with red border along the lower edge fits over the jacket (high waist-line), and is held by a girdle with a very artistic knot in front. Pointed shoes complete her ensemble. The figure is 40½ inches high.

The two ladies-in-waiting are alike, each in a firm and graceful pose. Both of these statuettes are 48 inches high. The coiffures of these ladies are arranged in serpent windings. They wear very high shoulder capes

or what we call Medici collars, evidently fashionable in China many centuries before the Medicis. Each is provided with two jackets, an inner one with tight sleeves and an outer one with long, drooping sleeves. Their girdles, artistically tied, consist of red and green silken cords. The tops of their shoes are cut out into lotus designs.

Women's feet were poetically compared by the Chinese to lotus flowers. It was said that lotuses sprout forth from the steps of beautiful women, and this thought may have inspired the fashion of shoes with lotus designs. Until the end of the Manchu dynasty it was customary in Peking to

Ladies of Fashion in Ancient China

Portrait statuettes of a princess and her ladies-in-waiting, on exhibition in George T. and Frances Gaylord Smith Hall, showing how high waistline, Medici collar, and modern décolleté were anticipated in China hundreds of years before they appeared in Europe.

embroider a lotus on shoes worn by the dead at burial. The lotus, being an emblem of purity, was intended to convince the

(Continued on page 2)

Three Noted Scientists Visit Museum

Three distinguished scientists were visitors at Field Museum last month. Dr. Alexander Wetmore, Assistant Secretary of the Smithsonian Institution, in charge of the National Museum, was here on June 3. Dr. Thomas Barbour, Director of the Museum of Comparative Zoology at Harvard University, and Chairman of the Executive Committee of the Institute for Research in Tropical America, came on June 7. Sir Henry Wellcome, founder of the Wellcome Foundation, famous research laboratories in London, visited the Museum on June 9.

MUSEUM MAKING COLLECTION OF RARE ELEMENTS

By Henry W. Nichols
Associate Curator of Geology

The Department of Geology is assembling with the cooperation of Herbert C. Walther of Chicago, a collection of the rare elements. A number of these are already exhibited i Frederick J. V. Skiff Hall (Hall 37), bu owing to the extreme scarcity of many them it may take years to complete th collection.

The crust of the earth is composed ninety-two known elements. Eight of thes account for more than 98.5 pe cent of the crust. Only iron an aluminum among the heavy meta are included in these eight. Of th other eighty-four elements, onl five are present in quantitie greater than one-tenth of one pe cent. The remaining seventy-nin elements together comprise les than one-half of one per cent of th mass of the earth's crust. Most the useful metals such as coppe zinc, and silver, are included i this half per cent, and most of ther in quantities of less than one one hundredth of one per cent. If thes metals remained uniformly distrib uted they would be so difficult t procure that mankind would b deprived of their use. It is onl because geological agencies caus these useful elements to segregat in ore bodies that their use become possible.

The quantity of gold present i the crust of the earth has bee estimated as one-half of on millionth of one per cent, yet gol is not included among the reall rare elements being assembled i this collection. These exist in eve smaller quantities. Deposits con taining them are few and the con tent of a rare element in a deposi is often low.

The rare elements not so lon ago were mere curiosities and o scientific interest only. But som of them have such remarkable prop erties that in spite of their rarit and necessarily high price the have come into regular use. Radium an helium are examples. Some, as for exampl tantalum, are used whenever a supply ca be obtained, in spite of their price, because o qualities which, while present in lesser degre in other elements, are so exceptionall developed.

The present and probable future utiliza tion of these elements is the reason fo beginning this collection. Besides the rar elements proper, a number of what may b called semi-rare elements are included. Als whenever the elemental form, either from lack of use or from use restricted to certain industries, is likely to be unfamiliar to mos people, the common elements are displayed

The collection of fossil and amber-lik resins at Field Museum ranks among th finest of its kind in the world.

Field Museum of Natural History

Founded by Marshall Field, 1893

Roosevelt Road and Lake Michigan, Chicago

FIELD MUSEUM NEWS

STEPHEN C. SIMMS, *Director of the Museum*......*Editor*

CONTRIBUTING EDITORS

BERTHOLD LAUFER............*Curator of Anthropology*
B. E. DAHLGREN............*Acting Curator of Botany*
OLIVER C. FARRINGTON............*Curator of Geology*
WILFRED H. OSGOOD...............*Curator of Zoology*

H. B. HARTE.............*Managing Editor*

Field Museum is open every day of the year during the hours indicated below:

November, December, January	9 A.M. to 4:30 P.M.
February, March, April, October	9 A.M. to 5:00 P.M.
May, June, July, August, September	9 A.M. to 6:00 P.M.

Admission is free to Members on all days. Other adults are admitted free on Thursdays, Saturdays and Sundays; non-members pay 25 cents on other days. Children are admitted free on all days. Students and faculty members of educational institutions are admitted free any day upon presentation of credentials.

The Library of the Museum, containing some 92,000 volumes on natural history subjects, is open for reference daily except Sunday.

Traveling exhibits are circulated in the schools of Chicago by the Museum's Department of the N. W. Harris Public School Extension.

Lectures for school classrooms and assemblies, and special entertainments and lecture tours for children at the Museum, are provided by the James Nelson and Anna Louise Raymond Foundation for Public School and Children's Lectures.

Announcements of courses of free illustrated lectures on science and travel for the public, and special lectures for Members of the Museum, will appear in FIELD MUSEUM NEWS.

There is a cafeteria in the Museum where luncheon is served for visitors. Other rooms are provided for those bringing their lunches.

Busses of the Chicago Motor Coach Company (Jackson Boulevard Line, No. 26) provide service direct to the Museum. Free transfers are available to and from other lines of the company.

Members are requested to inform the Museum promptly of changes of address.

PRAISE FROM EDUCATORS

Too often are the benefits made possible by philanthropists taken for granted by a public which, if appreciative, at least seldom expresses its appreciation. A philanthropic enterprise which does elicit voluntary commendation from a large number of individuals, therefore, may be regarded as unusually successful in performing the service for which it was established.

For this reason it is very gratifying to note the response which has been made to the activities of two separately endowed units of Field Museum—the Department of the N. W. Harris Public School Extension, founded by the late Norman W. Harris and further endowed by surviving members of his family; and the James Nelson and Anna Louise Raymond Foundation for Public School and Children's Lectures, founded by Mrs. James Nelson Raymond in memory of her late husband.

Shortly before the closing of Chicago's

schools last month for the summer holidays, Field Museum received more than 100 letters from principals and teachers of Chicago schools expressing appreciation for the services rendered to education by the Harris Extension and the Raymond Foundation. The writers emphasized that the traveling exhibition cases of the Harris Extension, and the lectures and entertainments provided by the Raymond Foundation, perform functions of the greatest importance in the general educational activities, and that they are a boon to teachers and pupils alike.

Many of the writers remarked their gratification over the fact that it had been possible for the Museum to maintain these services in spite of the difficulties imposed by present economic conditions, and indicated their strong desire for continuance of the work when the schools reopen next autumn.

Frequently the letters had praise for the efficiency and courtesy of the Museum representatives sent to their schools in connection with these activities. A number cited specific advantages which their schools had derived from the service. Some mentioned the fact that the children themselves had often expressed their appreciation of the exhibits and lectures.

At the Museum every effort is constantly made to improve and expand the value of these services, and the inspiration derived from these letters will strengthen continuance of these efforts.

HERRING ARE QUEER FISH

By ALFRED C. WEED

Assistant Curator of Fishes

Although quite an ordinary fish in appearance, the herring has attracted fisher-folk from as far back as we have any record, and probably as far back as men have gone to sea. It was always an event of importance when fish of any kind could be caught easily and in large numbers. When such fish could be caught year after year almost as regularly as the return of the seasons it became the occasion for the founding of a great industry. Such an industry has often been founded on a herring fishery and grown until whole communities depended on it for their existence. Suddenly, for no apparent reason, would come a year when the catch was small. Next year, no herring at all. No one could tell where they had gone. No one in the region would see them again for years. The whole fishery would have to be abandoned, or else the fishermen would have to travel far out to distant seas to get the fish for their people to salt, smoke, dry or pickle.

Such has been the history of herring in Labrador. When first commercially discovered, its waters teemed with cod and herring. The herring were even more valuable than the cod. Their quality was considered the best in the world. They were large and fat, and could be caught easily. Fishermen from all parts of Europe flocked to those waters. Danish, Breton and Basque fishermen crossed the Atlantic to dispute the waters with Yankee skippers from the New England coast, while the governments of France and England argued and fought over the ownership of those stony and forbidding shores. Suddenly the herring stopped visiting these waters. For a whole generation they were not seen there. Then no one wanted the land. It was kicked back and forth like a football between Canada and Newfoundland. The cod were not valuable enough to bother with, and the wealth of salmon and trout was not considered. Finally the herring came back and the cod fishery became important. Canada and Newfound-

land both wanted Labrador. That dispute was settled about five years ago, but meanwhile the herring have come and gone twice.

The herring is a fish difficult to prepare for museum exhibition. Like so many of the fishes of the open ocean, it is delicate and fragile. Its scales fall off at a touch. It shines with a wonderful, silvery and pearly sheen that is entirely impossible to retain and almost impossible to reproduce. The celluloid reproduction in Albert W. Harris Hall (Hall 18) however, shows well the appearance of the scales, with color and brilliance almost equal to the living fish.

Museum Methods Described

How Field Museum hardens and preserves fossil bones with their original color and texture by means of a new method of impregnating them with the material known as bakelite is described in an article by Henry W. Nichols, Associate Curator of Geology, and P. C. Orr, also of the geological staff, which appeared in the May issue of *The Museums Journal* (London). The method was first used on fossils by Dr. E. C. Case of the University of Michigan, but certain modifications and innovations have been introduced at Field Museum which make it more suitable for the special requirements here.

Origin of Modern Fashions

(*Continued from page 1*)

Judge of Purgatory that the wearer was a person of good moral standing. Incidentally, it is interesting to note that although the custom of artificially bound feet, also known as lotus feet, sprang up during the T'ang period, all women represented in the clay figures have naturally proportioned feet.

The woman credited with the invention of the Medici collar is Catherine de' Medici (1519–89), daughter of Lorenzo de' Medici, Duke of Urbino. In 1533 she married the Duke of Orleans, who subsequently became Henry II, king of France (1547–59). Catherine is said to have brought from Italy to France the fashion of wearing the high collar still named for her. In those days, and earlier, Italy carried on a lively trade with the Orient, and Marco Polo had reported the wonders of China. It is most likely that the impetus to the Medici collar was received in Italy from the Orient long before the Medicis, and that Catherine by wearing it merely made it fashionable and lent it her illustrious name.

The reinstallation of the exhibits mentioned in this article has been made in new cases with concealed lighting and improved labels, greatly enhancing their attractiveness. The exhibits are in George T. and Frances Gaylord Smith Hall (Hall 24).

216

宗教卫道士及其奇迹

THE
OPEN
COURT

OCTOBER

1932

Vol. 46   *Number* 917

THE OPEN COURT

Volume XLVI (No. 10)   OCTOBER, 1932   Number 917

A DEFENDER OF THE FAITH AND HIS MIRACLES

BY BERTHOLD LAUFER

Field Museum of Natural History, Chicago

AN exhibit recently installed in a case on the East Gallery of
Field Museum, Chicago, comprises a number of fine, carved,
wooden images of Buddhist and Taoist deities most of which were
obtained by me from ancient temples in and around Si-an fu, the
present capital of Shensi Province and erstwhile glamorous metropolis of the Han and T'ang dynasties of China. Negotiations for
such images were by no means easy and required a great deal of
pourparlers, tact, and diplomacy. The mere hint at a commercial
transaction would have been regarded as an insult by the abbot of
a temple both to himself and the gods. It was necessary first to gain
the confidence and amity of monks and abbots and then to try to
effect on this basis an exchange of gifts—the gods while not venal
might be presented to one who possessed an intimate understanding
of their nature and who promised to treat them with care and reverence. Under this formula, by making a present to the abbot and
an offering to the temple, which reconciled the gods, I became the
proud owner of a statue of Wei-to, who is the loyal protector of
Buddha's temples and a staunch defender of his faith.

Not that statues of this god are rare, nearly every Buddhistic
temple boasts of one, but most of these are made after a common-
place routine by artisans who copy their models mechanically. This
one was singularly beautiful, free and original in his conception,
well carved and finely lacquered, the lacquer coat being mellow in
tone with age. Last but not least, this particular Wei-to was glorified by a tradition which clothed him with a nimbus in the eyes of
his admirers and imbued him with a historical significance of the
first order. The story briefly is this. Long ago, during the seventh
century, in the glorious age of the T'ang emperors, there lived at
Si-an fu a Buddhist priest, Tao Süan by name, a profound thinker
and an eminent writer on subjects connected with his religion. Like
all members of the monkhood he was deeply devoted to self-con-

centration and contemplation, looked upon as the great means of self-perfection. Meditation would naturally lead to dreams putting him in contact with the supernatural world. Tao Süan wrote his memoirs, and has left to us many precious documents in which he records his conversations with gods and spirits that revealed themselves to him in his inspirational dreams. Among others Wei-to appeared to him in his visions and directed him to have his statue made exactly as his apparition. Tao Süan obeyed the command, and from this time onward images of Wei-to were set up as the guardians of Buddha's temples and clergy. It is a singular fact that all Buddhistic divinities of China are derived from types created in India where Buddhism was born, but that Wei-to is the only Buddhistic deity conceived in China. He has the appearance of a handsome Chinese youth with a smiling countenance and is a powerful general fortified by a heavy suit of mail, ever ready to strike demons and the foes of the faith. Now it happened that the temple from which the Wei-to, at present in the Museum, has come was erected on the very spot where Tao Süan, master of contemplation and original creator of Wei-to, lived and taught. According to a tradition of this temple, which appears to be well founded, this statue was a direct descendant of Tao Süan's work, traceable to his inspiration, permeated by his spirit. It was regarded as a palladium capable of innumerable blessings, and one may realize how hard it was for my friend, the abbot, to part with this treasure for the benefit of Chicago.

The statue of this god was a great miracle-worker. Wei-to, above all, was a good provider, an efficient money-raiser and bill collector. In some monasteries where the monks thought more of their temporal than spiritual welfare they placed his statue in the kitchen the supervision of which was entrusted to his care. It even occurred that when he was installed as a culinary purveyor, the monks recited incantations, threatening him with severe corporal punishments if he should ever neglect his duty to supply the kitchen regularly with provisions.

Whenever it happened that the temple buildings were in need of repairs or that the walls had to be repainted or that a pagoda was to be restored, good Wei-to was instrumental in raising the necessary cash for such labors. In a case like this the brotherhood would stage a solemn procession through the streets and lanes of the city. One of the monks carried a shrine harboring Wei-to's picture, beating

a wooden drum in the shape of a fish and soliciting funds or subscriptions from wealthy shopkeepers and well-to-do families. If
not successful in this venture, one of the monks would deposit
Wei-to's image on the threshold of the house of a very prominent
family, obstruct the entrance, and remain seated there cross-legged
like a Buddha, for days if necessary, and would patiently wait till
the expected contribution was turned over to him.

If the monks failed in this quest of charity, the racket was pushed
to extremes. A member of the brotherhood was locked up in a
cage just high enough to allow him to squeeze in, and was openly
exhibited to the crowd in the market place. The door of this cage
was carefully shut with several padlocks, and the news was broadcast that the fellow in the cage was on a hunger strike, doomed to
die, and that he would not be released until the necessary amount
was raised. The people were urged to have pity with the moribund
man and to surrender speedily their loose change. In order to play
their feelings up to a pitch, it was alleged that the prisoner's bare feet
rested on sharp iron spikes; this in a way was true, but the spikes
were so deeply sunk into a wooden plank that it formed a smooth
surface, and moreover he was always secretly released before any
harm could befall him.

It will thus be seen that rackets are not an institution of recent
origin, or peculiar to Chicago, and that rackets also have their history whose threads may take us back to the intricate mysteries of
the Orient.

217

中国古代的科举考试

Field Museum is open every day of the year during the hours indicated below:

November, December, January   9 A.M. to 4:30 P.M.
February, March, April, October   9 A.M. to 5:00 P.M.
May, June, July, August, September   9 A.M. to 6:00 P.M.

Admission is free to Members on all days. Other adults are admitted free on Thursdays, Saturdays and Sundays; non-members pay 25 cents on other days. Children are admitted free on all days. Students and faculty members of educational institutions are admitted free any day upon presentation of credentials.

The Library of the Museum, containing some 92,000 volumes on natural history subjects, is open for reference daily except Sunday.

Traveling exhibits are circulated in the schools of Chicago by the Museum's Department of the N. W. Harris Public School Extension.

Lectures for school classrooms and assemblies, and special entertainments and lecture tours for children at the Museum, are provided by the James Nelson and Anna Louise Raymond Foundation for Public School and Children's Lectures.

Announcements of courses of free illustrated lectures on science and travel for the public, and special lectures for Members of the Museum, will appear in FIELD MUSEUM NEWS.

There is a cafeteria in the Museum where luncheon is served for visitors. Other rooms are provided for those bringing their lunches.

Busses of the Chicago Motor Coach Company (Jackson Boulevard Line, No. 26) provide service direct to the Museum. Free transfers are available to and from other lines of the company.

Members are requested to inform the Museum promptly of changes of address.

SPLENDID COOPERATION GIVEN BY BOMBAY SCIENTISTS

A large shipment of accessory material required for the preparation of eight habitat groups of Asiatic mammals which it is proposed to install in William V. Kelley Hall (Hall 17) was received last month from the Bombay Natural History Society under a cooperative arrangement between that society and Field Museum. The Museum is especially indebted to Sir Reginald Spence, Honorary Secretary of the Bombay Natural History Society, and S. H. Prater, Curator of its museum, for the splendid spirit of friendship and scientific cooperation which they manifested in making possible this arrangement.

The accessories received consist of trunks of native trees, samples of bark and leaves, soil, rocks, various plants and flowers preserved in fresh state, and other material needed in reproducing landscapes as environ-mental backgrounds for such animals as the sambur deer, swamp deer, axis deer, sloth bear, nilghai or blue bull, blackbuck, and Indian leopard or panther. The Bombay society sent out a special expedition to gather this material, which was sent in such quantity as to fill eleven packing cases and crates.

CIVIL SERVICE EXAMINATIONS AS HELD IN CHINA

By BERTHOLD LAUFER

Curator, Department of Anthropology

Up to 1905, toward the end of the Manchu dynasty, China was the classical land of examinations. The country was always guided by an aristocracy of intellect, not of birth; in fact, there was no hereditary nobility. There was free competition for all in obtaining official positions, and a very complex but just system of civil service examinations held by the government recruited the best talent from all ranks of society.

The object of these examinations was not, as with us, a test merely of knowledge, but a test of culture and literary ability. Elegance of style, in conformity with ancient recognized models, was the primary condition of the essays to be submitted. Perfect mastery of literature, a formidable memory and a highly disciplined mind were necessary, as quotations from the Classics had to be given with rigid accuracy. Degrees were conferred upon the successful candidates by the state, without the intervention of school or college, and opened the way to official rank and service, but were bestowed on only a small percentage of those who competed, the average being 2 or 3 per cent out of a number of 2,000,000 candidates who reported during the year.

Examinations were held in the capitals of the provinces once in three years, in special huge buildings known as examination halls, consisting of many rows of thousands of small cells. Remains of these may still be seen at Canton and Nanking.

A rough plank served as a table by day and a bed by night. The candidates were virtually imprisoned in these cells for a nine-day session, undergoing a great strain of their physical and mental powers, to which older people—there was no restriction as to age—frequently succumbed. It has happened that father, son and grandson have appeared at the same time to compete for the same prize. Each candidate took along into his cell all food he required, likewise fuel, candles, and cooking utensils, for no one was allowed to accompany him. The doors of the cells were sealed up and carefully guarded. On entering the hall, everyone was received by four soldiers and searched through his wadded robes, pockets, and shoes for precomposed essays or other illegitimate aids that he might have been tempted to smuggle in. Fraud was severely punished, and he who was caught risked dismissal and the loss of all titles and degrees previously acquired.

A silk handkerchief tucked in the sleeve certainly looked harmless and did not cause any suspicion. Such handkerchiefs therefore became a medium to aid the candidate's memory. They were inscribed on one side only. An example of a "crib" or "pony" of this kind was recently presented to the Museum by Edward Barrett of New York, who had obtained it on a recent trip to China. It is a strip of yellow silk, thirty-five inches long and fourteen inches wide, containing 24,365 finely written characters in 443 lines, copied from the work of Mong-tse, the most gifted of Confucius' disciples, a pioneer of statecraft and a practic philosopher. It must have taken man months to make this copy, and, as show by a collation of several passages wit the original text, it was exactly made probably in the K'ang-hi period (1662-1722

MONKEYS AID SCIENTIST IN COLLECTING PLANTS

By PAUL C. STANDLEY

Associate Curator of the Herbarium

Many more or less serious suggestion have been made that it might be possibl to train monkeys to aid the human race i performing difficult tasks, one being to pic cotton. Botanists collecting plants in th tropics, where the trees are so tall that it i impossible for a man to reach their branches often wish for a trained monkey that woul climb trees and bring down samples of leave and flowers.

Apparently one scientist has almost solve this problem. The botanists of Field Mu seum have just named a hundred plan specimens sent by Dr. Ray Carpenter o Yale University, who spent several month in Panama studying the black howle monkeys. These are the largest of a American monkeys. Dr. Carpenter com mented on the state of some of the speci mens sent, many of which were edible fruits He declared their imperfect condition wa due to the fact that the monkeys perche in the treetops picked the leaves and fruit and threw them down to him. Such intel ligent cooperation between man and beas in scientific investigation is without paralle No other monkey-collected plant collectio has ever reached Field Museum, at least

Philip M. Chancellor Visits Museum

Philip M. Chancellor of Santa Barbara California, who sponsored and led tw zoological expeditions for Field Museum the Chancellor–Field Museum Expedition to the South Pacific in 1929-30, and th Chancellor–Field Museum Expedition t Aitutaki in 1930, was a visitor at the Mu seum on June 28. Mr. Chancellor stoppe in Chicago on his way home from German where for several months he had bee engaged in studies. He is a Patron an Contributor of the Museum.

Patron of Museum Dies

Dr. George Frederick Kunz, who was a Patron and a Corporate Member of Fiel Museum, died June 29 in New York. D Kunz, internationally known as a mineral ogist and gem expert, was in his seventy sixth year.

218

《中亚与俄罗斯》序言

THE
OPEN
COURT

New Orient Society Monograph
Second Series   Number Two

MARCH

1933

Vol. 47   *Number 921*

THE OPEN COURT

Founded by Edward C. Hegeler

Editors:

GUSTAVE K. CARUS   ELISABETH CARUS

SECOND MONOGRAPH SERIES OF

THE NEW ORIENT SOCIETY OF AMERICA

NUMBER TWO

CENTRAL AND RUSSIAN ASIA

EDITED BY

BERTHOLD LAUFER

Published

Monthly: January, June, September, December
Bi-monthly: February-March, April-May, July-August, October-November

THE OPEN COURT PUBLISHING COMPANY

337 EAST CHICAGO AVENUE   CHICAGO, ILLINOIS

Subscription rates: $3.00 a year, 35c a copy, monograph copies, 50c

Entered as Second-Class matter March 26, 1887, at the Post Office
at Chicago. Illinois, under Act of March 3, 1879.

CONTENTS

THE OPEN COURT

Volume XLVII (No. 2)	March, 1933	Number 921

NEW ORIENT SOCIETY MONOGRAPH: SECOND SERIES   NUMBER TWO

CENTRAL AND RUSSIAN ASIA

PREFACE

BY BERTHOLD LAUFER

IN PLANNING this monograph it has been my aim to secure the good offices of the best living authorities on the three countries which are here represented. I was particularly fortunate in obtaining the collaboration of Dr. Sven Hedin while he resided in Chicago last year, supervising the erection of the Jehol Lama temple on the grounds of the Century of Progress. Dr. Sven Hedin, incontestably the greatest geographical explorer of Tibet of all times, past and present, has devoted his lifetime to science and research with a stupendous productivity in books and maps to his credit, all of permanent value. No one is more qualified than he to write on Tibet, which is his second home. In the sketch here presented he has outlined with the hand of a master a magnificent fresco painting, tracing the development in the exploration and opening of the land of mysteries, characterizing its geographical features, surveying its history, setting forth and interpreting its hierarchical system with the complex machinery of this priest-government and its relations to China, India, and England.

Mr. Owen Lattimore, author of *The Desert Road to Turkistan, High Tartary,* and *Manchuria Cradle of Conflict,* has extensively traveled in northern China, Manchuria, and Chinese Turkistan, and has studied political and social conditions with an open mind and keen observational power. His sketch of present-day Chinese Turkistan is a brilliant and penetrating analysis of intense interest. One of his statements that furnishes food for reflection is that although the currency of the country is worthless, yet its economic condition is remarkably steady, compared not only with China proper but with almost any country in the world and that although backward in every respect, it is probably more stable and contented than any region of equal area in the world. The latest news from Turkistan is that it seeks complete independence from Chinese sovereignty; Mr. Lattimore's article gives a clear answer to the why of this movement.

Mr. Lopatin is a young and energetic Russian ethnologist, now living and studying in this country. He has successfully explored the Goldi and Tungusian tribes of the Amur region, and has published many scientific monographs. The vivid picture that he unrolls here before our eyes of the transformation of Russian Asia under Soviet rule will be especially welcome at this moment when our Government seems to be determined to grant official recognition to the U.S.S.R.

I wish to express my warmest thanks to the three eminent scholars for their excellent contributions to this monograph.

A chapter to be devoted to Mongolia was scheduled in the original plan for this monograph. The subject, however, proved too large to be included here. Both Dr. Hedin and Mr. Lattimore have briefly touched on Mongolian problems, and in the monograph pertaining to China the editor will discuss modern cultural movements among the Mongols.

219

东方和西方

THE OPEN COURT

Founded by Edward C. Hegeler

Editors:

GUSTAVE K. CARUS   **ELISABETH CARUS**

SECOND MONOGRAPH SERIES OF

THE NEW ORIENT SOCIETY OF AMERICA

NUMBER FIVE

CHINA

EDITED BY

BERTHOLD LAUFER

Published

Monthly: January, June, September, December
Bi-monthly: February-March, April-May, July-August, October-November

THE OPEN COURT PUBLISHING COMPANY

149 EAST HURON STREET   CHICAGO, ILLINOIS

Subscription rates: $3.00 a year, 35c a copy, monograph copies, 50c

THE OPEN COURT

Volume XLVII (No. 7) October-November, 1933   Number 926

NEW ORIENT SOCIETY MONOGRAPH : SECOND SERIES   NUMBER FIVE

PREFACE

BY BERTHOLD LAUFER

THE contributors to this monograph, with a single exception, are Chinese scholars of note. In interpreting for us the conditions and aspirations of modern China, the Chinese are naturally placed in a better position and are more competent than any of us, and whether right or wrong are entitled to a fair hearing. The December issue of The Open Court will contain an essay on art tendencies in present-day China by Teng Kwei, which was too long to be inserted in this number.

Greece and Rome are irrevocably dead, but China with a past of five millenniums is still alive and looms in our eyes like a giant. In the nineteenth century the western world was still dominated by the ideal of classical humanism based on the study of Greek and Roman civilizations and restricted to the Mediterranean. This is a thing of the past. We now live in the better and bigger era of the Pacific humanism that recognizes the Pacific ocean as the center of world history and is more broad-minded in embracing the study of all great Oriental civilizations.

The study of Chinese civilization is not merely a fad or a capricious hobby or just one out of many hundred specialties in which the modern scholar fondly indulges. We study the language, literature, and art of China because such study has a paramount educational and cultural value and is part and parcel of a truly humanistic education. We are confident that the study of China will contribute much to the renaissance of our own civilization and will mean an important step forward into the era of a new humanism and philosophy that is now in process of formation.

CONTENTS

CHINA

NORTH AFRICA

THE OPEN COURT

Volume XLVII (No. 8)	December, 1933	Number 927

NEW ORIENT SOCIETY MONOGRAPH: SECOND SERIES NUMBER SIX

EAST AND WEST

A FOREWORD BY BERTHOLD LAUFER

AN avalanche of platitude and blah has fallen on the subject of differences between the East and the West. Discussing the spiritualism of the East versus the materialism of the West is a favorite sport of grandiloquent orators. As if there had never been any materialistic philosophies in the East and any spiritual tendencies in the West! We must not take East and West in the sense of abstract ideas, which will inevitably lead to vague idealizations, but must sense them as living realities in their proper setting and perspectives. In the first place, all orientals taken individually are not radically different from ourselves, they are just as human as you and I, subject to the same human emotions and passions; all shades and grades of character are found among them. Armenians, Arabs, Persians, Indians, Chinese, and Japanese are as shrewd, keen, and enterprising industrialists and merchants as any in the western world. India has been a dreamland of mysticism, speculative philosophies, and good fairy tales, but this has not prevented her from cornering the world market in precious stones for two thousand years. China has been a land of thinkers and great poets and artists, but this has not prevented her mercantile class from dominating the world market in silk, porcelain, and tea.

The fundamental divergencies are not between individuals or classes of people, but are deeply sunk in the thoughts of the folk mind fostered by a different background of civilization. There are only two such fundamental differences between the East and the West, which may be tersely formulated as the difference between the ego and the non-ego and the difference between the definite and the indefinite article.

China has an impersonal or non-ego culture, while ours in consonance with our heritage received from classical antiquity is an egocentric or ego culture, largely obsessed by the glorification and over-valuation of the individual, which has resulted in a standard codification of the individual's rights, while the East keenly emphasizes the individual's duties toward the family and the state. The Chinese mental complex has always been detached from the ego, without much regard for the individual, focused on the cosmos, the joy and deification of nature, striving for the beyond and reveling in dreams of eternity and immortality. The same observation holds good, more or less, for India and Japan.

A few practical examples will clarify this distinction between the ego and the non-ego aspect of culture. The Chinese, and other Asiatics likewise, have never hit upon the idea of erecting personal statues or monuments to their emperors, statesmen, generals, and war heroes, such as decorate or disgrace the squares and parks of our cities. No portrait of an emperor appears on any Chinese coin— quite in opposition to Greek, Roman, and late European coins. The Chinese erected marble arches or gateways to commemorate moral and abstract ideas, for instance, an extraordinary act of filial piety on the part of a dutiful son, or to honor a virtuous widow who did not remarry after her husband's death. This contrast of the ego with the non-ego philosophy finds its most noteworthy expression in the field of art. With the majority of our artists (there are exceptions, of course) vanity, ambition, self-love, and an inordinate craving for fame and notoriety are the principal incentives to work. Ostentatiously they paint or carve their names in huge letters on their pictures or sculptures and are prone to ascribe their work to their own merit and genius, usually forgetful of what they owe to their milieu, to their predecessors and teachers, and to the inspiration of a long and time-honored tradition. Thoughts drift along different lines in the East. China has produced the most skilful bronze founders, potters, and lapidaries the world has ever seen. We know their works, but are ignorant of their names. These men were too modest and too sensible to mar their productions by affixing to them their signatures. Among more than a thousand carvings assembled in the Jade Room of Field Museum there is not one inscribed with an artist's name. Why is it? The Chinese master just because he was a superlatively great artist was not fool enough to believe, nor did he flatter himself into the belief, that he personally was the creator of his creations, but humbly attributed them to the action of a higher power, to the merits and benign influence of his ancestors, or to the will and decree of Heaven. The artist was a sort of high priest: practising an art was a sacerdotal and sacred function. He produced, not to make a living or to please his contemporaries, but to honor his ancestors and to attain his own salvation.

As to the difference between the definite and the indefinite article, it amounts to this that East and West have an entirely different attitude toward religion. All Asiatic nations, excepting those that profess Islam, look upon religion as *a* means of salvation, as *a* road possibly leading to salvation—in opposition to the religions of Semitic ancestry. Judaism, Christianity, and Islam, each of which claims to possess *the* road, *the* only possible, truthful, infallible, permanent, and unchangeable road to salvation. It is at the surface merely the difference between the definite and the indefinite article, but this difference is profound, vast, and far-reaching, and has shaped the trends of history in the East and the West in almost opposite directions. The exclusive Semitic attitude toward religion naturally made for intolerance and persecution: the inclusive, broad-minded Far-Eastern attitude resulted in liberality, moderation, and tolerance. The Chinese in particular have been the most tolerant people in matters of religion, and have willingly listened to and extended hospitality to all religions that knocked at their door. No Chinaman has ever launched a campaign for religious persecution or would ever go to war for the triumph of a religious dogma, nor does he long to die for the glory of his country. He desires to live for it to the greater glory of his ancestors.

The ancient Semitic idea of blood-sacrifice, and redemption of sin by blood has always been alien to the humanized and refined spirit of India and the Far East. These happier nations were spared the ordeals and terrors of religious struggles and persecutions, as staged in Europe, all emanating from the merciless and violent Semitic idea of a jealous, ever irate, and vindictive deity thirsty for a sacrifice. Hence we face the sorry spectacle of what is called the history of Europe—an eternal rivalry and strife between the Church and the State, the temporal and the ecclesiastic powers, an endless chain of religious wars and persecutions, massacres of heretics and dissenters, burning of witches, the tortures of the inquisition—all for the triumph of theological dogmas. The one word, Canossa, denotes the martyrdom of our medieval society; then the clash of the Cross with the Crescent in the Crusades, the clatter of creeds, Spain combating the Arabs, the only cultured nation during the middle ages, to which Europe, then owed everything in the line of philosophy, medicine, and sciences. At the same time arts and letters flourished in China and Japan, and the great Chinese masters developed their sublime landscape painting which is now a source of joy and inspiration to the entire civilized world.

China has always had plenty of religion and religions, but religions only; never, however, an organized Church or a hierarchy or priesthood that would have meddled with state affairs or interfered with social customs or the freedom of the individual. One of the wisest institutions of China is that marriage has always been strictly a matter of civil law, the exclusive business of the family without interference on the part of a priesthood. This is the more remarkable when we consider that in all European countries civil marriage is an achievement of recent date and that our ancestors were compelled to struggle for centuries until the separation of the State from the Church was brought about and the Church was assigned to its proper place.

Among us, an individual is definitely labeled like a wine-bottle, in his peregrination from the cradle to the coffin. We consist of a birth certificate, a baptismal certificate, a vaccination certificate, possibly a marriage certificate, and infallibly a death certificate; or, as a cynic once expressed it, when we are born, they pour water on us, when we marry, they throw rice on us, and when we are buried, they throw dirt on us. Moreover, we are definitely labeled in matters of religion: we profess a religion officially and publicly, we may be associated with a certain denomination and be registered by a church to which we belong for a lifetime. Nothing like that exists in the East. The question so frequently addressed to a Chinaman as to whether he is a Confucianist, Buddhist, or Taoist is irrelevant; for, as a matter of fact, he is nothing of the kind or may be everything of the kind. No one in the East makes a showing of religion or professes it *urbi et orbi* as we do, and no one is attached to a church, for the simple reason that there is no such institution as a church in our sense of the word. The temples of China, Japan, Tibet, and Mongolia are essentially for the benefit of the monkhood which resides in them. The layman may visit a temple for the purpose of seeking the advice of a priest or consulting the deity by resorting to some means of divination, and he may visit a Buddhist or a Taoist temple on the same day, but there is no community service. Contributions to the maintenance of religious buildings and the clergy are frequently made, but there is no obligation or coercion, and any service is voluntary. The main concern of a Chinese is to obtain long life in this life and salvation in the other life. To him religion is a vehicle carrying him into a better land of bliss, and he welcomes any religion that holds out any promises of salvation and offers the best guaranties. He has never been willing to believe exclusively in one infallible religion that alone might be capable of bringing this result about. Whatever we may think about this attitude toward religion, we are compelled to admit that it has made for tolerance toward all religions and for a large measure of personal religious liberty. One of the most curious features of this development then is that the East with its non-ego, anti-individualistic tendency has ended with granting greater personal liberty to the individual, while the West with or despite its theory of the pursuance of individual happiness has finally succeeded in fettering the individual and restricting his movements to a minimum.

In studying other nations outside our own culture sphere and especially oriental nations we awaken to know ourselves and to see our own limitations. We have a great deal to learn from India and the nations of the Far East. We have frequently reproached the Chinese for their lack of patriotism and national spirit and have thereby merely displayed poor judgment and sheer ignorance of the historical factors involved. The ancient Greeks were not nationalists, but merely aimed at being civilized. True they were swallowed by the Romans politically, but their superior civilization conquered the Romans and the entire Roman Empire. Like the ancient Greeks, the Chinese people were never united by the principle of nationalism, but solely by the consciousness of a common bond of a great civilization. In a similar manner the Germans of the eighteenth century were not nationalists; Goethe and Schiller, Lessing, Herder, and Kant were cosmopolitans whose home was the world. German nationalism dates from 1871 with the foundation of the German empire. Nationalism will always run to extremes, and that extreme supernationalism such as prevails at present is no blessing we have seen from the days of the World War and see more and more from day to day. The present Chinese government, in accordance with the teachings of Sun Yat-sen, inculcates and fosters the spirit of nationalism, which is alien to their people and never formed part of their traditional background. Unfortunately they are compelled to adopt it from our "civilization," together with militarism, bombing planes, and other instruments of warfare. In some quarters this may be hailed as "progress." We are confident that this will merely be a transitory stage evolving into a finer and bigger era of true culture in the near future.

The New Orient Society of America, as evidenced by this series of monographs, endeavors to promote public interest in the Orient and to diffuse accurate knowledge of the oriental nations of the present time. Its objects are both scientific and educational, for only by serious study and research can we hope to be a safe guide and to fulfill our mission. We have no ax to grind, we carry on no propaganda, we are not pro this or anti that, but we preach a gospel of good-will and understanding, of honest cooperation and friendly reconciliation. We are not interested in politics nor in the promotion of trade; we do not tell you that if you will study the Orient and the customs and manners of its peoples you will be able to extend your business connections. But we promise you something more than mere material gain, we promise you that if you will study the Orient you will enrich your intellectual and spiritual life, that you will gain a new soul and that you will make the greatest discovery of your life—discovering yourself by discovering others.

220

中美观点的交流

SINO-AMERICAN POINTS OF CONTACT*

BY BERTHOLD LAUFER

ABOUT a hundred and fifty years ago Americans first came in direct contact with Chinese when the American ship *Empress of China*, sailing from Boston and rounding the Cape of Good Hope, cast anchor in the harbor of Canton. This occurred in the year 1784, under the reign of the great Emperor Ch'ien Lung, who was a contemporary of George Washington. Thus Americans were late arrivals—in fact, the last of foreign peoples to enter into commercial and political relations with China. Europeans, first the Portuguese, then the Spaniards, Hollanders, British, and French, had preceded them by several centuries. It is no empty saying that from the first days of Sino-American intercourse the two great countries have been linked by bonds of sympathy which have not existed and do not exist between China and any European power. These bonds of sympathy and friendship have been strengthened from year to year, as witnessed particularly by the ever increasing number of Chinese students and scholars annually flocking to our universities athirst for knowledge.

What, then, have Americans and Chinese in common? I think, a goodly number of very fine traits. First, the spirit of democracy, which has pervaded China for more than two thousand years, ever since the First Emperor Ch'in Shi smashed the old feudal system. The principle of government for the benefit of the people certainly is American, but it is equally Chinese and goes back to the fourth century B.C., when Meng-tse (Mencius), the most gifted of Confucius' disciples proclaimed the doctrine, "The people are the most important in a nation, and the sovereign is the least important of all." Second, the spirit of religious tolerance. I know of no more tolerant nation than the Chinese. Third, the lack of a caste system and lack of a hereditary nobility. China was always guided and governed by an aristocracy of intellect, not of birth; the old system of free competition by civil service examinations recruited the best talent from all ranks of society. Fourth, Americans and Chinese do not suffer from the obsession of that great evil, the race superiority complex; they are averse to armed force; they are friends of

*Reprinted from THE SCIENTIFIC MONTHLY, March, 1932, Vol. XXXIV.

peace, and are animated by a deep sense of justice and fair play toward all, regardless of race, color, or creed. Fifth, and this is the greatest asset that the two nations have in common, they have an unbounded, almost religiously fanatic, faith in the power of education and knowledge as the best guarantors of progress, as the best possible safeguards of the permanence of their social structure and institutions. With this capital of a common historical tradition and mentality—democracy, tolerance, equality, justice, and education—we are well prepared to stand the test and storms of the time.

Aside from these ideals, there are culture elements inherent in the two civilizations that establish a common basis for a harmonious social life and sympathetic fellowship among representatives of the two nations. In reflecting on cultural similarities between Americans and Chinese, it is advisable to proceed from realities and direct observations. A white man who is in a good state of health is able to live in China in a house of Chinese style, in a purely Chinese surrounding, on Chinese food, in every fashion exactly like a Chinaman, not only for years, but a lifetime, without suffering impairment or injury to his health. Chinese houses are very much like our own; their plan of arrangement comes very close to that of the ancient Roman house. Rooms are airy, spacious and well ventilated, and comfortably stocked with tables, chairs, armchairs, settles, and sofas. There is no other nation in the world whose house furniture offers so complete and striking a coincidence with our own. In fact, it is one of the amazing points of culture history that of all nations of Asia the Chinese is the only one that takes its meals seated on chairs around a table, in the same manner as we do. This custom was acquired by the Chinese only in comparatively late historical times. The ancient Chinese, down to the epoch of the two Han dynasties, used to squat at meal times on mats spread over the ground, in the same way as it is still customary with the Japanese and the peoples of India. The remarkable step leading to the use of raised chairs and high tables was taken in the period between the Han and T'ang dynasties, as a sequel of many foreign influences that came from Central Asia at that time, and speaks volumes in favor of Chinese adaptability and readiness to adopt foreign institutions. The Japanese, with all their temperamental changeability, still adhere to the old primitive custom of sitting cross-legged on the mats covering the floors of their rooms; and while an American, for the sake of curiosity or experience, may enjoy living in a Japanese home for a few days or weeks, he will never acquire the Japanese mode of sitting, which is a source of physical discomfort to us.

The objection may be interposed that many travelers and adventurers in almost all parts of the world have conformed to the life of the natives whom they set out to explore. Such examples indeed are numerous. Any normal individual of good physique and temperate habits is able to live wherever other human beings of whatever race can exist, whether they be Eskimo, American Indians, South Sea Islanders, Pygmies or Negroes, Berbers or Beduins; but such adventures are usually transient, and the explorer will always be glad, once his task is accomplished, to return into the harbor of "civilization." Speaking of myself, it fell to my lot to live for many months among such primitive folks as the Gilyak and Ainu of Saghalin Island, the Golde and Tungusian tribes of the Amur region, sharing their huts or spending the night in the open, sleeping on a bearskin, living like them on salmon and game, even amid smallpox and trachoma epidemics, without any harm to my health, save a temporary discomfort from parasitic insects. I could not, however, have stood this sort of life for a number of years, and while I enjoyed studying these tribes and gathering data concerning their daily life, languages, folk-lore, and religion, I can not say that I felt at home with them, at least not so intimately as I do feel at home with the Chinese. It was also my good fortune to spend a year and a half among the Tibetans, both the nomads and the agriculturists, just living like one of them; and while the Tibetans have my unstinted sympathy, the time I should be willing to dwell in their midst will always be one of restricted duration. The lesson to be retained, therefore, is that a robust man with a definite object in mind may live anywhere without hazard of life and welfare within a limited period, whereas no such time limit is attached for us to China. Again, it can not be doubted that many white individuals have settled among Indians, Eskimo and other primitive peoples, taking native women as their wives, even adopting native speech, clothing and habits, and thus ending their days. Examples of this kind are not typical, however, and such individuals have usually been fugitives, castaways, tramps, derelicts or sailors cast adrift.

In order to settle among the Chinese, no foreigner need feel anxiety about his health, at least no more than if he stayed at home, nor does he require the explorer's physical fiber. China beckons to

the man of culture, and the more cultured he is, the more welcome
and the happier he will be there, since the Chinese are highly cultured, well-bred and well-mannered people. Even most Chinese
farmers and laborers are gentlemen, and from many of them many
a so-called gentleman in our midst could learn a useful lesson in good
manners or etiquette.

One of the most remarkable inventions ever made by the Chinese
is the chopsticks, "the nimble ones," as they are called in Chinese,
the invention of which goes back to the days of the Chou dynasty.
Chopsticks are not only characteristically Chinese but also set the
Chinese people clearly off from other nations of Asia that are still
in the habit of taking food to their mouth with their fingers, which
is even done by so highly civilized people as those of India. Annamese, Koreans, Japanese, and other peoples who came under the
spell of Chinese civilization adopted from the latter the use of chopsticks. It is self-evident that these make for good table-manners,
which are the first criterion of a civilized individual; and whatever
opinions we may hold on the Confucian system of ethics, it is un-
deniable that it has at least brought about the one good effect to
transform the majority of the people into a body of highly decent,
respectable and well-bred men. The sanctity of the home and the
purity of family life belong to the greatest achievements of Confucian social ethics. For all these reasons, official and personal intercourse of Americans with Chinese is easy and a source of pleasure.
Their sense of humor, their delight in story-telling, their conversa-
tional gifts and oratorical power are other qualities that will not
fail to make a strong appeal and endear them to us the closer we
get acquainted. At Chinese parties there is less formality and con-
ventionality than in our country.

Their eminent faculty of assimilating and absorbing foreign racial
elements has struck many observers. In fact, the Chinese no more
than any other nation represent a pure race. The northern Chinese
have a strong admixture of Tungusian, Mongol and Turkish blood;
the southerners have to a great extent intermarried with the aboriginal tribes which preceded the Chinese as owners of the country.
The question of intermarriages of Chinese and whites is naturally
a delicate one, and it would be futile to generalize on so vital and
large a problem; but if limited personal experience and observation
may count a little, I may say that many happy marriages of Euro-

peans and Americans with Chinese women have come within my notice. There is no gulf separating the two races, and there are no obstacles of a racial or cultural character in the way of such unions. The offspring of American fathers and Chinese mothers belong to the best citizenry of China, and commanding the two languages as they do, they make the best liaison officers to maintain and strengthen the bonds between East and West. Many of these Eurasians are splendid fellows, and I have found in them the most willing and enthusiastic helpmates in scientific investigations.

As an analyst of human nature I should be the last to deny that there are psychological differences between Chinese and ourselves. These, however, do not spring from a basically divergent mentality or psyche but are merely the upshot of a distinct set of traditions and education based upon the latter. As the grasp of ancient traditions upon their minds will gradually loosen and as the best in our institutions and inventions will be adopted (I advisedly shun the ambiguous and much misused word "progress"), these small divergences will gradually disappear or be reduced to a minimum. The abandonment of foot-binding and opium-smoking may be cited as relevant instances. The student of anthropology who has learned to fathom and to understand the customs and usages of every people knows only too well that the Chinese are not different from other peoples but are just human and humane. There is no custom in China that in one or another form would not appear among other peoples or even among ourselves. The Chinese worshiped their ancestors and to a large extent still do so; they are justly proud of their ancestors, and in their modesty attribute their own good luck and success to their ancestors' virtues and beneficent influence. We with our pride in ancestors and with our passion for genealogical quests, are no less ancestor worshipers; our "worship" has merely assumed a different form.

221

海龟岛

TURTLE ISLAND

BY BERTHOLD LAUFER

Dedicated to Dr. Moritz Winternitz, Professor of Sanskrit at the University of Prague, in honor of his seventieth birthday, December 23, 1933.

IN his monumental work *Geschichte der indischen Litteratur* Professor Moritz Winternitz has devoted an admirable chapter to a discussion of Indic stories and their migration eastward and westward, and observes wisely (II. p. 105), "Although many tales may have found their way from India into the West, yet it can hardly be doubted that also many a foreign tale has migrated into India. This, for instance, might be the case with reference to the mariners' tales which relate shipwrecks and various strange adventures at sea." On the other hand, Erwin Rohde (*Der griechische Roman,* 3d ed., 1914, p. 193) is inclined to trace to India the more important motives of the Arabic marine novels and sees Indic influence likewise in the Greek literature of marine romances.

. About a decade ago I read a paper before the American Oriental Society under the title "Tales of the Indian Ocean" of which I gave the following definition:

> There is a type of mariner's story or sailor's yarn which we meet in all countries bordering on the Indian Ocean—in Greek, in Syriac and Arabic, in Sanskrit or Pāli, as well as in Chinese. Where and how these stories originated is often difficult to decide, and it seems best to characterize them simply as tales of the Indian Ocean. The Indian Ocean, so to speak, functioned as the broadcasting station which sent these stories out to all ports. The Indian Ocean had a peculiar fascination upon the minds of Greeks, Arabs, Persians, Indians, Malayans, and Chinese; its many wonders stirred their power of imagination, and its marine animals even gave rise to new mythical conceptions.

Most of these tales appear to have originated in the circle of navigators and to have been spread by sailors from one port to another. This fact is clearly disclosed in the story of "The Capture of the Rhinoceros" which a Chinese physician of the T'ang period. as he states advisedly, recorded from the lips of a foreign sea-captain whom he had met in Kwang-tung (Laufer. *T'oung Pao,* 1913, pp. 361-364 and *Chinese Clay Figures,* pp. 145-147).

The story of Turtle Island or Whale Island belongs to this cycle. Zacher (*Pseudocallisthenes,* p. 147) characterizes it well as "one of those very ancient migratory tales coming down from an unknown period, which float between Orient and Occident from early times." Its distribution has often been discussed, but the Chinese versions have not yet been utilized.

The *Kin-lou-tse* ("The Golden Tower", chap. 5, p. 19) written by Yi, prince of Siang-tung, afterwards the emperor Hiao Yüan of the Liang Dynasty (A.D. 552-554), which contains several curious traditions pointing to a foreign origin, offers the following tale:

> Once upon a time there lived a huge turtle amidst sandy islets. The animal's back was covered with trees which made it appear like a regular island in the ocean. It happened that merchants came there, and believing that it was an island, gathered fuel with a view to prepare their food. The turtle was burned hot and dived back into the sea, whereupon several tens of men suffered death.

However terse and sober this account may be, it embodies all essential elements and represents the primeval version of the story which seven hundred years later appeared in the romance-like adventures of Sindbad the Sailor in the Arabian Nights.

About a century later the Chinese were treated to an Indic version of the story. Hüan Tsang, the illustrious Chinese pilgrim to India (Julien, *Mémoires*, I, p. 474; Beal, *Buddhist Records*, II, p. 125) tells in his Memoirs about a merchant prince from Jāguda, who worshiped the heavenly spirits and despised the religion of Buddha. With some other merchants he embarked in a ship on the southern sea and lost his way in a tempest. After three years their provisions became exhausted, and they invoked the gods to whom they sacrificed. All their efforts were futile when unexpectedly a great mountain with steep crags and precipices and a double sun radiating from afar was sighted. The merchants were overjoyed at the prospect of finding rest and refreshment on this mountain. But the merchant-master exclaimed, "This is no mountain, it is the fish makara (whale); the high crags and precipices are but its fins and mane; the double sun is its eyes as they shine." The master then remembered Avalokiteçvara as the savior from the perils of the sea, and they all invoked his name. The high mountains disappeared, the two suns were swallowed up, and the mariners were rescued from shipwreck through the intervention of a Çramana walking over the sky.

It is obvious that the Buddhists made use of an old story and adapted it to their own purpose. Rescue at sea through the intervention of the Buddha or his saints is a frequent motive in Buddhist hagiography and iconography. Foucher (*Etude sur l'iconographie bouddhique,* p. 82) points out a bas-relief of Bharhut where a makara devours a ship with its crew (Cunningham, *Stūpa of Bharhut,* plate 34,2). Further, an allusion to this motive is made in the Tibetan cycle of legends associated with the name of Padmasambhava of the eighth century (Grünwedel, *Buddhistische Studien,* p. 106).

An echo of Hüan Tsang's story, as pointed out by me in *Journal of American Folk-lore,* 1926, p. 89, occurs in *Liao chai chi i,* No. 82 (translated by H. A. Giles under the title *Strange Stories from a Chinese Studio*). I have no intention of covering the whole ground occupied by this legend or giving a complete bibliography of previous studies; suffice it to refer to Zacher, *Pseudocallisthenes,* p. 147; Runeberg, "Le conte de l'Isle-poisson," (*Mémoires Soc. néo-philol. à Helsingfors,* II, pp. 343-395), and Cornelia C. Coulter, "The 'Great Fish' in Ancient and Medieval Story" (*Transactions Am. Philol. Assoc.,* LVII, 1926, pp. 32-50). I do not agree with previous investigators in regarding the heroic exploit of Keresāspa in the Avesta as the earliest version of this tale. This, in my estimation, is entirely distinct: Keresāspa slays a horned monster on land, which is the principal motive of the story, while the feature that he cooked meat on the monster's back is merely an incidental accessory. There is no sea, no monster island, no casualty or rescue from this alleged island in the Avestan episode. If anything is clear, it is the fact that the monster island motive must have originated in a maritime setting and have grown out of marvelous incidents of a sea voyage. Thus remain as the oldest versions the cycle of the Greek romances of Alexander and the Physiologus. In one of Alexander's alleged letters to Aristotle, the monster is specified as a "giant turtle" (otherwise "sea-monster"; see Ausfeld, *Der griechische Alexanderroman,* p. 178).

A turtle appears in the *Wonders of India* (*Livre des Merveilles d l'Inde,* ed. Van der Lith and Devic, p. 37). Qazwīnī, in his Wonders of Creatures, likewise connects the story with a marine turtle, and such we find in the Chinese version of the *Kin-lou-tse.* The transformation of the turtle into a whale seems to be due to the

Physiologus. The Chinese version, I am inclined to think, was transmitted to China by oral tradition. I can cite a specific instance of the occurrence of our tale, where literary diffusion is out of the question. It was recorded about eighty years ago among the Karen, an illiterate tribe of Upper Burma, by Francis Mason ("Religion and Mythology of the Karen," *Journal As. Soc. of Bengal*, XXXIV, 1858). As it has escaped the previous writers on the subject, it may be cited here in extenso.

> The Elders among the Karen say there are fish in the sea as large as mountains, with trees and bamboos growing on them as on land. Voyagers have to be careful where they land to cook. They carry axes, and cut into the ground to try it. If juice springs up where it is cut, they know that they are on a fish; but if the ground seems dry, they are on land, and go to cooking. It is related that a man landing on an island went to cooking without trying his ground, and it turned out to be a fish which sunk with him into the sea and then swallowed him. When the man was in the fish's belly, he said to the fish: 'When males acquire large game, they shout and cry out in exultation, but you are silent. Are you not a male?' On hearing this, the fish opened his mouth to scream, when the man leaped out and escaped. The elders say that when people kill one of these fish, it is impossible for them to eat it all up, and they burn its fat. With its bones they can make beams and rafters for houses.

In this version the "island" motive is connected with the "swallow" motive, both of which, according to Miss Coulter in the article quoted above, were developed in India and spread westward, leaving their mark in turn on Greek, Arabic, medieval Latin, and the vernacular literatures. While I concede the possibility of an Indic origin of the "island" motive, I am not so sure of the "swallow" motive being specifically Indic (compare my article "The Jonah Legend in India," *The Monist*, 1908, p. 576). Allusions to the "swallow" motive occur in Chinese authors of the pre-Christian era when Indic influence is out of the question. Several ancient philosophers (Chuang-tse among them) use the phrase "the boat-swallowing fish" (*t'un chou chi yü*) as a well-known affair or a firmly established expression: thus, Shi-tse (*Chu tse wen sui sii pien*, ed. by Li Pao-ts'üan, chap. 9, p. 7) says, "Where water gathers, the boat-swallowing fish will arise" (cf. also Pétillon, *Allusions littéraires*, pp. 313, 497, and Pelliot, *T'oung Pao*, 1920, pp. 294, 351).

The *Kin-Lou tse* contains the expression in three passages (chap. 4, pp. 18, 19b, 25b). The only explanation of this phrase I have found thus far occurs in the *Chi lin sin shu* (*op. cit.*, chap. 5, p. 15): "In the southern region there is the alligator fish [thus literally: *ngo yü*] whose snout is eight feet long and which reaches its largest size in the autumn. The fish stretches its head out of the water and swallows the men near the border of the ship. Other men in the boat seize spears and try to keep the fish off." Granting that it might have happened in ancient times that a frail boat struck an alligator, a huge fish, or some species of whale and capsized, drowning some of the sailors, the report could easily gain ground that these men were swallowed by a marine monster. As long as we do not know more about the boat-swallowing fish of the ancient Chinese, I am rather disposed to credit it to an actual experience or several experiences than to an outside influence.

Ulrich Schmidt, in his *Voyage to the Rivers La Plata and Paraguai* (1567, Hakluyt Soc. ed., 1891, p. 86), relates, "Between S. Vicenda and Spiritu Sancto there are plenty of whales which do great harm: for instance, when small ships sail from one port to another, these whales come forward in troups and fight one another, then they drown the ship, taking it down along with the men." This is not a reminiscence of the tale of Whale Island, but the simple record of incidents or experiences within a well-defined locality of South America. Qazwīnī (*Kosmographie*, translated by K. Ethé, p. 268) remarks in his description of the whale that sea-going vessels have much to suffer from it and that it devours whatever it finds. Another kind of whale (p. 289) is defined by him as a very large fish which can smash a ship. See also *Livre des merveilles de l'Inde*, pp. 14-15.

222

热河金庙模型到展

Field Museum News

Published Monthly by Field Museum of Natural History, Chicago

Vol. 4   APRIL, 1933   N

THE BOWER BIRD, AN AESTHETE OF THE ANIMAL KINGDOM

By RUDYERD BOULTON
Assistant Curator of Birds

One generally thinks of an aesthetic sense, the appreciation of color, form, and sound for their own sakes, as an attribute solely of human beings. For example, there is no evidence that a bird will select for its mate one that is more brightly colored than its fellows, or that sings, to our ears, more beautifully than other birds. Detailed studies indicate that the songs of birds are expressions of physical vigor, and many songs are known to be warnings to rivals rather than invitations to mates.

The bower birds of New Guinea and Australia, however, definitely display a sense of beauty which makes them unique in the animal kingdom. It is the habit of these birds to build complicated structures, which they decorate in various ways and use as playgrounds during the period of courtship and mating.

A habitat group of the New Guinean fawn-breasted bower bird was recently installed in Stanley Field Hall. This is a species which builds a bower on the ground, with platforms at each end, constructed of twigs and sticks by the male bird. He devotes about two weeks to the task and performs a remarkable feat of architecture. One platform is plain, but the other is definitely decorated with fresh and colorful berries, fruits, and leaves. These are not eaten; they are purely for ornamentation. They are replaced at frequent intervals, and the bird carries the old withered decorations to a neatly maintained rubbish pile near-by, instead of scattering them about indiscriminately. On the decorated platform the male performs a courtship dance, while the female stands on the undecorated platform at the opposite end of the bower to watch. When the courtship is over, a nest is built high in a tree near-by, and the bower is used as a playground by the male.

The Museum group, consisting of both male and female birds, and a bower, illustrates the courtship stage. The birds and the bower were collected near the Sepik River in New Guinea by Assistant Curator Karl P. Schmidt and Walter A. Weber while members of the Cornelius Crane Pacific Expedition of Field Museum in 1929. The birds were mounted by Assistant Taxidermist John W. Moyer.

Another species of bower bird uses shells and shiny pebbles for decorations, while a third species builds a mossy roofed hut and distributes flowers and bright petals on the carefully leveled dooryard of moss, renewing them as fast as they wither. An Australian bower bird has recently adopted as cherished decorations for its playground pieces of broken china and glass bottles, showing that the selection of materials is not an iron-bound, inherent mechanical reflex. Here, indeed, is an artistic genius among bower birds! One can think of many parallels in human society, but conscious effort devoted to a non-utilitarian result is not common among animals. The hoarding of bright-colored objects by jays and crows is one of the few comparable instances known. Activity of this kind is, no doubt, an outgrowth of secondary sexual charac-

Bower Bird Group in Stanley Field Hall
The male bird is seen performing his courtship dance on the platform decorated with fruits and berries, while the female watches through the bower.

teristics, such as the drumming of the ruffed grouse and the dancing of the prairie chicken, which, as in the case of the bower birds, is performed by the males whether or not any females are present. It is dangerous to describe these actions in the anthropomorphic terms of human psychology and behavior, yet, in default of detailed modern studies, one is left no choice.

Gift from C. Suydam Cutting

Through the generosity of C. Suydam Cutting of New York, Field Museum has received an extremely interesting collection of birds and mammals from Upper Burma. The collections were made by Lord Cranbrook and Captain F. Kingdon Ward. Among the most interesting specimens are several rare water shrews and moles, and paratypes of two species of new babbling thrushes, recently described by N. B. Kinnear of the British Museum (Natural History). A pair of very rare blood pheasants is also included.

JEHOL PAGODA MODEL ON EXHIBITION

By BERTHOLD LAUFER
Curator, Department of Anthropology

An exact miniature reproduction of pagoda in the imperial palace of Je China, the region recently invaded by Japanese, is on exhibition in the Sc Gallery on the second floor at Field Muse The original of this pagoda, which octagonal in shape, contains nine sto and is 213 feet high. It is one of the fi pagodas in northern China.

Between the years 1751 1765, the Emperor K'ien-l made four journeys thro the midland provinces of empire. On his visits to N king and Hangchow he deeply impressed by the famed pagodas of these c —the Pao-en-ta ("Pagoda the Reward of Kindness") the Leu-ho-ta ("Pagoda of Harmonies"), models of wl are also shown in the l seum's collection. He des to have these reproduced his summer palace at Je where he had erected a ten in 1751. The plan was car out, but one of the two pago was destroyed by fire and other collapsed on its c pletion.

The geomancers counse and gave the verdict t southern monuments must be built in the north. emperor, however, scor their decision and orde new and more solid build material. After ten ye labor, the pagoda was c pleted. Its story is told in inscription engraved on stone tablet and composed the emperor. The tablet set up in front of the pagoda on a terr enclosed by a stone rail, and is reprodu in the Museum model. Five lion c playing with a ball are carved in high re on the top. Each side is adorned wit' dragon in clouds striving for the flam pearl. Each story has four doors and f windows. The pinnacle is in the shape an Indian stupa (tope).

The territory of Jehol formed part Chi-li Province under the Manchu dyna (1644–1911). Originally inhabited by rov Mongols, it was part of Inner Mongolia which it also belongs geographically. never was part of or in any way connec with Manchuria. For many centuries country has been settled by Chinese agric turists. The Mongols returned to th steppes, and through hard labor the Chin farmers transformed the inhospitable mo tain region into fertile land.

The Museum's economic botany coll tions include a display of oils, resins a lacquers.

223

中国镜子展

SIR HUBERT WILKINS, EXPLORER, TO LECTURE AT MUSEUM

Captain Sir Hubert Wilkins, famed explorer of the Arctic and Antarctic, will lecture at Field Museum on Saturday afternoon, March 4. His subject will be "What I Have Discovered in the Arctic and Antarctic," and he will relate his experiences on expeditions made by dog team, by airplane, and by submarine. The lecture will be illustrated with motion pictures. It will be given in the James Simpson Theatre of the Museum, and will begin at 3 P.M.

Sir Hubert's lecture is the first in the fifty-ninth course presented by the Museum, which will comprise eight other lectures to be given on successive Saturdays throughout March and April. The complete schedule for this course will appear in the March issue of FIELD MUSEUM NEWS.

No tickets are necessary for admission to these lectures. A section of the Theatre is reserved for Members of the Museum, each of whom is entitled to two reserved seats on request. Requests for these seats may be made by telephone or in writing to the Museum, in advance of the lecture, and seats will then be held in the Member's name until 3 o'clock on the day of the lecture. Members may obtain seats in the reserved section also by presentation of their *membership cards* to the Theatre attendant before 3 o'clock on the lecture day, even though no advance reservation has been made. All reserved seats not claimed by 3 o'clock will be opened to the general public.

NEW WORLD FOOD PLANTS

BY B. E. DAHLGREN
Acting Curator, Department of Botany

The Department of Botany recently installed in Hall 25 an exhibit showing the principal food plants of American origin.

On his first voyage to the New World, Columbus found the inhabitants using vegetables that were strange to him, especially some starchy tubers, probably sweet potatoes and cassava. He carried these back to Spain and presented them to Queen Isabella, together with other products of the newly found land. It is doubtful whether the queen was greatly impressed with the present. She would much rather have had a gift of cinnamon, cardamoms or sandalwood, which would have constituted proof of a new route to India. The incident, however, is noteworthy as marking the first introduction of American food plants into the Old World, an event of considerable significance to the world's dietary, which has America to thank for many important contributions.

It is interesting to note that the introduction of Old World food plants into America also dates from the voyages of Columbus, and has continued ever since.

After Columbus, the early explorers and conquistadores found other food plants in use and in cultivation among the New World inhabitants, especially the Aztecs of Mexico and the Incas of Peru. Cortez made the first acquaintance with chocolate and vanilla at the court of Montezuma. It is evident that the areas inhabited by the Mayas and Incas have been important centers of origin and dispersal of plants.

The settlers in North and South America soon learned to use many of the vegetable foods of the Indians, such as corn, beans, pumpkins and cassava. Certain of the newly discovered food plants spread rapidly over most of the world. This was true of the peanut, which was carried to Africa from the east coast of South America and to the Orient from the west coast, early in the history of world-wide navigation. Some American food plants, such as potatoes, were first carried to Europe and developed in cultivation there before coming into general use among the new population in the land of their origin. Others, such as tomatoes, were very slow in becoming adopted. The tomato was grown in Europe for several centuries as a curiosity and ornamental plant before it became, rather recently, the important article of food that it is today. A few valuable American food plants such as the avocado are only now becoming well known. Others, e.g., the chayote, are scarcely known at all in the United States in spite of efforts made to introduce them.

The recent discovery by a party of Russian botanists of more than a dozen potato-like plants and potato relatives cultivated by Indians, a few in the Maya area in southern Mexico, the rest in Bolivia on the margin of the former Inca region, may prove to be of importance for the development of new and improved sources of food at the hands of expert plant breeders.

The new exhibit shows only the principal native American vegetable products. Many tropical American fruits and some vegetables, little known in North America, are omitted. Also omitted are various small fruits such as strawberries, raspberries, blueberries, and plums, which belong to the circumpolar flora and have their counterparts in Europe. The display includes maize, or Indian corn, potatoes, sweet potatoes, tomatoes, pimentoes, Jerusalem artichokes, which are the roots of a western sunflower, pumpkins, squashes, lima and kidney beans, cassava, which in the United States is best known in the form of tapioca, peanuts, cranberries, persimmons, papaws, papayas, the avocado, the pineapple, cacao, vanilla, and others.

Chinese Mirrors Displayed

Two exhibition cases of metal mirrors from China, some dating back as far as 246 B.C., have been installed in George T. and Frances Gaylord Smith Hall (Hall 24).

Mirrors were important to the Chinese not only as aids to vanity, but also because of the belief that they dispelled evil spirits and goblins, according to Dr. Berthold Laufer, Curator of Anthropology. The common superstition that breaking a mirror brings bad luck prevails in China, and goes far back into antiquity, states Dr. Laufer.

Progress of Rockefeller Project

J. Francis Macbride, Assistant Curator of Taxonomy, in Europe for several years to obtain photographs of type specimens of Central and South American plants in European herbaria—a joint project of the Rockefeller Foundation and Field Museum carried on for the benefit of botanists generally—reports that he has completed 2,000 photographs at the University Museum of Copenhagen, and is continuing similar activity at Geneva. To date more than 23,000 photographs have been assembled in various European herbaria. The herbarium of Copenhagen contains some early and important Central American collections, especially from Costa Rica, and its curator, Dr. Carl Christensen, generously permitted Mr. Macbride to select a large amount of duplicate material for the herbarium of Field Museum.

A DEFENDER OF THE FAIT AND HIS MIRACLES

BY BERTHOLD LAUFER
Curator, Department of Anthropology

An exhibit of carved wooden imag Buddhist and Taoist deities was rec installed in George T. and Frances Ga Smith Hall (Hall 24). Most of these obtained from ancient temples in and ar Si-an fu. One of them is a statue of W the loyal protector of Buddha's ten and a staunch defender of his faith.

This statue, well carved and finely quered, is glorified by a tradition. D the seventh century there lived at Si- a Buddhist priest, Tao Süan by Like all monks he was devoted to conter tion, looked upon as the means of atta self-perfection. Meditation naturally to dreams, in which he had contact the supernatural. Tao Süan wrote memoirs, in which he records his conv tions with the gods. Among others W appeared and ordered his statue exactly like his apparition. Tao obeyed, and thenceforward images of to were set up as the guardians of Bude temples and clergy.

All other Buddhistic divinities are der from types created in India, where Budd was born. Wei-to is the only one conce in China. He has the appearance handsome Chinese youth with a sm countenance, yet is a powerful ge fortified by a suit of mail, ever read strike demons and foes of the faith.

The temple from which came the to now in the Museum was erected on spot where Tao Süan lived and tau According to tradition this statue w descendant of Tao Süan's work, perme by his spirit. It was regarded, theref as a great miracle-worker. Wei-to, al all, was a good provider, an efficient mo raiser, and bill collector. In some mo teries the monks placed his statue in kitchen, entrusting its supervision to care. Sometimes they even recited inca tions, threatening him with corporal put ment if he should neglect to supply th with provisions.

Whenever a temple was in need of rep or a pagoda was to be restored, Wei-to instrumental in raising the necessary c The brotherhood would stage a proces through the city. One monk, carryin shrine harboring Wei-to's picture, and b ing a wooden drum in the shape of a solicited funds from the wealthy. If was unsuccessful, a monk would dep Wei-to's image on the threshold of house of a prominent family, obstruct entrance, and remain seated there cr legged like a Buddha, for days if necess until the contribution was made.

If the monks again failed in this ques charity, they resorted to extreme measu One would be locked in a cage just h enough to allow him to squeeze in, and wo then be exhibited in the market pla The door of the cage was padlocked, and the news was broadcast that he was door to die of starvation unless the money raised. The people were urged to h pity. To arouse their feelings, it was s that the prisoner's bare feet rested on i spikes. This in a way was true, but spikes were so deeply sunk into a pla that it formed a smooth surface. Moreov the man was always secretly released bef harm could befall him.

it will thus be seen that "rackets" not of recent origin, but that they hav history whose threads may take us ba to the intricate mysteries of the Orient.

224

《原始人》绪言

Prehistoric Man
Hall of the Stone Age of the Old World

BY

HENRY FIELD

ASSISTANT CURATOR OF PHYSICAL ANTHROPOLOGY

———

FOREWORD BY BERTHOLD LAUFER

CURATOR, DEPARTMENT OF ANTHROPOLOGY

———

8 Plates in Photogravure and 1 Map

ANTHROPOLOGY
LEAFLET 31

FIELD MUSEUM OF NATURAL HISTORY
CHICAGO
1933

FIELD MUSEUM OF NATURAL HISTORY
DEPARTMENT OF ANTHROPOLOGY
CHICAGO, 1933

LEAFLET NUMBER 31
COPYRIGHT 1933 BY FIELD MUSEUM OF NATURAL HISTORY

FOREWORD

A century of progress is being celebrated in Chicago during this summer. While a century may be regarded a considerable span of time in the life of a community, it is just a drop in the ocean compared with the long history of mankind and just an atom in the history of the universe. In the opening of Hall C, which depicts in eight dramatic groups human prehistory from its incipient stages in the Chellean period down to the dawn of recorded history, we are privileged to celebrate 250 millenniums or 2,500 centuries of progress—a progress brimming over with the most unusual human interest and tinged with the luring colors of adventure and romance. This reconstruction of man's past extending over a period of 250,000 years is a spectacle never attempted before in any museum of the world.

We may bemoan the fact that no historian's pen has chronicled for us the doings and sayings of the Neanderthalers and Cro-Magnons and that we must laboriously restore their life and appearance from more or less fortuitous remains of mute bone and stone. What is to be regretted much more profoundly, however, is the fact that motion pictures have been invented too late. I would gladly sacrifice all mediaeval local chronicles of European towns and monasteries and throw the lives of the emperors and martyrs for good measure into the bargain in exchange for one contemporaneous motion picture reel taken of the life of the Neanderthalers and Cro-Magnons and a dozen dictaphone records of their speech and songs, not to speak of the gain that would have accrued to our knowledge of

3

history and anthropology if Alexander the Great, on his conquest of Asia, had been accompanied by an army of camera men. The next best to the motion pictures of which we unfortunately are deprived is the drama in eight acts represented by the eight groups of prehistoric man and his culture reconstructed by the incomparable talent of Frederick Blaschke under the direction of Henry Field in Hall C. These restorations are as accurate in detail as warranted by our present knowledge. They are not dogmatic nor doctrinal, nor visionary nor sentimental. On the contrary, one of their beauties is their restraint, their simplicity, and above all, their power of suggestion. The artist who created these enchanting scenes does not try to be original by being different, but is original by being sincere; he takes us into his confidence, he makes us pause, reflect, and speculate, and allows us to turn our thoughts longingly back to eons of time that were still a sealed book to the preceding generation. The man who is able to produce such an effect of inspiration is a real artist and master. Another test of a great work of art is the permanence of the impression which it is apt to leave on our minds. No one who will spend only a few minutes in front of each of these groups will ever forget them; they live and endure in our memory, and their memory will always urge us with irresistible force to return to them. A new world has been opened here to all of us with plenty of food for thought and study.

BERTHOLD LAUFER

225

人类种族厅于 6 月 6 日开放

Field Museum News

Published Monthly by Field Museum of Natural History, Chicago

Vol. 4   JUNE, 1933   No. 6

HALL OF THE RACES OF MANKIND (CHAUNCEY KEEP MEMORIAL) OPENS JUNE 6

BY BERTHOLD LAUFER
Curator, Department of Anthropology

Chauncey Keep Memorial Hall, which will be opened to the public on June 6, contains a series of statues, busts, and heads of bronze (with the exception of four which are of stone) by Malvina Hoffman, sculptor of international fame, intended to illustrate the principal racial types of the human species and depict their physical characteristics. This hall, unique among the museums of the world, is named in honor of the late Chauncey Keep, a highly esteemed member of the Museum's Board of Trustees from 1915 until his death in 1929. A legacy of $50,000 which he left to the Museum has been applied to the hall and its contents, and the balance of the cost has been met by generous contributions, totaling more than $150,000, from Marshall Field, Mrs. Stanley Field, and Mrs. Charles H. Schweppe.

Chauncey Keep

In the carrying out of its novel idea, Chauncey Keep Memorial Hall required special treatment, and new resources of museum technique had to be developed. A great deal of construction work had to be undertaken, alcoves built to provide a suitable setting for the bronzes, and careful studies made for the purpose of giving them the best possible display and lighting. Both in the formulation of the plan and in the solution of the many complex problems connected with the work, President Stanley Field has spent much of his time and energy, and has to a considerable degree contributed to the success of the hall. The plan was carried out after long and mature deliberation; as a matter of fact, its inception goes back to the year 1915 when it was first mapped out in the Department of Anthropology. In the course of years it was frequently modified and improved, and finally brought to fruition with the cooperation of Henry Field, Assistant Curator of Physical Anthropology.

Blackfoot Indian

The hall is divided into three sections, the central one being an octagon. The material is distributed by continents in geographical order.

Samoan

The section on the west side of the hall contains the races of Africa and Oceania; the octagonal section in the center is devoted to the races of Europe, Asia, and America, those of Asia being continued in the section on the east side. The center of the octagon is occupied by a monumental bronze group symbolizing the unity of mankind. It consists of a white, a yellow, and a black man of heroic size representing the three principal racial divisions. Each figure embodies the highest qualities of the race.

More bronzes will be added to the hall from time to time; also, colored transparencies of racial types will be installed, and special exhibits of a scientific character will be arranged at the east end.

Each sculpture is the result of careful selection of subject and long anthropological study. Malvina Hoffman was sent by the Museum on an expedition to Asia, visiting Japan, China, Java, Bali, Sumatra, the jungles of the Malay Peninsula, Ceylon, and India, studying racial types and modeling her subjects directly from life in clay; from clay they were transformed into plaster and finally cast in bronze. As far as possible, the patina of the bronze has been finished in conformity with the skin color of the race.

Mongol

Mangbetu Woman

Because of the rapid extinction of primitive man due to the white man's expansion over the globe many a vanishing race will continue to live only in the sculptures displayed in this hall. Both the racial and the individual character is grasped and portrayed in the bronzes with a rare insight into the mind of primitive man. Pose and action are chosen in consonance with the character of each particular tribe and permit the study of the physical functions which are more important for evaluation of a race than bodily measurements.

Only a few can be selected here for illustration and some brief comment. The Blackfoot Indian shows a perfect development of the body, which is intense with health and vigor; he is signaling to his friends in the distance that he has hit his quarry. The Samoan is a fine example of Polynesian stock.

Sara Dancing Girl

The graceful, fifteen-year-old dancer of the Sara tribe and the Mangbetu woman from the Belgian Congo well represent Negro types of beauty. The portrait bust of a powerful Mongol evokes memories of the Mongol empire, greatest in history.

One of the most attractive figures is that of a middle-aged Ainu, full of dignity and poise, an eloquent spokesman of the once glorious past and subsequent tragedy of his vanishing race. In a prehistoric age the Ainu were the original inhabitants of the Japanese islands. Clashing for centuries with the Japanese who were migrating from southeastern Asia, the Ainu finally yielded to forces superior in number, retreating into the northern island of Yezo.

Ainu

226

《马常奇（音译）的藏品——中国古代青铜器》导言

THE MA CHANG KEE COLLECTION

ANCIENT CHINESE BRONZES

Exhibited at the Galleries of

RALPH M. CHAIT

600 Madison Avenue
New York

"He who has money is fortunate; but he who has ancient bronzes is blessed by Heaven."—MA CHANG KEE.

FOREWORD

TSING FAH MA, the celebrated collector and connoisseur, is a Chinese gentleman of advanced years. For generations, his family has dealt in Chinese works of art of the highest importance. As intermediaries or agents representing the members and nobles of the courts of the Chinese Empire of the now defunct Ch'ing dynasty, which had ruled China for two hundred and sixty-eight years, many of the most cherished treasures, even those considered as the palladia of the empire, have been negotiated or acquired through them.

In China, Mr. Ma, or Ma Chang Kee as he is also known there, is called the "source" from whom great works of art and information concerning them can be acquired. No matter how serious their need for cash may be in these trying times, the princes, nobles, and fastidious members of the aristocracy, whose names are famous throughout China as collectors and connoisseurs, will not permit strangers or dealers to pass the gateways of their homes, but deal with and show their possessions to Mr. Ma. He is financially independent, and is considered one of China's foremost experts, especially in ancient bronzes, being the scion of a family preeminent in that field.

Mr. Ma belongs to the old school. It has been his custom to conduct his affairs from his home and gardens in Shanghai, where most of the important art deals are consummated only after weeks of discussions, tea-drinkings, and exchanges of salutations, as Chinese etiquette prescribes.

Two years ago, he was enabled to gratify an old desire to visit the Occident, to see again in his lifetime the many chefs d'oeuvres which he and his father before him had been so instrumental in procuring and collecting, and which since had left the shores of China. At that time, he had acquired an assemblage of rare bronzes which he considered the crowning glory of his career.

And so, on this, his first trip to America, he brought with him a few pieces from this rare group, which contained objects gathered from the imperial and most noted private Chinese collections, the like of which no Occidental had ever before seen. Upon his arrival here, he stated in his limited English that his reason for personally bringing these objects, rather than having entrusted them to agents as was his practice in the past, was his desire to learn at first hand, from his visit, whether the Americans "eyes have got," meaning,—appreciation, taste and knowledge.

In the unique collection which he had acquired, there were a group of gold bronze vessels, surpassing in quality and technique those known to Occidental experts, from which, in 1930, the Boston Museum of Fine Arts purchased one bronze jar. In describing it*, the curator of Asiatic art, Mr. Kojiro Tomita, states: "So exceptional in technique and quality is this bronze jar of the Han dynasty, that its acquisition is a great event."

About a year later, when Mr. Ma arrived here, he brought three other specimens of gold bronze which he had decided to part with. These consisted

* See "Bulletin of the Boston Museum of Fine Arts", No. 167, Volume 28, June, 1930.

of a similar gold bronze jar, differing from the one in the Boston Museum only in the handles; a covered ceremonial gold bronze vessel, and a gold bronze bowl, which the celebrated collector, Miss Kate S. Buckingham of Chicago, was fortunate in purchasing, and all of which are now proudly exhibited at the Art Institute of Chicago.

Regarding the jar, Mr. Charles Fabens Kelley, curator of Oriental art at the Art Institute, says that the **"Magnificent jar of ceremonial type, decorated with engraved patterns in silver on a groundwork of gold . . . is an almost exact duplicate of the one recently acquired by the Boston Museum of Fine Arts and described in a scholarly article by Kojiro Tomita."

Mr. Ma, having learned that we "eyes have got" went back to China, and returned to America, bringing with him the most important objects in the collection, because of the belligerent threats made by Japan to China, and the general chaotic state of affairs there. This group comprises the remaining and the most outstanding gold bronzes in the collection, among which is the third of the series of the only three gold bronze jars known, differing slightly in shape from the one in the Boston Museum of Fine Arts, and the one in the Art Institute of Chicago; also, the large gold bronze plate,—an extraordinary and unique specimen, distinguished by its beauty of design, perfect technique, and its size; as well as four other amazingly beautiful and rare gold bronze ceremonial vessels; and some finely patinated, rare sacrificial vessels differing from and surpassing those known to be in our most noted collections.

Mr. Ma, however, could not continue his stay in America because the Sino-Japanese war broke out, and it became imperative for him to return to China to attend to the care of his family and interests.

Therefore, it falls to my good fortune to present for exhibition, these, the rarest of ancient Chinese gold and sacrificial bronzes, used in the ritual ceremonies of Heaven and Ancestor-worship by the feudal lords of ancient China.

The bronzes exhibited have been authoritatively described, authenticated, and catalogued by the eminent scholar, Dr. Berthold Laufer, of the Field Museum of Chicago, who also made the epigraphical determinations and translations of the pictographic and ideographic inscriptions appearing on the vessels. The collection has also been inspected and examined by Dr. Colin G. Fink, Professor of Electro-Chemistry, Columbia University, and Consultant to the Metropolitan Museum of Art, who also rendered a report of his findings from his microscopic and other scientific analysis of one of the gold bronzes.

To both these gentlemen, I extend my grateful thanks for their assistance and scholarly contributions, and regret that the limited space of this booklet does not permit the printing of their full, minutely detailed descriptions and findings.

RALPH M. CHAIT.

December, 1932.

** See "Bulletin of the Art Institute of Chicago", No. 1 and 2, Volume 26, January and February, 1932.

CHRONOLOGY

INTRODUCTORY

MANY excellent bronzes have come out of China during the last two decades and may be seen in our museums and private collections, but the earliest and best are still retained in the hands of collectors in China. Owing to a lucky chance we are now offered an opportunity to view in Mr. Chait's Galleries a series of monumental bronzes that have come from the mansions of various prominent Chinese collectors.

The importance of this gathering of bronzes rests on the fact that it allows us to study the fundamental characteristics of Shang, Chou, and Ts'in (or Ch'in) bronzes. It is not so long ago that students of Chinese art were hesitant to admit the existence of Shang bronzes, and there were even those who denied the existence of Chou bronzes. These apprehensions, however, were elicited by lack of experience and proper knowledge of the subject. There is no longer any ground for this safety-first-attitude which relegates a Shang bronze to the Chou period, and a Chou and Ts'in bronze to the Han epoch or even later. Any authentic archaic bronze can now with perfect safety be assigned to its proper setting, for the characteristic features of the bronzes of each period are perfectly well-known.

During the last decade sound and systematic excavations undertaken by Chinese scholars have made us familiar with the culture of the Shang that represents a world in itself, fundamentally different from that of the succeeding Chou dynasty. The results of this painstaking work have been published in Chinese, and this accounts for the fact that the majority of museum curators has not taken notice of it, or has not been able to follow the progress of science made in China. It is now possible to outline a clear picture of what Shang civilization was; we know its antiquities, its peculiar script and the contents of its inscriptions,—all radically at variance with what constitutes the later Chou culture.

In Mr. Chait's exhibit there are three prominent objects which indubitably are offsprings of Shang civilization, Nos. 1, 6 and 12. No. 1, a twin owl wine kettle (type *yu*) is provided with an inscription cast both in the bottom and inside of the cover. This inscription consists of four characters of primitive pictographic style (note the bird caught in a net in the character *Lo*) such as was never used under the Chou or any subsequent period. This inscription reads: Lo tso fu Kwei, which means: "Lo had this vessel made in memory of his father Kwei." Kwei is one of the twelve cyclical signs used in the reckoning of years. These signs were employed as personal names under the Shang dynasty, and under this dynasty only, and are thoroughly characteristic of it. They were not utilized as personal names under any subsequent dynasty. The above inscription in its form and content could only have been indited by a man of Shang.

Again, in the bronze No. 6, a beaker of the type *tsun*, we find an inscription cast inside of the hollow foot and consisting of three characters, which read: Kia shi tso. This means: "The historiographer (*shi*) had this vessel

5

made (*tso*) in memory of his father Kia." The word Kia is again one of the twelve cyclical signs in vogue under the Shang as a personal name.

The complex cosmology personified in the large animalized wine-vessel (No. 12) is the outcome of Shang, not Chou mythology. On the other hand, the mythological scenes and creatures so profusely represented on the pair of gilded tazzas (*tou*), on the globular vase (*hu*) and on the marvelous plate (*p'an*) are emanations of the fantastic mythology evolved in the period of the Ts'in (or Ch'in) (246-207 B.C.), and the designs on the gilded wine jar (*lei*) are characteristic of the same epoch.

The evaluation of Chinese bronzes requires an intimate knowledge of history, palaeography, and mythology. The stylistic criteria usually relied upon by art students are deceptive because they are purely subjective,—a fact to which Dr. W. Percival Yetts has recently again called attention (in an article entitled "Problems of Chinese Bronzes" in *Journal of the Royal Central Asian Society*, August, 1931).

It is wholly unnecessary to say a word of praise on behalf of the bronzes displayed in Mr. Chait's Galleries. They plead their own cause, they justify themselves, they stand, like the pyramids of Egypt, as great monuments for all times.

BERTHOLD LAUFER.

6

HISTORICAL OUTLINE

THE salient bulwarks of the antiquity of nations, are witnessed in the majority of instances, by their architectural ruins. The pyramids of Gizeh and the temple ruins of Luxor and Karnac testify to the antiquity of the ancient civilization of Egypt, but when we approach China we find no architectural ruins of such antiquity, for the Great Wall of China is but an infant in comparison.

Instead, we meet with their ancient bronze sacrificial vessels which are the most valuable legacy of China's ancient civilization, dating back to the 2nd millenium B.C., representing a monument of national art the beginning of which is buried in the profoundest mystery and obscurity. This is also true of the race; where the Chinese originally came from and how they reached China are matters simply of speculation.

Bronze has been known in China since prehistoric times under the name of "*tung*" and the Chinese valued it more than any other media because of its durability and its admixture of metals which enabled it to withstand the vicissitudes of time and the ravages of the elements so that under ordinary conditions it suffered no noteworthy damage. Having this high regard for the metal, they fashioned from it their vessels for the ritual of their ceremonies, and ornamented them with designs and motives which, by their symbolism, tend to express an animistic and cosmic idea, or a quasi-religion, between man and the unseen powers supposed to influence his life. This form of ornamentation is quite characteristic of these early sacrificial bronzes. In many cases they also bear hieroglyphic inscriptions which now distinguish and help in determining their antiquity.

From such existing evidence as we have in the form of vessels and bells, we observe that the art apparently starts mature and there are cycles of decline, rebirth, and finally decadence. The starting point seems to be lost in legend, yet reason and logic tell us that no matter how distant there must have been a beginning. Whenever we may assume Chinese civilization to have begun, no one today, in the light of our advanced knowledge, meager as it may be, will deny a greater antiquity to the Chinese than any other living civilization, or a continuity which despite foreign influence, is reflected in its artistic forms of expression.

To attempt to research Chinese bronzes historically, is to discover with much dismay that the sources of information are scant, despite the fact that the study of ancient bronzes has been industriously pursued in China by generations and generations of scholars, and despite the many references to the bronze ritual vessels in canonical books and annals, some of which like the "Chou Li" date back to the 3rd century B.C. Of this work, the late Prof. Hirth said that "as an educator of the nation, the 'Chou Li' has probably not its like among the literatures of the world, not excepting the Bible." This remark refers especially to its minute details of public and social life because it throws considerable light on the constitution and culture

7

of the nation during the Chou dynasty.   Nevertheless, so far as their bronzes are concerned, their origin still lies hidden behind a nebulous veil of legend, and we are unable to ascribe them to an earlier epoch than the Shang dynasty (1766-1122 B.C.).

No example is known at this time, to exist of a period antedating this era, though in the "Shu Ching" we read of the Nine Tripod Cauldrons (*ting*) as having been cast of bronze in the Hsia dynasty (2205-1766 B.C.). These tripods are said to have been decorated with carved designs of maps and figures, illustrating the productions of the Nine Provinces forming the Empire, and were long preserved as the palladia of the kingdom.   These, we are further told, were lost in the internal troubles which ushered in the close of the Chou dynasty, (1122-255 B.C.).

The difficulties of establishing an antiquity for the bronze sacrificial vessels at an earlier era, are due to the fact that criteria vital to a fuller definite knowledge have as yet not become entirely available.   We have learned much as a result of the momentous discovery made in 1899, in the environs of An-yang in the west of Honan, of several thousand fragments of bone and tortoise-shell incised with archaic script, said to be the remains of archives left by a royal diviner of the Shang-Yin dynasty, which Dr. W. Percival Yetts[1] considers as "incontestable genuine additions to the stock of pre-Ch'in documents." However, this knowledge by reason of the fact that it is chiefly in Chinese, has escaped the notice of many and awaits translation by Western scholars.

For this knowledge, we are indebted to the labors of such eminent native scholars, to mention but a few, as, Sun I-jang[2] and to Liu O,[3] who in 1903 published a huge work illustrating a thousand of these fragmentary specimens; and especially to Lo Chen Yu,[4] the famous archaeologist, who in 1914 visited the site and recorded his findings, as well as to Wang Kuo-wei's[5] important contributions, from which much has been added to enrich our knowledge, not so much about bronzes alone, but as Dr. Berthold Laufer says, "It is now possible to outline a clear picture of what Shang civilization was; we know its antiquities, its peculiar script, and the contents of its inscriptions."

We however still await the final results of the research work in progress by Chinese, Japanese, and Western epigraphists and archaeologists concerning the bone finds.   These however, already have well established substantial testimony of an early Chinese civilization and culture of which the ancient bronzes in the possession of museum and private collections throughout the world, are living witnesses.

The early bronzes of China, beside serving as the palladia of the kingdom,

1 Catalogue of the George Eumorfopoulos Collection of Chinese and Corean Bronzes, Volume I.   W. Percival Yetts, 1929.
2 Ch'i wên chü li.   A pioneer study of inscriptions on the Honan finds, 1927, with a preface dated 1904.
3 T'ieh-yün tsang Kuei.   Reproductions of ink-squeezes of inscriptions on bone fragments from the Honan finds, 1903.
4 Yin sü ku ch'i wu t'u lû.   Illustrations of the notes on ancient objects obtained by the author's brother at the site of the Honan finds, 1916.
  Yin-shang chên pu wên tzu K'ao.   An initial study of divinatory inscriptions on the Honan finds, 1910.
5 Chien shou t'ang so tsang Yin hsü wên tzu K'ao shih.   Style of inscriptions on the Honan finds.

8

were used in the ritual ceremonies of ancestor-worship, and as a consequence are the dearest and most precious heirloom possessions of the Chinese who value them above everything else. From time immemorial, we read that these bronzes incited tribes to war and men to murder for their possession; caused the looting of palaces, temples, and tombs, not only for their sacred or artistic value, but as well for their intrinsic value when melted down and cast into cash by forgerers. This constant recasting of bronze has been going on for over two thousand years, at least since the beginning of the Ch'in dynasty (246-207 B.C.), depleting the vast number of ancient bronzes which had been cast in the preceding fifteen hundred years during the Shang and Chou dynasties.

For the ritual of their elaborate ceremonies for ancestor and other forms of worship, and to perform their sacrifices in a manner befitting their importance, we read that a vast number of especially designed vessels were required by the ancient Chinese. These varied in shape, size and decoration according to the demands of the tyrannical laws of their ritual which were always strictly adhered to. The emperor, we read, made sacrificial offerings to Heaven as the supreme ruler. His subjects sacrificed to beings of a lower order,—the sun, the moon, and the stars; the hills, rivers and forests, and to the departed souls of their ancestors. The manner in which sacrifices were made, was regulated and prescribed by myriads of petty rules; in fact, the sacrificial service, from all we read, leads us to believe that it was the leading feature in the spiritual life of the Chinese of the *San Tai* or Three Ancient Dynasties.

No enterprise was embarked upon, or decision made without the seeking of divine aid. Much like the ancient Greeks and their oracles, the Chinese depended upon divination, for many references appear in their records showing that the art of obtaining an omen from the unseen spirits was cultivated in every minute detail. The chief means of auguration was of course, the system of the *Pa Kua* or the "Eight Mystic Trigrams", though in many cases the scales of the tortoise scorched by fire were used as oracles, as well as numerous other means, even to the practice of observing the stars to ascertain man's fate. If the outcome prayed for was successful, the occasion was celebrated by the casting and presentation, with pompous ceremony, of a bronze vessel to the ancestral shrine.

The bronze vessels used in the ritual worship of ancestors, were divided, according to the scholars, into two classes: one for water and liquids, such as the fragrant *Chang*,—an alcoholic wine fermented from millet and mixed with odoriferous herbs. These were generally called *tsun* or wine vessels. The other group was intended for solids and called *yi*; they were used as sacrificial food vessels in which the offerings were made of cakes, fruits, vegetables and viands. These groupings however, have been added to considerably in later centuries. Chinese scholars likewise divided their ancient bronze sacrificial vessels into two chronological categories. The first consisted of the bronzes made during the *San Tai* or Three Ancient Dynasties, namely, the Hsia, the Shang, and the Chou. The second group included those bronzes of the Ch'in, Han and later dynasties.

9

During the Three Ancient Dynasties, China was made up of a congeries of tribes, much like the Feudal system of Europe, differing not only in dialect, but also in culture. All however shared equally in their fealty and obligations as vassals to the supreme lord of the suzerainty. Fealty however, did not deter or deprive them of the privilege of maintaining private miniature courts wherein was observed a religious ceremonial elaborately conducted with bronze vessels to befit its dignity. Bronzes were also cast for the emperor and feudal lords, who presented them as trophies or gifts to triumphant generals and others upon whom they wished to bestow their favour.

These ritual and honorific vessels adorned with designs of monumental grandeur gracing their plastic shapes, have intrigued the whole civilized world not only by their massive dignity and inimitable grand flow of lines, but as well by the mythological creatures,—quadrupeds, birds, and reptiles, (highly conventionalized and stylized it is true) who disport themselves on the surfaces of the bronzes, adding further to the mystery of their origin because of our inability through lack of records to comprehend their meaning and purpose.

And so, the artistic temperament of the Chinese found an outlet and an impetus in this gorgeous ceremonial which combined and utilized all forms of beauty. Authorities universally agree that this sacerdotal art of China is one of the highest summits of human art, and that even in those remote epochs the Chinese were daring artists and master-craftsmen in metal. All of which is further proof that the art, because of its maturity at this early period, must of a necessity be at least a millenium or more older than we know it to be.

This belief is somewhat substantiated by the startling finds of a Neolithic pottery made by Prof. Andersson[6] about a decade ago in Kansu at the foot of the hills and in the caves of the valley of the Upper Yellow River. A ware antedating even this was discovered at about the same time, in Honan in the lower valley of the Yellow River, known as the Yang-shao site. These pottery vessels dating back to at least the 3rd millenium B.C., indicate that even at that time a high state of civilization and culture existed, capable of producing vessels of earthenware, classic in form and decorated with geometric designs painted in pigments.

To produce these pottery vessels, it was necessary to fire them in kilns or ovens at about a 1000 or more degrees, and as George Soulie de Morant says, "in a good oven, with a forced draught, implying an advanced culture". Therefore some knowledge of furnaces, kilns and moulds must have been known about a thousand years earlier than the earliest properly identified bronze. Who can tell what excavations in China, scientifically conducted, might further bring forth?

The method of casting bronze vessels by the early Chinese, known as the "cire perdue" or "lost wax" process, was unquestionably limited to a family or guild of craftsmen to whom were entrusted their production, and who must have sacredly guarded the secrets and traditions of the processes

6 Preliminary Report on Archaeological Research in Kansu. Y. G. Andersson, Peking, 1925.

of their craft. In fact, this time-honored tradition of jealously guarding the secrets of a craft, as the late Prof. Hirth said, "is a feature of Chinese social life in that specialties in art and workmanship are treated as a monopoly of certain families on which no outsider is allowed to trespass, and these secrets are so well-guarded that a branch of art may die out with the last scion of the family that created it, as in the case of Fu Chow lacquer, the secret of which was lost during the T'ai Ping rebellion.

No contemporary account is known of the technical processes or of the furnaces, crucibles and moulds used, because of the secrecy that surrounded the production of these bronzes, and the general wanton destruction that took place in the Ch'in dynasty (246-207 B.C.) when Emperor Shih Huang Ti broke the ritualistic cult by ordering the melting down of all bronzes and the burning of all books having reference to past history of the empire.

His ambitious desire was to have the history of China commence with himself as its first ruler, under the pretext that the government should not be fettered by ancient usages. The literati who pertinaciously adhered to former usages were either imprisoned or burned alive. Likewise, many scholars and ministers were put to death for having failed in obedience to this edict. Then it was that the sacred vessels which had been handed down for centuries from father to son with their records, were hidden by those who treasured and cherished them. In this manner, early documentary evidence was lost or destroyed, or perhaps still lies hidden in the bowels of some mountain-side or river-bed.

Following the death of the tyrant Shih Huang Ti, and later upon the overthrow of the Ch'in dynasty which he founded, reverence for the relics so intimately connected with their ancestors came again into vogue. Great attempts were made to recover hidden bronze vessels and their records. Their traditional uses, secretly transmitted from memory recitals by scholar to scholar in the past centuries in order to keep within the edicts established by Shih Huang Ti, were now again openly recorded and discussed. Thus was established the nucleus of the literature which grew richer as new discoveries of ancient bronzes were made, and vaster with every succeeding century. Ever since, archaeologists, historians, and sinologues have been kept busy interpreting not only the vessels and their ornamentation, but also their pictographic and ideographic inscriptions. These had become a lost language as a result of the many sweeping changes introduced in the preceding dynasty when an attempt was made to unify and standardize the styles of writing by a compulsory change in script and character writing. Even today, diverse renditions are given by the few great scholars who are able to decipher and translate these inscriptions.

These dedicatory and honorific inscriptions met with on the bronze sacrificial vessels enable us to establish their antiquity by the epigraphical determinations made from them. To date, unfortunately, no specimen, even of a utilitarian character, fashioned from bronze, has been unearthed to which a date earlier than the Shang dynasty can be ascribed.

RALPH M. CHAIT.

11

"Consider again some thousand-year-old bronze, jealously preserved in the collections of emperors: its solemn and rude form, its powerful curves, its priestly scrolls betoken its high ritual destination: from the depths of the hard, resonant metal well up the sombre tawny suffusions of its glowing patina; a patina that spreads in smoky splendours, in living flushes, in sudden pallors, in poisonous greens beneath the outer skin of the mysterious substance. It lives before our eyes to the very core of its substance, where slumber the virtues of so many metals fused by the heat of the flame whose memory they keep."

from "China" by EMILE HOVELAQUE

CATALOGUE

No. 1. **DOUBLE OWL BRONZE WINE KETTLE AND COVER,**

One of the rarest bronzes in the collection. Superior and larger than the only other one known which measures but 8 inches, in the famous Baron Sumitomo's collection housed in Japan.

The kettle bears pictographic inscriptions cast in the interior bottom and inside of cover, reading "Lo tso fu Kwei" meaning "Lo had this vessel made on behalf of or in memory of his father Kwei". Kwei is one of the cyclical signs which were used as personal names under the Shang.

Type *yu*, Shang period.
Height, 13 inches.
Width, 9 inches.

No. 2. **BRONZE PHOENIX BOWL,**

Type *I*, completely gilded in the interior.
Chou period.
Height, 5¾ inches.
Diameter, 8 inches.

No. 3. **RECTANGULAR FOUR-FOOTED BRONZE CEREMONIAL VESSEL,**

An inscription consisting of five lines and twenty-five characters is cast in one of the long inner sides. It relates that an emperor of the Chou dynasty, pleased with the services of a certain Te, bestowed upon him five strings of cowrie-shells which were used by Te for the casting of this precious bronze vessel.

Type *ting*, Chou period.
Height, 9½ inches.
Sides: Length, 7 and 5¾ inches.

No. 4. **CRESCENT-SHAPED BRONZE SWORD,**

A bronze sword of exactly the same crescent-shaped type, is in the collection of the Field Museum. It was found in Southern Mongolia, and the conclusion is therefore warranted that this type of sword was peculiar to the Hiung-nu, corresponding to our Huns, with whom the Chinese were constantly at war for many centuries. These swords were cast, of course, by Chinese founders for the Hiung-nu.

Chou period.
Length from tip, 16¼ inches.

No. 5. **RECTANGULAR BRONZE VESSEL,**

Bears inscriptions which relate that "Ko-su kung had this vessel made for the offering of sacrifices to his deceased father and grandfather,

13

that he might obtain long life, and that this vessel might serve as an eternal treasure to his descendants for ten thousand years".

Type, double *fu*; Chou period.
Length, 11¾ inches.
Width, 8 inches.
Height, 7¼ inches.

No. 6. BRONZE BEAKER,
Originally this vessel was gilded. Bears inscriptions made in the mould, consisting of three characters of Shang style, reading "Kia shi tso", meaning "The historiographer made this vessel in memory of his father Kia". Kia is also one of the cyclical signs which under the Shang served as a personal name.

Type *tsun*, Shang period.
Height, 11 inches.
Diameter of opening, 8¼ inches.

No. 7. RECTANGULAR GILDED BRONZE VESSEL WITH ROOF-SHAPED COVER,
This ceremonial vessel is the only one of its kind known in gilded bronze. There is a similarly-shaped but ungilded vessel in the Freer Gallery of Art, Washington, D. C.

Type *fang-i*, Ch'in period.
Height, 13¼ inches.
Length, 7¼ inches.
Width, 6¼ inches.

No. 8. GLOBULAR GILDED BRONZE WINE JAR,
The *lei* was a receptacle for carrying wine to the shrine where ceremonial offerings were made to the God of Heaven, and to ancestors. The decorations on the vessel are all of geometric character with a rhythmic parallelism, and because of the symbolism attached to its triangular and other designs, the motif represents a sort of landscape in hieratic style,—heavenly powers controlling the phenomena of the sky and atmosphere framed by the mountain ranges of the earth.

Type *lei*, Ch'in period.
Height, 9¾ inches.
Diameter, 7 inches.

No. 9. PAIR OF GILDED BRONZE TAZZAS WITH COVERS,
Receptacle and cover are of globular shape, and are very beautiful and harmonious in form. In regard to this type of vessel, called *tou* and its development see Laufer's "Chinese Pottery of the Han dynasty". These two receptacles are the only two known in gilded bronze. The ceremonial gilt bronze vessel in the Lucy Maud

14

Buckingham Collection, Art Institute of Chicago, acquired from Mr. Ma, though comparable in quality and technique, is of a different though not dissimilar form, having a shorter foot giving it a squattier appearance.

Type *tou*, Ch'in period.

Height, 7½ inches.

Diameter, 9½ inches.

No. 10. GILDED BRONZE VASE,

An exceedingly rare vase of which type there are only three known. The other two, differing slightly in shape, formerly belonged to Mr. Ma; one is now in the collection of the Boston Museum of Fine Arts, and the other is in the Lucy Maud Buckingham collection, Art Institute of Chicago. The one above-described is the third of this rare set of three.

Type *hu*, Ch'in period.

Height, 13 inches.

Diameter at opening, 5¾ inches.

No. 11. LARGE GILDED BRONZE PLATE,

This extraordinary and unique specimen, distinguished by its enormous size, beauty of design, and perfect technique, is decorated all over on both sides, and is the outstanding and finest gilded bronze known in the Oriental or Occidental world, and far excels in beauty and technique the only other similar known plate, which is in the celebrated collection of the Marquis Hosokawa, in Japan.

Type *p'an*, Ch'in period.

Diameter, 20¼ inches.

Height, 3¼ inches.

No. 12. LARGE ANIMALIZED BRONZE WINE VESSEL AND COVER,

Of this great masterpiece of bronze casting, Dr. Laufer says: "No description can render justice to the beauty of this vessel, and the large variety and complexity of its design. It rises before our eyes like the lost Atlantis, as an attempt at depicting primeval ideas of the world's creation, or some of the favorite evolutionary theories entertained by China's ancient philosophers".

Of this type there is but one other (smaller) known in Occidental collections, measuring 30 cm. high, which belongs to Mr. and Mrs. Eugene Meyer of Washington, D. C.

Type *I*, Shang period.

Height, 19½ inches.

Length, 21 inches.

15

No. 13. UNIQUE TRIPOD LIBATION VESSEL,

    This bronze libation cup is of unusual size and decoration, and there are but two other examples like it known. One is in a famous collection in the Middle-West, and the other is in the collection of Baron Sumitomo, illustrated on Plate 88 of his catalogue. His however lacks many of the magnificent features of the one herein described.

    Type *kin*, with two spikes surmounted by crested birds; Chou period.
    Height, 18½ inches.
    Diameter, 8½ inches.

No. 14. GILT BRONZE SACRIFICIAL VESSEL AND COVER.
IN THE FORM OF A RECUMBENT OX,

    An inscription of eight characters is cast in the vessel to the effect that "Hien mei-sho made this sacred vessel in memory of his ancestor Ting". This vessel is cast in two parts, head and back of animal forming the lid, and covered with a network of frets forming the background, with superimposed designs of monsters on each side of the body.

    Type *I*, Chou period.
    Length, 9½ inches.
    Height, 6 inches.

No. 15. BRONZE BEAKER,

    With interesting, unusual light green patina.
    Type *tsun*, Chou period.
    Height, 10⅞ inches.
    Diameter at opening, 9⅞ inches.

No. 16. BRONZE ANIMAL, TIGER,

    Almost completely mineralized by its light green malachite patina.
    Chou period.
    Length, 8¾ inches.
    Height, 4 inches.

No. 17. BRONZE CEREMONIAL VESSEL WITH TWO HANDLES,

    The interior bottom bears an unusually long inscription cast in the mould of some eighty odd characters which at the time of the printing of this catalogue had not yet been completely deciphered, and the epigraphical determinations are still in preparation.

    Type *I*, Chou period.
    Height, 4⅜ inches.
    Diameter at opening, 7½ inches.

227

中国的秋千

LIBER
SEMISAECULARIS
SOCIETATIS FENNO-UGRICAE

Suomalais-ugrilaisen Seuran Toimituksia LXVII
Mémoires de la Société Finno-ougrienne LXVII

HELSINKI 1933
SUOMALAIS-UGRILAINEN SEURA

The Swing in China

by

BERTHOLD LAUFER.

The swing in its relation to magical and religious practices has been the subject of several studies. From the general ethnological viewpoint it has been treated by J. G. Frazer, »Swinging as a Magical Rite» (in The Golden Bough, 3d ed., pt. III, vol. IV, 1914, 277—285). The swinging ceremonies of the Indo-European nations have been comparatively set forth and ingeniously interpreted by Leopold von Schroeder in his work »Arische Religion», vol. II »Naturverehrung und Lebensfeste» (1916). A very critical investigation of the Greek swinging festival is due to F. Boehm, »Das attische Schaukelfest» (in Festschrift Eduard Hahn, 1917, 280—291), with whose conclusions I find myself entirely in accord. The famous swing festival of Siam has been described by Gerini (Encyclopaedia of Religions, V, 870), but this article is now superseded by the book of H. G. Quaritch Wales, »Siamese State Ceremonies» (London, 1931), who has devoted an entire chapter to the swinging festival of Siam (238—255) with a detailed analysis of the ceremonies, also in their relations with those of India. No one, however, has as yet made the faintest allusion to the swing in China; and as far as I am aware, no sinologist has ever dealt with the subject.

The first surprising fact about the Chinese notices of the swing is that a foreign origin is ascribed to it, while to my knowledge no other nation has a tradition to this effect; all others simply take the swing for granted and claim it as an old national property of their own. It is still more amazing that it is a tribe of »northern

barbarians» which receives credit for having transmitted the swing
to China. If it were credited to India or Siam, where from ancient
times the swing has played a prominent part in religious and other
ceremonies, we would easily understand and naturally regard this
as a logical consequence of the long and intimate cultural contact
between these three countries.

Almost all Chinese authors who have written on the swing quote
the *Ku kin i shu t'u* 古 夕 藝 術 圖 , a work now lost, which
seems to be the oldest source to make mention of the swing. This
book was still extant under the Sui dynasty (A.D. 590 —617), for
it is listed in the catalogue of Sui literature (*Sui shu*, chap. 34, p. 5)
as »*Ku kin i shu*, consisting of twenty chapters» in the division of
light literature called *siao shwo* 小 說 . The author's name is not
given. As far as I am aware, this work is not cited in the literature
of the Han (*Ts'ien Han shu*, chap. 30), so that the supposition may
be warranted that it must have been written some time between
the third and fifth century A.D. A certain difficulty arises from
the fact that various authors cite the text of this book in a different
manner and that it is thus not entirely clear what was contained
in the original text and what may be due to interpolation. Perhaps
the oldest source quoting the *Ku kin i shu t'u* is the *King ch'u sui
shi ki* 荊 楚 歲 時 記 by Tsung Lin 宗 懍 assigned to the
Tsin or the Liang dynasty (Wylie, Notes on Chinese Literature, 56).
Referring to football after Liu Hiang, this author continues, »According to the *Ku kin i shu t'u*, the swing is a sport of the Shan Jung
(Mountain Jung 山 戎) of the northern region, who practise it
for the purpose of gaining lightness and agility of the body» (ed.
of *Han Wei ts'ung shu*, p. 9).

The same sentence, and this one only, is likewise quoted in the
Sui hua ki li 歲 華 紀 麗 by Han Ngo 韓 鄂 of the T'ang
(ed. of *T'ang Sung ts'ung shu*, chap. 1, p. 17) and in the *Tsing k'ang
siang so tsa ki* 靖 康 緗 素 雜 記 by Hwang Ch'ao-ying 黃 朝 英
of the Sung (ed. of *Shou shan ko ts'ung shu*, chap. 8, p. 3b). K'ang-

hi's Dictionary quotes the same text from the *Tsi yün*, a dictionary of the Sung period, but curiously enough omits the Shan Jung, making the swing simply »a sport of the northern regions.»

The *Sui shi kwang ki* 歲時廣記 by Ch'en Yüan-tsing 陳元靚 of the Sung (chap. 16, p. 7b, ed. of Lu Sin-yüan; cf. Pelliot, *Bull. de l'Ecole française*, IX, 1909, 224—225) cites the *King ch'u sui shi ki* as follows: »At the spring festival they suspend long ropes from high trees. The daughters of the officials, in festive attire, are seated or standing on the rope and push it forward and backward. This is called the swing (*ts'iu ts'ien* 鞦 韆). In the vernacular of Ch'u 楚 it is named *t'o kou* 拖鈎 ('pulling the hook'); in the Nirvāṇa Sutra, *küan-so* 罥索.»

The same work quotes the *Ku kin i shu t'u* directly as follows: »The swing, used on the day of the cold provisions (*han shi* 寒食), in its origin was a sport of the Mountain Jung of the northern region for the purpose of gaining lightness and agility of the body. Later generations took advantage of it and on every day of the cold provisions amused themselves with this sport. Subsequently Chinese girls learned it. Colored ropes were suspended from a tree and a framework set up, called a swing (*ts'iu ts'ien*). Others say that since the time that Duke Hwan of Ts'i 齊桓公 annihilated the Mountain Jung, this sport began to spread in China.»

The *Shi wu yüan shi* 事物原始, by Sü Kü 徐炬 of the Ming, quotes the *King ch'u sui shi ki* as follows: »At spring time they suspend long ropes from high trees. The daughters of the officials, clad in colored dresses, are seated on the rope and push it forward and backward. This is called swinging (*ta ts'iu ts'ien* 打鞦 韆).» Again, this work is credited in the *Ko chi king yüan* (chap. 60, p. 4) with the following: »The swing in its origin is a sport of the Mountain Jung of the northern region, who practise it for the purpose of gaining lightness and agility of the body. Subsequently Chinese girls also learned it. They suspend a wooden board from colored ropes and erect a framework above it. The girls in bright attire mount the swing, seated or standing, pushing it forward and backward. This is called 'the swing' (*ts'iu ts'ien*). In the vernacular of Ch'u 楚 俗 it is also called *shi-kou* 施 鉤. The Nirvāṇa Sūtra calls it *küan-so* [as above].» It would lead me too far to enter here into a discussion of the game *shi-kou*, *shi* being probably a wrong reading for 拖 *t'o* (»pulling the hook») offered by Lu Sin-yūan's text. It has nothing to do, however, with the swing. It is a sort of tug-of-war. The confusion between the two was obviously brought about by the fact that they were practised on the day of the cold provisions (*han shi*) and on Ts'ing-ming day.

The term *küan-so* ascribed to the Nirvāṇa Sūtra means literally »suspended ropes» and apparently is not a transcription, but represents the literal translation of a Sanskrit term that appears likewise in Tibetan *dpyaṅ-t'ag* or *ap'yaṅ-t'ag*, which has the same meaning. What the Sanskrit equivalent is I do not know at this moment. The common Sanskrit words designating the swing are *dolā, dolikā, hindola, preṅkhā, preṅkholana*. None of our Tibetan dictionaries assigns to *dpyaṅ-t'ag* the meaning »swing», but it is defined by them merely as »a cord or rope by which a thing is suspended, as a plummet, a bucket, a miner.» The meaning »swing» is clearly indicated by the K'ien-lung dictionaries in four and five languages, where Chinese *ts'iu-ts'ien* is rendered by Manchu *čeku*, Tibetan *dpyaṅ-t'ag*, Mongol *degüdžing*, Turkī *ilänggü*; the verb »to swing» is *ta ts'iu-ts'ien* (as above), Manchu *čekudembi*, Tibetan *ap'yaṅ-mo nyug*, Mongol *degüdžingnemüi*, Turkī *ilänggü ujadu*.

Jung was a general term for barbarian tribes. The Mountain Jung (Shan Jung 山 戎) are identified by Legge (Chinese Classics, V, 904b) with the Northern Jung (Pei Jung), and inhabited the present department of Yung-p'ing in Chili Province (*ibid.*, 118). These were possibly tribes akin to the Hiung-nu (Chavannes, Mémoires historiques de Se-ma Ts'ien, I, 31 and Laufer, Sino-Iranica, 203). The *Ch'un ts'iu* and *Tso chwan* do not mention Duke Hwan's invasion of the Shan Jung, but an officer of Ts'i invaded them in the thirtieth year of Duke Chwang (Legge, p. 117). Se-ma Ts'ien (Chavannes, Mémoires historiques, IV, 136) refers to an invasion of the Mountain Jung in 664 B.C. when Duke Hwan of Ts'i routed them. In another passage (*ibid.*, 564, and III, 425), the Duke boasts of this victory.[1] This apparently is the event alluded to in the above tradition allegedly recorded in the *Ku kin i shu t'u*. It is no wonder, of course, that there is no contemporaneous record anent the introduction of the swing preserved in the historical books of-the period if such an introduction should then have really taken place. A swing was not a matter serious and important enough to arouse the interest of a dry chronicler of the Confucian school. If the swing should have gradually been diffused over China ever since 664 B.C., it is curious, of course, that we hear nothing about it during the centuries that follow. Another gap in the tradition is that no word for the swing is recorded for the time of its adoption and after. The common designation *ts'iu-ts'ien* is traced, according to another tradition, to the Han period (see below), so that the swing must have been known under a different name in pre-Han times. Be this as it may, devoid of chronological value as the above tradition may be, it remains, nevertheless, one of folkloristic value to us interested in culture-historical movements.

Personally I believe that Chinese memories are excellent and capable of retaining traditions through the centuries. Such traditions are not altogether to be discarded in dealing with a subject as the one under discussion where the sober historian naturally leaves us in the lurch. The author of the *Ku kin i shu t'u* was the first who took an interest in a minor affair such as a swing, and made inquiries among the people as to what was known to them about it. I do not believe that he invented his story. I do believe that he placed on record a popular tradition current among his contemporaries. There are two salient points to be retained in this tradition: first, that the swing was believed to be of foreign origin, and second, that it was received from some tribe of northern barbarians in a period anteceding the Han. The combination of this tradition with Duke Hwan's expedition against the Shan Jung (it is not even certain that this

---

[1] For concise biographies see Mayers, Chinese Reader's Manual, No. 211, and Giles, Chinese Biographical Dictionary, No. 841.

item was contained in the original text of the *Ku kin i shu t'u*)
savors strongly of a learned interpretation or afterthought, and in
all probability may not have formed part of the oral popular tradition.

It is clear that the text of the *Ku kin i shu t'u* has suffered
from the inevitable fate of interpolations. The *Shi wu ki yüan*
事物紀原 (chap. 8, p. 20), written by Kao Ch'eng 高承 of the
Sung and published in 1472 by Li Kwo 李果, quotes under the
heading »swing» solely the work in question, saying that the Jung
and Ti 狄 of the northern region, who were fond of practising
swinging in order to gain agility, made swings whenever the day
of the cold provisions appeared 每至寒食為之. It is manifest
that the latter clause is plain absurdity; for this holiday is a purely
Chinese institution which was assuredly not adopted by the barbarians. The Ti are likewise interpolated in the text. Again, the
T'u shu tsi ch'eng, in reproducing the passage of the *Ku kin i shu t'u*,
does not give the word *küan-so* from the Nirvāṇa Sūtra, but credits
the latter with a pompous definition of the swing, with the wrong
addition that it is a kind of *shi-kou*. Aside from such minor details
and the reference to Duke Hwan, however, we might say that we
have the text of the *Ku kin i shu t'u* before us in as good a form
as might be expected under the circumstances and that it imparts
to us a bit of useful information on the subject.

According to another tradition, swings are ascribed to the emperor
Wu of the Han dynasty (140—86 B.C.), under whose reign they
are said to have been popular in the apartments occupied by the
women. This is explained by the fact real or alleged that the emperor
was in the habit of praying for a lifetime of a thousand autumns
漢武祈千秋之壽. *Ts'ien ts'iu* 千秋 (»a thousand
autumns») then was a phrase used in praying for long life. This
phrase was made to serve as a designation for the swing, evidently
prompted by the belief that swinging as a wholesome sport was
apt to prolong one's years. It happened that for some reason or
other, which is not explained, the two words were interchanged,

and the swing was then called *ts'iu ts'ien* 秋 千. In fact, the term is thus written in the ancient texts, and is also entered in Giles' Chinese-English Dictionary. The original meaning of the term was subsequently forgotten, and the classifier 177 革 (»hide») was added to each character, then written as also at present 鞦 韆, although hide does not enter into the making of a swing. Thus far the Chinese.

The tradition that swings were in vogue at the court of Wu, however, is not contemporaneous, and is not contained in the literature of the Han, but first crops up under the T'ang in a *fu* 賦 written by the poet Kao Wu-tsi 高 無 際. Practically the same information is given in a *Ts'ien ts'iu fu* 千 秋 賦 by Wang Yen-shou 王 延 壽, cited in the *Sui shi kwang ki* (chap. 16, p. 9). Afterwards it was repeated in several books, e.g. in the *Tsing k'ang siang so tsa ki* (chap. 8, p. 3b) and in the *Fu ku pien* 復 古 編 by Chang Yu 張 有 (eleventh century).

The value of a T'ang tradition pointing to an event under the Han with a gap of several centuries between the two dynasties is difficult to determine. The proposed etymology, however, is suggestive and presumably correct. The swing was regarded as a life-prolonging instrument, and it is quite plausible that it received the designation *ts'ien ts'iu* (»a thousand autumns»), implying the wish that a person who would indulge in the sport of swinging might live a thousand (i.e. numerous) years. The same term, at a later time, was applied to the heir-apparent, in the same manner as *wan sui* (»ten thousand, i.e. numberless, years») referred to the emperor. An inscription on a cast-iron bell with date corresponding to A.D. 1595 (obtained by me in Shansi for the Field Museum) begins: 皇 帝 萬 歲 太 子 千 秋, »Numberless years to the emperor, a thousand autumns to the heir-apparent.» Giles registers *ts'ien sui* (»a thousand years») as title of a prince. The main point, however, is that the formula *ts'ien ts'iu wan sui* and the two phrases *ts'ien ts'iu* and *wan sui* separately appear on the roofing-tiles of the Han period (see A. Forke, Inschriftenziegel aus der Ch'in- und Han-Zeit, 72, 73, 96, 97).

Leaving aside the retrospective traditions, this much can be formulated from a purely historical standpoint. The swing is not mentioned in pre-Christian literature, nor is any term on record that might be thus interpreted. I feel almost certain that it is not referred to in the literature of the Han. The *Han tsien* 漢雋 does not list any term for a swing. The *Ts'e yüan* writes that »swings were much used in the palaces of the Han and T'ang»; however, this assertion can be positively made solely for the court of the T'ang. The *possibility*, of course, remains that the swing may have been known at the Han court; this depends upon the critical attitude we may adopt toward the tardy tradition to this effect. The earliest dictionary that registers the word *ts'iu-ts'ien*, according to K'ang-hi, is the *Yü p'ien* of Ku Ye-wang (A.D. 523). The *Sui shi kwang ki* quotes a dictionary, *Tse shu* 字書, the date of which is not known to me. The definitions of the swing by the lexicographers are simply »rope sport», 繩戱 *sheng hi*. The word *hi* designates any kind of play or amusement, a game as well as a stage play.

Under the T'ang dynasty swinging was an entertainment enjoyed by the ladies of the palace. The *K'ai yüan t'ien pao i shi* (chap. B, p. 19b) has this notice: »In the period T'ien-pao (A.D. 742—755), whenever the festival of the cold provisions 寒食節 [on the day preceding the Ts'ing-ming festival] arrived, they vied in the palace to erect swings 秋千, so that the ladies of the seraglio might have an occasion for pleasure and rejoicing. The emperor [Hüan Tsung] called it the 'game or play of the half fairies' 半仙之戱 — an expression adopted by the literati and people in the capital.» The idea underlying this poetic phrase is not hard to understand. The fairies or immortals of Taoism when their earthly career had come to an end were able to ascend heavenward. The ladies rising into the air in their swings looked to the monarch like fairies ready for their aerial journey, but swinging back they returned to earth — hence not full-fledged but merely »half» fairies.

As regards the Sung period, references to the swing are numerous. Mong Yüan-lao 孟元老, in his *Tung king mong hwa lu*

東京夢華錄 (p. 10 b of the edition in *T'ang Sung ts'ung shu*) points it out for K'ai-fung, capital of the Northern Sung. In a verse of Su Shi (A.D. 1036—1101) maidens swing in a garden on a moonlit evening of the spring to the accompaniment of music (Admiral Ts'ai Ting-kan, Chinese Poems in English Rhyme, 1932, No. 42). The praise of the swing is sung in many other poetical compositions, some of which are reprinted in the *T'u shu tsi ch'eng* XVII, chap. 804, and *Yüan kien lei han*, chap. 187, pp. 21—22.

Under the Sung, the swing makes also its appearance in pictorial art. In the Pictures of a Hundred Boys 百子圖 ascribed to the painter Su Han-ch'en 蘇漢臣, the sympathetic exponent of children's life, swinging as a pastime of lads is always in evidence. A passage in the *Sui shi kwang ki* (chap. 16, p. 9b) points out that people were in the habit of buying small swings to give pleasure to their boys; the swings were made of wood skilfully painted in bright colors.

The *Ming kung shi* 明宮史 (chap. 5, p. 3) by Liu Jo-yü 劉若愚 (cf. Hirth, *T'oung Pao*, 1895, 440—446) contains this notice: »On the fourth day of the third month the eunuchs set over the women apartments don Lo-han (Arhat) garbs. The Ts'ing-ming festival is the swing festival. They carry willow twigs in their hair. In the rear of the K'un-ning Palace 坤寧宮 and in every palace they erect a swing.» In Peking the women put willow-catkins in their hair on Ts'ing-ming day (Grube, Zur Pekinger Volkskunde, 64); in central China, the houses were decorated with willow branches on the same day.

»During the Ts'ing-ming festival it was formerly customary in the Peking palaces to erect swings with colored cords. Those who participated in the ceremony of swinging donned a special garment which consisted of a gold-embroidered jacket and a girdle provided with a smelling pouch. Those on the swing were grouped in pairs opposite each other. Specially palatable morsels were served at the banquet which surpassed other festive occasions. In the houses of the well-to-do arrangements for the feast were not inferior to those in the palace. In the mansions of the high palace officials, the eunuchs, and the nobles, swinging was regarded as capable of warding off evil spirits and simultaneously as a pastime» (W. Grube, Zur Pekinger Volkskunde, p. 65, quotation from *Si tsing chi* 柝 津 志).

At present swinging is still a pastime of Chinese youngsters as among us (compare, for instance, I. T. Headland, The Chinese Boy and Girl, 1901, with reproduction of a photograph of a lad standing on a swing suspended from the branch of a tree).

Chinese records attribute the swing also to barbarous tribes; e.g. to the land of the Se-lo 厮 囉 north of the Yellow River, whose customs in general were like those of the Tibetans; they lived in block houses, while the rich had tents of felt; they frequently practised the sport of swinging (*Sung shi*, chap. 492, p. 6b). An album depicting in water-colors the life of the aboriginal tribes of Yün-nan (雲南百蠻 圖, 2 vols., about K'ien-lung period) contains the picture of a swing consisting of a rope suspended from a wooden framework which is surmounted by three little flags. There is no seat, and a woman simply stands on the rope. According to the legend added to the picture swinging is practised on New Year's day in order to »gain lightness and agility of the body» (using the same phraseology as the *Ku kin i shu t'u*) and with the idea of praying for long life.

Like the Chinese, the Koreans delight in swinging, but do so on the festival of the fifth day of the fifth moon. H. B. Hulbert (The Passing of Korea, 1906, 279) writes, »Sometimes the lofty branch of a pine-tree is used, but more often two great poles are erected for the purpose. These are held in place by guys, and are variously ornamented. The Koreans are adventurous swingers, and accidents are not infrequent. The rough straw ropes break sooner or later, and some one gets a nasty fall, which terminates the sport for that season.» Again, he writes (p. 371), »Korean girls are very fond of swinging, and on a certain day in spring there is a swing festival in which men, women, and children participate. Huge swings are arranged in public places, but these are used only by men and boys.» As to the swing in Korea compare, further, Koike in *Int. Archiv f. Ethnographie*, IV (1891), 10 and Schlegel's note, *ibid.*, 121; Rockhill, *Am. Anthr.*, 1891, 185; S. Culin, Korean Games, 1895, 34—35. As the Korean designation for the swing is *chyu-chyen* identical with the Chinese word, there can be no doubt as to the derivation of the Korean swing from China.

Frazer, in The Golden Bough, writes, »The custom of swinging has been practised as a religious or rather magical rite in various parts of the world, but it does not seem possible to explain all the instances of it in the same way. People appear to have resorted to the practice from different motives and with different ideas of the benefit to be derived from it. . . . The Letts, and perhaps the Siamese, swing to make crops grow tall. . . . People swing in order to procure a plentiful supply of fish and game as well as good crops. In such cases the notion seems to be that the ceremony promotes fertility, whether in the vegetable or the animal kingdom; though why it should be supposed to do so, I confess myself unable to explain.»

In China, swinging was not associated with any magical rites or fertility ideas. It was chiefly practised to promote long life, a notion peculiar to China and not to be found among any other nation (therefore not registered by Frazer, who makes no reference to China). No other nation has expressed its yearning for longevity so fervently and intensely as the Chinese, and has devised so many hundred ways and means of obtaining it. The swing was added to the long catalogue of these recipes. Only one text, and a recent one (*Si tsing chi*), alludes to warding off evil spirits in connection with swinging. Frazer remarks, »Swinging is sometimes resorted to for the purpose of expelling the powers of evil.» Nothing like the mythological significance of the swing in India and Siam is met with in China. What China shares with other nations is the feature that swinging was restricted to, or rather with predilection practised on, certain holidays in the beginning of the spring. This, of course, is not due to historical contact, but is the outcome of conditions.

Swinging is an outdoor sport which naturally hibernates in the winter. and awakens in the spring, and on holidays there is more leisure and opportunity for play and games of all sorts. Thus the Chinese also had a »swing festival», a term used in the *Ming kung shi* with reference to the court of the Ming, but apparently no ritual was connected with it — our sources at least are silent as to this point. Swinging in China was a festal sport surrounded by splendor and merry-making, not, however, a festal rite.

Chicago.

228

戽水车或波斯水轮

ORIENTAL STUDIES

In honour of

Cursetji Erachji Pavry

EDITED BY

JAL DASTUR CURSETJI PAVRY

WITH A FOREWORD BY

A. V. WILLIAMS JACKSON

LONDON   *MCMXXXIII*

OXFORD UNIVERSITY PRESS

ONE THOUSAND COPIES OF THIS BOOK HAVE BEEN
PRINTED, AND THE TYPE HAS BEEN DISTRIBUTED

PRINTED IN GREAT BRITAIN
AT THE UNIVERSITY PRESS, OXFORD
BY JOHN JOHNSON

THE NORIA OR PERSIAN WHEEL

IN his *Systema Agriculturae, The Mystery of Husbandry Discovered* (2nd ed., London, 1675, p. 18), J. W. Gent has the following notice on the Persian wheel:

'The most considerable and universal is the Persian wheel, much used in Persia, from whence it hath its name, where they say there are two or three hundred in a river, whereby their grounds are improved extraordinarily. They are also much used in Spain, Italy, and in France, and is esteemed the most facile and advantageous way of raising water in great quantity to any altitude within the diameter of the wheel, where there is any current of water to continue its motion; which a small stream will do, considering the quantity and height of the water you intend to raise. This way, if ingeniously prosecuted, would prove a very considerable improvement; for there is very much land in many places lying near to rivers that is of small worth, which if it were watered by so constant a stream as this wheel will yield, would bear a good burthen of hay, where now it will hardly bear corn. How many acres of land lie on the declining sides of hills by the rivers sides, in many places where the water cannot be brought unto it by any ordinary way? Yet by this wheel placed in the river or current, and a trough of boards set on tresles to convey the water from it to the next place of near an equal altitude to the cistern, may the land be continually watered so far as is under the level of the water. Also there is very much land lying on the borders of rivers that is flat and level, yet neither doth the land-floods overflow the same, or at most but seldom; nor can the water be made by any obstruction thereof, or such-like way to overflow it. But by this Persian Wheel placed in the river in the nearest place to the highest part of the land you intend to overflow, therewith may a very great quantity of water be raised. For where the land is but little above the level of the water, a far greater quantity of water, and with much more facility may be raised, than where a greater height is required; the wheel easier made, and with less expence.'

Gent offers a good woodcut of the Persian wheel, and his account goes to show that in the latter part of the seventeenth century no better method of raising water for irrigation purposes was known in Europe. The term 'Persian wheel' is still used in English in the same sense, and is registered as such in the *Oxford English Dictionary*.

The Persian wheel is familiar to us under the name *noria*, which we adopted from the Spaniards. The latter, on their part, received it from the conquering Arabs. The consensus of opinion is that the word is based on Arabic ناعورة, *nāʿūra* or *nāʿōra*, with the article *an-nāʿōra*, which is still preserved in the Spanish forms *anoria* and *añoria*; in old Spanish it was *naora* and *alnagora*. Both the Arabic and Spanish word refers to the same hydraulic device. The Portuguese form *nora* is still nearer to the Arabic prototype, and the Arabic etymology is listed as early as 1830 by J. De Sousa (*Vestigios da lingoa arabica em Portugal*, p. 169: 'Maquina hydraulica, que serve de tirar agua dos poços, cisternas, e rios'). L. Marcel Devic (*Dictionnaire étymologique des mots français d'origine orientale*, 1876, p. 177) states that the Arabic noun is derived from the verb نعر *na'ar*, which means 'laisser jaillir le sang par saccades, en parlant d'une veine; ce qui s'applique assez bien aux norias, formées d'une série de seaux en chapelet qui se remplissent au fond du réservoir et viennent se vider l'un après l'autre à l'extérieur'. The Arabic etymology of *noria* has remained uncontested, and has been adopted by W. Meyer-Lübke (*Romanisches etymologisches Wörterbuch*, 1911, No. 5856; also by K. Lokotsch, *Etym. Wört. der europ. Wörter orientalischen Ursprungs*, 1927, No. 1561) and by all our English dictionaries. The *Oxford English Dictionary* defines *noria* as 'a device for raising water, used in Spain and the East, consisting of a revolving chain of pots or buckets which are filled below and discharged when they come to the top'. Another Spanish-Portuguese word has a bearing on the same contrivance: Spanish *arcaduz* or *alcaduz*, Portuguese *alcatruz*, which represent Arabic القادوس *al-qādūs* and signify 'the bucket of a noria' (Devic, p. 6; Meyer-Lübke, No. 1456).

From Italy the Persian wheel spread to Warsaw, Poland, where Peter Mundy (*Travels*, iv. 203, Hakluyt Society, 1925) saw one in operation in 1643 and described it under the title 'A Strange Water Work' as follows:

> 'The gardener's servant showed me a house in a garden near the palace, which by wheel works drew water out of a well of itself, which he gave me to understand after this manner, viz.: among other is one great principal wheel unto which are fastened a great number of pots. This wheel, having once motion given it, forces up a quantity of water through pipes by the help of pump holes, leathers, etc., as I have seen in other water works. Of this water, part runs to the palace and the rest runs back into the vessels fastened on the great wheel, which being of a great compass, a little weight on the circumference causes it to go about. Having once motion, it forces up so much water that supplies the king's house and itself again to continue the said motion of itself, so that if this be true, as I think it is, it may be rightly called a perpetual motion: the water in the well supplied by his own spring. It was contrived by an Italian, who dieing and the work coming out of frame, there has been none since can be found that can bring it into order again, so that at present there must be twelve Tartar slaves to supply the work which the wheel alone performed of itself.'

Sir Richard Carnac Temple, Mundy's editor, comments justly, 'Mundy is describing what is known in India as the 'Persian wheel' (*rahat*), in this case driven by water machinery supplied by itself. In India, Persia, and Mesopotamia it is driven by a bullock. Such wheels are still common in Italy and Portugal'. Mundy himself (*Travels*, ii. 228, Hakluyt Society, 1914) had observed them also in Spain and India. 'In Spain', he writes, 'we call them noraies' (noria).

A. von Kremer (*Culturgeschichte des Orients unter den Chalifen*, ii. 322) maintained that the water-wheel was brought by the Arabs to Spain, but added judiciously that he hardly believed that its invention was a merit of the Arabs; and I concur with him in this opinion. G. Staunton (*Account of an Embassy from the King of Great Britain to the Emperor of China from the Papers of the Earl of Macartney*, ii. 1797, p. 479) writes: 'Most Eastern nations seem to have been acquainted at an early period with the machine for raising water, known by the name of the Egyptian wheel, which was however un-known in Europe till the Saracens introduced it into Spain, in an imperfect state, and under a very awkward form.' Ibn al-'Auwām, an Arabic agri-cultural writer, who lived at Seville in the twelfth century, has given a technical description of the noria in his famous treatise on agriculture (*Le livre de l'agriculture d'Ibn al-Awam*, traduit de l'arabe par Clément-Mullet, i, 1864, p. 129). The most interesting contributions to our knowledge of water-raising devices from Arabic sources we owe to the eminent Arabist and physicist, the late Eilhard Wiedemann (*Beiträge zur Geschichte der Natur-wissenschaften*, vi, 'Zur Mechanik und Technik bei den Arabern', Erlangen, 1906, pp. 13, 50; x, 'Zur Technik bei den Arabern', p. 331). The water-wheels are also styled *zurnūq*, *daulāb*, *hannāna*, and in Egypt *sāqiya* ساقية. *Hannāna* means 'the constantly groaning ones' from the creaking sound pro-duced by the wheels, which even impressed the poets as beautiful. *Daulāb* or *dulāb* دولاب, a word of Persian origin (*dol-āb*), means a 'wheel', and speci-fically refers to a water-wheel drawn by oxen or horses and designed for drawing water from a well, while *gharrāf* غرّاف is a water-wheel drawn by oxen or horses and designed for drawing water from a river. The word *nā'ūra* (plural *newā'ir* نواعير) is chiefly used in the Maghreb, but also in Syria and Persia. The water-wheels are constructed by a special class of artisans whose craft descends from father to son; the most skilful ones live in Syria. Historically the noria can be traced to the earliest times of the Caliphs. Man-power also was then enlisted to operate it, sometimes for the purpose of inflicting a punishment. The earliest example cited by Wiedemann refers to the year A.D. 884–5, and is an account by Ahmad Ibn al-Tayyib.

There was a celebrated noria at Fez, Morocco, which raised the water of the river up to the royal garden and which became proverbial (Gaudefroy-Demombynes, *Masālik El Absār*, p. 156, Paris, 1927, who remarks that the norias and their sighs are a theme of Arabic poetry). The *sāqiya* of Egypt is described by E. W. Lane (*Account of the Manners and Customs of the Modern Egyptians*, 5th ed., 1871, p. 26): it mainly consists of a vertical wheel which raises the water in earthen pots attached to cords and forming a continuous series; a second vertical wheel fixed to the same axis, with cogs; and a large horizontal cogged wheel which, being turned by a pair of cows or bulls or by a single beast, puts in motion the two former wheels and the pots. T. Shaw (*Travels, or Observations Relating to Barbary and the Levant*, Oxford, 1738, p. 431) writes: 'Persian wheels, called *sakiah* in Egypt, were in general use along the banks of the Nile, from the sea to the cataracts.' Again, the term 'Persian wheel' is noteworthy.

In Syria and along the banks of the Tigris and Euphrates the norias are still in operation. At Basra the water-wheels were also driven by camels. Near Mecca norias were likewise employed. Yāqūt writes that 'the water of Ma'din al-Burm is well-water which irrigates the fields by means of the *zurnūq*.' Dimashqī, who wrote a cosmography about A.D. 1325, refers to the norias of Ḥamā on the river 'Aṣī (Orontes) as being of a construction as seen nowhere else, in order to maintain considerable streams of water for the irrigation of numerous gardens abundant in fine and excellent fruits such as the apricot of camphor and almond flavours not found in any other country (A. F. Mehren, *Manuel de la cosmographie du moyen âge*, p. 281, Copenhague, 1874; see also Gaudefroy-Demombynes, *La Syrie à l'époque des Mamelouks*, 1923, p. 106; S. Guyard, *Géographie d'Aboulféda*, ii, pt. 2, 1883, pp. 40, 138). Yāqūt likewise speaks of the irrigation of the gardens through the water of the 'Aṣī. Large water-wheels still exist at Ḥamā (R. Oberhummer and H. Zimmerer, *Durch Syrien und Kleinasien*, p. 92). Illustrations of such wheels may be viewed in the book of A. T. Olmstead, *History of Assyria* (New York, 1923), Figs. 66, 71, 76.

In the environment of Constantinople the Persian wheel was noticed by G. Jacob (*Altarabisches Beduinenleben*, 1897, p. 228), and it likewise occurs in Asia Minor (K. Kannenberg, *Kleinasiens Naturschätze*, 1897, p. 83, under the name *sakie*). In other words, it is widely diffused all over the Islamic world, including northern Africa.

The tendency of certain Egyptologists to draw retrospective conclusions from present-day conditions is well known, and it is even more drastic among Assyriologists. Thus we are treated to the gratuitous speculation that the ancient Egyptians 'perhaps utilized also the *sāqīya*' (F. Hartmann, *L'Agriculture dans l'ancienne Égypte*, 1923, p. 118, who, however, adds cautiously, 'an ancient design of which has not yet been found'). There is no tangible evidence for this assertion, and the fact remains that the *sāqīya* was introduced into Egypt during the Middle Ages by the conquering Arabs. Again, there are Assyriologists who from the modern water-wheels existing in Mesopotamia conclude naïvely that they must have been in existence in ancient times (cf. Handcock, *Mesopotamian Archaeology*, p. 369: 'What the larger machines were we do not know, but as Johns suggests, they may have very possibly consisted in a set of buckets fastened to a wheel, etc.: but whatever the machine was it must have been fairly elaborate, for it sometimes required as many as eight oxen to work it.' (B. Meissner, *Babylonien und Assyrien*, i. 192, in discussing the methods of irrigation in ancient Babylonia, says nothing about this alleged use of the noria). It might be confidently stated that if the noria had been known in ancient Egypt and Babylonia, it must have spread to Greece and Italy, or that at least some notice of it would have been preserved by Greek or Roman writers, neither of which, however, is the case. The passage in Vitruvius x. 5, as already pointed out by J. Beckmann (*Beyträge zur Geschichte der Erfindungen*, ii, 1788, p. 14), relates to water-mills, not to water-wheels for irrigation (cf. M. H. Morgan, *Vitruvius*, 1914, p. 294, and A. Neuburger, *Technik des Altertums*, 2nd ed., 1921, p. 232).

Abu 'Abdallah el-Maqdisi, called El-Muqaddasi, wrote in A.D. 985: ''Adud

R

al-Daula dammed the river which flows between Shīrāz and Iṣṭakhr by means of a gigantic wall whose foundations he closed with lead. Behind, the water is stowed, and is higher than the river. On both sides he set up ten water-wheels and beneath each water-wheel a mill—at present one of the wonders of Fars. There he built a city and conducted the water into canals and supplied three hundred places with water.' In another manuscript the following version occurs: 'And on each side he made arches from which the water flowed roaringly and impetuously into the lowest parts of the norias and set these in motion. Around the felloes of the wheels there are boxes which fill themselves with water. When they have performed a revolution, they pour the water into canals from which it is distributed among three hundred places' (Wiedemann, op. cit., p. 324).

According to Rashīd-ad-Dīn, when the Mongol forces laid siege to Baghdad in A.D. 1258, their lower camp was pitched at a place called Dūlāb-i Baqal, which means 'the Water-wheel of the Vegetable Garden' (G. Le Strange, 'Baghdad during the Caliphate', *JRAS.*, 1899, p. 882).

F. von Schwarz (*Turkestan*, 1900, p. 346) refers to the peculiar water-wheels used in Russian Turkestan, which he says are called *čigir* (in W. Radloff's *Wörterbuch der Türkdialecte*: *čikir*, given as Jagatai). He describes them 'as paddle-wheels crudely patched together from wooden rods and operated by the current of the channel water. To the periphery of these wheels are usually attached earthenware pots which draw water in the canal at the turning of the wheels and empty the water into a wooden gutter. Beside the garden of the Tashkent observatory a Sart had once set up such a water-wheel which had a diameter of about six metres and which irrigated a small cotton plantation. Owing to the frequent need of repair these wheels are but seldom utilized at Tashkent where there is no lack of water. Among the agricultural Kirgiz, who have an abundance of draught-cattle, water-wheels driven by oxen are also used' [as also in China]. F. von Schwarz concludes, 'The water-wheels used in Central Asia agree perfectly with those of Egypt, which goes to prove that the Central Asians derived them from the Egyptians' (which, of course, is an untenable conclusion).

H. Moser (*L'Irrigation en Asie centrale*, 1894, p. 266) states that the word *čigir* is Turkish in origin and that the Bukharians still bestow upon this apparatus the name 'Persian wheel' (*roue persane*). He observed these Persian wheels on the banks of the Amudarya and in Khiwa. They reminded him of the *sāqīya* of Egypt, and he thinks it probable that the latter is akin to the *čigir* in origin and that this origin is very ancient. His description is as follows:

'Il se compose essentiellement d'une roue en bois, de 3 à 4 mètres de diamètre, tournant dans un plan vertical au dessus d'un puisard de façon à ce que la circonférence de la roue plonge, en bas de sa rotation, au dessous du niveau de l'eau. Sur le pourtour de la circonférence sont fixés, à des intervalles convenables, des cruchons de poterie ou de bois évidé, obliquement au rayon de la roue, de façon à ce qu'ils s'emplissent dans le puisard, puis se déversent à la hauteur voulue dans un canal récepteur. A l'aide d'un engrenage à roues dentées en bois, le tchiguir est mis en mouvement par un cheval, chameau, bœuf, âne ou quelquefois plusieurs de ces bêtes de somme disparates accouplées ensemble. Cet appareil peut élever de 4 à 5 mètres cubes d'eau à l'heure.'

Peter Mundy (*Travels*, ii. 228) observed Persian wheels at Fatehpur Sīkrī in the palace of the Moguls: 'The water to water it is also to fill the tanks aloft, first into one tank and then from that into another higher, and so into four or five until it come aloft, by that which we in Spain call Noraies [noria].' Sir Richard Carnac Temple, Mundy's editor, annotates that 'the ruins of the series of Persian wheels and reservoirs, by which water from the lake outside the city was supplied to the palace, still exist.'

John Fryer, who travelled in India and Persia from 1672 to 1681, noticed the use of Persian wheels both in India and Persia. They have 'pans or buckets of leather hanging round about a wheel, some always in the water, others rising up, and at the same time others pouring out as the wheel turns round: and thus are their best gardens kept alive.' While in Persia, he speaks of 'the Indian wheel drawn up and let down by oxen, with as little intermission day or night, as Sysiphus's repeated trouble is reported' (*New Account of East India and Persia*, ii. 94, 171; iii. 156, Hakluyt Society, 1912, 1915).

A. Neuburger (in H. Kraemer, *Der Mensch und die Erde*, ix. 306) pleads for India as the home of the water-wheel, 'as may be concluded from various criteria'; but no evidence for this assertion is forthcoming.

'Persian wheels', N. G. Mukerji writes in his *Handbook of Indian Agriculture* (2nd ed., Calcutta, 1907, p. 142), 'are in use on the Malabar coast, in Rajputana, Kathiwar, and the Punjab. Some are of very simple and cheap construction. This type is used chiefly on the coast of Kathiwar, Gujarat, and the west coast of India generally. A bamboo or wooden drum of light framework turns on an axle which rests on two pivots. A sitting man turns the drum with his hands and feet. Round the drum is attached an endless garland of mud vessels which are brought up by the revolution of the drums carrying water in them, and discharging the water (from three mud vessels at a time), into a trough of stone whence it flows out to the field. With this implement one man can irrigate one-tenth of an acre a day. The Persian wheel of the Punjab pattern is the same as the Egyptian-Persian wheel.'

Mukerji further refers (p. 145) to the noria or bucket-pump as another form of improved Persian wheel, which consists of buckets chained one to another in an endless series and worked by hand or animal power.

Discussing the subject of irrigation in India, W. Crooke (*Things Indian*, New York, 1906, p. 282) writes,

'Lastly comes the curious machine known as the Persian wheel, of which the history is obscure. It does not seem to be used in Persia [this is erroneous], but it is represented by the Egyptian *sāqīya*, and it appears in Palestine. Possibly the Indian title merely implies that the idea came from the West. In Egypt it was probably a late invention, as it has not been recognized on the tomb frescoes. The Burmese have a somewhat similar machine, the *yit*, in which, as in the Persian wheel, the water is raised by a wheel, to which bamboo baskets are attached [as in China]. The Persian wheel is known as far south as Malabar, and it is purely a matter of habit or tradition whether the farmer uses the wheel or the leather bag,' &c.

'Pucka wells are usually worked by the *harth*, or Persian wheel. A broad-edged lantern wheel whose axis lies horizontally over the centre of the well's mouth, carries on its broad edge a long belt of *moonj* rope, made like a rope ladder, the ends of which joined in an endless band reach below the surface of the water. To this at every step of the rope ladder, an earthen pot called *tind* is fixed. As the wheel revolves, the large rope belt descends into the water with its pots, the pots become filled with water, and are drawn up: as they reach the top of the wheel, they are by the revolution of the wheel inverted, and their contents poured out into a trough, which is ready to receive them, and which leads to the water-course of the fields to be irrigated. The wheel bearing the belt and waterpots is caused to revolve by having on the same axle another wheel parallel to it, and cogged in one side, the teeth of which work into the cogs of another vertical lantern wheel, whose axis again rests in a bar supported between two upright brick or wood pillars at one side of the well's mouth; this vertical wheel is turned by a pair of oxen yoked to a pole, which is fixed into the axis of the wheel in question. The oxen by walking round and round on a tramway drag the pole with them, and cause the whole apparatus to turn' (B. H. Powell, *Hand-Book of the Economic Products of the Punjab*, i, 1868, pp. 207-8).

In his *Bihār Peasant Life* (Calcutta, 1885, p. 210) Sir George A. Grierson writes, 'The Persian wheel is not used in Bihār. Its name, *rahat*, is however known in Patna.' The sole allusion to the Persian wheel in Sanskrit literature I have encountered so far occurs in Bāṇa's *Harsha-Carita* (translated by Cowell and Thomas, 1897, p. 264): 'His right hand shook a rosary, like a Persian wheel containing the buckets for raising water from the well of all delightful motions.' It seems, therefore, that the Persian wheel was known in India at least in the sixth century of our era.

With reference to the Burmese methods of irrigation good information is contained in J. G. Scott's *Gazetteer of Upper Burma and the Shan States* (pt. 1, ii, Rangoon, 1900, pp. 342-3). The *yit*, referred to above by W. Crooke, is described there as 'an ordinary water-wheel with lengths of bamboo tied transversely opposite the floats. These act as buckets for lifting the water and, as the wheel revolves with the current, are tilted so as to empty themselves into a trough or channel, which carries the water into the fields. In some places in the Shan States where the rivers have a deep channel, these wheels are forty or fifty feet high and raise water enough to form quite a considerable rivulet.'

The noria is also widely distributed throughout the Far East, and is a conspicuous and indispensable adjunct of Chinese and Japanese agriculture. The Chinese water-elevators have attracted the attention of many travellers, and have frequently been depicted and minutely described, but no one seems to have noticed that they are identical in principle and construction with the water-raising devices of the West, nor has any one ever raised the question as to their origin and historical connexions. In an excellent study, entitled 'Westöstliche Landwirtschaft' (in *Festschrift P. W. Schmidt*, pp. 416–84), P. Leser has recently traced the interrelations of the East and West in matters of agricultural implements, but he has not dealt with machinery for irrigation. The older literature relative to the noria in China is listed by J. H. Plath ('Die Landwirtschaft der Chinesen', *Sitzungsberichte der bayerischen Akademie*, 1873, pp. 815–17). A brief summary of the subject is given by W. Wagner (*Die chinesische Landwirtschaft*, Berlin, 1926, pp. 189–99) and F. H. King (*Farmers of Forty Centuries*, Madison, 1911, pp. 300–3, 363, 411, with good illustrations; see also S. Syrski in *Anhang Berichte über österr. Exped. nach Siam, China und Japan*, 1872, p. 79; G. Schlegel, *Ouranographie chinoise*, 1875, p. 457; S. W. Williams, *The Middle Kingdom*, 1901, ii. 7; H. R. Davies, *Yün-nan*, 1909, p. 158; J. G. Anderson, *The Dragon and the Foreign Devils*, 1928, p. 27; and others).

As regards the water-wheel of Japan I shall refer only to E. S. Morse (*Japan Day by Day*, ii, 1917, p. 284), who gives a sketch of it, saying that it is a Chinese device, rare about Tokyo and farther north, but not uncommon in the southern provinces of Japan. According to G. Sarton (*Introduction to the History of Science*, i, 1927, p. 580), the introduction of the noria into Japan is ascribed to Yoshimine Yasuyo, a scholar and son of Kwammu-tennō, emperor from A.D. 782 to 805.

F. H. Nichols (*Through Hidden Shensi*, New York, 1902, p. 31) writes,

'Every quarter of a mile or so a donkey at the end of a long pole may be seen walking around a windlass. He is raising water from a well by a chain-pump, whence it is discharged into the furrows that cross the fields in every direction. Some of the wells are very deep, and are constructed on the Artesian principle, a series of hollow bamboo-rods taking the place of an iron pipe. A well-donkey is a thing essentially Chinese. No one drives him or apparently takes the slightest interest in him. He wears big straw blinders over his eyes, which prevent his seeing anything. He is oblivious of his surroundings. All the ordinary aims and ambitions of donkey life he seems to have forgotten. Hour after hour he walks slowly around the windlass, only a speck on the flat landscape, only a cog in the simple but vast system of agriculture which keeps millions of men alive.'

In the Chinese standard work on agriculture, the *Nung cheng ts'üan shu* 農政全書, published in 1640, seven years after the death of its author, Sü Kwang-k'i 徐光啟, the famous disciple of the early Jesuits, six illustrations of native hydraulic engines are given (reproduced also in *Shou shi t'ung k'ao*, ch. 37, pp. 4-5, and *T'u shu tsi ch'eng*, xxxii, ch. 244) under the generic name *shwi ch'o* 水車 ('water engines'). The first, *fan ch'o* 翻車, is the chain-pump (also described and figured by G. Staunton, *Macartney's Embassy*, ii. 481; J. F. Davis, *China*, ii, 1857, p. 258; *Chinese Repository*, v, 1837, p. 494).

The second of these, designated *t'ung ch'o* 筒 車 ('tube engine'), corresponds exactly to the noria or Persian wheel, save that, as implied by the term, the buckets are replaced with bamboo tubes, in conformity with the universal use of bamboo as a convenient material throughout China. Williams speaks of buckets, and Anderson of tub-shaped containers. This noria is also figured and described by Davis (ii. 260). The third, styled *shwi chwan fan ch'o* 水 轉 翻 車 ('revolving machine turned by water'), corresponds exactly to the *sāqīya* of Egypt. The fourth shows the same apparatus operated by a water-buffalo; the fifth, the same set in motion by two donkeys; and the sixth, called *kao chwan t'ung ch'o* 高 轉 筒 車, is a double chain-pump running over two sprocket-wheels, a lower one in the water, a higher one in the rice-field.

Those who have not access to Chinese publications will find the Chinese illustrations well reproduced and described in the excellent book of O. Franke, *Kêng Tschi T'u, Ackerbau und Seidengewinnung in China* (Hamburg, 1913, pp. 149–52). It is noteworthy that the chain-pump appears in the series of engravings prepared by Lou Shou about A.D. 1145 (Franke, Plate XXXVI; and compare Pelliot, *A propos du Keng Tche T'ou*, Plate XXIII). The principle in all of these machines is the same, the only difference being in the mode of applying the moving power; one is worked by the hand, another by the feet, of a labourer, and the third by an animal (R. Fortune, *Two Visits to the Tea Countries of China*, i, 1853, p. 230). J. Barrow (*Travels in China*, 1804, p. 540) remarks that 'the water-wheels still used in Syria differ only from those of China by having loose buckets suspended at the circumference, instead of fixed tubes.'

In regard to the inventor of the water-wheel, Chinese records refer to the names of two individuals, Pi Lan 畢 嵐 and Ma Kün 馬 鈞 (see *Shi wu ki yüan* 事 物 紀 原, ch. 9, p. 3*b*, by Kao Ch'eng 高 承 of the Sung period, ed. by Li Kwo 李 果 in 1472, original edition in Gest Chinese Research Library, McGill University, Montreal). Pi Lan is said to have lived under the reign of the Emperor Ling (A.D. 156–189) of the Later Han dynasty and to have constructed a *fan ch'o* used for sprinkling the streets. Ma Kün (Giles, in his *Chinese Biographical Dictionary*, p. 565, mentions him as 'a famous mechanic who constructed a variety of ingenious machines') lived in the third century A.D. under the Wei dynasty, and it is on record that 'when he lived in the capital, he owned a plot of arid land suitable as a garden, but there was no water to irrigate it; thus he made a turning wheel (*fan ch'o*) which he caused to be revolved by boys, and this wheel conducted the water for the irrigation of his garden.' This wheel, as also added in the *Shi wu ki yüan*, is identical with the one now used by farmers for field irrigation and took its beginning from Ma Kün of the Wei. This text is quoted in the *Shi wu ki yüan* from the *Wei lio* 魏 略, an historical work now lost, written by Yü Huan 魚 豢 and covering the period from A.D. 239 to 265; portions of it are preserved in the commentary of P'ei Sung-chi to the *San kwo chi* (Chavannes, *T'oung Pao*,

1905, p. 519). As the *Wei lio* is a reliable work, the passage cited from it in the *Shi wu ki yüan* may correctly reflect the tradition of the Wei period (see also *Ko chi king yüan*, ch. 48, pp. 7–8; *Shi wu yüan hui* 事 物 原 會, ch. 23, p. 3; *T'u shu tsi ch'eng*, and others). There is no doubt that Ma Kün was a good hydraulic engineer, for the *Shi wu ki yüan* (ch. 9, p. 26b) has preserved another tradition according to which he made for the Emperor Ming of the Wei artificial fountains in the shape of animals, fishes, and dragons.

The *Shi wu yüan shi* 事 物 原 始 ('Origin and Beginning of Things') connects the invention with the name of the Emperor Ling, but attributes its inception to another engineer, called K'o Mien 渴 兊 , who used the device of bamboo tubes attached to the wheel. Fang I-chi 方 以 智 , in his *Wu li siao shi* 物 理 小 識 (written toward the end of the Ming dynasty, ch. 8, p. 31), refers solely to the Emperor Ling and concludes that the method of irrigation by means of water-wheels was known in China from under the Han; he also points out a passage in a poem of Su Shi or Su T'ung-po (A.D. 1036–1101) relative to the raising of water by means of bamboo tubes from the wells of Se-ch'wan.

In the palace of the Mongol emperors at Peking there was east of the Wan-sui Hill a stone bridge in the middle of which there was an aqueduct leading the water of the Kin-shwi to the top of the hill; the water was pumped to the top of the hill by means of machines, and was poured from the jaws of a stone dragon into a square basin (E. Bretschneider, 'Arch. and Hist. Researches on Peking', *Chinese Recorder*, vi, 1875, p. 319, after *Ch'o keng lu*).

The history of the water-wheel for irrigation cannot be dissociated from that of the water-mill which is based on the same mechanical principle. As formerly pointed out by me (*Chinese Pottery of the Han Dynasty*, p. 33), the invention of mills driven by water is attributed by the Chinese to Tu Yü 杜 預 (A.D. 222–284); and this event coincides exactly with the time of Ma Kün, the inventor of the irrigation water-wheel. I also drew attention at that time to the curious coincidence that water-mills sprang up in China at about the same time as in the Roman empire. Strabo is the first to mention a water-mill with reference to Mithridates who had one in his residence at Cabira, and this hints at the fact that water-mills were first known in the Orient (J. Beckmann, *Beyträge zur Geschichte der Erfindungen*, ii, 1784, pp. 12 et seq.; O Schrader, *Reallexikon*, 2nd ed. by A. Nehring, ii, 1923, p. 27). They gradually spread in Italy in the early days of the Imperium. In view of the fact that the noria did not conquer Europe at the same time it would follow that the use of the water-wheel for mills is older than its use for irrigation purposes, or at least that the Persian wheel had not yet advanced beyond the western boundaries of Iran in Roman times; and this is in harmony with my conclusion that it was the Arabs who brought it from Persia to the empire of the Caliphs and to Egypt. Now, if in the third century of our era Tu Yü constructed a water-mill on the same principle as the water-mills of western Asia, and if simultaneously Ma Kün constructed a noria or what ultimately resulted in this device, we cannot believe in the miracle that these two Chinese engineers should have independently evolved what pre-existed in the West and achieved the same result; the only possible conclusion is that the two worked out their ideas on plans and models brought to China from some locality of central Asia, which presumably was Sogdiana.

The ancient territory of Sogdiana appears to have the best claim to the invention of water-raising devices by means of wheels. From ancient times Sogdiana enjoyed an unrivalled state of prosperity: its fecund valleys, the wealth of its soil, industry, and commerce, and its powerful cities fired the imagination of the ancients to such a pitch that they styled it Paradise of Asia. It is well known how highly agriculture was esteemed in the ancient Persian religion, how the farmer's good deeds and irrigation of the fields in particular are extolled in the Avesta, and how highly developed the art of gardening was in Persia and how the Persian garden became the model for all gardens of Asia. In Sogdiana (K'ang), it is stated in the Annals of the T'ang Dynasty (*T'ang shu*, ch. 221 B; cf. Chavannes, *Documents sur les Tou-kiue occidentaux*, p. 135), 'they have very ingenious machines'. Soon after General Chang K'ien's famous expedition to the West, Chinese engineers wended their way to central Asia (*Shi ki*, ch. 123; cf. Hirth, 'Story of Chang K'ien', *JAOS.*, 1917, pp. 111, 113). There is no doubt that such engineers, on their return to the homeland, brought back plans and specifications of water-raising engines.

Barthold and Petrow have written in Russian detailed monographs on the history of artificial irrigation in Russian Turkestan, which goes back to the times of the first colonizers of the country (F. Machatschek, *Landeskunde von Russisch Turkestan*, 1921, pp. 141, 327). Another method, that of subterranean irrigation, the so-called *kyärise*, is also due to Persia whence it was introduced into the adjoining countries (Machatschek, p. 144, and Vavilov and Bukinich, *Agricultural Afghanistan* [in Russian], Leningrad, 1929, pp. 140, 547). Of this method a Chinese record is also preserved. Ch'ang Te, who in A.D. 1259 was delegated by the Mongol Emperor Mangu to his brother Hulagu in Persia, noticed that the people of Persia dig wells on the summits of mountains and conduct the water several tens of miles down into the plain for the purpose of irrigating their fields (E. Bretschneider, *Chinese Recorder*, v, 1874, p. 326, who annotates: 'This is still the custom all over Persia. The aqueducts are all subterraneous in order to prevent the evaporation of the water. As in Persia it never rains in the summer, agriculture would be impossible there without this artificial irrigation.').

It is possible that the water-mill also is of Iranian origin. Dimashqī (translation of Mehren, p. 254) describes a water-mill at Merend in Aderbeidjān Kedrenos, a Greek monk of the eleventh century, who compiled a Synopsis of History beginning with the creation of the world and terminating with the year 1057, says that under the reign of the Emperor Constantine, Metrodoros a Persian by birth, went off to India and constructed for the Brahmans water-mills and baths—things previously unknown in the country (McCrindle, *Ancient India as Described in Classical Literature*, p. 185).

According to Gauthiot (*Journal asiatique*, 1916, i, p. 251) a water-mill is called *āsyāw* in Pahlavi, *āsyā* in Persian; in the dialects of the north-east of Iran: *yāva-rĕnĕm* ('grain-mill'). The Iranian tribes of the Pamir are familiar with the water-mill (Wakhī *khadorg*, Sariqoli *khadorj*, Minjan *khairgha*, Sanglikh *khadari*).

Another type of mill, the windmill, is likewise of Oriental origin. For a long time this was denied. The learned Beckmann (op. cit., p. 32) wrote in 1788, 'It has often been asserted that wind-mills were first invented in the Orient and became known in Europe in consequence of the crusades, but this is improbable. First of all, they do not now occur in the Orient or but seldom, not in Persia (with reference to Chardin), not in Palestine, not in Arabia. Second, wind-mills occur prior to the crusades or at least right in their beginning,' &c. Windmills therefore were believed to be of medieval European origin, and even in the second edition of Schrader's *Reallexikon* (ii, 1923, p. 28) all that is said about the subject is that windmills seem first to be mentioned in an Anglo-Saxon document of A.D. 833; this account, however, according to F. M. Feldhaus (*Technik der Vorzeit*, col. 1326), is a forgery. This opinion of a European origin is no longer tenable. Dimashqī (translation of Mehren, p. 246) describes wind-mills in a country west of Sejistān (Seistān), where the winds and floating sands are very frequent: 'therefore the inhabitants avail themselves of the winds to turn their mills and to transport the sand from one place to another, so that the winds are subject to them as they were formerly to Solomon.' The construction of these windmills is then described in detail (see also Barbier de Meynard, *Dict. géogr. de la Perse*, 1861, p. 301). Above all, we owe valuable information on the subject to the learned researches of E. Wiedemann ('Zur Mechanik und Technik bei den Arabern', *Sitzber. phys.-med. Soz. Erlangen*, 1906, pp. 44–9). The Caliph Omar I (A.D. 634–644) ordered a Persian, Abū Lulua, to make a windmill for him. More information is given by Wiedemann on the wind-mills of Sejistān.

On a former occasion (*Chinese Pottery of the Han Dynasty*, p. 19) I have set forth that windmills are unknown in China. F. H. King (*Farmers of Forty Centuries*, 1911, pp. 332–5) has figured and described a sail windmill pumping sea-water into evaporation basins at the Taku government salt-works, but this device is entirely different from our windmill and independent of the development of the latter. The only example of a windmill that has come within my experience during my travels is the windmill employed by the Tibetans for driving their prayer-wheels, and a specimen obtained by me may be seen in the collections of the Field Museum, Chicago. I have no doubt that the Tibetan notion of the windmill is traceable to Tibetan contact with Iranian regions. The fact that the windmill is practically absent in China and in the Roman empire and that it appears in Europe only during the Middle Ages goes to prove that the windmill is a late invention, much later than the water-mill and the water-wheel.

The position of the noria in the history of agriculture remains to be determined. On several occasions I have emphasized the fact that the entire economic structure of ancient Chinese civilization rests on a common foundation with the other great civilizations of Asia and ancient Egypt. The whole agricultural complex—intensive farming by means of the plough drawn by an ox, cultivation of wheat and other cereals, artificial irrigation, pottery shaped by means of the wheel—is fundamentally the same everywhere (see my *Beginnings of Porcelain in China*, 1917, p. 176). The elements of this economic basis certainly go back to a prehistoric age unfathomable by dates. A system of canals and the use of the well-sweep are features of this prehistoric irrigation. The noria, however, does not belong to this ancient complex. It is plainly an invention of historical times made in an age when mechanical engineering had reached a high stage of development. It is not one of those subconscious or semiconscious gropings of primitive or prehistoric man, but it is the outcome of a volitional, judiciously conceived plan of a thinking engineer well versed with the laws of mechanics. There are good reasons for the conviction that this engineer was an Iranian, probably a Sogdian. The designation 'Persian wheel' is valid. It cannot be fortuitous that this name appears in Europe, in Bukhāra, as well as in India.[1]

BERTHOLD LAUFER

Field Museum of Natural History, Chicago.

[1] After the above article was completed in July, 1929, I received a copy of *A History of Mechanical Inventions*, by Abbott Payson Usher, just published by McGraw-Hill Book Co., New York. On pp. 80–1 the noria is briefly dealt with in a descriptive manner without any reference to Asia. The subject of windmills and water-mills is treated somewhat more critically, and the Oriental origin of the former is admitted (p. 128).

In regard to Japan, reference should be made to the article of J. Troup, 'On a Possible Origin of the Waterwheel,' *Transactions of the Asiatic Soc. of Japan*, xxii, 1894, 109–114.

229

伊特鲁里亚语

Field Museum is open every day of the year during the hours indicated below:

Nov., Dec., Jan., Feb., Mar.	9 A.M. to 4:30 P.M.
April, September, October	9 A.M. to 5:00 P.M.
May, June, July, August	9 A.M. to 6:00 P.M.

Admission is free to Members on all days. Other adults are admitted free on Thursdays, Saturdays and Sundays; non-members pay 25 cents on other days. Children are admitted free on all days. Students and faculty members of educational institutions are admitted free any day upon presentation of credentials.

The Museum's natural history Library is open for reference daily except Saturday afternoon and Sunday.

Traveling exhibits are circulated in the schools of Chicago by the N. W. Harris Public School Extension Department of the Museum.

Lectures for schools, and special entertainments and tours for children at the Museum, are provided by the James Nelson and Anna Louise Raymond Foundation for Public School and Children's Lectures.

Announcements of free illustrated lectures for the public, and special lectures for Members of the Museum, will appear in FIELD MUSEUM NEWS.

A cafeteria in the Museum serves visitors. Rooms are provided for those bringing their lunches.

Chicago Motor Coach Company No. 26 buses go direct to the Museum.

Members are requested to inform the Museum promptly of changes of address.

MEMBERSHIP IN FIELD MUSEUM

Field Museum has several classes of Members. Benefactors give or devise $100,000 or more. Contributors give or devise $1,000 to $100,000. Life Members give $500; Non-Resident Life and Associate Members pay $100; Non-Resident Associate Members pay $50. All the above classes are exempt from dues. Sustaining Members contribute $25 annually. After six years they become Associate Members. Annual Members contribute $10 annually. Other memberships are Corporate, Honorary, Patron, and Corresponding, additions under these classifications being made by special action of the Board of Trustees.

Each Member, in all classes, is entitled to free admission to the Museum for himself, his family and house guests, and to two reserved seats for Museum lectures provided for Members. Subscription to FIELD MUSEUM NEWS is included with all memberships. The courtesies of every museum of note in the United States and Canada are extended to all Members of Field Museum. A Member may give his personal card to non-residents of Chicago, upon presentation of which they will be admitted to the Museum without charge. Further information about memberships will be sent on request.

BEQUESTS AND ENDOWMENTS

Bequests to Field Museum of Natural History may be made in securities, money, books or collections. They may, if desired, take the form of a memorial to a person or cause, named by the giver.

Cash contributions made within the taxable year not exceeding 15 per cent of the taxpayer's net income are allowable as deductions in computing net income under Article 251 of Regulation 69 relating to the income tax under the Revenue Act of 1926.

Endowments may be made to the Museum with the provision that an annuity be paid to the patron for life. These annuities are tax-free and are guaranteed against fluctuation in amount.

A MESSAGE FROM THE DIRECTOR

For the loyal support they have given during the difficult years of depression, Field Museum thanks its thousands of Members. The Museum has been fortunate in the number of Members who have continued on its rolls in spite of adverse economic conditions. While there has again been some decline during 1933 in the total number of Members, the rate of decrease has been markedly less than in the two preceding years.

In expressing appreciation to those who have retained their memberships it may be considered pardonable, perhaps, to urge that they continue further this support in the new year now beginning, and in succeeding years. The Museum is still confronted with severe financial problems in carrying on its mission in educational and scientific fields of endeavor, and the contributions received in the form of membership fees are an important part of the institution's revenues. The help of every present Member is needed, and many new Members must be sought. An appeal is made to all present Members not only to continue their own direct assistance, but to propose the names of other persons of their acquaintance who might be interested in becoming Members.

—STEPHEN C. SIMMS, Director

THE ETRUSCANS
BY BERTHOLD LAUFER
Curator, Department of Anthropology

The origin of the Etruscans has long been a mystery that has aroused many lively controversies. It is now certain that the Etruscans were not natives of Italy; they were Orientals or semi-Orientals who came from some part of Asia Minor, sailing across the Mediterranean and landing in northern Italy about 800 B.C. The new immigrants soon developed into a powerful nation which possessed a strong navy and carried on a considerable maritime commerce.

The Oriental origin of the Etruscans is revealed by many features of their religious beliefs and worship, but above all by their earliest art, which shows close contact with Mesopotamia, Syria, and Cyprus on the one side and with Egypt on the other. Sphinxes, human heads, and lotus designs of true Egyptian style are found on early Etruscan pottery and sarcophagi. From the seventh century onward Greek influence began to be felt in Etruria. More than any other Italic people the Etruscans appreciated the beauty of Greek art and made it their own by importation and imitation.

Field Museum has a remarkable Etruscan collection, part of which is the result of excavations carried on in 1895–96 at the ancient cemetery of Narce about ninety miles north of Rome, in Etruria, under the direction of A. L. Frothingham. This was one of the Museum's earliest expeditions, and was sponsored by the late Charles L. Hutchinson. The period over which the tombs of this locality extend is the first half of the seventh century B.C. The tombs were laid out in the form of trenches and characterized by the change from incineration to inhumation. Numerous objects found in the tombs reveal an advanced stage of culture and include many fine examples of bucchero ware, pottery painted red and polished, pottery decorated with designs in red on yellowish ground, and quantities of bronze ornaments which were concealed in jars. Bucchero is the designation of a black pottery for which the Etruscans were famous. The deep lustrous

black surface was produced by fumigation in a closed furnace, or by covering before baking with a coating of charcoal, whereupon the vase or dish was carefully polished.

The Etruscans had a peculiar kind of stove. A pottery jar in which the food was cooked was placed on a slender, hollow pottery support usually cut out in openwork. These so-called vase carriers were placed over a fire which carried the heat upward into the jar on top.

A curious feature of many bronze bracelets is that fibulas are attached to them. A fibula is a clasp, usually ornamented, and it was indispensable to the ancients for fastening their garments; it is the precursor of our safety pin. As we carry spare tires on our cars to be used in case of emergency, so the Etruscans carried spare fibulas on their bracelets to have them handy in case one was lost, which could easily happen.

Cremation of the dead was much practised in some parts of Etruria and resulted in the making of urns for the ashes of the deceased. These urns are made of pottery, tufa, or alabaster. In shape they are miniature sarcophagi, the cover being decorated with a recumbent figure intended to represent the deceased person, while the front of the chest is carved with a scene in high relief and painted in colors. Most of the subjects have some allusion to death, either directly when a dying man is represented, or indirectly in mythological scenes of fatal combats.

A burial urn of marble shows a fine relief, with considerable remains of color, representing Achilles, the hero of Homer's Iliad, holding the head of a Trojan youth. A very fine alabaster burial urn from Chiusi shows the effigy of the deceased on the cover. The front of the chest is carved with a vivid battle scene in high relief: an armored horseman attacking a Gaul with a lance, while the latter, kneeling on the ground, plunges his sword into the horse's belly.

The treasures of the Museum include three large painted sarcophagi, made from a volcanic tufa, which are unique. They have been described in detail by the late Professor F. B. Tarbell in Field Museum Publication No. 195. Prominent among the designs painted in colors are marine monsters or sea dragons, creations of Etruscan mythology elicited by the people's fondness of the sea and maritime enterprise.

The Etruscan collections have recently been reinstalled in Edward E. and Emma B. Ayer Hall (Hall 2), in a more systematic arrangement, and are now interpreted by more informative labels.

Oysters a Foot in Diameter

Had there been men on earth to eat them, one oyster of the Oligocene or Miocene period, nineteen to thirty-nine million years ago, would have been sufficient to provide a feast for an entire family and their guests, as against a half-dozen to a dozen modern oysters on the half-shell for each individual.

Fossil shells of some of these giant prehistoric mollusks are on exhibition in Ernest R. Graham Hall (Hall 38). They range from six to twelve inches in diameter, and the shells alone weigh as much as sixteen pounds. They were obtained in southern Argentina, from ledges of ancient sandstone and gravel in which they had been buried so many millions of years. The prehistoric oyster beds from which they came occur over several hundred miles along the coast. Many of the shells are also found far inland, and are thus among the indications that, eons ago, a sea covered the continent.

230

中国和其他地方的柠檬

JOURNAL

OF THE

AMERICAN ORIENTAL SOCIETY

EDITED BY

W. NORMAN BROWN
University of Pennsylvania

JOHN K. SHRYOCK
Philadelphia, Pa.

E. A. SPEISER
University of Pennsylvania

VOLUME 54 · NUMBER 2 · JUNE, 1934
Published June 15, 1934

CONTENTS

————————*————————

PUBLISHED BY THE AMERICAN ORIENTAL SOCIETY

ADDRESS, CARE OF

YALE UNIVERSITY PRESS

NEW HAVEN, CONNECTICUT, U. S. A.

*Books for review should be sent to one of the Editors (addresses on the inside of this cover)
Annual subscriptions and orders should be sent to the American Oriental Society,
care of Yale University Press, New Haven, Conn., U. S. A.
(See last page of cover.)*

*Entered as second-class matter June 1, 1916, at the post office at New Haven, Connecticut,
under the Act of August 24, 1912. Published quarterly.*

THE LEMON IN CHINA AND ELSEWHERE

BERTHOLD LAUFER

FIELD MUSEUM

OF THE NUMEROUS useful fruits that we owe to India the lemon is the most democratic and the most widely known. It has become a denizen of this world and, with its Indic name, has penetrated even into the darkest parts of Africa and the tropical jungles of South America. Next to the word " tobacco," the word type " lemon," of Indic origin, is the most universal, reverberating from every tongue of the globe. To cite a few examples—Tukano *erimoá* and Tuyuka *uīnimoá* (of the Betoya group in South America on the Upper Rio Negro) are derived from Portuguese *limão*. Tupi, the lingua franca of Brazil, which has adopted many loan-words from the Portuguese, upholds the word for lemon in the form *limaw,* although the liquid *l* is foreign to the language (Tatevin, *La langue Tapihiya,* 1910, p. 142). Along the east coast of Africa we hear *limao* or *ndimu* for the fruit, *mlimao* for the tree.

It was heretofore supposed that the lemon is of recent origin in China, introduced by " foreigners." It will be shown that this conception of the matter is erroneous and that Chinese acquaintance with the lemon dates from the middle of the twelfth century under the Sung dynasty.

Dr. W. T. Swingle, in his revision of the genera Citrus and Poncirus in China (in C. S. Sargent, *Plantae Wilsonianae,* 1914, II, pp. 127-137), has restored for the lemon the name *Citrus limonia* Osbeck on the ground that this is the oldest available name for the lemon (1765), i. e. in our botanical literature. He says that the lemon is still commonly grown and sold in pots as in Osbeck's day (see also his article " Citrus " in Bailey's *Cyclopedia of Horticulture,* 1914, p. 781).[1]

---

[1] Lemons are pointed out in our literature on China prior to Osbeck's time. We read in Du Halde's *Description of the Empire of China* (London, 1738, I, p. 317) : " Limons and citrons are very common in some southern provinces, and extraordinary large; but these are scarce ever eaten, being only made use of for ornaments in houses, where they put seven or eight in a china dish, to please the sight and smell; however, they are exceeding good when candy'd. Another sort of limon, not much larger than a walnut, is likewise in great esteem; it is round, green, and sharp, being reckon'd

143

I

F. P. Smith (*Contributions towards the Materia Medica* etc. *of China,* 1871, p. 131) cites *ning-mung* 檸檬 ("lemon") and observes, "No mention is known to be made of the lemon in the *Pen ts'ao*. The characters here given are from English dictionaries." He further gives *ning-mung chi* 汁, "lemon-juice" as "a name introduced by foreigners and applied to lime juice as well," and *ning-mung-shwi* 水 ("lemonade").

De Candolle (*Origin of Cultivated Plants,* p. 179) doubts whether the area originally covered by the lemon includes China or the Malay Archipelago, and continues, "Loureiro mentions *Citrus medica* in Cochinchina only as a cultivated plant, and Bretschneider tells us that the lemon has Chinese names which do not exist in the ancient writings and for which the written characters are complicated, indications of a foreign species. It may, he says, have been introduced." In his article "The Study and Value of Chinese Botanical Works" (*Chinese Recorder,* 1870, p. 178), Bretschneider wrote as follows: "The common lemon-tree at Peking is frequently raised in a dwarf form in pots as an ornamental shrub and also on account of the lemons which it produces and which do not differ from our European lemons. It is called 香桃 *siang t'ao* ['aromatic peach'] [2] and may have been introduced. This name is not in Chinese books. The name 檸檬 *ning-mêng* given to the lemon in Bridgman's Chrestomathy (p. 443) is not to be found either in Chinese books. Perhaps by these sounds the Hindustan name of the lemon, being *nee-moo,* is rendered."

In *Mesny's Chinese Miscellany* (Shanghai, IV, 1905, p. 8) we read, "Lemonade, *ning meng shui, Ho-lan shui* 荷蘭水,[3] *ch'ang*

---

excellent for ragous. The tree that bears them is sometimes put in boxes, and serves to adorn the outward courts or halls of houses."

[2] MacGillivray, in his *Mandarin-Romanized Dictionary* (2nd ed., 1907, p. 261) lists 香圓 or 檬 as vernacular names of the lemon, but these, properly speaking, refer to the Buddha-hand citrus (*Citrus sarcodactylus*) 香櫞。俗作圓 (see *Pen ts'ao kang mu,* chap. 30, p. 13). Giles No. 4256 renders *hiang yüan* by "lemon," but under No. 13,738 defines it as "the Chinese citron—a variety of *Citrus medica* L." Perrot and Hurrier (*Matière médicale et pharmacopée sino-annamites,* p. 137) explain *hiang yüan* as *Citrus decumana.*

[3] This, as a matter of fact, is a general term for soda water and proves nothing in favor of an introduction of the lemon through Hollanders, as possibly might be inferred from the name.

sheng kuo shui 長生菓水. In western China the true lemon grows and the fruit remains on the tree for years, hence its name *ch'ang sheng kuo,* i. e. long life fruit. This name is, however, given to peanuts or groundnuts at Shanghai."

G. A. Stuart (*Chinese Materia Medica,* 1911, p. 117) writes, " The lemon has been called by the same name by foreigners in China, as well as by the names *ning-meng* 檸檬 and *li-meng* 黎檬. But it is pretty certain that the lemon does not grow in China proper, or at least has been but lately introduced, and therefore it is not named." All this turns out to be erroneous.

It is correct, as stated by F. P. Smith, that the lemon is not mentioned in the *Pen ts'ao kang mu* of Li Shi-chen. This is accounted for by the fact that Li Shi-chen was not a botanist, but a herbalist and pharmacologist, his interest in plants and fruits being limited to their use in the pharmacopœa, and as the lemon was not medicinally employed up to his time, it failed to receive a place in his work. What F. P. Smith and his successors did not note, however, is that the lemon is clearly described in the *Pen ts'ao kang mu shi i* (chap. 7, p. 60b) and in the *Chi wu ming shi t'u k'ao* (sect. 果, chap. 16, p. 82). The former work gives extracts from the *Ling nan tsa ki and Yüe yü;* the latter cites the *Kwei hai yü heng chi, Ling nan tsa ki,* and *Nan yüe pi ki.* Yet neither points out the fundamental text of the *Ling wai tai ta.*[4]

The earliest reference to the lemon in Chinese records is made by Fan Ch'eng-ta 范成大 (A. D. 1126-93), in his *Kwei hai yü heng chi* 桂海虞衡志 (preface dated A. D. 1175; ed. of *Chi pu tsu chai ts'ung shu,* p. 25b), who writes as follows: " The *li-mung* fruit 黎朦子 has the size of a large plum; again, it resembles a small orange, and is exceedingly sour to the taste." No further information is given. The *Kü lu* 橘錄, a treatise on oranges, edited in A. D. 1178 (translated by Kiang Kang-hu and Hagerty in *T'oung Pao,* 1923, pp. 63-96) is reticent as to the lemon.

The earliest important description of the lemon is contained in the *Ling wai tai ta* 嶺外代答, written by Chou K'ü-fei 周去非 in A. D. 1178 (ed. of *Chi pu tsu chai ts'ung shu,* chap. 8, p. 8b). In a notice on the fruits cultivated in southern China at this time,

---

[4] The present case goes to show conclusively that a plant cannot be assumed to be unknown to the Chinese simply for the reason that it is not mentioned in the *Pen ts'ao* literature.

the author mentions the *li-mung* fruit 黎朦子 as "being of
the size of a large plum; again, it resembles a small orange, and is
exceedingly sour to the taste" (same definition as in *Kwei hai yü
heng chi*), and continues, "Some people say that it has come to
us from the southern barbarians 或云自南蕃來. The people
of P'an-yü (Canton) do not use vinegar in large quantity, but
avail themselves in particular of the juice of this fruit, which is
well known for its sourness, squeezing the juice out with a spoon.
They also boil it in honey, soak it in a brine, and dry it at the sun
when it is ready for consumption." [5]

There is no doubt that the lemon is visualized in this text and
that the Cantonese made a sensible use of it. The Arabs also preserved lemons in salt (Ibn Baṭūṭa, transl. by Defrémery and Sanguinetti, III, p. 126). The lemon, we may conclude, was introduced into what is now Kwang-tung Province under the Sung
dynasty, probably in the first part of the twelfth century, possibly a
little earlier, since Fan Ch'eng-ta and Chou K'ü-fei met it in the
south as a well established cultivation. The tradition that it came
to China from the "Southern Barbarians," vague as it may be, is
entirely credible, and is confirmed by the non-Chinese name *li-mung*
which was received with the fruit from some foreign people. The
Shanghai Medical Dictionary classifies it "among fruits of the
Barbarians" 夷果. Prior to the twelfth century the lemon must
therefore have migrated from India to Indo-China and possibly the
Malay Archipelago. The form *li-mung* is phonetically too simple,
and its congeners of almost identical structure are too widely diffused to afford a clue as to the particular nation or country from
which the southern Chinese might have derived the fruit. One
fact stands out clearly, and this is that *li-mung*, unlike numerous

---

[5] There is good reason to believe that this text has been pointed out
here for the first time. It has remained unknown to the editors of the
T'u shu tsi ch'eng, to the *Pen ts'ao kang mu shi i*, *Chi wu ming shi t'u k'ao*,
Ts'e yüan, and the *Shanghai Botanical and Medical Dictionaries*. The *T'u
shu tsi ch'eng* has devoted no article to the subject, being content to quote
Fan Ch'eng-ta's definition of *li-mung tse* without any additional text and
placing it among "miscellaneous fruit-trees" (XX, chap. 313, *tsa kwo mu
pu hui k'ao*, p. 4) and again among "fruits" (XX, chap. 15, *kwo pu hui
k'ao* 2, p. 10), in this case without citation of the source. Even J. Matsumura
(*Shokubutsu-mei-i*, pt. 1, p. 86) cites the Chinese names for *Citrus limonia*
merely from recent works such as *Ling nan tsa ki*, *Kwang-tung sin yü*,
Kwang k'ün fang p'u, and *Hwa i k'ao*, but has neglected the *Ling wai tai ta*.

other Indic plant names, is not a bookish transcription based on Sanskrit, but was orally received together with the plant through an intermediate tongue from a mediaeval Indic vernacular. Sanskrit *nimbū* or *nimbūka* is probably based on the vernacular forms: Bengali *lebu, nebu*; Konkani *limbo, nimbo, nimbu*; Oriya *nembu*; Hindī *nībū, limbu, limu*; Panjabi *nimbu*; Marathi *nībū*; Gujarati *lïbu*; Nepali *nibu* or *nïbu*, Assamese *nemu*.[6] Old Javanese and Bali *limo*; Malay *limon, lïmau, lïmaw*; Dayak *liman*; Sunda and Makasar *lemo*; Nias *dima*; Formosa *rima*.

It should be emphasized right here that the Buddhists had nothing whatever to do with the propagation of the fruit or its name. Buddhist texts and lexicographical literature are reticent as to both, and the conclusion to be drawn from this and other facts inevitably is that at the time Buddhism was diffused from India to China the lemon was not yet cultivated in India. It is noteworthy also that the Chinese have never been aware of the fact that the lemon is a native of India; India is never referred to in connection with it. Only recently have they learned this fact from us. Both the *Ts'e yüan* and the *Shanghai Botanical Dictionary* (*Chi wu hio ta ts'e tien,* p. 516) point out that the lemon originally grows in India, but foolishly do not say a word about its cultivation in China. It is clear that this cultivation was firmly established in Kwang-tung Province in the second half of the twelfth century when Chou K'ü-fei wrote, for the fruit was then extensively used for culinary purposes at Canton. Neither at that nor at any later time do we hear of any importation of lemons into the country.[7]

---

[6] Mr. Edwin H. Tuttle has kindly favored me with the following note on the Dravidian names of the lemon: "I find Kanara *iliminci* 'lemon,' *nimbe* 'lime'; Tamil *elumiccai* 'lemon' or 'lime'; Telugu *nimma* 'lemon' or 'lime'; Tulu *nimbe, limbe, limbi* 'lemon.' Kanara regularly has *i* for *e* before *i* or *u*: *iliminci* comes from a form with initial *e*. Weak *i* and *u* interchange often in Tamil; *elu-* may come from **eli-*. Telugu regularly has *mm* for *mb*, and often *n* for *l* or *r* near a nasal. Kanara final *e*, Telugu final *a* and Tulu final *ę* (very open *e* or *œ*) probably represent *-as*. I think a basis **limbas* might be assumed for all of the words given above. Native initial *l* is unknown in Tamil and unsual in Kanara; *iliminci* and *elumiccai* may have come from Telugu **limma*; the ending looks like Sanskrit *icchaka*. Sanskrit *nimbū* may have come from Dravidian **limbūs*, which might be an older form of **limbas*."

[7] I have searched through the *Mong liang lu* (A. D. 1274) of Wu Tse-mu, the *Tung king mong hwa lu* of Mong Yüan-lao, and the *Wu lin kiu shi* of Chou

The *Ling nan tsa ki* 嶺南雜記, a record of the geography and productions of Kwang-tung Province, written by Wu Chen-fang 吳震方 in the seventeenth century (Wylie, *Notes*, p. 63), gives the following information (*Siao fang hu chai*, IX, p. 194; also in *Lung wei pi shu* and *Shwo ling*):

"The fruit *i-mu* 宜母 resembles the orange, but is sour. It is much used as a condiment to food. It improves the breath, and is grateful to the stomach. Women, who during the time of pregnancy feel uneasy, will be comfortable after eating this fruit. Hence it has received the name *i-mu* ('beneficial to the mother'). It is also called *i-mung-tse* 宜濛子. It is prepared in the form of a liquid sweet or sour, that dispels the heat (i. e. it is cooling) and that is styled *kie k'o shwi* 解渴水 ('thirst-allaying water,' i. e. lemonade). Wu Lai 吳萊 of the Yüan period is the author of a song entitled 'lemon hot water song' (i. e. a song in praise of hot lemonade)." [8]

The *Nan yüe pi ki* 南越筆記, written by Li Tiao-yüan 李調元 in the eighteenth century, contains the following text: [9] "The fruit *li-mung* 黎檬子 is also called *i-mu* 宜母. It resembles the orange (*ch'eng* 橙, *Citrus aurantium* L., now *Citrus sinensis* Osbeck), but is smaller in size. It ripens in the second or third month when it is yellow in color. It is exceedingly sour of taste. Pregnant women, when their liver is empty, have a craving for this fruit, whence its name *i-mu* ('beneficial to the mother'). At the time of the Yüan dynasty, Li-chi-wan 荔支灣 in Kwang-chou (Canton) was an imperial fruit orchard, where eight hundred large and small lemon-trees (*li-mu* 里木) had been planted for

---

Mi in the hope of lighting upon the use of *li-mung tse* in the Hangchow of the Sung, but so far in vain. I am looking for further evidence before hazarding the conclusion that the lemon was unknown in central China under the Sung.

[8] Wu Lai, styled Yüan-ying 淵頴, lived during the thirteenth century. He is the author of the *Nan hai ku tsi ki* 南海古蹟記. His writings were collected under the title *Yüan-ying tsi* 淵頴集.

[9] The text in question is reprinted in the *Chi wu ming shi t'u k'ao*, sect. 果, chap. 16, p. 82. The *Pen ts'ao kang mu shi i* (chap. 7, p. 60b) ascribes the same text (with a few insignificant variants) to the *Yüe yü* 粤語. The *Nan yüe pi ki* is reprinted in the *Han-hai* collection and in *Siao fang hu chai*, IX (the above text on p. 277).

the purpose of making lemonade (*k'o shwi* 渴水).[10] The word *li-mu* designates the same fruit as *i-mu tse* 宜母子, also called *li-mung tse* 黎濛子. In the poem of Wu Lai it is said that the officials in charge of the gardens of Kwang-chou sent lemonade (*k'o shwi*) as tribute to the imperial court. When weather and wind are hot during the summer, a wine made from lemons and various flowers makes a 'sweet dew beverage' (*kan lu tsiang* 甘露漿).[11] In the countries [other reading: gardens] of the south they boil 'red dragon marrow' 赤龍髓[12] and cover this with lemons, squeezing the water out and boiling it with sugar. The Mongols call lemonade *she-li-pie* 舍里別 [a transcription of Arabic *sherbet*]. It is also styled 'medicinal fruit' (*yao kwo* 藥果). During the hot season people endeavor to buy lemons up for storage purposes; they keep for several years and still yield juice, which is a good substitute for vinegar." [13]

The *Hwa i hwa mu niao shou chen wan k'ao* 華夷花木鳥獸珍玩考 (chap. 10, p. 2) of 1581 [14] contains a note of the

---

[10] According to the *Kwang-tung sin yü* 廣東新語 (chap. 17, p. 12b), Li-chi-wan was the name of one of the famous gardens of Canton, situated five *li* west of the city.

[11] The term *kan lu* ("sweet dew") denotes (1) a heavenly dew as a symbol of universal peace (under the Han); (2) the nectar of the gods, rendering Skr. *amṛta*; (3) the manna furnished by *Hedysarum alhagi* and other manna-like substances (*Sino-Iranica*, pp. 343-350), hence also used as translation of the Biblical manna; (4) the tuber of *Stachys sieboldi*; (5) hard sugar (*Hwa i k'ao*, chap. 5, p. 29).

[12] I do not know what this vegetal substance is; it is not listed in any of the relevant sources.

[13] The *Yüe yü* contains a notice to the effect that lemons put in a brine keep for years and change their color to black; juice from such lemons can heal wounds and "fire resulting from cold phlegm" 寒痰火. The same clause is found in the *Kwang-tung sin yü* (chap. 25, p. 33), which for the rest offers the same text as the *Nan yüe pi ki*.

[14] The title means "Researches into the botany, zoology, and mineralogy (including some art crafts) of China and foreign countries." The various chapters are grouped under subtitles as they deal with plants, animals, or precious stones. This book, written by Shen Mou-kwan 慎懋官 whose preface and postscript are dated 1581, is a mine of curious information, although most data are quoted from earlier works. A copy of the original edition of this now very rare book is in the Library of Congress to which I am grateful for its loan. It is usually cited under the abbreviated title *Hwa i k'ao*. See also Wylie, *Notes on Chinese Lit.*, p. 168.

lemon under the heading *i-mu tse* and begins by saying that in an ancient record it is also called *li siang tse* 梨橡子. The first element of this compound means " pear "; the second refers to a species of oak (*Quercus bungeana* or *chinensis*). It is difficult to see how a combination of these two plant names could be used for designating the lemon tree. I believe that *siang* is an error for *yüan* 橡 (above, note 2) and *li* (" pear ") for 黎 used in *li-mung*. The text of the *Hwa i k'ao* then continues, " In shape the lemon is like a sweet orange, but in taste it is sour. In the third year of the period Ta-te 大德 (A. D. 1299) of the Yüan dynasty, the officials in charge of sugar manufacture in the Ts'üan-chou circuit [in Fu-kien] reported that they used lemons (*li-mu tse,* as above), by a process of boiling the juice, in the preparation of sherbet (*she-li-pie*), which is the Mongol word for lemonade (*k'o shwi* 渴水). Of course, all fruit juices can be prepared in this manner, but only the lemon is sour in flavor and remains unchangeable for a long time. The word *li-mu tse* is identical with *i-mu tse*. At the time of the Yüan there was to the east of the city of P'an-yü (Canton) a lotus pond called Nan Hai (' southern sea '), and to the west of the city there was Li-chi-wan, with an imperial fruit orchard, where eight hundred lemon trees of various sizes had been planted. In the seventh year of the period Ta-te (A. D. 1303) the tribute gift (of lemonade) came to an end. At present this garden is the dwelling-place of common people."

This text is given as a quotation from the *Kwang chou chi* 廣州志 (" Records of Canton "), not to be confounded with the two *Kwang chou ki* 記 listed by Bretschneider (*Bot. sin.,* pt. 1, No. 377).

The form *li-mu* 里木 of the Mongol period is obviously based on Persian *līmū* ليمو, Persian being the lingua franca of the Far East during that memorable epoch. This form is not registered in any of our dictionaries, not even by Palladius, nor in the *Ts'e yüan.* Solely the *Shanghai Botanical and Medical Dictionaries* list it as a synonym of *i-mu,* but without any reference to the source. It would be interesting to trace this *li-mu* in the *Yüan shi* and other his-torical sources concerning the Mongol period.

At the time of the Yüan dynasty *she-li-pie* (as above) or *she-li-pa* 舍里八 (Arabic *sharbat* or *sherbet*) was a beverage favorite with the Mongol emperors, who appointed a special official charged with its preparation and called *she-li-pa-chi* (in Mongol probably *šarbači*). Mar Sergius, a Nestorian Christian, who founded a Nestorian church at Chen-kiang in A. D. 1281, was reputed, as were also his ancestors, for his ability to prepare sherbet, and the emperor bestowed upon him a diploma in form of a golden tablet, granting to him the privilege of specially applying himself to that occupation. In A. D. 1268 the emperor Kubilai ordered Mar Sergius to come to Peking post-haste, in order to present sherbet, and he received ample reward for this service. *She-li-pa* is defined as " a beverage made of fragrant fruits boiled in water and mixed with honey " in the Chinese text in question, a chronicle of Chen-kiang fu written in the period Chi-shun (A. D. 1330-32), and *she-li-pa-chi* as the name of an office. Mar Sergius was obliged to send annually to the court from Chen-kiang forty jars of sherbet prepared from the juices of grape, quince, and orange, as the beverage was believed to have curative power. In 1272 Mar Sergius, together with the minister Sai-tien-chi, traveled to Yün-nan Province; in 1275, to the provinces of Che-kiang and Fu-kien, always for the purpose of preparing sherbet (Palladius, " Traces of Christianity in China and Mongolia," *Chinese Recorder,* VI, 1875, pp. 108-110; Palladius identifies *she-li-pa* with the Persian *sherbet,* but the word is of Arabic origin; cf. also Moule and L. Giles, *T'oung Pao,* 1915, pp. 633-635, 647, 653).

It follows from the above texts of the *Hwa i k'ao* and *Nan yüe pi ki* that lemon is to be added to the fruits which entered into the making of sherbet under the Yüan and that the word *sherbet* was then used principally in the sense of lemonade. Lemons were likewise so used in the Near East. Peter Mundy (*Travels,* I, p. 63, Hakluyt Soc. ed.), in 1620, describes the Turkish sherbet as " a drink made of sugar, juice of lemons and water." Sir Thomas Herbert (about 1630) wrote, " Their liquor may perhaps better delight you; 'tis faire water, sugar, rose-water, and juyce of lemons mixt, called sherbets or zerbets, wholsome and potable." John Fryer (*New Account of East India and Persia,* III, pp. 137, 149), who traveled in the East from 1672 to 1681, writes with reference to Persia that " the usual drink is sherbet made of water, juice of lemmons, and ambergreece [ambergris] " and that " sherbets are made of almost all tart pleasing fruits as the juice of pomegranets, lemmons, citrons, oranges, prunellas."

A. Bergé (*Dict. persan-français,* p. 237) gives for شوربت *shar-bat* the meaning "limonade, sorbet." As is well known, the series *sharbat, sherbet, sharāb* represents the ancestor of our words *sherbet, syrup,* and *shrab* (Osmanli *shorbet* migrating into Italian as *sorbetto,* hence French *sorbet,* Spanish *sorbete,* Portuguese *sorvete*). In the same manner as we learned the use of water-ices from the Near East, the Chinese adopted it from Persians and Arabs, as witnessed by their word *she-li-pa* and the prominent role played by the Nestorians in this industry.[15] After the fall of the Yüan dynasty, the word *she-li-pa,* which perhaps never was popular, sank into oblivion, but the preparation and use of sherbets have persisted in China to this day. In Peking they are known as *shu t'ang* 暑湯 (lit. "heat beverages," i. e. beverages to ward off the heat, cooling beverages), and during the summer months are sold by hucksters in the streets (at least this was the case under the Manchu dynasty 1900-10 when I lived in Peking).

"Among summer drinks there is the *swan mei t'ang* 酸梅湯, a decoction of a certain kind of green plum obtained from the south, which is taken during the hot months with ice as a cooling pleasant drink. It is sold everywhere in the streets. The plum is mixed with sugar and made into a dry paste, and is so sold in the dry fruit shops. It is also mixed with some *kwei hwa* 桂花, the flowers of the *Osmanthus fragrans* of Loureiro" (J. Dudgeon, *The Beverages of the Chinese,* Tientsin, 1895, p. 17; see also W. Grube, *Zur Pekinger Volkskunde,* p. 76).

Matsumura cites also the name *lo-mung-tse* 羅蒙子 from the *Yang-ch'un hien chi* 陽春縣志 as a synonym of the lemon. In the *Ling wai tai ta,* however (chap. 8, p. 9b), *lo-mung-tse* (*lo* being written 蘿) is given as a distinct fruit, described as " being yellow,

---

[15] A strange confusion has been brought about by Hirth (*Chau Ju-kua,* pp. 115, 120, 121, 127) in regarding *se* 思 and *sha* 沙 as transcriptions of Arabic *sherbet.* Aside from the fact that this is phonetically impossible and that *she-li-pie* or *she-li-pa* are the correct transcriptions of the Arabic word, there is no question at all of sherbets in the text, but of "wines (i. e. alcoholic beverages) which are heating and stimulating." A sherbet is just the opposite, a non-alcoholic, cooled and cooling beverage. Peter Mundy says advisedly that sherbet is the ordinary drink of great men among the Turks, their law forbidding them wine. Ch'ang Te mentions orange juice mixed with sugar as the beverage of the caliph without giving the name sherbet (Bretschneider, *Med. Res.,* I, p. 140).

of the size of an orange or pumelo" 橙 柚 (the former character is identical with 橙).

The origin of the form *ning-mung* (or *meng*), as given by F. Porter Smith, Bretschneider, and others, remains obscure. As far as I have been able to ascertain by interrogating Chinese, it is chiefly used in Kwang-tung and Fu-kien, while Shanghai and Peking men prefer *li-mung*. There is to my knowledge no authority for the characters 檸檬, as given by Bridgman, F. Porter Smith, and successors, although entered in all current dictionaries and even in the *Ts'e yüan*. The *Shanghai Botanical Dictionary* (*Chi wu hio ta ts'e tien,* p. 516) winds up its discourse on the lemon, which is poor enough (only the *Yüe yü* and *Ling nan tsa ki* are laid under contribution), by saying that "in recent times the lemon is generally called *ning-mung*" (same characters as those of F. Porter Smith). K'ang-hi's Dictionary does not give them; above all, however, there is no literary source that gives them, and I have searched for them long and patiently. The only work in which I found them is one of recent origin, the *O yu ji ki* 俄遊日記 (*Diary of a Journey to Russia*) by Miu Yu-sun 繆祐孫 (*Siao fang hu chai,* III, p. 416).

The Chinese nomenclature of the lemon may now be tabulated as follows:

黎朦子 *li-mung* or *li-mong tse* (Sung).—*Kwei hai yü heng chi* and *Ling wai tai ta.*

里木 *li-mu* (Yüan).—*Hwa i k'ao, Nan yüe pi ki.*

黎檬子 *li-mung tse.*—*Nan yüe pi ki* and K'ien-lung's *Polyglot Dictionary,* Appendix, chap. 3, p. 15b, with the following equivalents: Manchu *jušuči* (lit. "sour fruit"), Tibetan *li-meṅ* or *li-mōṅ siu* (transcription of Chinese),[16] Mongol *küjiltäi jimin* ("aromatic fruit").

---

[16] As to the acquaintance of the Tibetans with lemons, I have no personal experience. Jäschke, in his *Tibetan-English Dictionary,* cites *gam-bu-ra* ("citron, lemon") as West-Tibetan; in his *Tibetisches Handwörterbuch,* which preceded the English edition, he has added Sanskrit *gambhīra,* which according to Boehtlingk and Roth denotes "lemon-tree, lemon." Chandra Das, in his *Tibetan-English Dictionary,* copied Jäschke's *gam-bu-ra* and joined to it Sanskrit *jambīra; jambīra,* of course, could never be transformed into a Tibetan *gam-bu-ra;* it denotes not the lemon, but *Citrus medica.* Jäschke, further, gives " *spyod-pad, dpyod-pad* (spelling uncertain), pronounced *čö-pe* " as a designation of the lemon; this is quite enigmatic

宜 母 *i-mu,* by way of popular etymology with reference to an alleged medicinal virtue of the fruit.—*Ling nan tsa ki.*

宜 濛 子 *i-mung tse,* a compromise or missing link between the correct form *li-mung* and the popular *i-mu.*—*Ling nan tsa ki.*

含 里 別 or 八 or *she-li-pie* or *pa,* sherbet, lemonade (Yüan).

解 渴 水 *kie (chieh) kʻo shwi,* or merely *kʻo shwi,* lemonade.— *Ling nan tsa ki, Nan yüe pi ki.*

ning-mung shwi or *li-mung shwi,* lemonade (modern colloquial).

The lemon is still cultivated in the provinces of Kwang-tung and Se-chʻwan. Mesny (above, p. ——) refers to its cultivation in western China. A. Hosie (*Report on the Province of Se-chʻwan,* 1904, p. 17) specifies the district of Kin-tʻang 金 堂 in the prefecture of Chʻeng-tu as the seat of lemon cultivation. Rockhill (*The Land of the Lamas,* p. 303) mentions a Catholic mission, near the famous Lu-ting suspension bridge in Se-chʻwan, where there was a fine vegetable garden around the vicarage, and he noticed in it pomelo and lemon trees laden with fruit, but he was told that it never matured. J. Anderson (*Report on the Expedition to Western Yün-nan,* 1871, p. 64) noted lemons at Bhamo. "The lemon is not grown in China as a fruit tree but only as a dwarf pot-plant, bearing as many fruits as can be got on it" (S. Couling, *Encycl. Sinica,* p. 410, after F. Meyer).

The "foreigner," who in the "introduction" of the lemon into China loomed so large in the minds of sinologists of the preceding generation that the Chinese sources were not even consulted, may have had his share in giving a fresh impetus to the cultivation of lemons in consequence of his greater demand for lemonade and lemon slices for tea, salads, and other dishes. The *Industrial Handbook of Kiangsu Province,* just issued by the Bureau of Foreign Trade (Shanghai, 1933, p. 220), contains the statement that "the import of lemons from the United States to Shanghai has increased from Haikwan Taels 96,523 in 1925 to H. Tls. 126,812 in 1928."

---

to me. Lama D. Kazi (*English-Tibetan Dictionary*) lists *čos-pad* as a Sikkim word for the lemon. If the lemon is known to Tibetans, it must be due to importation from India, Kashmir, or Sikkim. According to Risley's *Gazetteer of Sikkim* (p. 76), the lemon is cultivated there.

II

After this reconnaissance I determined to follow the trail of the lemon in the Chinese records relative to the countries of the Indian Ocean, in the expectation of lighting upon data that might enable us to trace the gradual stages of its migration back to its native home, India. This attempt proved disappointing, however. Works such as the *Ying yai sheng lan,* the *Tung si yang k'ao* (of which I have the original edition of 1618), the *Si yang ch'ao kung tien lu* (in *Pie hia chai ts'ung shu*) do not mention the lemon anywhere. Those who have not access to the Chinese sources may convince themselves by consulting the relevant translations of Phillips, Mayers, Groeneveldt, Pelliot, and Rockhill. Nor is the lemon listed as an article of import into China; Chao Ju-kwa and others maintain silence about it. This, however, is surprising only at first sight, but considering the fact that everywhere in the Far East and in India the lemon is merely planted in gardens here and there for local needs and that it is nowhere cultivated on a large scale, this situation becomes easily intelligible. Only in southern France, Italy, Spain, Portugal, California, and the West Indies has lemon culture developed into an industry of such a magnitude that it pays exportation. The Chinese, being matter-of-fact people, in visiting foreign countries were interested to know, first, on what the inhabitants subsisted (the inquiry as to whether they cultivated rice and other cereals was uppermost in their minds) and, second, what agricultural and other products lent themselves to exportation. The lemon did not come within this category and therefore remained unnoticed.

In the Philippines the lemon was established long before the times of Spanish colonization. Pigafetta, who accompanied Magellan on his circumnavigation of the globe (1519-22) mentions lemons (*limoni*) among the fruits of the island Zubu or Cebu (Blair and Robertson, *The Philippine Islands,* XXXIII, pp. 133, 187, 231). Miguel de Loarca (*Relacion de las Yslas Filipinas,* 1582) reports, " There are also many good oranges and lemons " (*op. cit.,* V, p. 171) ; they are likewise referred to by Antonio de Morga (*Sucesos de las Islas Filipinas,* 1609 ; *op. cit.,* XVI, p. 87). This excludes the notion held by some scholars of the preceding generation that the Malayan words for the lemon are derived from the Portuguese (e. g. W. Joest, *Das Holontalo,* 1883, p. 74).

4

H. Kern, in a brief article entitled "Limoen" first published in 1897-98 (reprinted in his *Verspreide Geschriften,* XII, pp. 151-153), regards Skr. *nimbū* as a sanskritization of Hindustani *nimbū,* which on its part should be a corrupted pronunciation of *līmū.* He further points to Old Javanese *limo* occurring in the *Rāmāyaṇa,* developed from an older *limau,* and to the cognate words in other Malayan forms of speech. In Samoa, Fiji, and Mota there is a word *moli* meaning "orange," which according to Kern is the same word as *limo.* His conclusion is that Dutch *limoen* in its origin is a Malayo-Polynesian word which by way of Hindustan, Persia, and Arabia has found its way to Europe. This linguistic somersault is made without any regard to the botanical facts. If Kern's speculation were correct, the lemon tree would have to be regarded as a native of Malayo-Polynesia and as having been introduced from there to India. The reverse, however, is the case. According to G. Watt (*The Commercial Products of India,* 1908, p. 325), who calls the lemon *Citrus medica* L., var. *acida,* it is "undoubtedly a native of India." It grows wild in the forests of northern India, on the southern slopes of the Himalaya, especially in the valleys of Kumaon and Sikkim. In the valley of Nepal lemons grow most luxuriously and are of very fine flavor (*Imp. Gazetteer of India,* XIX, p. 47). How the plant spread from India to Malaysia we have no means of ascertaining; there are two possibilities—either by way of Indo-China or from southern India or Ceylon directly across the sea, possibly by both ways. A. de Candolle already emphasized the fact that nowhere in the Archipelago does the lemon occur in the wild state, but is only cultivated. The occurrence of the word *limo* in the Javanese version of the epic Rāmāyaṇa is merely an example of the application of the word in literature, but does not go to prove that the lemon was anciently known in Java, not to speak of cultivation; nor is it by any means certain that the word refers to our lemon (cf. W. Marsden, *History of Sumatra,* p. 100, where it follows from the names for various citrus fruits cited that *limau* is a general term covering all members of the Citrus family).

The task of elaborating a history of the lemon in India if such is possible must be left to competent Sanskrit scholars. The fact that it appears on the horizon of the Chinese as late as the twelfth century and that at about the same time it starts on its westward migration leads me to think that the beginnings of its cultivation in India may fall in the early middle ages, say the fourth or fifth to the ninth century. The earliest reference to the lemon of India is made by the Arabic geographers of the tenth century (below, p. 158). It is significant that the Chinese Buddhist pilgrims to India, while they describe many plants of the country, are reticent as to the lemon and that it is not mentioned in Buddhist literature. It is not contained in the Bower Manuscript, but according to Watt in the work of Suśruta. The Petersburg Dictionary refers under *nimbū* to *Rājanighaṇṭu* (11,176) and *Bhāvaprakāśa* (2,38); the term *nimbūkaphalapānaka* goes to show that lemonade was known in India. François Bernier (*Travels in the Mogul Empire 1656-68*, transl. by A. Constable, 2d ed. by V. A. Smith, p. 253) refers to the excellent lemonade to which a wise man will here accustom himself and which costs little and may be drunk without injury.

The earliest references to lemons in India on the part of European travelers are by the two friars, Odoric of Pordenone and Jordanus. Odoric (1286-1331), on his visit to the island of Sillan (Ceylon), describes a pool full of precious stones and abounding in leeches. The king, he relates, allows the poor to search the water for the stones once or twice a year and to take whatever they can find. But that they may be able to enter the water in safety they bruise lemons and copiously anoint the whole body therewith, and after that when they dive into the water the leeches do not meddle with them (Yule, *Cathay,* 2d ed. by Cordier, II, pp. 171, 306, 347). As Yule annotates, Ibn Baṭūṭa writes that the people of Ceylon take care to keep ready a lemon and to squeeze its juice upon leeches that may drop upon them. Knox and Tennent corroborate Odoric's notice of lemon juice as the remedy for leech bites. Hence it is quite certain that the lemon is intended in Odoric's text and that the medicinal properties of lemon juice were anciently known in India. Another early mention of lemons in Ceylon is by Gabriel Quiroga de San Antonio (*Brève et véridique relation des événements du Cambodge,* ed. A. Cabaton, p. 178), who paid a visit to Ceylon in 1600. Friar Jordanus, in 1328, wrote that India, as regards fruit and other things is entirely different from Christendom, except that there be lemons in some places, as sweet as sugar, while there are other lemons sour like ours (Yule, *Hobson-Jobson,* p. 514).

It is said that the so-called Nabatean Agriculture, written in
A. D. 903 by Ibn Wahshiyah (regarding this work see Carra de
Vaux, *Les Penseurs de l'Islam,* II, pp. 296-300), contains an allusion to the lemon (Flückiger and Hanbury, *Pharmacographia,* 2d
ed., 1879, p. 115, after Meyer, *Gesch. der Botanik,* III, p. 68). If
this be true, it would be the earliest reference to the fruit in the
literatures of the world. I note from E. Seidel (*Mechithar,* p.
216), however, that the word in the text thus translated is ﺣﻤﺴﺎ
which Seidel regards justly as a transcription of Khasia, a district
in India known for Citrus cultivation. This being the case, it is
not certain that the lemon is intended; it may be one of the many
other species of Citrus as well.

The geographers Iṣṭakhrī and Ibn Haukal (toward the middle
of the tenth century) are the first Arabic authors who attribute to
Sind a fruit as large as an apple and very sour, called *līmūnah.*
This information has been copied by Edrīsi of Cordova and Abu'l-
Feda (Guyard, *Géographie d'Aboulféda,* II, pt. 2, p. 113; A. von
Kremer, *Culturgesch. des Orients unter den Chalifen,* I, p. 312).
According to von Kremer, the migration of the lemon from India
to the Near East took place under the caliphate. The Arabs apparently transmitted it to Persia, Iraq, Syria, and Egypt. In regard
to Persia see, for instance, G. Le Strange, *Description of the Province of Fars in Persia,* pp. 39, 47. In Syria the lemon was cultivated under the Mamluks in the thirteenth century (Gaudefroy-
Demombynes, *La Syrie à l'époque des Mamelouks d'après les
auteurs arabes,* p. 26).

Ibn al-Baiṭār of Malaga (A. D. 1197-1248; Leclerc, *Traité des
simples,* III, pp. 255-262) gives a lengthy description of the lemon,
its properties and uses, and it is noteworthy that he does not cite,
as in most cases, his predecessors; but he evidently describes the
plant and fruit from personal experience. He gives a recipe for
the preparation of lemon syrup or lemonade as then was customary
in Egypt: three or four ounces of lemon juice were mixed with a
pound of sugar; this mass was heated, and water was added to it
according to individual taste. Lemons are frequently mentioned
in the Arabian Nights, and lemon trees in a garden of Egypt are
described poetically in the story of Nūr ed-Dīn and Maryam (Night
846). In Morocco lemons were known in the fourteenth century,

according to Ibn Faḍl Allah al-'Omarī, 1301-49 (*Masālik el Absār* etc., transl. by Gaudefroy-Demombynes I, 1927, p. 175).

Documentary evidence as to how and when the lemon was introduced from the Near East into southern Europe is lacking. It is supposed that the Crusaders took it along from Palestine and that the Arabs transmitted it to Spain. The former supposition is based on the fact that Jacobus de Vitriaco (or Jacques de Vitry, about 1200) describes the lemon which he had seen in Palestine, but he does not say that he was instrumental in taking it to Europe.

Ibn el-'Awam, who lived at Seville in the twelfth century (Clément-Mullet, *Le Livre de l'agriculture,* 1864, I, p. 300), in his great work *Kitāb-el-felāhah,* mentions the lemon or citron tree (*limonier ou citronnier* in Mullet's translation), but does not say that it was cultivated in the Spain of his time, nor does he refer to lemonade; the chances are that the lemon is not visualized in his text. Perhaps it was to Sicily and southern Italy that the lemon was first transplanted through Arab agency. The fact of the transmission itself cannot be called into doubt, for it is upheld by the migration of the Arabic word *limūn, leimūn* ليمون into Italian *lima, limone* (Old Italian *lumia, lomia*); Spanish *lima, limon*; Portuguese *lime, limão*; Provençal and French *limon*; Rumanian *lemej, alemej, alimon.* The early English travelers to India also have preserved the vowel *i*: thus William Finch (in India 1608-11) spells *limmons*, Edward Terry (in India 1616-19) *limons* (see W. Foster, *Early Travels in India,* pp. 166, 297).

By the sixteenth century lemon culture was well established in Italy. Castore Durante (*Herbario nuovo,* Roma, 1585, p. 259) writes that lemons grow in great quantity in Calabria, in Puglia, and in the kingdom of Naples and are found in many gardens in Rome and neighboring places. From Italy lemons became known in Germany in the first half of the sixteenth century, and then and in the seventeenth century were still called *limone, lemone,* subsequently superseded by *citrone.* Around 1700 Germans became acquainted with lemonade (Kluge, *Etymol. Wörterbuch*). From about 1630 the *limonadiers* began to play a prominent role in France, subsequently taken over by the cafétiers—a subject treated in detail by Larousse (Grand Dictionnaire).

In England lemon trees were cultivated as early as the reign of James I (1603-25) as Lord Bacon mentions lemons, oranges, and myrtles housed in hot country plants. In some parts of Devonshire lemon trees were trained to the walls, requiring no other care than to be covered with straw or mats during the winter. Being of a much hardier nature than the orange, the lemon was brought to greater perfection in England than the latter fruit (H. Phillips, *Pomarium Britannicum,* p. 229).

During the seventeenth century the lemon had completed its triumphal procession around the world. The great traveler Peter Mundy, in 1634 and 1638, found lemons in St. Helena, where there was a " Lemmon Valley because it leads to the place where lemmon trees are " (*Travels in Europe and Asia,* III, pp. 330, 412, Hakluyt Soc. ed.). Although St. Helena never had a native population, it has played a great role in the diffusion of cultivated plants. H. Phillips (*op. cit.,* p. 230) wrote in 1821 that " the lemons of St. Helena are the most esteemed, growing larger, and of a milder flavor than other kinds." In 1613 Rodrigues da Costa found citrons and lemons in Madagascar (*Collection des ouvrages anciens concernant Madagascar,* II, p. 12). Sir Thomas Roe (*Embassy to India 1615-19,* ed. W. Foster, pp. 9, 13) reported lemons on the Comoro Islands. In 1638 Mundy encountered lemons in Mauritius, Madagascar, and Mohilla, one of the Comoro group (II, pp. 14, 319; III, pp. 350, 369). On the island of Bourbon (then Mascaregne) lemons were observed by the Sieur D. B. (P. Oliver, *Voyages made by the Sieur D. B. 1669-72,* p. 86). On the east coast of Africa lemons were known much earlier: Ibn Baṭūta mentions lemon trees on the island Manbasa, two days' voyage from the land of the Swahili (Defrémery and Sanguinetti, *Voyages d'Ibn Batoutah,* II, p. 191).

231

稀有的中国毛笔笔杆

A "FAMILY TREE" OF MAN AND THE APES

A "family tree" of man and other animals belonging to the same order, the Primates, has been placed on exhibition at the entrance to the Hall of the Stone Age of the Old World (Hall C).

On the background of the exhibition case is represented a branching tree, and attached to the branches are reconstructions of the skulls of primitive monkeys and apes, of types of prehistoric men, and finally skulls of modern men of various races, and modern apes. The exhibit graphically demonstrates the theory that man, while not the descendant of any living type of ape, had, from many lines of evidence accepted by scientists, a common ancestry with the apes; and that while apes were evolving from primitive types to those living today, a parallel evolution was taking place through various primitive human types and culminating in present races of man.

The exhibit begins with reconstructions of the skulls of a primitive lemur, a tarsier, an ancestor of the anthropoid ape, a hypothetical intermediate type between the apes and man, and an anthropoid ape. Emanating from the same original sources, but taking separate lines, are found branches with reconstructions of the most famous types of prehistoric men of which scientists have found evidence—the Trinil or Java Man (Pithecanthropus), the Piltdown Man (Eoanthropus), the Peking Man (Sinanthropus), Heidelberg man, Neanderthal man, and Cro-Magnon man. Other branches indicate the relationship to these and to each other of the four principal racial types existing today—the Australian, the Negro, the Mongolian, and the White, of each of which a skull is displayed. Likewise, from the lower branch is indicated the parallel development of the outstanding modern apes—gorilla, chimpanzee, orang, and gibbon. The exhibit performs the double function of indicating the relationships between the various branches of the primate order, and of providing material for studying the physical similarities and differences between the head structures of the various monkeys, apes, and men.

PLANTS COLLECTED IN 1778-88 RECEIVED AT MUSEUM

By PAUL C. STANDLEY
Associate Curator of the Herbarium

A collection of plant specimens gathered while the Revolutionary War was in progress in the North American colonies has been added recently to the Herbarium of Field Museum. The collection was made not in North America, however, but in Peru, in 1778-88, by the first botanists who visited that country. They were Hipolito Ruiz and José Pavón, two men sent by the King of Spain to make a botanical survey of the country.

Some of the work of these men was performed under great hardships. A large part of their collections, after being assembled most laboriously, was destroyed by accident. After the two botanists returned to Spain they prepared a partial account of the plants. This was published in lavishly illustrated folios, of which there is a set in the Museum Library.

This valuable material, almost every specimen of which represents one of the new species given Latin names by Ruiz and Pavón, was received by Field Museum through exchange with the Botanic Garden of Madrid. Among the plants represented are several orchids, and numerous palms,

including the Panama hat palm, from whose leaf fiber the so-called Panama hats are manufactured. This plant, first discovered by Ruiz and Pavón, was named by them *Carludovica*, in honor of Carlos IV of Spain and Queen María Luisa, the royal patrons of the expedition to Peru.

It is a noteworthy fact that these pressed and dried herbarium specimens are in a perfect state of preservation, and in some of them the colors of the leaves are preserved as well as if they had been collected only a year ago.

The Museum's collection of Peruvian plants is probably the largest in the world. These newly acquired specimens of authentic material will be immediately useful in the preparation of the *Flora of Peru*, with which Assistant Curator J. Francis Macbride has been engaged for several years.

Rare Chinese Brush-holder

An unusually fine Chinese writing-brush holder was recently acquired with funds given to Field Museum by the American Friends of China. It was immediately placed on exhibition in a case illustrating writing materials in Hall 32. This brush-holder is carved from a rare Burmese tropical wood, known in trade as padouk. It is decorated with bamboo, rock, and a magpie perching on a plum tree, these designs being carved out of ivory, jade, rose quartz, chalcedony, carnelian, lapis lazuli, tiger-eye, spinel ruby, and mother-of-pearl. The magpie was sacred to the Manchu dynasty. An inscription carved in ivory means, "May you have white eyebrows (i.e. long life) and may your years be prolonged!" This is followed by the date "first year of K'ien-lung," corresponding to A.D. 1736, when the emperor K'ien-lung succeeded to the throne. The brush-holder was presented to him in commemoration of this event.—B.L.

Gift of Lamaist Paintings

An interesting collection of fourteen Lamaist paintings illustrating the Buddhist pantheon of Tibet, the largest in the world, was recently presented by Leon Mandel and Fred L. Mandel in memory of their deceased mother, Mrs. Blanche R. Mandel. These pictures are painted in bright watercolors on cotton cloth, and were executed by monks in the Lama monasteries. There are pictures of Buddha surrounded by a thousand imaginary or celestial Buddhas and of goddesses in the various forms of their numerous incarnations. Of especial interest is the portrait of one of the Dalai Lamas in his yellow and red priest's robe, holding a lotus in each hand.

A Camel from Wyoming

A slender, stilted skeleton from the hills of Wyoming, on exhibition in Ernest R. Graham Hall (Hall 38), tells the story of a family of animals once quite common in North America. The slender head, long neck, slender legs and spreading foot betray relationship to the camels and llamas.

This skeleton was collected from a sandstone formation of early Miocene age. Bones of these animals are fairly abundant as the animals must have been in the Great Plains region twenty million years ago.

This animal is taller and more slender than the llama of South America, but related to it. In fact the southern members of the family are known to have branched off a little after the skeleton of this individual was buried in the wind-blown sands of Wyoming.—E.S.R.

JUNE GUIDE-LECTURE TOURS

Conducted tours of exhibits, under the guidance of staff lecturers, are made every afternoon at 3 P.M., except Saturdays, Sundays, and certain holidays. Following is the schedule of subjects and dates for June:

Friday, June 1—Halls of Animal Life.

Week beginning June 4: Monday—Prehistoric Life; Tuesday—General Tour; Wednesday—North American Indians; Thursday—General Tour; Friday—Geology Exhibits.

Week beginning June 11: Monday—Primitive Peoples; Tuesday—General Tour; Wednesday—Egyptian Exhibits; Thursday—General Tour; Friday—Halls of Plant Life.

Week beginning June 18: Monday—Moon and Meteorites; Tuesday—General Tour; Wednesday—Chinese Exhibits; Thursday—General Tour; Friday—Jade and Crystals.

Week beginning June 25: Monday—Reptiles, Past and Present; Tuesday—General Tour; Wednesday—Man Through the Ages; Thursday—General Tour; Friday—Plants of Unusual Interest.

Persons wishing to participate should apply at North Entrance. Tours are free and no gratuities are to be proffered. A new schedule will appear each month in FIELD MUSEUM NEWS. Guide-lecturers' services for special tours by parties of ten or more are available free of charge by arrangement with the Director a week in advance.

Gifts to the Museum

Following is a list of some of the principal gifts received during the last month:

From C. E. Tober—a stone effigy pipe and a pottery vessel, Illinois; from Mrs. George H. Martin—2 carved horn spoons, Sitka Indians, Alaska; from Van Cleef Brothers—4 samples of rubber; from von Platen-Fox Company—a board of tamarack, Michigan; from Dr. Earl E. Sherff—71 herbarium specimens, Hawaii; from E. I. Du Pont de Nemours and Company —2 samples of synthetic rubber; from Franklin G. McIntosh—8 specimens of minerals, California; from Colonel W. H. Surgbnor—an Alaskan mountain sheep head; from Karl Plath—a South American tanager and 2 bird skeletons; from Miss Bertha Cramer—2 bird skeletons, Illinois; from Mrs. E. Walton—a golden-crowned kinglet and skeleton, Illinois; from A. G. and Raymond B. Becker—81 specimens of invertebrate fossils, Florida; from Leon Mandel— 5,000 feet of motion picture film taken during the Leon Mandel-Field Museum Zoological Expedition to Guatemala; from Lieutenant Seeley A. Wallen—a wild boar and a jungle fowl, Philippine Islands.

NEW MEMBERS

The following persons were elected to membership in Field Museum during the period from April 16 to May 15:

Contributors
Mrs. Sarah S. Straus

Associate Members
Mrs. Frederick C. Gifford, Miss Helen K. Gurley, Mrs. Elmer A. Howard, Raymond J. Koch, Mrs. William S. Mills, Sigurd E. Naess, James P. Soper, Jr., H. E. Wills.

Annual Members
Charles S. Babcock, Mrs. Anna K. Brown, J. Amos Case, Dr. Fremont A. Chandler, Edmund J. Claussen, Henry Towner Deane, Harry L. Diehl, David W. Edgar, Earl E. Enos, Dr. Alexander Gabrielians, Mrs. John L. Gardiner, Robert N. Golding, Joseph B. Hawkes, Garner Herring, Haven Core Kelly, Arthur W. Nelson, Joseph J. Nevotti, Dr. Edward H. Ochsner, C. N. Owen, Dwight S. Parmelee, William D. Price, L. J. Quetsch, T. E. Quisenberry, John Glen Sample, Rev. Dudley S. Stark, William O. Trainer, F. K. Vial, Charles Weiner, Elmer Zitzewitz.

Zinc and Lead Ores

A group of zinc and lead ores from the Embree Mines of Tennessee, recently presented by Charles P. Wheeler, is of unusual interest. These ores have the appearance of cave deposits. Stalactites, stalagmites and other cave formations, which in ordinary caverns are composed of carbonate of lime, are in this deposit composed of the carbonates of lead and zinc and the silicate of zinc. Many of the specimens are pure white with a good luster which is exceptional in ores of this kind. They may be seen in Frederick J. V. Skiff Hall (Hall 37).

PRINTED BY FIELD MUSEUM PRESS

232

中国乾隆皇帝的黄金藏品

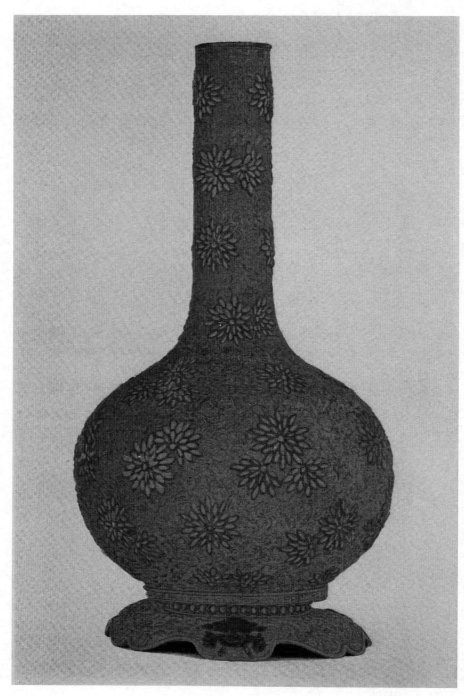

17 (see page 32)

The Gold Treasure

of the

Emperor Chien Lung of China

by

BERTHOLD LAUFER

EXHIBITED

by

PARISH-WATSON & COMPANY, INC.

44 EAST FIFTY-SEVENTH STREET

NEW YORK

A CENTURY OF PROGRESS

CHICAGO 1934

PREFACE

IT may be recalled by many of my friends that I have been interested and intimately associated with the Chinese Imperial Gold Collection since its arrival in America some years ago. It was brought to New York where it has remained up to this time. Meanwhile, the House of Parish-Watson Co., Inc., with whom I am now associated, has taken over the collection.

In order to make the collection accessible to a larger public, Mr. Parish-Watson recently consented to its exhibition at Chicago's Century of Progress. Mr. Parish-Watson, broad-minded and public-spirited as he is, has spared no expense in making the exhibition of the collection as attractive and efficient as possible, and furthermore, has honored me by placing me in charge of the arrangement and direction of its public showing.

The obstacles to overcome have been many. It was essential that the safeguarding of the collection be assured; suitable safety devices consequently had to be set up; effective installation was imperative. This, of course, necessitated considerable experimentation with lighting, spacing, background, and numerous other details which would be of no interest to the reader.

Mr. Rufus Dawes, President of A Century of Progress, fully approved our enterprise with enthusiasm, and met all our requirements in a spirit of helpful sympathy and co-operation. I wish herewith to express my thanks to him, his associates, and in fact all other officials of the Fair for the assistance kindly rendered and for the many and various courtesies extended to me.

We believe to perform an act of public duty and an educational mission by showing at A Century of Progress this magnificent gold treasure, which is certainly unique in this world. It is our fervent hope that it will serve to interest the general public in the glory of the ancient civilization of China, and to further the understanding and appreciation of its great art.

[5]

It affords me considerable pleasure to lay before the interested public an account of this collection written by Dr. Laufer as a result of a long and profound study. It consists of two parts, an historical introduction and a complete descriptive catalog of the objects, which, in turn, gives much information on the significance and symbolism of the designs.

In conclusion, I wish to proffer my most profound thanks to Dr. Berthold Laufer, indisputably America's foremost sinologist, for his kindly co-operation and his consenting to write the text of this catalog. His spontaneous enthusiasm has acted as a positive inspiration to me and has helped me over the many problems with which I was confronted.

HERBERT J. DEVINE

Chicago, June the first, 1934

[6]

HISTORICAL INTRODUCTION

I N A. D. 1783 the great Emperor Chien Lung, who
ruled over China from 1736 to 1795, and who was a
contemporary of George Washington and a character
not unlike Frederick the Great, was made the recipient
of a set of gold objects consisting of altogether eighteen
articles designed for use on the imperial desk. This presentation
was made by a Manchu official of high rank, named Pao Tai,
imperial envoy to Tibet. This fact is clearly indicated on the
white silk labels pasted in every box, which was especially made
for each gold object. The interesting point is that the inscriptions
on these labels are not merely written in Chinese, but in the four
principal languages which at that time dominated in the empire
of China—Chinese, Manchu, Mongol, and Tibetan. The Emperor himself had studied these four languages and mastered them
to such a degree that not only could he fluently express himself
in each of them, but also was able to write essays in all of them
in excellent style. Under his patronage a comparative dictionary
of these four languages was published—a work which is still used
by us as an indispensable and fundamental source-book.

The labels in question are inscribed in a beautiful calligraphic
style. Each inscription contains the identical data pertaining to
the dedication of the set and gives the name of Pao Tai, the exact
date of the presentation, and the name for the particular object.
The official term used in making this gift signifies "a gift of supreme importance and value," including the notion that a man has
invested in it his entire fortune, that he has put his heart and soul
in it with reverence and devotion and is intent on serving his
sovereign to the bitter end.

Each of the eighteen objects intended for use or decoration on
the imperial desk is wrought from pure gold, the total weight of
which amounts to 5,966 grams (gold and silver were always used
unalloyed in China), and is inlaid with beautifully carved and

[7]

polished plaques of turquois and lapis lazuli. Now, both turquois and lapis lazuli are the favorite precious stones of the Tibetans and Mongols and represent an allusion to the great colonial possessions of the empire in Central Asia. Tibet has likewise been celebrated since times immemorial for its wealth of gold. The famous tradition of the Gold-digging Ants related by Herodotus and the Indian epic Mahābhārata, as shown by me in an article written in 1908, refers to gold-digging Tibetan and Mongol tribes in the region of the upper Yellow River valley. The felicitous and artistic combination of gold with turquois and lapis lazuli is an intentional and distinct allusion to Tibet. In this connection it is noteworthy that the Chinese imperial envoy sent from Peking to Lhasa, the holy capital of Tibet, was styled in Tibetan "the gold-letter bearer" (gser-yig-pa). In the same manner as the Chinese speak of the sovereign's dragon face, the Tibetans refer to it as "the royal golden face" and call the emperor "the golden king" and Peking "the golden castle."

In 1725 the Chinese Government appointed two High Commissioners to control the political affairs of Tibet. Several attempts at revolt in 1750 led to the entire suppression of the temporal sovereignty in Tibet, and the government of the country was placed thenceforward in the hands of the two eminent spiritual rulers, the Dalai Lama of Lhasa and the Panchen Lama of Tashilhunpo, who were aided by a council of four laymen, called Ministers of State, under the direction in chief of the two Imperial Commissioners or Residents appointed from Peking. In consequence of the British advance in India the possession of Tibet was jealously guarded by the Chinese, and the Emperor Chien Lung did everything to attract and to please the high and powerful dignitaries of the Lamaist Church, who on their part controlled the masses of the population, not only in Tibet, but also in Mongolia. The Emperor was deeply interested in Buddhism and Buddhistic teachings, but it was rather political motives that prompted him to promote and to maintain sumptuous Lama tem-

[8]

ples in his capital and in his summer residence Jehol, since in this manner he remained in direct personal contact with the Living Buddhas and Incarnations.

In 1793 the Emperor, as though he desired to reciprocate, transmitted to Lhasa a golden urn to be used in selecting the new incarnations. When a Living Buddha is about to die, that is, to transmigrate or to change his form of existence, he tells beforehand of the place where he will reappear, while at his birth he can without difficulty recite the events of his former existence. Little slips of wood, each bearing the name of one of the candidates, were thrown into the golden urn sent by the Emperor Chien Lung, which was then placed in the principal temple of Lhasa in front of the statue of Tsongkhapa, the reformer of Lamaism. A slip was then drawn from the urn, and the child whose name was inscribed on it was declared the new Living Buddha—subject to the Emperor's approval. It is thus manifest that the Chinese, while outwardly respecting Tibetan beliefs and customs, exercised unrestricted control of the political machinery and had a direct influence over the election of the high ecclesiastic dignitaries, including the Dalai Lama and the Panchen Lama. This is the historical background from which the presentation of this group of gold objects is set off. No doubt there was a hidden significant political intention behind this gift, the full import of which escapes our knowledge, but it is obvious that it was intended as a greeting from Tibet to the dragon throne of Peking, as a homage to the emperor of China in his function as the invisible ruler and real protector of Tibet.

In ancient times turquois was not much appreciated by the Chinese. Among archaic bronze vessels and bronze implements found in southern Mongolia (wrongly labeled Scythian art) there are some with turquois inlays. Turquois was a stone which enjoyed popularity among the nomadic tribes of ancient Turkish and Iranian stock living in central Asia. Persia has always been famed for its beautiful turquoises mined at Nishapur, which be-

[9]

came known in China during the fourteenth century in the Mongol period. I believe that the turquois employed in our gold set comes from the mines of Nishapur and was especially ordered there for the imperial treasure. The turquoises mined in Hupeh Province, China, are not of superior quality. Under the Tang dynasty (A. D. 618-906) we find Chinese gold and jade ornaments inlaid with plaques or beads of turquois.

Tibet, however, has always been the classical land of turquois. In Tibet a general national passion for this stone prevails among all classes of people, high and low, as the result of a centuries or millenniums old training. What jade is to the Chinese turquois is to the Tibetans. Turquois is used in the copper and bronze statues of the gods, in swords, daggers, and knives, in charm boxes, rosaries, and jewelry, and in the head-dresses of women. There is hardly any object used in Tibet, into which turquoises would not enter in some fashion. Turquois is the great medium of exchange throughout Tibet. Numerous articles now on exhibition in Field Museum were acquired by me from Tibetans through barter with turquoises. To the Tibetans a turquois is a symbol of their country comparable to the azure-blue of their beautiful lakes and flowers. As we speak of the blue of the sky, the Tibetans say poetically "the turquois of heaven." In a Tibetan poem the Himalaya is described thus: "This mountain range spreading like a thousand lotuses is white and like rock-crystal during the three winter months; during the three months of summer it is azure-blue like turquois; during the three autumnal months it is yellow like gold, and in the vernal moons, striped like the skin of a tiger. This chain of mountains, excellent in color and form and of perfect harmony, is inexhaustible in auspicious omens." This passage reveals the innate nature love of the Tibetan people and the parallel which they like to draw between the colors of their favorite gems and those of their natural surroundings in the course of the seasons. In another poem it is said, "On the plain where diamond rocks glitter is a lake with a mirror like turquois and gold."

[10]

Gold and turquois belonged to the most ancient offerings made to gods and demons, and ranked among the most precious gifts bestowed on saints and Lamas by kings and wealthy laymen. The thrones occupied by kings and church dignitaries were adorned with gold and turquoises, which likewise ornamented the cloaks worn by them. Unusually fine and large turquoises were known under poetical names such as "the resplendent, divine turquois" and had the value of a good race-horse. Marco Polo, speaking of the province of Caindu, which is identical with the western part of Sechwan Province, a territory largely inhabited by Tibetan tribes, refers to a mountain in that country "wherein they find a kind of stone called turquois, in great abundance, and it is a very beautiful stone; these the emperor does not allow to be extracted without his special order." For more information on the history of turquois in India, Tibet, and China the reader may be referred to Laufer's "Notes on Turquois in the East" (Field Museum Publication 169) and "Sino-Iranica" (pp. 516-520).

Lapis lazuli is mined in several localities of eastern Tibet and together with rubies was included by the ancient kings of Tibet among the presents sent to the sovereigns of China. The principal supply of the finest kind of lapis, however, comes from the mountains of Badakshan, north of the Hindu Kush, which produces two precious stones—lapis and the balas ruby or spinel. From this source the ancient Persians and Assyrians derived their lapis, and during the Tang dynasty (A. D. 618-906) it was exported from there to China. Marco Polo visited the mines, calling the stone *azure* and saying that it is the finest in the world and is obtained in a vein like silver. This exportation to China has persisted through the middle ages down to the present time. Lapis lazuli is called in Chinese "essence of gold" *(kin tsing)* or "dark-blue gold stone" *(ts'ing kin shi)*, and was chiefly enlisted for inlaying, occasionally also for jewelry and carving of figurines and snuff-bottles. The emperor used to wear a rosary of lapis beads when performing worship on the altar of Heaven, and a

[11]

rosary of turquois beads when officiating in the temple of the Moon.

It is difficult to outline briefly a history of gold in China. One point must be emphasized, and this will always speak in favor of the Chinese, that they were never obsessed by the hunger and greed for gold (the *auri sacra fames*, "the accursed hunger for gold," of the Roman poet), which characterizes the Semites, Greeks, Romans, and all other European nations ancient and modern. Gold was never coined into money in China, gold was not amassed and hoarded just for the love of it and, although found in many parts of the country, was never intensely or systematically exploited. We never hear in China of a "gold rush" or "gold fever," symptoms so characteristic of Europe and America. There is good evidence for believing that the oldest metal known to the ancient Chinese was silver, then copper, and lastly gold. Among the relics of the Shang dynasty, as far as I know, gold or objects of gold have not yet been traced, and probably were still unknown.

It seems that only in the late Chou period (about 300 B. C.) did gold come into prominence when gold foil was applied to bronze vessels as a coating or was used as an inlay in the surface of ritual bronzes. Gold was then considered as the most precious metal which also was the object of barter, and among precious substances was considered as ranking next to jade. Its chief use was for magico-religious purposes in that it was interred with the dead, the belief being entertained that gold, in the same manner as jade, was capable of preventing the body from decay, preserving it, and promoting the resurrection and immortality of the individual.

It is an interesting fact that there is no genuine word for gold in the Chinese language; there is only the descriptive term "yellow metal" *(huang kin)*, on the same level as "white metal" which describes silver or tin, "red metal" referring to copper, and "black metal" to iron. In course of time, the general designation *kin*

[12]

("metal") was reserved for gold, but the occurrence of this word in ancient texts presents a stumbling-block, as it is by no means certain in each and every case whether it refers to metals in general, or to gold, or to another specific metal, and there is no consensus of opinion among the ancient commentators.

Under the Han (206 B. C.-A. D. 220) the production of gold increased considerably. The philosopher Wang Chung writes that "gold and jade are considered the choicest omens; the sound of gold and the color of jade are most appreciated by man; gold is produced in the earth, and the color of the earth is yellow; the ruling element of the Han dynasty is earth, which accounts for the production of gold." Gilded bronze vases and vessels with gold incrustations belong to the finest achievements of Han art. Jade was still extensively used for personal ornaments, and for this purpose was more favorite with the people than gold.

Under the Six Dynasties gold and silver were lavishly employed for Buddhist statues and statuettes and for animal figurines, also for bowls, dishes, and boxes. We stand on firmer ground in coming to the Tang dynasty, whose productions are clearly recognizable. During that memorable epoch the Chinese began to appreciate the qualities of gold for artistic purposes. It is amazing that the art of the goldsmith was then fully developed and had apparently reached its climax: all manners of technique in treating gold were known and practiced in the glorious age of the Tang, such as beating out and cutting gold foil, gold filigree, repoussé work, work à jour, beading, making fine gold threads and gold wire and twisting it into spirals, and a marvelous combination of various processes into one harmonious work of art. The succeeding dynasties have adopted the technical lessons of the Tang without adding much that is new.

While many fine gold ornaments and even gold crowns of the Tang, Sung and later dynasties have come out of China during the last decade or so, nothing like the gold treasure of the Emperor Chien Lung has ever appeared before. This set, both from a

[13]

technical and artistic viewpoint, is absolutely unique in the world and the most perfect achievement of the goldsmith's craft that has ever been attained anywhere by human hands. The gold objects found in Tutankhamen's tomb are dwarfed and eclipsed by this production of a master mind, which baffles description. It is useless to attempt to describe the processes of its workmanship, which is so microscopically fine and so fairy-like delicate that its proper appreciation is only possible when studied under a powerful magnifying lens. We can but admire the enchanting color harmony of the gold with the charming blues of turquois and lapis lazuli, the simplicity and purity of style, and the exquisite choice of decorative elements.

Let us not be oblivious to the fact that period means but little in the history of art. A work of art is not necessarily great or good because it is old, and not necessarily inferior or poor because it is more or less recent. It is artistic merit and quality and the spirit pervading a work of art which is the decisive factor. Some scholars regard the Chinese art of the eighteenth century as one of a purely retrospective and imitative character and one of mere technical perfection. This sweeping generalization is not correct, however. True it is that ancient forms and designs were then perpetuated and reproduced, but not slavishly; it was, in the main, a new spirit cast into ancient molds, a new soul breathed into the bodies of the past, which rose again to a better and bigger life. In all lines of artistic endeavor we recognize a great amount of progress, improved taste, and novel ideas — in porcelain, textiles, embroidery, lacquer ware, jewelry, bronzes, sculpture, and painting. In many cases the artists of the Chien Lung period were more original than the originals taken by them as models, in the same manner as Shakespeare was more original and greater than the writers from whom he derived the plots for his dramas. Confucius, China's great sage, said, "Everything has its beauty, but not every one sees it." There are Chien Lung bronzes more artistic and therefore more desirable than many Han, Tang,

[14]

and Sung bronzes; and there are painters of the same period endowed with a striking originality of mind and power of brush. In fact, the reign of Chien Lung signals China's golden age in art and literature, a great epoch of renaissance, and the craft of the goldsmith must then have reached the climax of perfection, as witnessed by the exhibition of these superlative examples.

There is another important point to which attention must be drawn. While each piece individually merits careful study and analysis and must elicit our admiration for its beauty of form and mastery of execution, the whole set must also be viewed synthetically and examined as a unit in its totality. As every one will readily recognize, it was conceived by a single artist according to a well-devised and premeditated plan. It is this unity of plan and thought that lends another attractive charm to this group of desk paraphernalia. The set was first designed by the hand of a guiding genius who was endowed with a vision, a profound artistic sense, a refined taste, and a keen appreciation of the beauty of line and form. His was the mind of a master; assuredly he was the leader of his art during his days, another Benvenuto Cellini. His name unfortunately is unknown. In my essay "East and West" (*The Open Court,* December, 1933) I have set forth the reasons which prompted Chinese artists not to sign their masterpieces. They were too modest and too sensible to mar their productions with their signatures, and did not flatter themselves into the belief that they personally were the creators of their creations, but humbly attributed them to the action of a higher power, to the merits of their ancestors or to the will of Heaven. The artist was a sort of high priest; he produced, not to please his contemporaries, but to honor his ancestors and to attain his own salvation.

The artist, who designed this group of gold objects, did not work for the acclaim of the multitude or with a view to an exhibition and obtaining a *grand prix.* He had a finer and nobler ambition; his chief inspiration was the thought that his work was to be seen and judged by just one man — the Son of Heaven. All

[15]

his efforts were bent on this one objective. We may well realize how many years he must have toiled over his plans and designs in his study, how many sleepless nights he must have spent over them, how many years he must have anxiously watched his staff of artisans who were entrusted with the task of bringing his ideas to life. The result of his painstaking labors which without exaggeration we may estimate at ten years or even more is now happily before our eyes; surely it was worthy of the name of the great emperor. We can read from the superhuman efforts expended on the workmanship that the men who devoted to it their time and energy constantly had the thought of the Son of Heaven on their minds and were actuated by the earnest desire of service and loyalty to his majesty—loyalty, the cardinal virtue of a good citizen inculcated by Confucius.

On the other hand, solely a character and personality of the greatness of Chien Lung was capable of inspiring a masterpiece like this one. In other words, it has two focuses of radiation; on the one hand, the human, altruistic, and wonderfully devoted spirit of the artist and his staff; on the other hand, the highminded, art-loving, and generous spirit of the sovereign. There is, accordingly, a symphonic unit pervading this set of eighteen pieces. Each object has its definite place and significance, and bears a relation to every other object. It may not be an exaggeration to name this treasure a glorious symphony in gold, a triumph of the spirit over matter.

A word should be said as to why Pao Tai presented his lord with a series of objects just to adorn his writing desk. Here we have to remember that writing in China is calligraphy, an art on the same par as drawing and painting and the first and essential prerequisite and characteristic of a scholar. The written word was always worshiped as a fetish, and any materials and utensils devoted to the art of writing were given the most careful attention on the part of scholars. The celebrated calligrapher Wang Hi-chi (A. D. 321-379), whose handwriting is said to have been

[16]

"light as floating clouds and vigorous as a startled dragon," is credited with the dictum, "Paper symbolizes the troops arrayed for battle; the writing-brush, sword and shield; ink represents the soldier's armor; the ink-stone, a city's wall and moat, while the sentiments of the heart symbolize the chief commander." In this saying the mental attitude of the Chinese toward the arsenal of the learned is perfectly crystalized; paper, brush, ink, and ink-slab are the four great emblems of scholarship and culture; all of these are inventions which the Chinese may justly claim as their own, which constitute fundamentals of their civilization, and which have largely contributed to make them a nation of studious, well-bred and cultured men.

Chinese paper, brushes, and ink are the best ever made, and have achieved a world-wide reputation. Chinese ink is the only true black ink ever produced and that will last permanently. The presentation to the Son of Heaven of magnificent paraphernalia for his desk was a tribute to his standing and reputation as a scholar, a homage to his literary achievements. And a scholar and poet he was, and a very gifted and distinguished one. His collected essays and poems written in Chinese and Manchu fill several hundred volumes. One of these, a eulogy of Mukden and its environment, was translated into French by the Jesuit Father Amiot (published in Paris, 1770) and excited the admiration of Voltaire, who addressed an appreciative ode to the emperor. We may imagine that this gold set was capable of firing his imagination and inspiring him to many a composition.

He was a lover of books and literature and gathered in his palace the greatest library ever assembled. The catalogue of this library is the fundamental source for our knowledge of the history of Chinese literature. The works edited by him or issued under his direction and patronage are legion, and many of these belong to the masterpieces of typography of all times.

[17]

He was a liberal patron of art and artists. Although he disliked the Jesuit missionaries and forbade the propagation of the Christian religion, he appreciated their erudition and retained two Catholic artists, Castiglione, an Italian, and Attiret, a Frenchman, in his service.

His personal name was Hung Li. Chien Lung, the name by which we are accustomed to call him, was not his real name, but the slogan which he adopted for the period of his reign. Thus, the year 1736, in which he ascended the throne, in Chinese reckoning, is the first year of the period Chien Lung. He was the fourth son of the Emperor Yung Cheng (1723-35) and grandson of the great Emperor Kang Hi (1662-1722). Born in 1710, he was twenty-five years old when he succeeded his father, and soon rivaled his grandfather's fame as a ruler and a patron of letters. Gifted with an insatiable thirst for knowledge and a conspicuous talent for administration and statesmanship, he was an indefatigable worker until his last days. The year 1793 is memorable for the fact that the Earl of Macartney as ambassador of the King of Great Britain was received in two official audiences by the Emperor in the gardens of the palace of Jehol. On completing a cycle of sixty years of power and a successful reign, he abdicated in the year 1795 in favor of one of his sons (he had seventeen altogether), spending his last years in seclusion and study. The empire then was at peace, and the people were enjoying prosperity. He died in 1799 at the age of eighty-nine—the last great emperor of China and certainly one of the greatest who ever graced the dragon throne. He was canonized with the title Kao Tsung Shun Huang-ti.

A word remains to be said in regard to the manner in which this gold treasure found its way to America. It remained in the possession of the imperial house during the nineteenth century. In 1900 when the Boxer uprising lured the armies of Europe to Peking, the Empress Dowager and the court took refuge in the ancient capital of China, Si-an fu, and lived there in exile for some

[18]

time. According to a statement made by a member of the retinue, the gold set in question was regarded as of sufficient importance and value to be taken along with other treasures on this flight. Afterwards it was brought back to the palace of Peking, and in 1908 Prince Pu Yi, who now reigns as Emperor Kang Te over Manchukuo, fell heir to it. When the Republican Government withdrew its annuity from the young prince, he gradually became impoverished and was compelled to solicit loans from Chinese banks of Peking and Tientsin, placing with them numerous art treasures, among these the gold set, as collaterals. Unfortunately he was unable to meet his obligations though several extensions on the loans were granted. Finally the banks were forced to foreclose and to dispose of the art treasures to collectors. The gold set was the first to be segregated from the other collections which included bronzes, jades, and court paraphernalia, and was sent to the United States. It is now in charge of the well-known art firm, Parish-Watson & Company of New York, which has placed it on exhibition in Chicago's Century of Progress. All art lovers will be grateful to Parish-Watson & Company for the opportunity of viewing this unique gold treasure, which is like the embodiment of all beauty and splendor of the Arabian Nights.

BERTHOLD LAUFER

Chicago, June the first, 1934

[19]

2

1

DESCRIPTIVE CATALOG

1-2

A pair of censers of spherical shape, of solid gold wrought in open work, filigree, and beadwork. The surfaces are decorated with parallel rows of quatrefoils à jour, laid around in circles and overlaid with clusters of plum blossoms formed by turquoises. These are scattered around freely and inobtrusively as though a sudden blast of wind had shaken them off a tree. Like sketches of bamboo, the drawing of plum blossoms has developed into a special branch of Chinese art and with some artists into a veritable passion. The covers, in harmony with the shape of the vessels, are surmounted by finials formed by a knob of turquois, and these are enclosed by petals of turquois mounted on gold; the knobs function as handles. Either censer is equipped with a gold stand bordered by rows of turquoises which enclose a most graceful and beautiful palmette or palm-leaf design. The surfaces of these stands are embossed with an ornamental form of the character *shou* ("long life") surrounded by five bats. The bat *(fu)*, by means of punning, is an emblem of happiness *(fu)*, and the group of five bats symbolizes the five blessings—old age, wealth, health, love of virtue, and an easy natural death.

Height: 5 inches. Weight: 355 and 365 grams, respectively.

3

An arm-rest or wrist-support of gold, the designs being brought out in turquois. They represent cracked ice over a pond upon which the first plum blossoms of the spring have fallen, alluding to the awakening of the spring. The veins in the petals are delicately carved. This pattern reminds one of the famed motive of Chinese pictorial art—the poet astride a donkey in a snow-laden landscape searching for the first plum blossoms. The edges of the support are adorned with key patterns in lapis lazuli and turquois alternating. The ordinary supports of this kind were made of bamboo lacquered and carved, or of ivory; they were fashionable in the eighteenth century, but are no longer in use. They served as support for the wrist while writing with the brush. Specimens of such bamboo supports may be seen in a case devoted to Chinese writing materials in Field Museum (in center of West Gallery), where all sorts of writing-brushes, inks, inkstones, brush-holders, and other utensils are also assembled. He who will study this material will obtain a clearer idea of the meaning and scope of this desk set.

Length: 4 inches. Width: 2 3/16 inches. Weight: 145 grams.

4

A desk ornament of rock-crystal, in the shape of an ancient jade chime, mounted on gold and studded with turquoises. All figures and designs are skilfully carved from turquois in high relief and alike on both sides. They are carefully matched and registered on the foundation of rock-crystal. The picture illustrates a subject celebrated in Chinese painting: representatives

[21]

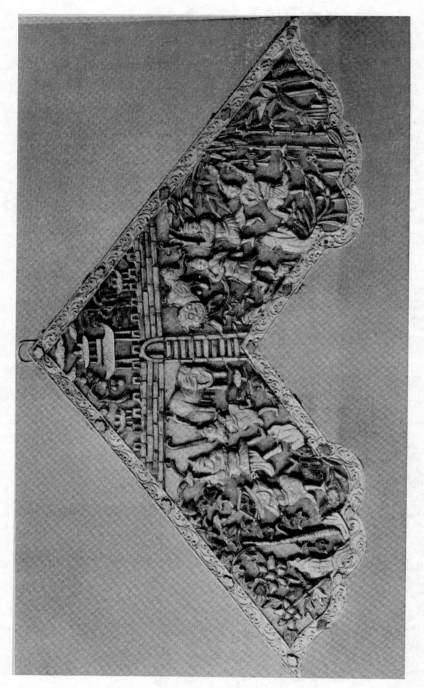

4

of foreign nations bringing gifts and tribute to the imperial court. The tribute-bearers are congregated in front of a crenelated city wall to which a staircase leads up. The top of the wall is crowned by a two-storied pavilion overshadowed by trees and overtopped by clouds. Two smaller pavilions are on the sides. Near the staircase are an elephant and a lion. The lion is chained and guided by a man from Central Asia, well characterized, like the other men, by his pointed conical cap. Live trained lions were frequently sent to Chinese sovereigns as gifts by the kings of Persia and Central Asia, the first lion tribute being recorded in A. D. 87 under the Han. Lions are not natives of China, and were always a source of wonder to the Chinese; in art, of course, the king of beasts is a familiar figure. When made in bronze, the tribute-bearers are usually astride the lion and the elephant. The mahout who guides the elephant is equipped with a hooked stick. At his feet there is a *ling-chi,* a fungus of immortality and symbol of long life, and a tribute-bearer carries such a fungus in a vase, while another on the opposite side carries coral branches, a very favorite tribute gift, and another holds an ingot of silver. Bamboo trunks, banana plants, a Wu-tung tree, and rocks complement the picture.

In the Jade Room of Field Museum chimes of rock-crystal and jade are on view. Their shape is derived from a carpenter's square. At an early date the ancient Chinese recognized the sonorous qualities of jade and used it as a musical instrument, suspending it in a wooden frame by means of a silk cord passing through a perforation at the apex. It was still so employed under the Manchu dynasty during the ceremonies performed in honor of Confucius. In the Chien Lung period, resonant stones were much favored as birthday presents or congratulatory gifts, as their name *ch'ing* is punned upon another word *ch'ing* of the same sound and tone, which means "good luck, happiness, blessings, felicitations." Formerly they were also part of a bride's dowry in Peking, and served as ornaments in the parlors of high officials.

Length: 9½ inches. Weight: 636 grams.

5

A desk ornament, of gold, in the shape of an ancient jade chime, being the mate to No. 4, but different in style and technique. It is entirely wrought from gold, while fruits, foliage, rock, and bats are carved from turquois against a delicate ground of gold filigree. The patterns are identical on both sides. The fruit is the peach, and it is the renowned peach of immortality growing in Paradise, believed to be located in the west, and conferring immortality on him who eats it. It is a symbol of longevity, also an emblem of marriage. The five bats, symbolizing again five blessings as in Nos. 1-2, are artistically scattered, one at the apex, two at the base and one each at the ends where the peaches terminate. Two rows of sea waves, of turquois, are represented at the two bases, jewels appearing on the crest of the waves and two rows of genuine pearls above them. It was believed that the Dragon King dwelling in a palace on the bottom of the ocean was the owner and dispenser of pearls and precious jewels. The de-

[23]

10

signs are therefore expressive of the following wish: "Felicitations! May you partake of the five blessings, may you obtain the peach of immortality and all treasures hidden in the sea!" The object is bordered by cloud patterns carved from turquois.

Length: 9½ inches. Weight: 360 grams.

6-7

Two paper-weights of gold, rectangular in shape. No. 6 rests on a solid gold plaque on which bats and clouds are engraved alternately (altogether six bats and seven clouds). On the upper surface, the center is occupied by an ornamental form of the character *shou* ("long life") carved from turquois; three bats and clouds alternating are to the left and right. The two sides are each decorated with five bats. The wish expressed by these designs is: "May longevity be yours, and may the clouds or the sky be the limit of your happiness!"

The other paper-weight, No. 7, rests on a gold plaque engraved with sprays of peonies and tendrils, and has a very unusual beaded design of geometrical character bordered by two bands of delicate key patterns. This gold filigree ground is overlaid by larger key patterns or meanders carved from turquois in relief. With the application of this pattern Pao Tai intends to thank his majesty for favors received; for the meander has been from ancient times a symbol of thunder and clouds which send fertilizing rain and produce affluence and wealth to the farmer. The meander was called "thunder pattern" or "rain-cloud pattern." Rain was conceived as a favor conferred by Heaven or the Dragon on mankind. For this reason the meander was applied to objects in allusion to the acknowledgment of favors received. Moreover, there is a rebus involved. The meander was also styled *hui wen,* i. e. returning or revolving pattern, and *hui* means also "to respond, to return a favor." Pao Tai therefore intends to express the message, "Thanks to the imperial dragon for favors showered upon me in the past!"

The center of the upper surface is occupied by the character *hi* of ancient seal form, which signifies "joy."

No. 6. Length: 9¼ inches. Weight: 270 grams.

No. 7. Length: 9 3/16 inches. Weight: 285 grams.

8

A scepter of happy augury or good luck, called Ju-i, which means "as you desire, according to your wish." On the occasion of a birthday or New Year, such scepters were bestowed by the emperor upon high dignitaries and courtiers and by these on the sovereign, with the idea of expressing good wishes. Queen Victoria received one from the emperor of China in honor of the fiftieth anniversary of her reign. Such scepters were made of jade, sandalwood, bamboo, amber, and even iron inlaid with gold and silver wire, but I have never before seen one of gold. The broad plaque atop is adorned with peaches and foliage, the peach of immortality and symbol of long life, and two bats carved from turquois and encircling a beautiful turquois

[25]

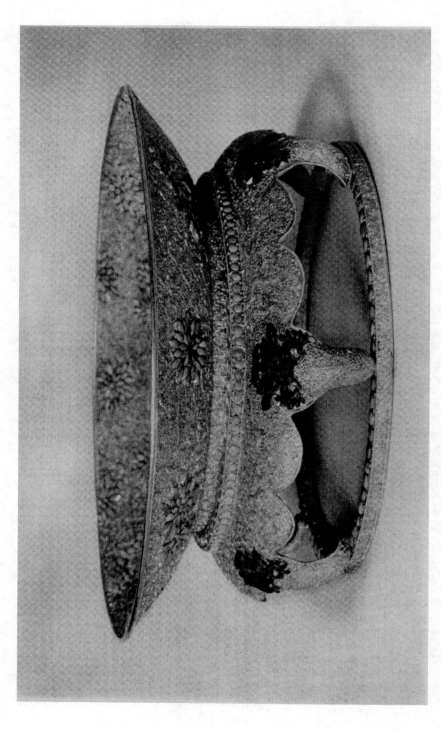

12

plaque with beaded mounting. The designs on the handle, likewise carved from turquois, represent the eight symbols of happy augury or eight precious objects, adopted by the Chinese from Buddhism and frequently represented on porcelains, in tapestries and embroideries. From top to bottom they run as follows: the wheel, an ancient emblem of the kings of India whose war chariots rolled over the universe, and of Buddha, who turned the wheel of religion; the conch, which was blown in announcing victory and calling the pious to prayer; the state umbrella, an emblem of royalty; a canopy, likewise so; lotus, emblem of purity; a holy-water vase, symbolizing abundance and peace; the double fish; and sacred knot, originally a mystic emblem on the chest of Vishnu, in China emblem of luck. The reverse is covered with an exquisite all-over pattern of gold filigree, while bands of meander decorate the sides.

These scepters have had a very curious development. Originally made of iron, they were a kind of blunt sword for self-defence or a symbol of command in the hands of a general. Buddhistic notions of magic were subsequently transferred to this weapon, and it was considered a magical wand capable of transforming the body, securing victory in battle, and enabling one to attain every wish. As a symbol of conquering power it was frequently placed in the hands of Bodhisatva statues.

Length: 9½ inches. Weight: 210 grams.

9

A box designated in Chinese on the accompanying label as a "circular twin box." Such "twin boxes" were commonly made in lacquer, and twin vases in porcelain and jade also are known. This twin box is wrought from pure gold in open work, the unit of design being a quatrefoil. This background is overlaid both on the cover and the sides with elegant plum blossoms carved from turquois, some single, others in pairs and clusters of three —a highly artistic principle. The concentric zones on the cover are filled with cloud patterns of turquois. The surfaces of the cover form two interlaced circles, twins grown together as it were.

Length: 4 5/16 inches. Height: 1½ inches. Weight: 195 grams.

10

A censer of compressed globular shape, of gold in open work, except the base which is solid, adorned with turquois and lapis lazuli. The central register of the vessel contains the eight Buddhistic symbols of happy augury, described under No. 8, carved from lapis lazuli of exquisite color. The eight symbols are separated by sprays of chrysanthemums, the petals being represented in turquois. Above and below this central register there is a band of quatrefoils of turquois alternating with lapis lazuli. The edge and the base are decorated with a design derived from lotus petals. The base is finished with a circle of turquois beads. The cover is crowned by a perforated knob of turquois mounted on petals of gold, and is laid out in three concentric zones. One of these is also occupied by the eight symbols in lapis lazuli.

Height: 2¾ inches. Weight: 280 grams.

[27]

13A

13B

11

An ink bed, of fine old turquois in the matrix, mounted on gold filigree of great delicacy and refinement. The turquois rests on a solid gold plaque the lower side of which is embossed with sprays of peonies and foliage of elegant style. The bed is posed on four feet of animalized style, dragon masks of archaistic style being brought out in high relief. A cake of ink was placed on top. Ink-cakes are gotten up in a very artistic manner, landscapes and portraits being stamped on their surfaces from molds, or designs being painted on them in gold and silver. The Chinese term used on the white silk label means literally "ink bed"; the Tibetan term means "a stand or support for ink."

Length: 4 inches. Width: 3½ inches. Height 1½ inches. Weight: 220 grams.

12

A dish or plate on a separate stand, a reverie somewhat reminiscent of the mysteries surrounding the Holy Grail. The gold filigree of the plate is overlaid with chrysanthemum blossoms carved in turquois, both inside and outside. The six feet of the stand are carved in the shape of elephant trunks, a tradition inherited from archaic bronze vessels of the Shang and Chou periods, and are surmounted by ogre heads of archaistic style, exquisitely cut from lapis lazuli of excellent quality and color. The surface of the stand is decorated with a key pattern or meander band in gold. A circle of turquois beads is laid around the edge of the stand and around the ring forming the base. This is a masterpiece, the climax of refined taste, a personification of pure beauty, combined with the highest possible qualities of technical perfection.

Diameter of bowl: 6 5/16 inches. Diameter of stand: 5½ inches. Total height: 2½ inches. Total weight: 435 grams.

13

Desk screen of gold, placed in a teakwood stand, with pictorial representations carved from turquois on both sides, which speak an eloquent symbolic language. In the upper part we see cranes on the wing soaring in clouds—emblematic of reaching immortality. Next to the phoenix, the crane is the most celebrated bird in Chinese legend. It is reputed as the patriarch of the feathered tribe and as the airship carrying heavenward the saints who have attained salvation. In fact, the crane itself is regarded as the incarnation of an immortal fairy. Cranes at sunset, e. g., depict their flight over the sea to the Fortunate Isles of the Blest. The crane is believed to reach a fabulous age; when six hundred years old, it drinks, but is no longer in need of taking food. As on our screen the cranes fly from heaven downward to earth, they are intended to confer immortality on the Son of Heaven, the recipient of the screen. On the left, a large Wu-tung tree (*Sterculia platanifolia*), one of the most beautiful trees of China, planted in many temples and parks, and sacred to the phoenix. The pine to the right is a symbol of strength, endurance, and permanence. The two deer

[29]

16

are of the spotted variety, called *mei hua lu* ("plum-blossom deer") or *kin ts'ien lu* ("gold coin deer"). The stag or deer, by means of punning, symbolizes prosperity or a good income. One of the deer holds a *ling-chi*, the fungus of immortality, alluding to long life, and deer is likewise believed to be a long-living creature.

The opposite side of the screen shows the five bats as the emblem of five blessings (explained in Nos. 1-2), coming down from the clouds. To the left, the peach tree of Paradise (see No. 5), at its foot a fungus of immortality. Beneath, a pomegranate tree and three tall bamboo stalks. The pomegranate fruits are ripe, displaying their seeds. Because of their exuberant seeds they are regarded as emblems alluding to numerous progeny. The pomegranate, so to speak, is an anti-race-suicide symbol. The fruit is still a favorite marriage gift and appears in the wedding feast, implying the wish for many sons and grandsons. The motives displayed on this side of the screen present a rebus which can be read. *Chu,* the bamboo, stands in the rebus for another word *chu,* which means "to pray." Thus, the donor intended to express to his majesty this wish: "I pray that you may obtain the five blessings, that you may be reborn in Paradise and eat the peach of immortality, and that you may leave numerous progeny behind!"

6¼ x 5¼ inches. Weight: 480 grams.

14

A quadrangular vase, of gold, in the shape of the archaic jade image of the deity Earth. Heaven was conceived by the ancient Chinese as circular, and earth as flat and square outside and rounded in the interior. The religion of the ancient Chinese was mainly nature worship; the great cosmic powers, Heaven and Earth and the Four Quarters, were the principal deities to whom worship was paid. Their philosophy was dualistic and classified all phenomena as male and female, light and darkness, heat and cold, positive and negative. These two primeval forces were seen active in Heaven and Earth; the union of the two and their constant interaction were believed to have resulted in the creation of nature and man. The deity Earth was represented in the form of a jade tube rectangular in cross section and round inside. Yellow was the color of Earth, and yellow jade whenever possible was chosen for this emblem. It therefore was a felicitous idea to select this emblem for a vessel in gold. The gold filigree background of the four panels is overlaid by turquois carvings in high relief, which represent the flowers of the four seasons. In one of the panels is displayed a flowering plum tree on which a magpie is perching. The magpie is a bird of good omen and was regarded as the protector of the Manchu dynasty. Another panel shows two dragon-flies, a banana, and orchid leaves growing from under a rock. The base is decorated with two meander bands in gold.

In this case it is especially true, as has been pointed out in the Introduction, that artists of the Chien Lung Period often surpassed the originals which served them as models and revived them with a new soul.

Height: 6 inches. Weight: 435 grams.

[31]

15

Water-receptacle, of gold, for the desk, consisting of vessel, cover, ladle, and stand, to contain the water which had to be poured on the ink-stone when a cake of ink was to be rubbed against the stone. The Chinese prepare only as much ink as is necessary at a time, since the liquid ink rapidly dries up. The ladle, used for pouring the water on the ink-stone, has the shape of a Ju-i scepter (see No. 8) gracefully curved, and is studded with turquoises. The shape of the vessel is exceedingly graceful and refined. The pattern—plum blossoms spread over cracked ice, of turquois—is identical with that in No. 3. Lapis lazuli is listed for the bands laid around the mouth and foot of the vessel and the stand.

Height: 3 inches. Weight: 200 grams.

16

Hexagonal gold vase, intended as a flower vase, of exquisite shape, such as we see in the finest examples of porcelains. Key patterns are laid around the mouth, middle, and base of the vase. The foot is adorned with a continuous swastika pattern, an emblem of good luck. The six upper panels are decorated with tendrils, emanating from a flower in the center. The six lower panels are ornamented with floral sprays, all carved from turquois. The vase consists of solid gold plaques which are overlaid with filigree patterns. Height: 7 inches. Weight: 290 grams.

17 (Frontispiece)

Flowervase of gold, with gold stand, called by the Chinese a hammer-shaped vase, in imitation of a porcelain vase which was a great favorite in the Kang Hi and Chien Lung eras. The delicate gold filigree background is overlaid with chrysanthemum blossoms of sky-blue turquois, artistically arranged single and in groups of two and three. The vase is posed on a stand of gold filigree, the circular top of solid gold being embossed with arabesques. The stand is adorned with monster heads of lapis lazuli; each head has two protruding fangs and arms equipped with four claws. This motive is well known in Tibetan art.

This subject presents a rebus which can be read and which is expressive of a wish. The word for the vase, p'ing, is punned on another word p'ing, which means "peace." A name for the chrysanthemum is chiu hua, "nine flower," because it blooms in the ninth month; chiu, "nine," is punned with chiu, "long." Therefore the wish is expressed, "May you enjoy peace for a long time!" Height: 8¼ inches. Weight: 450 grams.

18

Writing-brush holder, of gold, decorated with the same pattern—plum blossoms spread over cracked ice, of turquois—as in Nos. 3 and 15, an allusion to the early awakening of the spring. The veins in the petals of the blossoms are finely carved. The edge and the base of the vessel are decorated with a band of plain geometrical patterns carved from lapis lazuli of a clear blue. Height: 4 9/16 inches. Weight: 355 grams.

[32]

The Gold Treasure of the Emporor Chien L'ung.

<u>Foreword.</u>

On June 18th, I had an opportunity of inspecting the famous gold treasure of the Emperor Chien L'ung in the vault of the National City Bank, New York. While I have seen many rare gold objects from China, among others, many gold ornaments made under the T'ang and Sung dynasties, I must confess that I have never before seen anything like these precious gold objects, which, to my knowledge, are perfectly unique and a rare and priceless treasure. I had no idea that anything like this was ever made in China, and was surprised and thrilled at the sight of these costly and venerable relics which are one of the most remarkable productions of China's golden age under the reign of one of her ablest and most glorious rulers. These objects are the perfection and climax of the gold craftsman's art, and permit us to presuppose a long development of this craft that must have extended over many centuries. The workmanship is so fine and perfect, that it baffles description and the process of inlaying with turquoise and lapis lazuli is equally admirable, the beautiful colors of these stones matching harmoniously with the golden hue of the yellow metal.

I have carefully examined each object and read and partially copied the inscriptions written in Chinese, Mongol, Manchu, and Tibetan

on the paper labels pasted in the boxes. These inscriptions reveal the
fact that this set of eighteen gold objects was presented to the throne
by Pao-t'ai in the year 1783 (the forty-eighth year of the reign of
Chien L'ung), and these labels are perfectly authentic and contemporary.
Each label contains the same statement and date, and at the end gives
the name or term of the particular object. The entire set was made
for use on the Imperial desk, some of the objects being intended for
practical purposes such as the brush-holder, the support for the
ink-cakes, the water receptacle, and the censers; others being
devised as purely ornamental and esthetic objects.

It is not merely the beauty, elegance, and gracefulness of
each single article that merits our admiration, but no less, and
perhaps to a still higher degree, the unity of plan and harmony
of design that pervades the entire set. In other words, the set was
designed by the hand of a guiding genius possessed of a profound
artistic sense and a keen appreciation of the beauty of line and
form. No doubt it was a man of a master mind, a rival of Benvenuto
Cellini, assuredly the leader in his craft during his day. How
many years he may have toiled over his plans and designs and for
how many more years he may have anxiously watched over his staff of
artisans who were entrusted with the task of carrying his ideas into
execution, unfortunately, we do not know. The result of his painstaking labors is a truly royal gift worthy of the name of the great
Emperor to whom it was dedicated. It is plainly visible and readable
from the superhuman efforts expended on the workmanship that the men

who devoted their time and energy to this masterpiece constantly had the
thought of the Son of Heaven on their minds and were actuated by an intense desire of service and loyalty to him.   Solely a personality and
character as great as Chien L'ung's could have inspired a chef d'oeuvre
like this one.   Each object in this set has its definite place and significance, and bears a relation to every other object.   The whole set
presents a symphonic unit, a coherent story, an harmonious composition
as impressive and inspiring as an old Gothic Cathedral.

To testify that this unique treasure is genuine would seem to me
a frivolity.   It needs no apology and no defense; it can take the
witness stand on its own behalf, it speaks a language more eloquent
than all discourses of learned art-historians could lend it.   It is
the embodiment of a refined spirit and consummate mastery of the mind
over matter, such as only the Chinese could have acquired in
consequence of their century-old traditions of artistic skill and
training; it is supreme perfection in itself, a brilliant triumph
of the gold-craftsman's art, a live witness of what Chinese genius
is able to accomplish.

New York, June 22, 1929.   *Berthold Laufer*

The Gold Treasure of the Great Emperor Chien L'ung.

The Imperial Collection, comprising eighteen (18) golden objets d'art, was presented to the Emperor Chien L'ung by Pao Tai in the 48th year of the reign of the Emperor Chien L'ung, or in 1783. Pao Tai was a Mongolian prince and envoy from the Imperial Court to Tibet.

It was the custom of envoys so favored, to express their appreciation through the presentation of the finest works of art obtainable. Officials vied with one another in order to thus gain favor with the Emperor; each endeavored to present the most unique gift - unique as to conception, design, execution and form; magnificent as to material, and superb as to craftsmanship. The collection made before the year 1783, (the date of its presentation to his Majesty and therefore at least 150 years old,) is one of the most important and significant imperial collections that has ever been allowed to leave China. The set was intended for use on the Emperor's desk and consists partially of objects of utility, such as a brush-holder, support for the ink-cakes, a receptacle for holding the water to be poured on the ink-slab, and censers; partially of objects designed purely as decorative works of art. The question may be raised as to why turquoise and lapis lazuli have been utilized for inlaying, instead of jade, the most precious stone in the eyes of the Chinese. The answer to this question is that Pao Tai, being selected as the imperial envoy to Tibet, evidently made a gesture at Tibetan taste and customs, for turquoise has been the national and most popular stone of the Tibetans as well as Mongols throughout all ages, as may be read in detail in a monograph of B. Laufer, entitled "Notes on

Turquoise in Asia" (published by Field Museum, Chicago). This publication also gives full information on the turquoise mines, sources
of supply, trade and carvings in turquoise in Asia. In view of the
hardness of turquoise and the difficulty of cutting and carving it,
the fine ornaments cut out of this stone are worthy of highest
admiration. The gold used in the making of these objects is 22 Karat.

The inscriptions in the cover of the original boxes are in
four languages, - Chinese, Mongol, Manchu, and Tibetan. In every
inscription occur the two Chinese characters "Li Yi". The translation of this means "gifts of the highest importance and greatest
value". To make a gift of "Li Yi" represents one's "entire fortune"
in the presentation of the same; and to perform an act of "Li Yi"
means to serve with one's entire heart and to embody one's spirit
in the gift.

In 1900, at the time of the Boxer uprising, the throne
was threatened, and it therefore became imperative that the Imperial family and Court take flight into the interior. It has been
alleged by responsible members of the Court retinue that this
particular collection of golden objets d'art was considered to
be of such value as heirlooms that it was among the treasures
which were taken along during the flight. This undoubtedly accounts for some of the missing original boxes.

After the death of the Empress Dowager in 1908, these
golden objets d'art automatically came into the possession of
the young Emperor.

When the pseudo-court was abolished by Feng Yu-hsiang, these eighteen art objects were among the treasures turned over by the young Emperor Suan Tung to his bankers as collateral on a loan estimated at $6,000,000, Chinese Currency. Owing to circumstances, he has been unable to meet his obligations, and, therefore, the collateral has been foreclosed by the bankers after several extensions had been made on the loan. These golden objects have been the first group to be segregated from the famous Collection, which has been solicited by collectors and museums all over the world.

Representatives from several foreign countries were dispatched to China for the purpose of negotiating the purchase of the entire Imperial Collection, or parts of it.

Mrs. Eric Clarke's success in obtaining the Collection was entirely due to her familiarity with the country, her knowledge of the Chinese language, and her high repute in diplomatic and official circles.

Description.

No. 1 - Ti Tsung (Earth Symbol). The shape of this rectangular vase is derived from the ancient jade symbols representing the deity Earth, first conceived under the Chou dynasty and perpetuated under the Han, revived in pottery under the Sung. The pictures on the four sides represent the four seasons, the floral decoration being done in jeweled turquoise. The edges are finished after the style of the Chou and Han bronzes. Gold weight 435 grams.

No. 2 - Hexagonal Vase: After the well-known form of Kang Hsi porcelain - exquisite workmanship. Each panel is decorated with
floral designs inlaid in turquoise and lapis lazuli. 290 grams

No. 3 - A symbolic instrument made in the shape of a resonant jade.
At a very early time the Chinese recognized the resonant
qualities of jade and carved jade into this shape to be used
as a musical instrument. A series of such jade plaques were
suspended in wooden frames and struck with a wooden mallet.
There is a famous representation in one of the stone sculptures of the Han period, showing Confucius practising on
the jade chimes. These are called in Chinese k'ing or ch'ing,
and as there is another word of the same sound, but written
with a different character, which means "congratulations,"
the design of a sonorous jade is used symbolically in the
sense of "I congratulate you" or "with all good wishes."
This, apparently, is also the significance of the ch'ing in
this set. Set in jeweled turquoise, with a touch of lapis
lazuli in the wave of the sea and the spray thereof in pearls.
The design on both sides is identical. 360 grams

No. 4 - Resonant Instrument of Rock-crystal: Appliqued turquoise on
each side, set in gold; design of the Imperial palace, above
in the center. Tribute-bearers taking gifts to the Imperial
court like lion, elephant, coral branches, and others. All
these people represent foreign nations. 535 grams

No. 5   Sleeve Painting Holders: Paper weights executed in gold,
and   one following the design in block dragon and the character
No. 6   hai for joy as represented on the old Han bronzes; the other
is a study of clouds and bats. The character in the center
is the ornamental form of shou "Long Life". No. 5-285 grams,
No. 6-270 grams.

No. 7 - Screen: Decorated on both sides in jeweled turquoise.
The four trees represent the four seasons - the peach,
bamboo, maple, (wu-tung,) and pine. In the center of one
side are shown the cranes descending from the heavens. This
symbolizes the announcement of advancement in rank and conveys that wish. The crane is also an emblem of longevity.
In wishing advancement the artist placed the spotted deer
in order to make perfect his wish; for, in case the desired
rank was not achieved, the aspirant was to be contented with
his lot, - as the spotted deer always indicates contentment.
The deer holds a branch of ling-chi (fungus of immortality).
On the reverse side are shown Bats as symbols of good luck.
They are known as the animal of happiness because in pronunciation the words "bat" and "happiness" are exactly the
same, although each of the words is written with a different character; therefore, because of the pun, the bat has always stood for the symbol of happiness in Chinese art and
lore. There are also shown the pomegranate emblematic of a
wish for progeny on account of its numerous seeds, and
peaches emblematic of immortality in allusion to the peaches
in the Western Paradise. 480 grams

No. 8   Pair of Incense Burners: Made of delicate filigree work
and - studded with a simple design in jeweled turquoise, represent-
No. 9   ing hawthorne. The pieces of turquoise used in the knobs
        are of unusually fine quality and represent the sacred lava
        stones used to decorate Imperial gardens. The stands are
        separate pieces in themselves, mounted on carved teakwood.
        The top of the stand is embossed with the five bats sup-
        porting the characters of Long Life and Happiness. The 5
        bats symbolize five kinds of blessings most desired by man.
No. 8-365 grams   No. 9-355 grams

            1 - Happiness   3 - Wealth
            2 - Long Life   4 - Love of Virtue
                                 5 - A Peaceful End

No. 10- Plate: This superb piece is composed of two parts - the
        stand, which is decorated with chrysanthemums in turquoise,
        and six heads of the Imperial dragon in super-fine lapis
        lazuli; the plate, which is a separate piece, as mentioned,
        is of the finest gold filigree interspersed with bouquets of
        chrysanthemums. It is made in imitation of an early Kang-Hsi
        porcelain. 435 grams

No. 11- Bottle called in Chinese "hammer-shaped vase". Is of bulbous
        form and the conventional type developed at the Imperial
        kiln during the reign of the Emperor Kang-Hsi. It is of
        beautiful form and makes an exceedingly great appeal to the
        art student. 450 grams

No. 12- This is a "Figure-8" Box: The Chinese name on the label is
        "round twin box." Executed in filigreed gold with delicate
        floral designs around the sides and top. The top design is
        composed of interlaced circles. 195 grams

No. 13- Ink-cake Support: Cracked ice and hawthorne design, deli-
        cately executed with inlaid turquoise and jeweled lapis-
        lazuli. Exceptionally fine example of workmanship. 145
        grams.

No. 14- Brush Holder: Cracked ice and hawthorne design, deli-
        cately executed with inlaid turquoise and jeweled lapis-
        lazuli. Exceptionally fine example of workmanship. 335 grams

No. 15- Ju-i or scepter of good augury: This is the baton of pre-
        sentation. In other words, it is always presented, on both
        hands outstretched, as wishing "may you have your heart's
        desire". The reverse side of the Ju-i is superbly finished
        in feathery filigree. The 8 Buddhistic symbols of good
        luck are delicately designed and executed; the three large
        pieces of jeweled turquoise-matrix used in this particular
        piece are of unusually fine quality and polish. 210 grams

No. 16- Ink Bed: This is a most interesting block of uncut tur-
quoise-matrix, highly polished, and simply mounted on a gold
throne. The reverse side is beautifully decorated with the
Imperial chrysanthemum design. 220 grams

No. 17- Waterholder, with ladle, Stand and Cover separate: Cracked
ice and hawthorne patterns, delicately designed, with inlaid
turquoise and jeweled lapis lazuli. Exceptionally fine
example of workmanship. 200 grams

No. 18- Incense Burner: Open work done in finely-matched jeweled
turquoise and jeweled lapis lazuli. This is a symphony in
color and an exceptionally fine example of craftsmanship.
Decorated with the Eight Buddhistic Emblems of Happy Augury
in lapis lazuli. 280 grams

233

中国的穆罕默德青铜器

ARS ISLAMICA

PUBLISHED SEMI-ANNUALLY BY THE RESEARCH SEMINARY
IN ISLAMIC ART · DIVISION OF FINE ARTS · UNIVERSITY
OF MICHIGAN AND THE DETROIT INSTITUTE OF ARTS

MCMXXXIV

VOLUME I

UNIVERSITY OF MICHIGAN PRESS

ANN ARBOR

CHINESE MUḤAMMEDAN BRONZES   BY BERTHOLD LAUFER

WITH A STUDY OF THE ARABIC INSCRIPTIONS BY MARTIN SPRENGLING

During my travels in China (1901–04, 1908–10) I was always interested in the Muḥammedan population and made it a point to collect any objects that are characteristic of Muḥammedan life and culture and that are apt to distinguish the followers of Islam from the surrounding Chinese. I had occasion to visit the mosques of Peking, T'ai-yüan, T'ai-an, Ho-nan, Si-an, Hang-chou, Ch'eng-tu, and others, and had rubbings made of the inscription stones bearing on the history of these mosques, both in Arabic and Chinese,[1] and collected numerous Islamic books in Arabic, Chinese, and Chinese-Arabic. I likewise made a collection of such literature for the Newberry Library, Chicago.

The fact that the Muḥammedans of China, aside from their religion, are completely sinicized becomes patent to the most casual observer. They speak and write the Chinese language; only the Ākhūns and Mollās have a certain knowledge of Arabic, which I suspect is not very profound. Both men and women have adopted Chinese dress; only in Kansu Province did I notice a black veil worn by Muḥammedan women when appearing in public. They take the same food as their neighbors, save that they abstain from pork and do not eat together with infidels or anything cooked by them. Rosaries, prayer caps embroidered in gold and silver threads with Arabic aphorisms, and bronze vessels with and without Arabic inscriptions are the essential feature of my collection.

Much has been written about the history of the Muḥammedans in China, some of their inscriptions also have been translated, but much critical work remains to be done. Muḥammedan art, however, has but incidentally been treated, presumably for lack of good material which is not easy to procure; a great deal of it was undoubtedly destroyed during the formidable Muḥammedan rebellions in the nineteenth century (1855–73 and 1861–77) and presumably even earlier during the two Muḥammedan insurrections in Kansu (1648–49, 1781–83).[2] Much may still be retained in the possession of Muḥammedan families which are loath to part with their heirlooms. During the period of the Ming dynasty, notably in the fifteenth and sixteenth centuries, Muḥammedan art was in a flourishing condition, and

---

[1] M. Broomhall, *Islam in China*, 1910, p. XIII, asserts that the Rev. F. Madeley discovered the earliest Islamic inscription dated A.D. 742, hitherto not seen by any European or American. Again, on the plate reproducing a rubbing of this inscription, he parades the announcement that "it has never been found by any European before." The fact is that when I lived at Si-an fu in 1902 and was on the most friendly terms with the Muḥammedans residing in the city I was the first who actually saw and examined this inscription stone and had rubbings made of it as well as of all other inscriptions found in this and the second mosque of Si-an. My account was embodied by E. H. Parker in his article "Islam in China," *Asiatic Quarterly Review*, 1908; cf. also my report in *Anzeiger der phil.-hist. Klasse der Wiener Akademie*, 1905, No. 2. I was also the first who prepared a critical translation of this inscription based on the stone record, while the previous translation of Devéria was made from the unreliable text as printed in the *T'ien fang chi sheng shi lu*, Chap. 20, pp. 7–8.

[2] Cf. C. Imbault Huart, "Deux insurrections des Mahométans du Kan-sou," *Journal Asiatique*, 1890.

it was still active in the course of the eighteenth century, but the scarcity of the material that at present is at our disposal prevents us from evaluating the extent and effect of this art properly. It seems advisable to defer judgment and to refrain from generalized conclusions until all accessible material in our museums and private collections has been published. To encourage others to make known what they have and what they know is the main scope of this article.

Another reason why in the present state of our knowledge it is difficult to determine what the Muḥammedans have contributed to Chinese art, and to culture in general, is that they have always worked quietly, noiselessly, unostentatiously and have even purposely eschewed any sensations and publicity.[3] Buddhism always understood the art and value of advertising and never tired of announcing *urbi et orbi* the benefits that would accrue to its votaries. We are clearly conscious of the contributions made by Buddhism to the civilization of China, but who can give a positive answer when the question is raised as to what Islam has contributed? Muḥammedan literature fails us in this respect. Muḥammedans themselves cannot enlighten us on this point. We have to work out the facts for ourselves. We have to study the art-crafts of the Ming in bronze, pottery, enamels, glass, and textiles to ferret out any possibilities of derivations from Muḥammedan sources of art forms or ornamentation.[4]

M. Paléologue, who wrote the first book on Chinese art,[5] was also the first to discuss Muḥammedan bronzes. He illustrates three of these with Arabic inscriptions, but without giving translations, from the collection of C. Schefer, then director of the Ecole des langues orientales vivantes of Paris, adding that they date in the first years of the fifteenth century as testified by the marks engraved on the bottom. Paléologue holds that three bronze pieces forming a set are used in the Muḥammedan cult—a box to hold incense sticks, a censer to burn them, and a vase for the bronze spatulas by means of which the incense is taken. This affair, however, is not peculiar to the Muḥammedans, but is generally Chinese, nor do the Muḥammedans, as far as I know, employ it in their cult, but merely for domestic purposes.

S. W. Bushell [6] has illustrated a bronze censer with Arabic writing and the date-mark Süan-te (1426–35). The information given by him on Islam in China is almost literally copied from Paléologue and contains nothing new save some errors not contained in his predecessor's

---

[3] Vasilyev, "Der Mohammedanismus in China," in his *Die Erschliessung Chinas* by R. Stübe, p. 100, says justly, "Islam never applied to Government with request for privileges; on the contrary, it appears to have always endeavored to be forgotten from time to time. Everywhere do we see the minarets rise above other buildings; in China, however, they disappear between the other one-storied houses."

[4] As to cultural objects I may refer to the Chinese water-pipe the origin of which remained obscure for a long time until I found a passage in a Chinese text to the effect that the water-pipe made its first appearance at Lan-chou, capital of Kansu Province, in the beginning of the eighteenth century, and came from there together with the finely shredded tobacco used in connection with this peculiar smoking apparatus, which took its origin in Persia. Since Kansu is densely populated by Muḥammedans, especially those who emigrated from Turkestān, I have concluded that Kansu was the home of the Chinese water-pipe which was spread all over Asia by Musulmans. See my booklet *Tobacco and Its Use in Asia*, 1924, pp. 26–28.

[5] *L'Art chinois*, 1887, pp. 69, 72, 73.

[6] *Chinese Art*, I, 1921, p. 57 and fig. 43.

book. To these belongs the assertion that "young Muslims in China are taught the elements of their religion from books printed in Chinese Turkestan"; the books seen and collected by me have been printed at Ch'eng-tu, Sechwan Province, and in other Chinese mosques. The mosques of Yünnan Province are well known to have been active in the production of literature. Bushell's reference to "Muslim inscriptions in debased Arabic" is not justifiable on the part of one who is not an Arabic scholar.

Bushell[7] has also illustrated two Chinese vases of opaque glass with Arabic scrolls, one with date-mark of the Yung-cheng period (1723–35).

The most recent writer on Islamic art in China is Professor P. Kahle of Bonn University[8] in a discussion of the *Khitā'ī Nāme* of 'Alī Akbar, which was first made known by C. Schefer. Kahle shows that 'Alī Akbar spent the years 1505 and 1506 in China and that according to his account Islam appears to have played an important part in the China of the Ming. The emperor Čīn Khwār (Hiao Tsung, 1488–1505) is said by him to have been strongly inclined toward Islam and to have had an entourage of noble Moslems serving him as officials. The emperor Ch'eng-te (1506–21) is even credited with having adopted Islam clandestinely or publicly—doubtless an exaggeration. This rumor probably arose from the fact that the emperor studied Arabic, but he also studied Sanskrit and Mongol. Many Moslems were found among the eunuchs whose influence was all-powerful and to whom 'Alī Akbar owes much of his information. His descriptions of eunuchism are very circumstantial; he reports in detail whence the eunuchs came, how they got into the palace, how they gradually advanced in rank until they exerted a far-reaching influence. 'Alī Akbar writes, "The fact that the country of China is well populated and enjoys welfare and safety is accounted for by the existence of these eunuchs who are the agents of the emperor of China and who are treated by him like his sons; most of these eunuchs are Moslems." Professor Kahle writes me that he is planning to publish the entire *Khitā'ī Nāme* in text and translation—a work in which all orientalists will rejoice. In the article in question Professor Kahle proceeds to describe and figure blue and white porcelains with the date-mark Ch'eng-te (1506–21), partly with Arabic inscriptions or Persian-Arabic designs. Many of these are found in the Saray of Istanbul, and were made for the imperial court in the kilns of King-te chen.

Professor Kahle has justly pointed out that under the Ming numerous eunuchs were Muḥammedans and held important positions. There were likewise many Muḥammedans, who were not eunuchs, employed in high offices. One example out of many may suffice. When Shāh Rokh's embassy (1419–22) arrived at the court of Peking, it was received and advised by Mawlānā Ḥādji Yusīf the kādī, who was one of the Amīrs of Tūmān (commander of ten thousand men), one of the officers attached to the Chinese monarch and who was at the head of one of the twelve imperial councils. He was accompanied by several Moslems versed in languages, who admonished the Persians to make the kotow before the emperor.[9]

[7] *Chinese Art*, II, p. 68, figs. 84–85.

[8] "Eine islamische Quelle über China um 1500," *Acta Orientalia*, XII, pp. 91–110.

[9] Quatremère, "Histoire des deux sultans Shah-Rokh et Abou-Said," *Notices et Extraits*, XIV, 1843, p. 405.

The flourishing period of Islam in China was in the age of the Ming dynasty (1368-1643) when Muḥammedans occupied high offices in the government service and the emperors took a friendly and sympathetic attitude toward Islam. Outwardly this finds its expression in the fact that the majority of Islamic inscriptions were composed in the Ming period; thus, the first mosque of Si-an fu, called Ts'ing chen se, shelters three inscribed stone tablets dated 1405, 1526, and 1575; the second mosque of Si-an, called Ts'ing kiao se, has an epigraphic monument dated 1545; that of Hangchow one of 1493, and so on. In 1392 the first emperor of the dynasty, Hung-wu, issued an edict in favor of Islam, granting permission to found two mosques, one at Nanking and one at Si-an and to repair mosques whenever threatened with ruin; he also conceded to Muḥammedans the right to settle, travel, and trade in any part of the empire. This liberal and tolerant imperial attitude naturally resulted also in a high development of Muḥammedan art in which, as will be seen, the emperors themselves seem to have taken an interest.

All objects described in this article were obtained by myself at Si-an fu in 1908-9 during the Blackstone Expedition to China on behalf of Field Museum, Chicago. All photographs are due to the courtesy of Field Museum.

While there are Muḥammedan bronzes with the Süan-te date-mark (1426-35), as pointed out by Paléologue, and while there are others devoid of this date-mark, which, however, may be assigned to this period because of the technical character of the bronze, I am in a position to describe two censers (*Figs. 1-4*) which I have reason to believe are unique in that they are inscribed with a definite year in the Süan-te period and, more than that, bear each an inscription of sixteen characters of great historical significance. These two censers were acquired by me for Field Museum, Chicago, at Si-an fu, capital of Shensi Province, in 1908 from Mr. Su, a Muḥammedan, who at that time was the most prominent antique dealer of that city. The commerce in works of art then was entirely in the hands of Muḥammedans, and but for them our museums would be without Chinese art.

The two tripod censers are identical in shape, as may be seen from the illustrations, and were produced in the imperial foundry of Peking in the years 1430 and 1431, respectively. They are of exceptionally fine workmanship. One of the censers (Cat. No. 117602; 18x16.5 cm.; 11.2 cm. high), dated 1430, two views of which are shown in Figs. 1a and 1b, is coated with a lustrous brownish polish said to be characteristic of the Süan-te period, which is celebrated for its artistic work in bronze, and is entirely plain save three countersunk medallions or panels filled with Arabic script in flat relief.

Professor Sprengling, to whom I am very grateful for his kindly cooperation,[10] reads this inscription as follows:

'Afḍalu al-dhikr | lā'ilāha 'illā Allāhu | Muḥammadu rasūlu Allāh.

"The most excellent of confessional invocations (is): | There is no god whoever beside Allāh; | Muḥammed is the apostle of God."

---

[10] Arabic inscriptions on Chinese porcelains, to my knowledge, have not yet been translated. A blue and white porcelain vase with designs of Persian style and with an Arabic inscription is illustrated in the book of O. du Sartel, *La Porcelaine de Chine,* plate XIX, No. 91.

FIG. 3—CHINESE BRONZE TRIPOD CENSER, DATED 1431

FIG. 4—LOWER SIDE OF CENSER IN FIG. 3, INSCRIPTION

FIG. 5—CHINESE MUḤAMMEDAN BRASS PLATE

FIG. 6—CHINESE MUḤAMMEDAN COPPER PLATE

Fig. 2 shows the inscription cast on the lower side of the vessel, the sixteen characters being traced in an archaic style of seal script in four rows, four in each row, arranged in a square. The reading of these ancient forms is greatly facilitated by the analogous inscription in the other censer (*Fig. 3*), which is in modern style. Transcribed in modern form the inscription is as follows:

大 五 工 吳
明 年 部 邦
宣 監 官 佐
德 督 臣 造

"Made in the fifth year of the period Süan-te (A.D. 1430) of the great Ming dynasty by his majesty's servant Wu Pang-tso, who held office in the Board of Public Works with the title of superintendent or director (*kien-tu*)."

Before discussing the contents of this inscription it is desirable to examine the seal on the dragon censer (*Fig. 4*), which reads as follows:

"Made in the sixth year of the period Süan-te (A.D. 1431) of the great Ming dynasty under the supervision of his majesty's servant Wu Pang-tso, president of the Board of Public Works (*Kung pu shang shu*)."

We see that Wu Pang-tso, within a year or possibly less, had been promoted in 1431 to one of the highest offices in the capital, to the presidency of the Ministry of Works, while in the preceding year he was still holding a subordinate position in the same ministry. The fact that Wu Pang-tso had these two censers made for the imperial court becomes evident from the word *ch'en* ("subject, servant, minister of state") prefixed to his name and in smaller size in the second inscription. *Ch'en* was the word by which an official in addressing the emperor designated himself, so that in many cases we may simply translate it with the personal pronoun. It is well known that the court painters, whenever a certain picture was ordered by his majesty or when for certain reasons a picture was intended for him, signed it with their name preceded by the word *ch'en*.

The two pairs of five-clawed dragons represented in relief on the censer in Fig. 3 (Cat. No. 117601; 20.1 x 17.7 cm.; 9.6 cm. high)—there are three smaller four-clawed dragons on the bottom of the bowl and a pair on each handle, altogether eleven dragons on the entire vessel—likewise speak in favor of imperial patronage; for the dragon was the coat of arms of the house of Ming, and the one with five claws was an imperial prerogative. The flaming jewel or pearl is placed between the beards of two dragons and rests on a cloud pattern held between their claws—as far as I know, a rather unusual conception; thus it is the cloud and thunder dragon which is here intended, a symbol of fertilizing showers. It is conceivable that Wu Pang-tso had this vessel designed and cast in commemoration of his new appointment as president of the Board of Public Works and transmitted it to the sovereign as a token of his appreciation and gratitude. Unquestionably Wu was a Moslem, otherwise it would be unintelligible why he should have sent to the throne a censer adorned with an Arabic tenet, and just one expressing a cardinal doctrine of Islam. What was the motive prompting him to this act? We may assume that it was calculated either to render the monarch favorably disposed toward Islam, or if he was so inclined, to strengthen his sympathies and to give him a testimonial in return for favors he might have shown the cause of Islam. Or the case may be much simpler: Wu Pang-tso may have been actuated by the desire to impress upon his sovereign the fact that the Muḥammedans also possessed an art and a writing capable of ornamental treatment no less than Chinese; in other words he desired merely to present his lord with a specimen of Arabic calligraphy, with an example of Moslem art. This would by no means have been an unprecedented case; we need not invoke here Chinese tolerance in matters of religion, but we may emphasize their curiosity about exotics of all sorts such as plants, animals, minerals, strange foreigners, their manners, customs, and tales, and what is there that could not be found in China? There is nothing odd or amazing about the fact that a censer with an Arabic maxim should have found its way into the imperial palace which was a museum-like storehouse, where curiosities from all corners of the world were garnered.

The large plate of hammered brass (Cat. No. 117610; 76 cm. in diameter), illustrated in Fig. 5, is made in seven pieces, six forming the margin and one the central countersunk portion. The margin is engraved with an interlaced band filled with chrysanthemums and cloud patterns, the latter being enclosed in a crescent and surrounding a medallion which contains Arabic script. The center of the plate is occupied by an interlaced band arranged in the form of an eight-pointed star and laid around a circular band. Of the nine spaces thus made five are occupied by Arabic; the four small ones at the corners, with an acanthus-like pattern, which is decidedly Muḥammedan. The outer zone is decorated with an elaborate band of floral designs, evidently derived from textiles and probably an adaptation of Chinese decorative motives to Muḥammedan taste. This and the following plate, which were apparently produced by the same artisan, belong to the late Ming period (sixteenth to first part of the seventeenth century) and came to me from the possession of a wealthy Muḥammedan family of Si-an fu. All I could learn was that they were used for the decoration of walls in rooms.

The copper plate in Fig. 6 (Cat. No. 117609; 43 cm. in diameter), is adorned with the same interlaced star design as in the preceding plate, and has identical inscriptions. There are six five-clawed dragons engraved along the edges, one with head forward alternating with one with head turned backward. On the opposite side in the center there is a unicorn monster with the trunk of an elephant, the *makara* of India (*chu srin* of the Tibetans), which I have only encountered in Tibetan Buddhist art; attached to the monster is a conch-shell (Sanskrit *çaṅkha*, Tibetan *dung-dkar*). How this Indic motive came to be applied to this Islamic plate, which on the other side glorifies the Chinese dragon, I do not know, nor can I explain its significance in this connection.

Professor Sprengling has kindly supplied the following information on the inscriptions of the two plates:

"The copper plate and the brass plate are closely related to each other and, judging from the inscriptions, seem to be from the same workshop. The inscription or inscriptions on the sunken surface of No. 117610 are, excepting for minor variations in the forms or deformations of letters, exactly the same as the sum total of inscription on No. 117609. At this moment I have not the time to solve all of the intricacies inherent in a Chinese metal-worker's use of Arabic script for the ornamentation of his wares. The first word in the inner circle is most probably *Allāh*. The second word seems to be *walī*, "the patron of," perhaps *al-dīn*, "religion," or "judgment." In the upper left hand space we may read an attempt to write *ḥayāt* ("life"). The material in the four facets surrounding the inner circle is evidently intended for some such form of *ḥadīth* concerning the fundamental duties of a Moslem, as are found in 'Alā al-Dīn 'Alī al-Muttakī's *Kanz al-'Ummāl*,[11] *Ḳāla al-nabī a(laihi) (al-salā)m* in the right hand facet is perfectly clear. The next facet to the left of this one is extremely troublesome; I have found no satisfactory solution for it despite the fact that the following facets are again fairly clear. The third facet is probably to be read *wa-ṣūmū shahrakum;* the fourth: *wa hudjdjū baita rabbikum.* In English this would be: 'Said the prophet, hail to him' . . . (possibly *ṣalātī dhikrukum*, 'my prayer is your confessional invocation' 'and? . . . *al-zakāt?* 'the alms') . . . and fast your month and make pilgrimage to the house of your Lord.'

"On the rim of the large plate *al-ḥamdu lillāhi*, "praise to God," is repeated three times by itself, and once as the final portion of another phrase with *wa*, 'and' before it. At the very beginning of this lozenge is found the abbreviated form of 'hail to him,' found after the word 'prophet' in the first of the inner facets. This is odd, to say the least, as is the following involved and, to me, unintelligible word or words. In the facet just preceding this one may read *awwalan wa ākhiran*, 'first and last.' The facet or lozenge between two independent 'Praise to God' facets is at this writing quite insoluble to me; it seems to contain some reference, perhaps an injunction 'to us' (*lanā*), having to do with spending of money or wealth (*nafakā . . . al-māl?*). I am quite content to leave further, perhaps ultimate solution to some one else, who may be more intimately acquainted with the labyrinth of *ḥadīth*, or more fortunate in his search, or more expert in the Chinese manner of dealing with Arabic. Several of my assistants in collaboration with me have not in the time at their disposal been able to carry the matter further than here stated."

A circular brass box (Cat. No. 117605; 13.8 cm. in diameter and 5.4 cm. in height),

---

[11] Vol. III, p. 62, Nos. 1107-10.

Fig. 7, is decorated only on the top of the cover—in the center Arabic writing in flat relief laid around in a circle; in the outer zone there is a band of arabesques, the unit of design being repeated four times, the four units being divided by a flower with large spreading petals, and bird's heads being combined with the spirals. This object was acquired at Si-an fu in 1908, and is probably of the later Ming period. The inscription, according to Professor Sprengling, consists of four names, which read "Muḥammed, Maḥmūd, Ḥāmid, Aḥmed" and which are probably intended as variant names of the prophet Muḥammed.

Three flower-vases are grouped together in Fig. 8 (Cat. Nos. 120982, 117603, 117604; 16.5, 14.1, 17.3 cm. high, respectively), all presumably of the later Ming period. Each is provided with a pair of elegant loop-handles, and two of them are decorated with a neatly engraved band of floral designs. The first and third (Nos. 120982 and 117604), according to Professor Sprengling, contain two pious exclamations widely used in the Islamic world: *subḥān Allāh*, "glory to God," and *wal-ḥamdu lillāh*, "and praise to God." The vase in the center (No. 117603) bears the same inscription as the preceding brass box (No. 117605).

Two curiosa may be added here. Fig. 9 is an impression taken from a cast-iron seal in Chinese style with conical handle. The seal (Cat. No. 117606) is rectangular in shape, 6 x 3½ cm., 1.4 cm. thick; the handle is 5.2 cm. in height. Professor Sprengling reads it:

FIG. 9   FIG. 10

bi-mamnūnīyatī, "with my compliments." This seal was presumably stamped on packages containing presents sent to friends. Fig. 10 represents an impression taken from a wooden block engraved in Chinese fashion (Cat. No. 117608; 6.1 x 3 cm.). It was given me by a Chinese Muḥammedan with the explanation that it should contain a man's name to be printed on his visiting card which in the China of the Manchu dynasty was a rectangular sheet of paper dyed red. Professor Sprengling reads it, *al-salaam 'alaikum*, "hail to you!"—the commonest form of Moslem greeting. All Chinese Muḥammedans have Chinese names, and their name cards do not differ from those of the Chinese. It is conceivable that the stamp in question was added to the card of a Muḥammedan or imprinted on a separate sheet, so that in making a call on a coreligionist he identified himself as one of the faithful.

It seems that Muḥammedan art preferred expressing itself in Chinese forms to inventing new ones or perpetuating old ones inherited from Western Asia and acquiesced rather in the

FIG. 7—COVER OF CHINESE MUḤAMMEDAN BRASS BOX

FIG. 8—CHINESE MUḤAMMEDAN BRONZE VASES

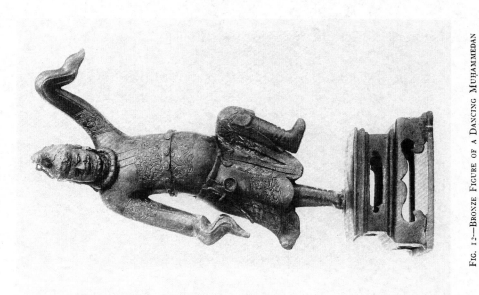

FIG. 12—BRONZE FIGURE OF A DANCING MUḤAMMEDAN

FIG. 11—CAST-IRON FIGURE OF A MUḤAMMEDAN

CHICAGO, FIELD MUSEUM

application of Arabic calligraphy to Chinese types of vessels of bronze, porcelain, and glass. In my estimation there is artistic merit to this idea, and even he who does not read Arabic must be struck with the beauty, the ornamental quality, and esthetic value of these scrolls and flourishes. At any rate it is a memorable fact that the two most calligraphic systems of the Orient—Chinese and Arabic—have met in a pleasing and peaceful rendez-vous on Chinese soil and that Arabic calligraphy has proved itself adaptable to Chinese art and acceptable to Chinese taste. I do not enter here into a discussion of the question as to whether new forms of vessels were introduced to the Chinese by Muḥammedans; there are such forms indeed, but the problem is complex; there are forms, for instance, inherited from Sasanian art by both Arabs and Chinese and subsequently perpetuated almost anywhere in Asia, and there are others whose history cannot yet be exactly traced. I hope to take up this problem some day when publishing some Chinese Muḥammedan bronzes which are not characterized by Arabic inscriptions.

On the other hand, it is noteworthy that the presence on Chinese soil of a considerable Muḥammedan population gave rise in Chinese art to a figure type which we may tersely dub "the Muḥammedan." This type is a creation of the Ming period. As is well known, the T'ang artists, with their love of the exotic, were fond of representing foreign nations, especially the inhabitants of Central Asia, in both painting and sculpture (pictures of tribute-bearing nations, *kung chi t'u*, and clay figurines), but the T'ang artists were chiefly interested in racial types, and in naturalistic manner stressed peculiar physical characters which were prominently brought out, sometimes even exaggerated. In Ming art the emphasis on physical type vanishes behind the interest in costume and action. We meet a turbaned man hailing from somewhere in Central Asia, dancing, kneeling, squatting, usually carrying something on his head, without being able to tell whether he is an Iranian, Turk, or what; but according to Chinese definition he is merely a Hui-hui, a Muḥammedan. Here again, the idea of tribute-bearer is still in evidence, especially in connection with lions, e.g. a Muḥammedan astride a lion and carrying a candlestick on his head.

A cast-iron figure is shown in Fig. 11 (Cat. No. 120256; 56 cm. high). A well-modeled nude man from Central Asia is kneeling on a circular base, with the function of a caryatid and supporting on his head and hands a tray on which a lamp was placed. The lamp was lost when I obtained this object, but it is referred to in the inscription cast in raised characters on the base. It appears from this inscription that this figure was placed in a Buddhist temple in front of a statue representing the Bodhisatva Kshitigarbha and was cast by an iron-founder, Ch'en Ying-kü by name, in A.D. 1618 (46th year of the period Wan-li).

The Ming bronze figure of a dancing Muḥammedan (Cat. No. 117697; 38.1 cm. high) is illustrated in Fig. 12. He is equipped with a hood terminating in a lion's head, and wears ear-rings, a necklace, and a belt. He is clad in a long, embroidered coat with long sleeves; a short sword and a bowl are suspended from his girdle. A Muḥammedan is guiding a lion on which the Bodhisatva Avalokiteçvara is seated—a unique conception in bronze (*Fig. 13;* Cat. No. 120179; 29.8 cm. high). It is reminiscent of the fact that live lions were frequently sent as gifts to the Chinese court by rulers of Persia and Central Asia, being transported overland. The conception is the same as that of a Muḥammedan leading a giraffe.[12]

Three small clay figurines of Muḥammedans (*Fig. 14*) are selected from a large number (Cat. Nos. 119946, 119947, 119943; 9.1, 10.1, 10.2 cm. high). These were used as candle holders, being hollow and having a vertical perforation run through their heads into which the stick or straw supporting the candle was inserted. They were formed in molds and apparently turned out in large numbers. Two of the Muḥammedans play the lute called *p'i-p'a*, an instrument of Central-Asiatic origin. The third is shown in the pose of a Bodhisatva with bare, protruding belly which he duly emphasizes by petting it with his right hand. He is equipped with a conical hat, a broad nose, and a generous moustache.

The preceding representations of Muḥammedans are all works of the Ming period. I could illustrate many more examples of this kind, but those given will suffice to demonstrate the fact that Muḥammedans played a prominent role in the life and culture of China in the age of the Ming dynasty.

[12] Cf. my booklet *The Giraffe in History and Art*, plate IV and fig. 13.

FIG. 1—SCHALE MIT RUBINLÜSTER, UM 850, PARIS, MUSÉE DU LOUVRE
(NACH M. PÉZARD)

FIG. 2—NAPF MIT KAIRUWĀN-LÜSTER, UM 860, BERLIN, STAATLICHE MUSEEN
(NACH FR. SARRE)

FIG. 13—CHINESE BUDDHIST BRONZE IMAGE
A LION GUIDED BY A MUḤAMMEDAN

FIG. 14—CHINESE CLAY FIGURINES OF MUḤAMMEDANS USED AS CANDLE-HOLDERS

CHICAGO, FIELD MUSEUM

234

远东的黑麦和跟我们的"Rye"相关的词来自亚洲的词源

通報

T'OUNG PAO

ARCHIVES

CONCERNANT L'HISTOIRE, LES LANGUES, LA GÉOGRAPHIE,
L'ETHNOGRAPHIE, ET LES ARTS DE L'ASIE ORIENTALE

REVUE DIRIGÉE PAR

PAUL PELLIOT

Membre de l'Institut
Professeur au Collège de France

VOL. XXXI

LEIDE
E. J. BRILL
1935

RYE IN THE FAR EAST AND THE ASIATIC ORIGIN OF OUR WORD SERIES "RYE"

BY

Berthold LAUFER.

I

All that has heretofore been known about rye in sinological literature is limited to two brief notes of E. Bretschneider. "The rye (*Secale cereale*), as far as I know, is nowhere cultivated in China. M. Perny, however, in his Dictionnaire français-latin-chinois (article production) mentions rye (*seigle*) as a product of China. I am very curious to know where he found rye" (*Chinese Recorder*, 1871, p. 225). Again, he writes (*ibid.*, p. 286), "I have expressed some doubt whether rye occurs in the Chinese dominions. Since writing this I read an article of Mr. Simon (Carte agricole de la Chine, *Journal North China Branch Royal Asiatic Society*, No. 4), in which he states that rye is cultivated in the province of Shensi. He does not say whether he speaks from his own observation; he does not give the Chinese name of the plant. It was in vain that I looked through Chinese works to make out a cereal which could be identified with rye. But perhaps the *Hei-lung-kiang mai* (wheat from the Black Dragon River, Amur), mentioned in the Memoirs of the emperor K'ang-hi (quoted in the *Shou shi t'ung k'ao*, chap. 26, p. 10) refers to rye. It is said there that this kind of corn was brought from Ao-lo-se (Russia). Rye is largely cultivated in Siberia."

In his Botanicon sinicum (I, pp. 84—86) Bretschneider has inserted a notice of the *Shou shi t'ung k'ao*, which was published in 1742, and in enumerating the plants treated in this work points out "rye (from Russia)". I shall come back to the text of the *Shou shi t'ung k'ao* below, and it will be shown that the cereal referred to by Bretschneider cannot be determined as rye.

Two years after Bretschneider, J. H. Plath (Landwirtschaft der Chinesen, *Sitzber. Bayer. Akad.*, 1873, p. 781) wrote, "Of cereals, rice, wheat, rye, barley, millet, etc., are cultivated". Again, "rye and maize from the north occur but rarely in Canton". The source of Plath's statement is an article in the *Chinese Repository* (III, 1835, pp. 457—471) entitled "Diet of the Chinese" by an anonymous author, who writes, "The grains which are cultivated include all those used for food, as rice, wheat, rye, etc., but the extent to which they are grown varies" (p. 458). "Rye and maize, at certain seasons of the year, are found in the market of Canton; but their use is mostly limited to the parts of the country where they are raised" (p. 459). Although rye is principally a cereal of temperate climes, it can be grown (and is grown) in warm and even tropical regions (e.g. in Brazil and in the Dutch Indies); it would therefore not be impossible that rye occurs here and there in southern China, but thus far this has not been confirmed.

Most writers on Chinese agriculture pass rye over with silence as, for instance, W. Wagner (Die chinesische Landwirtschaft, 1926), while others deny categorically its occurrence in China; thus, F. Koernicke (Arten und Varietäten des Getreides, 1885, p. 125), J. Hoops (Waldbäume und Kulturpflanzen, 1905, p. 447), T. H.

Engelbrecht (Festschrift Eduard Hahn, 1917, p. 18, who regards the origin of rye culture in Turkestan as highly improbable on the ground that it would have spread from there only to the west, not, however, to China where rye is unknown), and even N. I. Vavilov (On the Origin of the Cultivated Rye, *Bull. of Applied Botany* [in Russian], 1917, pp. 570, 588), who states positively that "rye is not cultivated in India, China, or Egypt"; thus likewise in his "Origin of Cultivated Plants" (Leningrad, 1926, p. 200): "It is cultivated neither in Syria and Palestine, nor in India and China"[1]).

A. Schulz (Geschichte der kultivierten Getreide, 1913, I, p. 83) asserts that "aside from Siberia rye is cultivated in Asia in Japan, Korea, Armenia, and Asia Minor, but nowhere, it seems, to a great extent", without further specifications or citations of sources.

As regards Japan, J. Matsumura, the eminent Japanese botanist (Shokubutsu-mei-i, pt. 2, p. 369), gives two Japanese terms for *Secale cereale -- natsu-komugi* and *lai-mugi* (in Kana only). The former apparently represents 夏 (*natsu*) 小麥 (*ko-mugi*), literally "summer wheat", while *lai* in the term *lai-mugi* is simply our "rye"; in other words, the two terms are learned productions of present-day Japanese botanists who are certainly familiar with *Secale* as a plant, but do not go to prove that rye was (or is) actually cultivated in Japan.

In lexicographical literature rye figures as *hadaka-mugi* 裸麥 (lit. "naked wheat"), colloquially simply *mugi* ("wheat") and *mugi-pan* ("rye-bread"). *Hadaka-mugi* with this meaning is entered

---

1) Its non-occurrence in India and Egypt is correct. Watt (Dictionary of the Economic Products of India, VI, pt. 2, p. 495) states that rye does not appear to exist in India to any extent either wild or cultivated, but grows wild in Afghanistan, where its grain is reaped with the wheat and ground up with it into flour. Pickering (Chronological History of Plants, 1879, p. 513): "Has not been met with in Hindustan by European observers." No book on the agriculture of India makes mention of rye.

in the Japanese-English Dictionaries of Hepburn, of Inouye, and of Nitobe and Takakusu, the last-named adding the botanical term *Secale cereale*. Above all, it is contained in the official publications of the Japanese Government with the French translation "seigle". The Résumé statistique de l'Empire du Japon (Tokio, 1899, p. 18) gives for the year 1897 an area of .651, 448. 5 *cho* 町 cultivated with rye, while in the same year wheat occupied only 458, 239. 2 *cho* and barley 639, 884 *cho*. The production of rye therefore would seem to be considerable in Japan, even considerable to such an extent that it must arouse suspicion. K. Rathgen (Japans Volkswirtschaft und Staatshaushalt, 1891, p. 325) states that *hadaka-mugi* means "naked barley" and that "rye does not exist in Japan, while oats (*karasu-mugi*) is very insignificant in quantity". M. Fesca also (Die landwirtschaftlichen Verhältnisse der Kai-Provinz, MDGO, IV, 1886, p. 174) renders *hadaka-mugi* by "nackte Gerste" (*Hordeum nudum*). The rendering of *hadaka-mugi* as "naked barley" is doubtless correct, since there is no such thing as "naked rye", but there is a naked or beardless barley. The Encyclopaedia Japonica gives as translation of 裸麥 "barley" (*Hordeum sativum*). This is half correct; it ought to be "naked barley" (*Hordeum gymnodistichum*) [1]. According to the Encyclopaedia, this cereal is cultivated chiefly in the western districts Kansai 關西 including Shikoku 四國 and Kyūshū 九州; but there is no genuine Japanese term for rye. Rye is grown

---

[1] In *T'oung Pao*, 1916, p. 90, I have pointed out *ts'ing k'o* 青稞 as the Chinese term of this species of barley. In the district of Hien-yang in Shensi "the naked barley is called *ts'ing k'o*" 有露仁者爲青稞 (*T'u shu tsi ch'eng* XX, *ts'ao mu tien*, chap. 32, *mai pu hui k'ao*, p. 17b). Potanin (Tanguto-Tibetan Borderland of China [in Russian], I, p. 357) writes that *ts'ing k'o* is cultivated by the Shirongol Mongols, rendering it "naked barley" (голый ячмень). [Le 漢回合璧 *Han-Houei ho-pi* (*circa* 1880) rend de même *ts'ing-k'ouo* en turkī par *yalāng arpa*, "orge nue", et j'ai noté aussi *yalāng ayāγ arpa*, "orge à pied nu". — P.P.]

only in small and limited areas in the northern part of Japan, the name given in the Encyclopaedia being ライ麥 *rai-mugi* (*Secale cereale*), i.e. *rai*-wheat, *rai* being a reproduction of our rye. When and under what circumstances it was introduced into Japan seems not to be known, at least is not stated in the Encyclopaedia; but, judging from the adoption of the English name, the question seems to be of a more or less recent introduction into Japan from the United States.

As to Korea, we have trustworthy evidence from a highly trained agronomist and eye-witness, F. H. King, formerly professor of agricultural physics in the University of Wisconsin, who in his book "Farmers of Forty Centuries or Permanent Agriculture in China, Korea and Japan" (1911, p. 367) informs us, "Here in Korea, too, as in China and Manchuria, nearly all crops are planted in rows, including the cereals, such as wheat, rye, barley, and oats". The cultivation of rye in Korea has been confirmed to me by a number of Korean students in Chicago, who give as name for it *pĭ* 稗 (Chinese *pai*) and say that it is cultivated only in isolated mountain regions. The term *pai-tse* 稗子 in China refers to a wild-growing panic grass occurring as a weed in wheat, millet, and rice fields (*Panicum crus galli* L. — Bretschneider, Bot. sin., II, No. 352; Stuart, Chin. Mat. Med., p. 304; O. Franke, Jehol, p. 72; *Chi wu ming shi t'u k'ao*, chap. 2, p. 1, with woodcut of the plant; Shanghai Bot. Dict., p. 1226; Matsumura, Shokubutsu-mei-i, pt. 1, p. 250). K'ien-lung's Polyglot Dictionary (chap. 28, p. 44*b*) gives as equivalents Manchu *hife*, Mongol *usun khonok*, Tibetan *k're č°us-ma*. It will be shown below that *pai-tse* in the sense of rye is also used in Chinese documents.

Evidence has increased to discredit the opinion that rye is lacking in China. It occurs not only in Manchuria and Mongolia, but sporadically also in China proper. Here in the *T'oung Pao*

(1909, p. 608) Dr. A. F. Legendre (Le Far West chinois) stated that the Lolo, in addition to other cereals, cultivate rye (*seigle*). The China Year Book of 1923 (p. 204) contains the statement, "Oats are found in Mongolia, Kansu, Kweichow; rye, only in Kansu". Above all, we have the testimony of an eminent botanist, E. H. Wilson (A Naturalist in Western China, I, 1913, p. 58), who noted in the mountainous region of eastern Sechwan "wheat, rye (*Secale fragile*), Irish potato, maize, and pulse". Again, in a chapter devoted to the agriculture of the province of Sechwan (II, p. 53) he writes, "In the mountains rye (*Secale fragile*) is sparingly grown and the grain eaten". The same information is given in the author's book "China Mother of Gardens" (1929, pp. 61, 335). P. Klautke (Nutzpflanzen und Nutztiere Chinas, Hannover, 1922, p. 23) writes, "Rye (*Secale fragile*) is cultivated only here and there in the mountainous regions of the west [of China], and its grain is used as food. The Chinaman esteems rye but little and strangely believes that it is unfit as an article of food"[1]. Pliny (XVIII, 141) calls rye flour ingratissimum ventri, and the belief that it is injurious to health still prevails in Greece.

This sporadic distribution of rye in isolated spots of China, in my estimation, presents an interesting scientific problem worth investigating. While the cultivation of rye in Manchuria and Mongolia is partly due to Russian influence, as expressly conceded in our source material, the Russians cannot be held responsible for the isolated phenomena in the mountainous regions of Sechwan, Kansu, and Shensi, which I shall try to explain in a different way. These various points will be taken up in proper order.

Bretschneider, as noted above, was not correct in recognizing

---

1) The author was an instructor in biology at the Tung Chi Medical and Engineering School for Chinese at Wusung. His book is valuable, being mainly the result of personal observations.

rye in the "wheat of Hei-lung-kiang", thus dubbed by the emperor K'ang-hi (1662—1722). The statement made by the latter in the *Shou shi t'ung k'ao* (chap. 26, p. 3*b*, ed. 1902) is as follows [1]: "The wheat produced in Hei-lung-kiang is very excellent and pure white in color; its nature also is beneficial to man. According to a tradition, the best of Chinese wheats was brought to China along from the western region 西域. The territory of Russia is more than ten thousand *li* distant from our western frontier, and the upper course of the Amur (Hei-lung-kiang) originally belonged to Russia. Thus, this cultivation has come from the west, being the finest of cereals, superior to those of other places". A woodcut entitled 黑龍江麥 ("wheat of Hei-lung-kiang") accompanies the text; it is exceedingly crude and looks somewhat like a species of millet; it may be anything, but does not represent rye.

The characterization "pure white" in the above text can only refer to wheat, not, however, to rye. The "wheat of Hei-lung-kiang" is simply what its name implies, a variety of wheat introduced by the Russians and made known to the Chinese from the province Hei-lung-kiang. There are hundreds of types and varieties of wheat and a great number of these cultivated in Russia (cf. M. A. Carleton, Russian Cereals, U. S. Dep. of Agr. Bull. 23, 1900, pp. 12—20).

The wheaten flour of Aihun 艾渾 (Aigun) on the Amur, opposite Blagoveshchensk, is praised by Fang Shi-tsi (*Lung sha ki lio*, ed. of *Siao fang hu chai* I, p. 377, see below) as sweet and fragrant and surpassing the flour of China; it renders cakes loose and pleasant of taste. This also must be a Russian variety

---

[1] This text is not included in the Observations de physique et d'histoire naturelle de l'empereur Kang-hi, inserted in the Mémoires concernant les Chinois, IV, 1779, pp, 452—483.

of wheat, in all probability the so-called Polish wheat (*Triticum polonicum*), the heads of which are of enormous size, the grains when perfect being very long, also hard and vitreous, in Europe used mainly for macaroni, occasionally for bread.

Sü Tsung-liang 徐宗亮, in his *Hei lung kiang shu lio* 黑龍江述略 (author's preface dated 1889; preface by Li Hung-chang 1890, printed 1891, chap. 4, pp. 2*b*, 3*a*), states that "forty bags of wheaten flour" were part of the annual tribute or taxes sent from the province Hei-lung-kiang to the court of the Manchu dynasty, and goes on to explain that this wheaten flour is very excellent and that according to an old tradition this cultivation has come from Russia long ago (進其中麥麵最佳。舊傳其種由俄羅斯國來已久). In another passage (chap. 6, p. 17*b*) Sü Tsung-liang writes that rice is not grown in the province, but is imported from Fung-t'ien 奉天 (Mukden) and that the principal food of the Manchu bannermen consists in millet and wheat mixed with beef or mutton.

The real Chinese designation for rye is *lao k'iang ku* 老羌穀, i.e. grain of the old K'iang (properly, designation of the ancient ancestors of the Tibetans; in this case, however, the Russians) and corrupted into 老鎗穀 *lao ts'iang* ("musket, gun, rifle") *ku*. The Manchu Si-ts'ing 西清, in his *Hei lung kiang wai ki* 黑龍江外紀 (ed. of *Siao fang hu chai*, I, p. 406)[1]), speaks of "a cereal from abroad, more than a foot tall, like blood in color, called grain of the old K'iang". Fang Shi-tsi 方式濟, in his *Lung sha ki lio* 龍沙紀略 (*ibid.*, p. 377), with reference to the second name given above, describes rye thus: "Its stalks and leaves are like cockscomb (*Celosia cristata*) more than ten

---

1) The original edition of this work in two volumes was procured by me for the library of the American Museum in New York. The author's preface is dated 1810 (Kia-k'ing 15th year).

feet high. The fruits are like those of the coir-palm *ping-lü* (*Chamaerops excelsa*). The seeds are deep red in color. The grain makes a fragrant and pleasant gruel". The same author describes also the Russian cabbage (*O-lo-se sung* 俄羅斯菘) under the name 老鎗菜 *lao ts'iang ts'ai* ("vegetable of the old guns"); others call it 老羌白菜 *lao k'iang pai ts'ai* ("cabbage of the old K'iang"); see the *Pei k'iao fang wu k'ao* cited below.

Sü Tsung-liang (*Hei lung kiang shu lio*, chap. 6, p. 25) makes the following statement: "The Russians in their origin are the survivals of a tribe of the K'iang. In every town of Hei-lung-kiang there are 'Old K'iang melons', 'Old K'iang cabbage', 'Old K'iang baskets' [1]) — all names for articles which have come from Russian territory. The character *ts'iang* 鎗 is an erroneous substitution for *k'iang* 羌 (俄本羌種之遺。各城有老羌瓜老羌菜老羌斗諸名皆自俄來者訛羌爲鎗). He says also that the Russian paper-notes 俄帖 used in the trade of the Chinese with the Russians on the Amur are called *K'iang t'ie* 羌帖, that he does not know at what time these *K'iang t'ie* sprang into circulation, but that in his time they were legal tender in all territories of Hei-lung-kiang, the Russian paper being tough and durable and no counterfeit bills being in circulation.

The name Lao K'iang for the Russians appears as early as the beginning of the Shun-chi period, i.e. the beginning of the Manchu dynasty, being mentioned in the *Ning-ku-t'a ki lio* 甯

---

1) *Lao k'iang tou* 老羌斗. These are explained in the *Hei lung kiang wai ki* (p. 407*b*) as vessels made from birch-bark 樺皮斗, also used by the Russians; the smallest are skilfully carved with ornaments and are used (by the Chinese) for storing betel-nuts and snuff, being called *lao k'iang tou*. They were made by the aboriginal tribes (for illustrations see Laufer, The Decorative Art of the Amur Tribes, plates XIX—XXI) and traded by the Russians to the Chinese. The name *tou* (lit. "grain measure, peck") is quite appropriate, as many of these baskets resemble in shape a Chinese *tou*. In regard to ancient trade in birch-bark see Sino-Iranica, p. 552.

古塔紀略 written in 1721 by Wu Chen-ch'en 吳振臣
(*Siao fang hu chai* I, p. 344b; also in *Hwang ch'ao fan shu yü ti ts'ung shu*), who lived at Ninguta in the Shun-chi period (1644—61) and also uses the curious transcription 邏車國 Lo-ch'e for Russia (instead of Lo-ch'a, as below).

The use of the tribal name K'iang for the Russians, of course, has no historical basis; it is merely a popular nickname, and it seems restricted to compounds designating certain Russian agricultural products. While the designation K'iang is not especially flattering, it is by no means offensive, nor do I believe that any offence is intended with it. It is somewhat surprising that this meaning of K'iang has remained unnoticed by the eminent Russian sinologist Palladius, who has not registered it in his Chinese-Russian Dictionary, and I believe that it is pointed out here for the first time in our sinological literature. The term Lo-ch'a 羅 剎 (transcribing Sanskrit Rākṣa, a class of men-devouring demons), however, in its application to Russia, certainly is offensive. This term appears in the beginning of the Shun-chi period (1644—61) — see *Ta Ts'ing i t'ung chi* [1]), chap. 423, § 7, p. 1 — chiefly in official documents, and was still used in the K'ang-hi period (1662—1722); many examples may be found in the *Ni pu ch'u* [Nerchinsk] *k'ao* 尼布楚考 by Ho Ts'iu-t'ao 何秋濤, in

---

1) I hope that my learned friend Pelliot with his stupendous knowledge of Chinese bibliography will give us some day a detailed bibliographical history of the official Geography of the Ts'ing Dynasty. Wylie (Notes, p. 43) says but vaguely, "First published about the middle of last century". His predecessor, W. Schott (Entwurf einer Beschreibung der chin. Lit., p. 81), remarks tersely, "Appeared 1744". A. Vissière (*J. As.*, 1901, sept.-oct., p. 323) speaks of "éditions de 1743 à 1784", while Wieger (La Chine à travers les âges, p. 524) gives the date 1764 (evidently the date of the second edition). M. Courant (Cat. des livres chinois, I, p. 92) says, "Edited at the end of the K'ang-hi period by a commission of officials; imperial preface, 1744; other edition with imperial decree, 1764". However, Giles (Cat. of the Wade Collection, p. 83) assigns the date 1745 to K'ien-lung's preface. I myself (Notes on Turquois, p. 65) said, "First printed 1745, second edition 1764". I shall abide by Pelliot's verdict.

Siao fang hu chai, I, pp. 424—430. Among the native tribes of Siberia and the Amur region the Russians are known as Luča, Luči, Lutsa, or Lutse — a term discussed in detail by L. von Schrenck (Reisen und Forschungen im Amur-Lande, III, 194—195), who holds that it is based on Rosiya, ruskiy. This problem merits a renewed and more profound investigation (see also Klaproth, Mém. relatifs à l'Asie, p. 453)[1].

Fang Shi-tsi (*Lung sha ki lio*, p. 370*b*) holds that Lo-ch'a is the correct name and that both 老鎗 and 老羌 are erroneous. Ho Ts'iu-t'ao (*Ni pu ch'u k'ao*, p. 427), with reference to this passage, contradicts him; while admitting that 老鎗 is wrong, he holds that 老羌 is not wrong on the ground that the Wu-sun 烏孫 in their origin were a tribe of the K'iang. Unfortunately he fails to explain what the alleged kinship of the Wu-sun to the Russians should be. Sie Ts'i-shi 謝濟世, in his *Si pei yü ki* 西北域記 (p. 7*b*, in *Hwang ch'ao fan shu yü ti ts'ung shu*), gives Lo-li 羅利, wrong reading for 羅剎 Lo-ch'a, as another name of O-lo-se, adding that it is the ancient country Ting-ling 丁零.

It should be noted that in the seventeenth and eighteenth centuries the term O-lo-se 俄羅斯 (based on Mongol Oros) principally applied to Asiatic Russia, while European Russia was known as Moscovia (莫斯哥未亞 Mo-se-ko-wei-ya). The latter is placed in the *Ta Ts'ing i t'ung chi* (chap. 423, § 4, p. 2) under the heading Europe (Si yang 西羊), while O-lo-se occupies a separate section (chap. 423, § 7). As far as I know, the name Mo-se-ko-wei-ya with the same characters first appears in

---

1) [Je ne puis me défendre de penser que Lao-k'iang (avec *k* palatalisé en pro-nonciation moderne devant *i*), Lao-ts'iang, Lo-tch'ö et Lo-tch'a sont autant de trans-criptions de Luča et Luca (*c* = *ts*). — P. P.].

Giulio Aleni's 艾儒略 *Chi fang wai ki* 職方外紀 (chap. 2, p. 18), 1623.

The fact that the aforementioned term *lao k'iang ku* refers to rye can also be established by a process of elimination, for all other cereals grown in Manchuria and the adjacent Russian territory are accounted for in the books quoted and in others as well. Oats are referred to in the *Lung sha ki lio* (p. 376) under the name *kung mai* 穬麥 or *yen mai* 燕麥 with the colloquial designation *ling tang mai* 鈴鐺麥 (explained: "because the fruit hangs down like a bell"), as coming from Morken and solely used for feeding cattle and horses (in the *Hei lung kiang wai ki*, p. 406). Moreover, the fact is established that rye is grown in the Amur country. According to Grum-Gržimailo (Description of the Amur Province [in Russian], 1894, p. 534), the principal crops there are summer wheat, oats, and summer rye which occupies 26 per cent of the cultivated area (for more information see below, p. 250).

Ho Ts'iu-t'ao 何秋濤 is the author of a small treatise entitled *Pei kiao fang wu k'ao* 北徼方物考 (reprinted in *Siao fang hu chai*, III, pp. 217—223). This contains a classified list of products of the northern border states (plants, minerals, animals, and products of industry) chiefly extracted from the *I yü lu* (see below) and other recent books, occasionally also from the T'ang and Yüan annals. It is a useful compilation, although very uncritical and without any attempt at interpretation or identification. Under the heading *lao k'iang ku*, the author quotes the texts given above of the *Hei lung kiang wai ki* and the *Lung sha ki lio*, and adds that the *Sheng-king t'ung chi* 盛京通志 calls this cereal *Kao-li ku* 高麗穀 ("Korean grain") or simply *Kao-li* ("Korean"). Indeed, the statement occurs in the *Sheng-king t'ung chi* (chap. 27, p. 2*b*) that "the Korean grain is fiery red

in color resembling the cockscomb flower" (cf. Fang Shi-tsi, above
p. 244, who has the same simile) — and this can only refer to
rye. The "wheat of Hei-lung-kiang" is treated by Ho as a separate
item, merely with reference to the notice of K'ang-hi. In view
of the fact that rye has been reported from Korea (above, p. 241)
it is credible that rye was introduced from Korea into Sheng-king
Province of Manchuria. The above work contains also notices of
the Old K'iang or Russian cabbage and the Old K'iang melon
老羌瓜.

The *Ta Ts'ing i t'ung chi* (chap. 42, p. 6) alludes to the
"Korean grain" among the products of Fung-t'ien fu 奉天府
in the following passage: "Among the kinds of grain 穀 there
are those which bear the names Si-fan 西番 and Kao-li all
cultivated in Fung-t'ien fu and other places. Of buckwheat, oats,
and *pai* 稗 [i.e. rye] those cultivated in Hei-lung-kiang are
the best."

Yang Pin 楊賓, in his *Liu pien ki lio* 柳邊紀略 (*Siao
fang hu chai*, I, p. 359), enumerates ten cereals as being cultivated
in the territory of Ninguta the first of which is 稗子 *pai-tse*,
obviously used in the sense of rye, followed by wheat, barley,
various species of millet, sorghum, buckwheat, and oats. Rye is
the most abundant of this series, according to Yang Pin, and on
account of its high price is only within the reach of wealthy
families. Of all cereals it calls for the highest price, 5 taels a
picul, while wheat is quoted at 3 taels, barley at $2^1/_2$ taels, millets
and sorghum at 2 taels, and oats at 1.3 taels a picul. In the
Ning-ku-t'a ki lio (p. 343*b*, see above, p. 246), written in 1721,
it is said, "Rye (*pai-tse*) is eaten by the better classes 貴人,
while the lower classes subsist on millet, on the ground that millet
gives strength, and do not drink tea". It is, of course, impossible
to assume that in the preceding cases *pai-tse* should refer to

Panicum crus galli, a wild Graminea with a bitter grain that is sometimes used in time of scarcity as a substitute for other cereals. The question in this case is of a cereal of high quality, priced more highly than wheat and therefore available only to the rich, and this can be no other grain than rye.

The Manchu Tulichen or Tulishen 圖理琛 (T'u-li-ch'en), in his diary *I yü lu* 異域錄 covering the period from 1712 to 1715, repeatedly enumerates the cereals and other plants cultivated in Siberia and Russia: wheat, barley, buckwheat, *yu mai* 油麥 (ed. of *Siao fang hu chai* III, p. 236*b*, thus likewise in *Ta Ts'ing i t'ung chi*, chap. 423, § 7, p. 2*b*), or barley, buckwheat, *yu mai*, and hemp (*ibid.*, pp. 238, 243); again, wheat, barley, buckwheat, *yu mai*, and beans (p. 245), and wheat, barley, buckwheat, *yu mai*, beans, and hemp (p. 246). The 油 in *yu mai* is a popular way of writing 莜 commonly signifying the naked oats (*Avena nuda inermis*), but since rye is the most important cereal of Russia, we cannot presume that it should have escaped the attention of a good observer of Tulichen's type. I am therefore led to the conclusion that the term *yu mai* in his account covers both oats and rye [1].

The Manchu do not sow rye because, as they say, "it is not profitable" (S. M. Shirokogoroff, Social Organization of the Manchu, p. 135). If this be correct, the Manchu must have made the acquaintance of rye somehow.

As to Hei-lung-kiang, the cultivation of rye in that territory

---

[1] The *I yü lu* has been translated into English by George Thomas Staunton under the title "Narrative of the Chinese Embassy to the Khan of the Tourgouth Tartars by the Chinese Ambassador" (London, 1821); the above passages on pp. 45, 109, 126, 131. There is also a Manchu translation of this book. For a full bibliography, etc., see G. Cahen, Histoire des relations de la Russie avec la Chine sous Pierre le Grand, 1911, pp. 115—133. A new, interesting treatise on Russo-Chinese relations is due to B. G. Kurtz, Russo-Chinese Relations in the 16th, 17th, and 18th Centuries (in Russian), State Publication of Ukraina, 1929.

is confirmed by J. Beckmann in his recent book "Heilungkiang, Land, Leute, Mission (Missionshaus Bethlehem, Immensee, Switzerland, 1932, p. 20), who enumerates soy bean, millet, kao-liang (sorghum), wheat, rice, barley, rye, hemp, and cotton as agricultural products of the province. I also notice from the same book (p. 18) that in the vernacular of the territory the term Lao-mao-tse ("Old Caps") is used for the Russians, said to be somewhat derogatory.

The following data are taken from the excellent treatise of V. I. Ogorodnikov, Native and Russian Agriculture on the Amur in the Seventeenth Century (in Russian, Memoirs of the State Far-Eastern University, Vladivostok, 1927). When the Russian conquerors reached the Amur about 1636, they encountered two powerful nations, called by them Dahurs or Daurians and Dyucheri, settled in the basin of the upper and middle Amur with a highly developed system of agriculture. In a report addressed to the czar by the voyevodes Golovin and Glebov in 1641 it is stated that "on the Shilka [a tributary of the Amur] there are many Daurian agriculturists who have every kind of bread fruit like the Russians and that the prince Ladkaya and some other princes on the Shilka raise cereals such as rye (rož), barley and some other seeds and that they send cereals for sale to the river Vitim where Tungusians buy them for sables, that bread is produced in quantity and that the people on the Shilka are sedentary agriculturists, but not warlike". Ogorodnikov (p. 31) arrives at the conclusion that in the seventeenth century Daurians and Dyucheri in various localities of the Amur basin cultivated rye, wheat, barley, oats, millet, buckwheat, hemp, and peas. Although Perfilyev (1658), Tomkanei, and the author of the "Descriptions of the Siberian Kingdom" testify to the occurrence of rye on the Amur, only one of Khabarov's companions, S. A. Andreyev (1652), affirms that "much

18

bread is produced in the Daurian country, but with the exception
of rye". In Ogorodnikov's opinion this statement refers not to the
entire area of aboriginal agriculture, but merely to the land of
the Dyucheri, especially those in the basin of the Sungari, where
according to the author of the "Descriptions of the Siberian
Kingdom" rye really did not grow, "because the soil was not
suitable for its cultivation"; and he thinks that possibly the cultivation of rye then was comparatively rare among the Dyucheri
of both the Amur and the Sungari. The Russian colonists of the
seventeenth century on the Amur cultivated rye, wheat, oats,
barley, buckwheat, hemp, and peas, i.e. almost the same plants
that were cultivated by the aborigines of the same territory and
all over Siberia generally. A document of the year 1682 mentions
summer rye, oats, and wheat. Winter rye, however, did not succeed
on account of the hardfrozen soil and severe winds, and after a
few failures the Russian settlers abandoned it entirely.

The fact that the Russian peasants when they began to colonize
the Amur country in the seventeenth century brought rye along
with them is thus well established. The fact that the Chinese
learned of rye from the Russians is evidenced by our Chinese
sources. I cannot decide the question as to whether the Daurians
and Dyucheri of the Russian accounts, as intimated by Ogorodnikov,
may have possessed rye prior to the advent of the Russians. This
would by no means be impossible, especially in view of Vavilov's
announcement that a native rye has been discovered in Mongolia
by Pisarev's expedition (see exact quotation below, p. 254). The
results of this expedition have not yet been published, or if so,
I have not yet seen them and must therefore reserve any conclusions until a later date.

In the middle of the nineteenth century the Russians were
still distributors of rye flour to the Gilyaks and probably other

native tribes of the Amur region. At the time of L. von Schrenck's travels (Reisen und Forschungen im Amur-Lande, 1854—56, III, p. 442) the Gilyak received rye flour from the Russians, which they liked on account of its cheap price, as they were able to buy only small quantities of wheaten flour from the Chinese at a high rate. The Gilyak word *oba* or *owa* given by L. von Schrenck for rye flour is a loan-word based on Manchu *ufa* ("flour, bread"). So likewise are the other Gilyak names of cereals cited on the same page: *budá* (*Setaria italica*), Manchu *buda*; *ssjussj* ("sorghum"), Manchu *šušu* from Chinese *shu shu* 蜀黍 [1]); and *mudi* ("barley"), Manchu *muji*.

At the time of my sojourn in the Amur country (1898—99) I was told that the most energetic agriculturists were Koreans and that the Russian peasants had abandoned the cultivation of cereals, as they could purchase flour from Chinese traders at a considerably lower cost than by raising their own crops. I. A. Lopatin (Etnografiya, a course of lectures delivered at the State Far-Eastern University, in Russian, Vladivostok, 1920—21, p. 145) writes: "The Russian colonists in eastern Siberia and in the Far East were compelled by unfavorable local climatic conditions (cold and snowless winters of the monsoon type) to abandon the cultivation of the most typical and favorite Russian grain, rye. Winter cereals cannot be cultivated here. Instead of rye, *yáritsa* has been introduced". Yáritsa (the same as *yarovaya rož*) is summer rye, which is sown in the spring; the flour made from it is not so dark as that of winter rye.

Rye is an important if not the most important crop of the Russian farmer of Siberia. In the district of Minusinsk, at least

---

1) The relation of this word to Persian شوشو *šušu* ("millet") remains to be studied in connection with the history of millets a rather complex problem on account of the number of different species involved and the relatively high antiquity of their cultivation.

during the last century, the cultivation of winter and summer rye occupied the first place among the crops, 35.2 per cent of the entire cultivated area (A. Jarilow, Ein Beitrag zur Landwirtschaft in Sibirien, 1896, p. 248).

In regard to the Mongols, P. S. Pallas (Samlungen historischer Nachrichten über die mongolischen Völkerschaften, 1776, I, p. 175) — and if any one ever knew the Mongols, it was old Pallas — informs us thus: "In the southeastern part of Mongolia entire tribes are impoverished to such a degree that they have been compelled to subsist on agriculture which China seeks to foster among the Mongols as much as possible. They cultivate wheat (*zagan buda*), barley (*arbai*), and rye (*oros*)". The word *oros* is the Mongol designation for Russia (Rus + *o* or *u*, as Mongol does not tolerate initial *r*, so that rye was aptly called by them "the Russian (scil. wheat)", and it seems permissible to conclude that the Mongols in question had obtained their rye from the Russians.

On the other hand, there is also rye in Mongolia of apparently indigenous origin. Vavilov (Origin of Cultivated Plants, p. 205) gives the following information: "Spring rye in eastern Siberia, in Transbaikalia, and in the Far East [in the Russian sense] has evidently also come out of the sowings of barley and wheat, and has formed a pure crop. It was observed by the Mongolian expedition of V. E. Pisarev that spring wheat and barley showed strong admixtures of rye in northern Mongolia. As shown by the data of this expedition, the cultivated field plants of Siberia and the Far East were borrowed from Mongolia" [1]).

The *Ts'e yüan* (sub 麵 *mien,* flour) calls rye 燕麥, explaining that Europeans have two kinds of bread, white and black, the

---

1) I believe that Mr. Vavilov's conclusion is correct. Paradoxical as it may seem, the Mongols have played a prominent role in the history of agriculture, as I hope to show in my history of buckwheat (complete in manuscript) and history of oats (in preparation).

former being made from wheat, the latter from *yen mai*. The term *yen mai* properly applies to oats (*Avena fatua*), but since Europeans do not make bread from this cereal, it can in this case refer solely to rye [1]).

The texts assembled permit us to tabulate the nomenclature of rye as follows:

老羌穀 *Lao K'iang ku* ("grain of the Old K'iang", i.e. Russian grain) — *Hei lung kiang wai ki*.

老鎗穀 *lao ts'iang ku* (same meaning) — *Lung sha ki lio*.

高麗穀 *Kao-li ku* ("Korean grain") — *Sheng-king t'ung chi*.

油麥 *yu mai.* — *I yü lu*.

稗子 *pai-tse.* — *Ning-ku-t'a ki lio*.

燕麥 *yen mai* (properly *Avena fatua*, oats). — *Ts'e yüan*.

Japanese *rai-mugi, lai-mugi*, based on "rye". Colloquially *mugi* ("wheat"), *mugi-pan* ("rye bread").

Korean 稗 *pi*.

Mongol *oros*.

As regards lexicographical literature, we find in H. C. von der Gabelentz' Mandshu-Deutsches Wörterbuch (1864) the word *niyanggu* with the meaning Roggen, Korn ("rye, corn"). K'ien-lung's Polyglot Dictionary (chap. 28, p. 43*b*) has the series Manchu *niyanggu je bele* ("millet" + "rice"), Tibetan *drus ajam* (lit. "soft millet"), Mongol *liyangku konok amu* ("millet" + "rice"), Chinese *liang ku mi* 涼穀米 (lit. "cold grain rice"). It follows from this series, as also remarked by Zakharov, that Manchu *niyanggu* (as well as the Mongol form) is nothing but a transcription of Chinese *liang ku*; it should properly be romanized *nyanggu* or *n'anggu*, as the Manchu mode of writing *niya* denotes merely the palatalization of the initial *n*. Zakharov defines the term, apparently

---

1) According to E. H. Wilson (China Mother of Gardens, p. 335), *yen mai* denotes *Avena nuda* and is the species preferred by the Chinese, while the Tibetans prefer *Avena fatua*.

after some Manchu dictionary, as "a white millet from which a grain soft and pleasant of taste is obtained". This, of course, is vague. I can neither confirm nor refute the notion that it may denote rye.

I. J. Schmidt (Mongolisch-deutsch-russisches Wörterbuch, p. 233*b*) lists a Mongol word *talkha* (*talxa*) with the meaning "grobes oder geschrotenes Roggenmehl" (coarse or rough-ground rye flour). Kovalevski also (Dictionnaire mongol-russe-français, p. 1635) renders this word by "farine de seigle".

In the modern vernacular rye is called *siao mai* 小麥 ("wheat"). This is listed as the word for rye in the English-Chinese Dictionary published by the Commercial Press of Shanghai, also in Taranzano's Vocabulaire des sciences [1]), and corresponds to Japanese colloquial *mugi* (above, p. 239). Inquiries made by me among Chinese living in America or passing through Chicago have confirmed this use of *siao mai* in the sense of rye. This is not surprising in view of the close botanical relationship of rye to wheat. Many other peoples to whom rye was a foreign product, on becoming acquainted with it, simply styled it "wheat" or "black wheat". In the Dutch Indies rye is but rarely cultivated; the Malayan designations of it are based either upon the Portuguese word for "wheat", *trigo*, or upon an Indian vernacular form for "wheat": (1) Malay and Sunda *tarigu, terigu, trigu*; Bugi and Makasar *tarigu*; Madura *darigu*; (2) Java, Madura, Sunda, Menangkabau *gandum* (Persian *gändum*, Sanskrit *godhñma*), Bugi and Makasar *yendong* (Nieuw Plantkundig Woordenboek voor Nederlandsch-Indië, p. 324). In Gypsy *biālo gīb* ("white cereal") signifies "wheat"; *gālo gīb* ("black cereal"), "rye" (R. Liebich, Die Zigeuner, p. 233).

---

1) The names for rye given by K. Hemeling (English-Chinese Dictionary, 1916) are artificially reconstructed, but not actually used. Taranzano gives also *hei mai* ('black wheat'), *yüan mai* ('round wheat'), and *lo mai* ('naked wheat') as terms for rye.

II

In order to understand the investigation that follows, it is necessary to have some knowledge of the history of rye and especially of the epochmaking researches of Vavilov.

We are all acquainted with rye as the typical cereal of the Russians, the North-Germanic and Finnish peoples. While it is a characteristic field crop of northern, central, and eastern Europe, yet the curious fact remains that botanists and linguists alike are agreed on the one point that its cultivation did not originate in Europe, but has come from Asia, while in Asia it plays only a subordinate or insignificant role. It is still more amazing that some scholars look for the original home of rye cultivation in territories of Asia where at present rye is not at all cultivated and where it is not even known that it was anciently cultivated. It has always remained unknown in the Egypto-Semitic culture area as well as in India and southeastern Asia including the Malay Archipelago. It is a plant of the north and of temperate climes. It is, further, the most recent of all our cereals the cultivation of which is considerably later than that of wheat and barley. Whereas the cultivation of barley, wheat, and millet in middle and northern Europe can be proved for the more recent neolithic period, the earliest hitherto known specimens of prehistoric rye do not antedate the earliest iron age. It is doubtful whether Greeks and Romans cultivated rye. Galen (A.D. 131—200) is the first Greek author who mentions it under its Thracian name βρίζα. I review some of the older speculations regarding the origin of rye, but the reader should keep in mind at the outset that these are now superseded by the matter-of-fact investigations of Vavilov.

In 1900 F. von Schwarz (Turkestan, p. 353) wrote, "Rye was unknown to the natives of [Russian] Turkestan until the advent

of the Russians, nor is it cultivated by them at present. Rye is
grown solely by the Russian colonists in the northern part of
Turkestan, on the northern slopes of the T'ien-shan, in the Ili
valley, and in the territory of Kasalinsk and Aulie-Ata. But
although the modern Central Asiatics do not engage in the cultivation of oats and rye, there can be no doubt that Turkestan
has been the primeval home (*urheimat*) of these two cereals; for
both rye and oats still grow wild in the mountains of Turkestan,
and even on the Pamir the mountain slopes are covered with
extensive meadows of wild rye and oats".

A. Schulz (Geschichte der kultivierten Getreide, 1913, I, p. 73)
has developed the following hypothesis in regard to the origin of
the cultivated rye. According to him, it was presumably domesticated in Russian Turkestan from *Secale anatolicum*, a wild-growing
species of rye (one of the three subspecies of *S. montanum* Gussone),
which has been observed in Asia Minor, Syria, Armenia, Kurdistan,
Afghanistan, the Turkmen steppe, Turkestan, Dzungaria, and the
Kirghiz steppe. "At present rye is but little cultivated in Turkestan,
but obviously it was formerly raised to a great extent, as may
be recognized from its present wide dissemination as an escape
from cultivation. It is abundant principally in the territory of
Tashkent. There, large tracts of the middle mountain range and
the fertile plains are occupied with fugitive, large-grained rye so
luxuriously that one might believe one self in the midst of a country
carefully cultivated with rye. In Turkestan rye serves merely for
the making of hay. The domesticators of rye were members of
the Turkish nation. From this stock, other Turkish peoples, as
well as the Finnish and Baltic-Slavic nations must have received
it. From the Slavs it must have spread to the Germanic peoples.
The names for rye in use among these peoples speak in favor of
this supposition". These are derived by him from Turkish: Tatar

areš, *oreš*, Finnish *ruis*, Lithuanian *rugiaĩ*, Russian *roži*, Old High German *rokko*, likewise Thracian βρίζα, alleged to go back to *ṷruĝi̯ā* evidently belonging to a Turkish language. So far Schulz.

J. Hoops, botanist and linguist (Waldbäume und Kulturpflanzen im germanischen Altertum, 1905, pp. 443—453, 635—637; and article "Roggen" in Reallexikon der germanischen Altertumskunde, III, 1916, pp. 508—514, the most lucid summary of the subject) holds in opposition to Schulz that the home of cultivated rye was not in Turkestan, but somewhat farther west, in the vast plains of southern Russia extending as far as Turkestan. The history of rye in Europe and the linguistic facts are well expounded by Hoops, and there can be no doubt that there is no Indo-European word for rye, but that the European forms represent a migratory word which with the plant passed on from tribe to tribe: Old Russian *rugĭ, Old Church Slavic *ružĭ*, Russian *rož'*; Lett *rudzu*; Esthonian *rukkis*, Lapp *rok*, Finnish *ruis*; Lithuanian *rugỹs* (plur. *rugiaĩ*); Old High German *rokko*, Anglo-Saxon *roggo*, Old English *ryge*, Old Norse *rugr*. Further, Syryän *rudžeg*, Permian *ružeg*, Wotyak *žeg*; Samoyed *ariš*, Ostyak *ariš*, Wogul *oroš*, Cheremissian *arša*.

T. H. Engelbrecht (Über die Entstehung des Kulturroggens, in Festschrift Eduard Hahn, 1917, pp. 17—21) contains nothing principally new, save that he looks for the origin of rye in Asia Minor whence he thinks it was brought across the Black Sea into southern Russia; as in anterior Asia wild rye occurs as a weed in wheat fields, it was easily carried into the Pontus region. While his suggestion of a possible connection of weed rye with cultivated rye is correct, he committed the usual error (exposed by Vavilov) in regarding the weed rye as *Secale montanum* Guss., while in fact it represents *Secale cereale* L.

A serious blow to all previous opinions and speculations was

of the Russians, nor is it cultivated by them at present. Rye is
grown solely by the Russian colonists in the northern part of
Turkestan, on the northern slopes of the T'ien-shan, in the Ili
valley, and in the territory of Kasalinsk and Aulie-Ata. But
although the modern Central Asiatics do not engage in the cultivation of oats and rye, there can be no doubt that Turkestan
has been the primeval home (*urheimat*) of these two cereals; for
both rye and oats still grow wild in the mountains of Turkestan,
and even on the Pamir the mountain slopes are covered with
extensive meadows of wild rye and oats".

A. Schulz (Geschichte der kultivierten Getreide, 1913, I, p. 73)
has developed the following hypothesis in regard to the origin of
the cultivated rye. According to him, it was presumably domesticated in Russian Turkestan from *Secale anatolicum*, a wild-growing
species of rye (one of the three subspecies of *S. montanum* Gussone),
which has been observed in Asia Minor, Syria, Armenia, Kurdistan,
Afghanistan, the Turkmen steppe, Turkestan, Dzungaria, and the
Kirghiz steppe. "At present rye is but little cultivated in Turkestan,
but obviously it was formerly raised to a great extent, as may
be recognized from its present wide dissemination as an escape
from cultivation. It is abundant principally in the territory of
Tashkent. There, large tracts of the middle mountain range and
the fertile plains are occupied with fugitive, large-grained rye so
luxuriously that one might believe one self in the midst of a country
carefully cultivated with rye. In Turkestan rye serves merely for
the making of hay. The domesticators of rye were members of
the Turkish nation. From this stock, other Turkish peoples, as
well as the Finnish and Baltic-Slavic nations must have received
it. From the Slavs it must have spread to the Germanic peoples.
The names for rye in use among these peoples speak in favor of
this supposition". These are derived by him from Turkish: Tatar

areš, *oreš*, Finnish *ruis*, Lithuanian *rugiaĩ*, Russian *roži*, Old High German *rokko*, likewise Thracian βρίζα, alleged to go back to *u̯rug̑i̯ā* evidently belonging to a Turkish language. So far Schulz.

J. Hoops, botanist and linguist (Waldbäume und Kulturpflanzen im germanischen Altertum, 1905, pp. 443—453, 635—637; and article "Roggen" in Reallexikon der germanischen Altertumskunde, III, 1916, pp. 508—514, the most lucid summary of the subject) holds in opposition to Schulz that the home of cultivated rye was not in Turkestan, but somewhat farther west, in the vast plains of southern Russia extending as far as Turkestan. The history of rye in Europe and the linguistic facts are well expounded by Hoops, and there can be no doubt that there is no Indo-European word for rye, but that the European forms represent a migratory word which with the plant passed on from tribe to tribe: Old Russian *rugĭ*, Old Church Slavic *ružĭ*, Russian *rož'*; Lett *rudzu*; Esthonian *rukkis*, Lapp *rok*, Finnish *ruis*; Lithuanian *rugỹs* (plur. *rugiaĩ*); Old High German *rokko*, Anglo-Saxon *roggo*, Old English *ryge*, Old Norse *rugr*. Further, Syryän *rudžeg*, Permian *ružeg*, Wotyak *žeg*; Samoyed *ariš*, Ostyak *arüš*, Wogul *oroš*, Cheremissian *arša*.

T. H. Engelbrecht (Über die Entstehung des Kulturroggens, in Festschrift Eduard Hahn, 1917, pp. 17—21) contains nothing principally new, save that he looks for the origin of rye in Asia Minor whence he thinks it was brought across the Black Sea into southern Russia; as in anterior Asia wild rye occurs as a weed in wheat fields, it was easily carried into the Pontus region. While his suggestion of a possible connection of weed rye with cultivated rye is correct, he committed the usual error (exposed by Vavilov) in regarding the weed rye as *Secale montanum* Guss., while in fact it represents *Secale cereale* L.

A serious blow to all previous opinions and speculations was dealt with by N. Vavilov, the Russian De Candolle, a brilliant scholar and thinker, who has placed the history of many cultivated plants on a new basis and has illuminated it with new and fertile ideas. In his study О происхожтеніи кульдурной ржи (On the Origin of the Cultivated Rye, *Bull. of Applied Botany*, Petrograd, 1917, pp. 561—590), he rejects completely the old theory that *Secale cereale* is derived from *Secale montanum*, the wild mountain rye characterized by a fragile ear, small seed, and perennial life. Vavilov states that all previous writers on the subject have merely said so, but that no one has ever given any actual proof for this assertion. The mountain rye may be cultivated for a long time, but will not change into the ordinary rye, and such a transition is genetically very improbable. Vavilov looks for the progenitor of the cultivated rye among the forms of *Secale cereale* itself. He emphasizes the fact that rye (*Secale cereale*) is widely distributed as a weed, sometimes in large quantities, in the wheat and barley fields of Persia, Afghanistan, Russian Turkestan (until the time of the Russian colonization), Bukhara, Syria, and Palestine-countries where the cultivation of this cereal is not known or almost unknown. This fact has been observed by many botanists and by Vavilov himself during his travels in Persia, Turkestan, and Bukhara. In Afghanistan, according to the observations of the botanist Aitchison, *Secale cereale* occurs as a weed among wheat, in some fields in such quantity that there is as much rye as wheat, and considered by the natives as a weed; it is not purposely sown along with the wheat, and is even considered by the people as very harmful to the system when a large amount of it is mixed with wheat flour. This common rye is perfectly wild in Afghanistan, and is not grown anywhere as a distinct crop. Vavilov points out that the Persians, Sarts, Arabs, Afghans and Turks designate rye *chu-dar* or *ju-dar*, which means "a herb growing among barley", or *gändum-dar*, which means "a herb growing among wheat". This, in his opinion, furnishes evidence for the fact that in these parts of Asia rye was merely known as a weed among wheat and barley crops, not, however, as a distinct cultivation. This rye weed is generally not distinguished from the cultivated rye of Europe, and like the latter is represented by a series of different hereditary forms or varieties differentiated by the color of the seed and structure of the ears; but in the rye weed are found also endemic forms, for instance, red-eared varieties, which are not known in Europe.

From a weed, rye began to be sown intentionally mixed with wheat, barley, and other plants (for instance, *Leguminosae*). This is demonstrated by the wide distributions of mixed sowings in the past and even at present in western Asia, the Caucasus, Russia, Crimea, and France (under the name *méteil*). Climatic conditions also exerted their influence: the farther northward, the more favorable were the conditions for rye, the more rye predominated in the wheat fields. The next step was the separate cultivation of rye.

In the opinion of Vavilov, the beginnings of rye cultivation should be looked for in Asia Minor, Turkish Armenia, Shugnan and Roshan (west of the Pamir). Ten years later he published an article entitled "Geographical Regularities in the Distribution of the Genes of Cultivated Plants" (*Bull. of Applied Botany*, XVII, Leningrad, 1927, pp. 411—428), in which he writes (p. 420), "The most probable center of the origin of cultivated rye (*Secale cereale*) as well as of the whole genus *Secale* are the eastern part of Asia Minor and Transcaucasia. In these countries are concentrated all species of rye as well as the diverse characters distinguishing the botanical varieties of rye. The most important fact, however, is that in these localities not only a great number of forms have been found, but also many dominant forms such as red-eared, brown-eared, and even black-eared forms, as well as varieties with strongly marked pubescence". In "Geographische Genzentren unserer Kulturpflanzen" (Verh. des V. internat. Kongresses für Vererbungswissenschaft, 1927, pp. 342—369), Vavilov has summarized and expanded his views on the subject: thanks to its endurance during the winter the weed rye begins to supersede wheat in high altitudes; in Turkestan winter wheat disappears at an altitude of 2.300—2.400 meters, and is replaced by winter rye; it is a struggle between the two cereals; wheat has introduced rye into cultivation; without the culture of wheat the genesis of cultivated rye remains unintelligible. This article is accompanied by a useful map (p. 361) showing the geographical distribution of the wild and cultivated ryes over Asia and Europe.

Finally, in his fundamental work Центры происхожденія культурных растеній ("Centers of origin of cultivated plants", or with the English title "Studies on the Origin of Cultivated Plants", Leningrad, 1926, pp. 72—90 Russian, English translation pp. 196—209), Vavilov has developed again his views of the origin and expansion of rye with renewed vigor and supported them with fresh data and observations. Of especial importance to the historian are his conclusions, with which I am entirely in accord, that "the cultivation of winter rye has evidently entered Europe by two principal ways — from Transcaucasia on the one hand and from Turkestan, Afghanistan, and adjoining regions on the other hand, a view supported by the fact that there are two distinct geographical types of weed rye"; and that not one center, but several must be assumed for the origin of rye cultivation; it was simultaneously and independently grown in many localities, as may be observed up to this time.

I feel that Vavilov's hypothesis in its main outlines is perfectly correct, explaining as it does the essential facts in a satisfactory manner. Above all, his opinion that cultivated plants have been developed from weeds is a fertile idea, and I hope to have occasion in further studies of Chinese cultivated plants to point out that it is valid in a number of other cases (cf. *T'oung Pao*, 1916, p. 89, note 2). In this connection it is interesting, though it is a fortuitous coincidence, that the word *pai* 稗, which, as mentioned above, has in recent times assumed the meaning "rye", also denotes "tares among wheat, weeds".

According to J. L. Schlimmer (Terminologie médico-pharmaceutique française-persane, Teheran, 1874, p. 510), rye is cultivated in Persia, but its cultivation seems to be restricted to five localities each of which has a different word for it, apparently of local value. The word *čodār* جودار referred to by Vavilov, according to Schlimmer, is restricted to Azerbeijan; but surely it must cover a wider area as it has been adopted by Osmanli (*čavdār*), and Steingass (Persian-English Dictionary, p. 402) lists it as a common Persian word (*čūdār*, "rye"). Schlimmer, further, notes for the designation of rye the words *diwek* for Suldeh Larijan, *kārnaware* for Kheregan, *dileh* for Na-inĕ, and *bārenj* for Chehar Mehell. Vavilov's *chu-dar* or *ju-dar* is given by Steingass (p. 377) as جودره *jaudara*, "herb growing among weed", and it may well be that the first part represents جو *jau*, *jaw* ("barley", Skr. *yava*). Steingass (p. 1101) also has گودر *gaudar* ("a plant growing among wheat and barley").

According to K. Kannenberg (Kleinasiens Naturschätze, 1897, p. 136), rye is cultivated in Asia Minor, but only to a small extent. Here, in the same manner as in Persia, we have unfortunately no historical documents that would enable us to judge how old this cultivation may have been. Another problem that remains to be studied more closely is the distribution and importance of rye among the peoples of the Caucasus who grow it in considerable quantity and distil from it their national brandy (see, for instance, G. Merzbacher, Aus den Hochregionen des Kaukasus, 1901, I, pp. 380, 492, 495, 522, etc.). The words *sekil* and *sukul* ("rye") which occur in Caucasian languages are suggestive of a connection with Latin *sécale*.

To come back now to the sporadic occurrence of rye in the mountains of Sechwan and possibly in Kansu and Shensi, our first difficulty is that we do not know the local names of the cereal and that we have a botanical determination only for Sechwan in Wilson's *Secale fragile*. One of the reasons which induced me to write this article is to call Mr. Vavilov's attention to the existence of rye in Chinese territory so that he might extend his investigations to this field also; and when these have been carried out, no one will be better qualified to explain this curious phenomenon than he. Meanwhile I restrict myself to submitting a few suggestive observations from a purely historical viewpoint. As rye is of no importance in Chinese agriculture, it cannot be assumed that it is spontaneous in the few isolated localities where it now occurs, although I believe that an intensive search will bring to light many more areas of its occurrence. The species *Secale fragile* grows also in Shugnan, and Vavilov has illustrated a specimen from this region in the Pamir (plate 166, No. 4, of his article On the Origin of the Cultivated Rye; cf. also the woodcut p. 566). Now the Pamir, according to Vavilov, has been one of the ancient centers of rye cultivation[1]), and the Pamir countries were in close contact with China under the T'ang dynasty (A.D. 618—906). It is not improbable, therefore, that this direct intercourse may have resulted in carrying rye to China.

Under the T'ang dynasty when Chinese power extended all

---

1) According to F. Machatschek (Landeskunde von Russisch Turkestan, 1921, p. 149), a variety of rye cultivated in Shugnan in the Pamir is probably indigenous there.

over Central Asia, the Pamir (Po-mi 播密) was well known to the Chinese. The country Wakhān (Hu-mi 護密) in the Pamir, according to the T'ang annals, produced beans and "wheat", and in A.D. 728 sent local products as a gift or tribute to the court of China (Chavannes, Documents sur les Tou-kiue occidentaux, p. 165). Again, in the following year, a high dignitary of the kingdom of Wakhān visited the court to pay homage (Chavannes, T'oung Pao, 1904, p. 50). In A.D. 730, the king of Wakhān came to render homage to the court and to offer products of his country (ibid., p. 51); another visit of the king in A.D. 733 (ibid., p. 55). In A.D. 753, Wakhān sent an envoy to China to pay homage and to bring tribute (ibid., p. 88). In the first month of the year 758, the king of Wakhān sent a high dignitary to China, and in the sixth month of the same year the king made a personal call at Ch'ang-an (ibid., pp. 93, 94). There was, accordingly, a rather lively intercourse between Wakhān and China in the first part of the eighth century. As we see from the documents of the Manchu dynasty, rye was simply called "wheat" or considered a species of wheat by the Chinese, and in view of the undoubtedly ancient rye cultivation by the nations or several nations inhabiting the Pamir, I am inclined to think that the "wheat" of Wakhān referred to in the T'ang shu is intended to designate rye or at least to include rye and wheat [1]). Further, if Wakhān

---

I) It must be borne in mind also that the original meaning of mai 麥 is not simply "wheat". The typical and specific ancient word for wheat is 秾 lai, loi, anciently *glo, *gro, identical with Tibetan gro ("wheat"), in the same manner as 來 lai ("to come") corresponds to Tibetan a-gro ("to walk, to go, to travel") and 賚 lai ("to give, to reward") to Tibetan gla ("pay, fee, wages"). Lai ("wheat") without the classifier appears in the Shi king (Legge's edition, pp. 580, 582) in combination with mou, 來 牟 lai mou ("wheat and barley"). The distinction made between ta 大 mai ("barley") and siao 小 mai ("wheat") first recorded in the Pie lu, as well as the numerous other compounds with mai forming names for other gramineous plants, would go to show that in its origin the term mai denoted a certain class of cereals, above all including wheat and barley.

repeatedly sent local products to China, it is possible that rye as
a cereal peculiar to the region and unknown to the Chinese was
included among these gifts. This assumption would plausibly account
for the sporadic occurrence of rye in Sechwan, Kansu, and Shensi.

The *Pei shi*, as quoted by Chavannes (*Bull. de l'Ecole française*, III, p. 401, note 9), reports that the inhabitants of Wakhān
subsist only on cakes and roasted grain and drink a brandy made
of grain.

The statement made in the T'ang Annals that Wakhān produced "beans and rye" obtains another significance in the light
of Vavilov's researches (pp. 578—579), who observes that the
inhabitants of Shugnan and Roshan consciously sow rye together
with beans.

The assertion of the T'ang Annals that Shignān (*Shi-k'i-ni*)
does not produce the five kinds of cereals (Chavannes, Doc. sur
les Tou-kiue occidentaux, p. 163) must rest on an error; for Hüan-
tsang (*Ta T'ang si yü ki*, chap. 12, p. 11b) says just the opposite
and attributes to Shignān "an abundance of pulse and wheat, but
little grain" (多菽麥少穀稼). The term "wheat" almost
certainly refers here to rye [1]), but the question is what is to be

---

1) [L'interprétation de *mai* par "seigle" dans le présent passage remettrait en
question la valeur des noms de céréales employés par Hiuan-tsang pour toute cette
région de l'Asie Centrale. J'en ai parlé à diverses reprises, à propos du terme de
宿麥 *sou-mai*, et en dernier lieu dans *Etudes d'oriental. publiées par le Musée
Guimet à la mémoire de Raymonde Linossier*, 423—425; j'y ai montré que Hiuan-tsang
employait *sou-mai* au sens du sanscrit *yava*, "orge", et je ne vois pas de raison sérieuse
pour que *mai* seul ne puisse ici désigner l'orge et le blé aussi bien que le seigle. Je
profite de l'occasion pour ajouter quelques mots sur *sou-mai*. Je ne connaissais l'expression, à part Hiuan-tsang, qu'à l'époque moderne, au sens littéral de "blé [ou orge]
semé à l'automne"; mais M. P. A. Boodberg, *Hu t'ien Han yüeh fang chu*, no. 5 [janv.
1933], p. 26, a signalé que l'expression se trouve déjà sous 120 av. J.-C. dans *Ts'ien
Han chou*, 6, 6b, et sous 106 ap. J.-C. dans *Heou-Han chou*, 5, 1b (cf. aussi *Library
of Congress, Orientalia added 1932—1933*, p. 9), et au même sens qu'aujourd'hui. Il
resterait à savoir si, sous les T'ang, c'est surtout l'orge qu'on semait à l'automne, pour
justifier l'emploi de *sou-mai* au sens restrictif d'"orge" chez Hiuan-tsang, — P. P.].

understood by *ku*. Beal (Records, II, p. 295) translates, "Much wheat and beans are grown, but little rice"; Watters, "There is a little vegetation" [1]).

III

Under the heading 山穀 *shan ku* ("mountain grain", or "wild grain"), the *Pen ts'ao kang mu shi i* (chap. 8, p. 33) cites the following text from the *Hwan yu pi ki* 宦游筆記 [2]): "It grows in Mongolia [Sai wai 塞外, lit. 'outside the passes'], where the natives call it *urkona* or *urkon* (烏爾格納 *wu-er-ko-na*). The stalks of this plant are more than a foot tall, thin like grass, with joints like those of bamboo. The leaves also are like those of bamboo. There is one leaf to every two joints. The ears of this grass resemble a *Polygonum* flower (蓼花 *liao hwa*, various species of *Polygonum*). The grains produced by this grass resemble those of the cultivated cereals (穀 *ku*), and are red in color. The Mongols gather these grains and dry them in the sun; after the removal of the skin, they are boiled into a gruel, which has a pleasant odor like a cereal. The Mongols use this gruel to satisfy hunger; they likewise make flour of the grain and mix it with tea. The Chinese traders 商民 (living among the Mongols)

---

1) I am now under the impression also that the mysterious cereal *ts'ing lo mi-tse* 青稞庶子 mentioned in the Annals of the Liao for the Si-hia country and pointed out by me in *T'oung Pao*, 1916, p. 95, may refer to rye, unless two names for different cereals are hidden in this compounds. [Je pense qu'il y a en effet deux noms, et que 青稞 *ts'ing-lo* est une faute de texte pour 青稞 *ts'ing-k'ouo*, "orge" (*Hordeum gymnodistichum*). Quant à 庶子, c'est là une forme abrégée de 糜子 *mi-tseu*, "millet"; cette forme abrégée se rencontre dans les documents retrouvés par Stein en Asie Centrale et que Chavannes a étudiés. — P. P.]

2) [Il semble que le *Houan-yeou pi-ki* soit un recueil de notes du XVIIe siècle, mais je n'en connais pas l'auteur, ni n'en puis indiquer d'édition. Le *Pen-ts'ao kang-mou che-yi* le cite assez souvent, et pas seulement pour la Mongolie. J'ai traduit dans *JA*, 1920, I, 170, d'après une citation de seconde main, la description que le *Houan-yeou pi-ki* fait de la préparation des alcools mongols. — P. P.]

19

mix this flour with millet and thus consume it. It is red in color, luscious, and very pleasing, in its taste not different from cultivated cereals, whence the name 'mountain cereal' has been conferred upon it. It grows on the banks of streams or along water courses in mountains. — As regards the root of this grass, it is called *mokor* (墨克爾 *mo-k'o-'r*) by the Mongols. Its outer skin is very fine, its interior is solid and contains a starch (粉 *fen*, 'flour'), which is white and has a sweet and pleasant flavor. The Mongols eat it raw, the Chinese traders consume it boiled and mixed with meat during the interval from autumn to winter. The Mongols search for it by digging in the burrows of rodents, where they obtain eatable things, and store these in baskets. This root is from two to three inches long. All wild rodents gnaw and cut things, drag them away, and store them".

The *Pei kiao fang wu k'ao* (above, p. 248) contains this notice of the plant 默克爾 *mo-k'o'r* (*mokor*): "It is the name of a grass and fruit which grows in the wilderness of the northern part of the Gobi. Rodents carry it off and store it in their burrows to feed on it. There are many impoverished people among the Khalkha Mongols who dig it up to satisfy their hunger. However, they leave a little in the burrows and do not wish to exhaust the animals' supply completely". A note is added to the effect that the emperor Kao Tsung Shun 高宗純 (K'ien-lung), on reading the biography of Su Wu 蘇武, came to the conclusion that this was the plant on which Su Wu subsisted during his nineteen years' captivity among the Hiung-nu[1]).

1) Sie Tsi-shi 謝濟世 (first part of eighteenth century), in his *Si pei yü ki* 西北域記 (p. 6 of the ed. in *Hwang ch'ao fan shu yü ti ts'ung shu*), speaks of "a country of rodents more than two hundred *li* east of T'o-lo-hai 陀羅海 and north of the Khangai Mountains 杭靄山, where rodents live in large communities in burrows and have store-houses with supplies in them". Li Wen-t'ien 李文田 comments on the word "supplies" that these are like 香附子 *hiang*

The truth of the Chinese story is corroborated by G. N. Potanin (Tanguto-Tibetan Borderland of China and Central Mongolia, in Russian, I, p. 111), who gives the following interesting account: "The Ordos Mongols are acquainted with a mouse who gathers supplies; they call it simply 'gray mouse' (*boro khuluguna*). It gathers exclusively from the fields, especially millet and hemp seeds, and picks only the very best. According to popular belief, the mice go to work in companies, climbing up to the top of the plants and plucking the best grains; then one, for instance, lies on its back, while others place the grains on its chest and belly and drag it by its tail into the burrow. Poor people, who in the opinion of the more well-to-do have lost all shame, plunder the mice during the winter. In the eighth month they find the storage magazines of the mice (called in Mongol *urgai* or *urgá*) filled with hemp seeds, and in the ninth month all seed is removed.

---

fu-tse, but larger and edible, called by the Mongols *mo-k'o-lo* 墨克勒 and by the Chinese "ginseng fruit" 人参果 *jen-shen kwo*. The term *hiang fu-tse* is one of many synonyms of *so ts'ao* 莎草, which is *Cyperus rotundus*; *jenshen* is *Aralia quinquefolia* (formerly *Panax ginseng*), but the term *jen-shen kwo* appears to be a different plant, which, however, is not listed in any of the relevant publications accessible to me. No proof is given for these identifications, and they seem altogether dubious. It seems to me that the root *mo-k'o-lo* has nothing to do with the plant *urkon*, but is a distinct species. [*Mo-k'o-eul* et *mo-k'o-lo* supposent théoriquement un mongol *mäkär, *meker. Il n'est pas douteux que ce soit là le *Polygonum* appelé en mongol "mykir" selon Pallas (*Samml. hist. Nachrichten*, I, 179—180) et *mekir* par Kovalevskiï. Pallas confirme que des rongeurs l'amassent dans leurs terriers, où des Mongols le recueillent, et indique pour ces terriers le nom d'"urgan", identique à celui d'urgaï que M. Laufer indique ci-dessous d'après Potamin ("urgun" dans Rudnev, *Materialy*, 133); cette plante *mekir* n'a donc rien à voir avec le seigle. Mais il est possible que le *wou-eul-ko-na*, malgré ce que dit le *Houan-yeou pi-ki*, soit une plante distincte du *mekir*. Il reste un peu surprenant que, d'après Li Wen-t'ien, les Chinois donnent au *mekir* le nom de "fruit de ginseng", alors que nos dictionnaires mongols indiquent, pour nom mongol de la racine de ginseng, *oryodoi*, phonétiquement assez analogue au *wou-eul-ko-na* du *Houan-yeou pi-ki*. Mais la restitution théorique de *wou-eul-ko-na*, si nous avions affaire à une transcription un peu rigoureuse, serait *ürgänä, autrement dit représenterait un mot de la classe palatalisée, tout différent par suite de *oryodoi*. — P. P.].

They assert that this mouse makes two storage burrows; at first it conceals the seeds in one of these, then it cleans the seeds, and transfers them to the other burrow."

The fact that the Mongols gather the grains of a wild Graminea and use them as food in the same manner as those of a cultivated cereal is interesting in itself, revealing as it does the primeval steps preceding the cultivation of cereals. What species this Mongol wild grain represents I do not pretend to know, but I wish to draw the attention of Mr. Vavilov and his collaborators to it, who will have no difficulty in procuring specimens and identifying them. But whether it is a wild rye or another gramineous species (it probably denotes wild cereals in general regardless of species), the Mongol designation of it, *urko-n* (*-n* being a suffix), is suggestive as bearing a close affinity with the Slavic, Germanic, Finnish, and Lapp names of rye. First of all, Mongol *urko-n* is related to Manchu *orho*, Golde *orokta* or *oroxta*, Niŭči or Jurči *urho* ("grass, weed"), Barguzin Tungus *orōkte*. The initial *ur* may be due to the fact that Mongol, as is well known, does not tolerate an initial *r*, so that *urko-n* may be evolved from an older **ruko-n*, which approaches closely Old Russian **rugi*, Lithuanian *rugỹs*, old High German *rokko*, Old Saxon *roggo* from Germanic **roggan-*, **ruggn*, **rug-n*; Anglo-Saxon *ryge*, Old Norse *rugr*, etc. And the Mongol form *urko-n* comes near to **uruĝiā* or **wrugiā* which has been traced from Greek-Thracian βρίζα. Hoops (*Waldbäume*, p. 448) has pointed out that the latter form connects with the series: Samoyed *ariš*, Ostyak *arüš*, Wogul *oroš*, Cheremissian *arša*, Chuvash *iraš*, Tatar *areš*, *oroš*, which in his opinion may represent loans from a Scythian or other Iranian language. If this be correct, these words must originally have referred to the weed rye as understood by Vavilov. Radlov (*Wörterbuch der Türk-Dialecte*, I, col. 278) assigns the word *aryš* (*y* = Russian ы) to the following dialects: Altai,

Teleutic, Shor, Lebed, Küärik, Baraba, Kazan, and Krym, which goes to show that the word covers a vast area, and gives it the meaning "rye, winter rye" [1]). It is clear that this word cannot be a loan-word from the Russian, but that it is an old Turkish word (although this would not exclude the possibility of an ancient derivation from some Iranian language); this being the case, it is equally patent that anciently this word did not denote the cultivated rye, but solely the weed rye and was subsequently transferred to the cultivated rye. The weed rye thrives abundantly in the Turkish area, while it is not known and also is highly improbable that the ancient Turks ever cultivated the plant (cf. also the remark of Vámbéry, *Primitive Cultur des turko-tatarischen Volks*, p. 216: "It is beyond doubt that the Turk in the interior of Asia has never cultivated rye and oats").

But there is still more. Finnish *ruis* ("rye") is regarded as a loanword from the Russian, but the fact has been overlooked by Schrader, Hoops, and others that Finnish has an older form of the word—*rukiit*, registered by Godenhjelm (*Deutsch-finnisches Wörterbuch*, 2nd ed., Helsingfors, 1906, p. 981) side by side with *rantavehnä*, which I cannot explain, both words being translated "wild rye". There is, however, in Finland no species of truly wild rye, in the botanical sense; the loosely used definition "wild rye" can denote only one of two things—either an originally cultivated rye which has escaped from cultivation ("rye occurs along roadsides and in waste ground in Hungary and Transylvania", Pickering, *Chron.*

---

1) [A tout hasard, je signale qu'il n'est pas exclu que le *segara* du *Codex Comanicus* (p. 131), rendu en persan et en turc par *"ous"*, ne soit pas *siligo*, "farine de froment", comme il est dit pp. 256 et 312, ni ne signifie *ūs*, "premier lait", comme l'a cru Radlov (cf. *T'oung Pao*, 1930, 256), mais soit pour *segala* ou *segale*, forme italienne de *secale*, "seigle", comme il est dit dans le même *Codex Comanicus*, p. 378; *"ous"* serait-il mal lu pour une forme apparentée à *oroš*, etc.? — P.P.].

History of Plants, p. 513), or the rye weed in Vavilov's sense. I believe the latter interpretation is involved in the word *rukiit*, which would go to show that the Finns first made the acquaintance of rye as a weed and in a later period as a truly cultivated plant; but this is a minor point. Finnish *rukiit*, first of all, is related to Esthonian *rukkis*, and second explains the Lapp form *rok*. The close affinity of this series with Mongol *urko-n*, **ruko-n* or **roko-n* is manifest. It is therefore not necessary to conclude, as was previously done, that the Finnish tribes had received their rye exclusively from the Slavs. They may have been acquainted with it, at least in the form of the weed rye, in an earlier period of their history when they were still in lively contact with Turkish-Mongol tribes. In the same manner as the Mongols, also Turkish and Finnish groups may have subsisted on wild cereals. Those who have occasion to visit Turkish tribes of Turkestan and Siberia should make inquiries as to whether there are distinctive words among them for the weed rye and the cultivated rye. If this information were available, it would assuredly clarify the situation. The oriental origin of our word series "rye" has intuitively been foreshadowed by O. Schrader (*Reallexikon der indogerm. Altertumskunde*, 2d ed., p. 266) and Hoops (*Waldbäume*, p. 448); it now becomes a certainty. The fact that rye is of Asiatic origin has been established beyond cavil. The rapprochement of the Finnish and Mongol terms signally confirms the views and conclusions of Vavilov.

I regret that I have but little material on the Pamir languages at my disposal and that the vocabularies I have do not contain words for rye. Vavilov (*Origin of Cultivated Plants*, p. 201) writes that the Tajiks of the Pamir and Badakshān call rye *kalp* and *loshak*. If this *loshak* could be equalized with *rošak*, we have a form approaching Syryän *rudžeg* (see above, p. 259). Now H. Paasonen (Über die Benennung des Roggens im Syrjänisch-Wotjakischen und im Mordvinischen, *J. S. Finno-Ougr.*, XXIII) is inclined to trace the Syryän-Wotyak words to a Scythian ancestor corresponding to Thracian **vriza*, which in the manner of Ossetian received the suffix *-äg*. I hope that an Iranian scholar familiar with the Pamir languages will enlighten this problem some day.

———

[Laufer n'a revu en épreuves que la première feuille du présent article. Le 30 août, il me la renvoyait avec un mot où il me disait que, relevant d'une opération sérieuse, il était en voie de rétablissement. Le 13 septembre, il périssait tragiquement à Chicago. Autrefois collaborateur assidu du *T'oung Pao*, Laufer lui revenait après une assez longue interruption. Je dirai dans une notice spéciale ce que la science perd à sa disparition prématurée, mais au terme des dernières lignes qu'il nous ait données, je veux du moins apporter à sa mémoire le tribut d'une vieille amitié. — P. Pelliot.]

235

柠檬的历史

HISTORY OF LEMONADE

With the season for iced drinks here, it is interesting to find that lemonade has a long and honorable history in the Orient. According to an article by the late Dr. Berthold Laufer, former Curator of Anthropology, lemonade was a favorite beverage of the Mongol emperors in China, and they were so fond of it that they appointed a special official of high rank to take charge of its constant preparation. Dr. Laufer wrote, in part:

"Mar Sergius, a Nestorian Christian, who founded a Nestorian church at Chenkiang in A.D. 1281, was reputed, as were his ancestors, for his ability to prepare sherbets (including lemonade), and the emperor bestowed upon him a diploma in the form of a gold tablet, granting him the privilege of especially applying himself to that occupation. Mar Sergius was obliged to send to the court annually forty jars of sherbet prepared from the juices of lemons, grapes, quinces and oranges. These beverages were believed to have curative powers. On various occasions this official lemonade maker was ordered to make special journeys post haste to various points in the empire to prepare the drinks for special functions.

"Of the numerous useful fruits that we owe to India the lemon is the most democratic and most widely known. It has become a denizen of the world, and, with its Indic name, has penetrated even into the darkest parts of Africa and the tropical jungles of South America. Next to the word 'tobacco' the word 'lemon' is the most universal, reverberating with only slight modifications from every tongue of the globe.

"The earliest references to lemons in India on the part of European travelers are by a Friar Odoric of the fourteenth century, who on a visit to Ceylon described a pool full of precious stones, and abounding in leeches. The king, he related, allowed the poor to search the water for the stones once or twice a year, and to take whatever they could find. But in order that they might be able to enter the water in safety they bruised lemons and copiously anointed their bodies with the juice to keep the leeches from biting them."

SUMATRAN WEDDINGS

Among the Menangkabau tribe, of central Sumatra, marriage is a really serious matter. Contrasted to the spur-of-the-moment weddings contracted in perfunctory ceremonies at some of our Gretna Greens, where a few hours' or even a few minutes' acquaintance may be a couple's only preliminary to matrimony, the Menangkabau go through eight solid days of elaborate and solemn ceremonials, culminating in a grand finale of feasting and dancing on the final day.

In Hall G (devoted to ethnology of the Malay Peninsula and Malay Archipelago) there are exhibited life size models of a Menangkabau bride and groom, dressed in the elaborate garments used on such an occasion. The trappings for these figures were collected for the Museum by the Arthur B. Jones Expedition to Malaysia.

A Menangkabau wedding is strictly an affair of the matriarchal family. A representative of the family negotiates the match, sets the time, and prepares the wedding feast. The garments worn by the bride are family possessions, and are used for generations.

The dresses shown on the Museum's models are typical of those worn by the bride and groom on the final day. The bride wears skirt, jacket, and shoulder

cloth of silk with designs in gold and silver thread. On her head and about her neck she wears the typical ornaments of a bride, while her wrists support huge bracelets covered with thin gold plate in design. On the fifth finger of her hand is a long golden fingernail protector—a result of Chinese influence.

The groom is dressed less elaborately than the bride, but his garments show some of the best weaving of the tribe. The lower

Menangkabau Bride and Groom
Life-size models of native Sumatrans in the elaborate trappings worn on the wedding day, exhibited in Hall G.

borders of his jacket and sleeves have designs woven in gold thread, while similar designs appear on the trousers and belt. Thrust into his belt is the kris or fighting knife, traditional weapon of the Malay.

SPHERICAL CONCRETIONS

Visitors sometimes inquire why some of the concretions in the large collection in Clarence Buckingham Hall (Hall 35) are spheres. Concretions assume many fantastic forms, but when grown under ideal conditions they are spheres as is illustrated by a recent addition to the collection presented by Mr. A. F. Sitterle, of Chicago. A study of this sand-calcite concretion, partially embedded in its sandstone matrix, may make the reason for the ideal form easier to understand.

This concretion was formed in a sandstone bed by growth, from the center, of a mass of minute calcite crystals which fill spaces between the grains of sand. The sandstone is of the variety called freestone, made up of uniform grains with the porosity equal in all directions. The concretion grew by the deposition of carbonate of lime from hard water which slowly percolated through the porous stone. This deposit from hard water is not unusual—it accounts for the scale formed in steam boilers and tea kettles. The concretion started as a single crystal of minute size or by the coating of a small nucleus and grew outwards. The reason for its spherical form is merely the absence of any reason for another shape. With conditions uniform on all sides of the growing mass there is no reason why it should grow faster in one direction than another. If it grows equally in all directions the shape is necessarily that of a sphere. The reason why more concretions are not spheres is that ideal conditions are as seldom encountered where concretions are growing as they are elsewhere.
　　　　　　　　　　　　　　　　—H.W.N.

GUIDE-LECTURE TOURS

During July and August the conducted tours of the exhibits under the guidance of staff lecturers will be given on a special schedule, as follows:

Mondays: 11 A.M., Halls Showing Plant Life; 3 P.M., General Tour.

Tuesdays: 11 A.M., Halls of Primitive and Civilized Peoples; 3 P.M., General Tour.

Wednesdays: 11 A.M., Animal Groups; 3 P.M., General Tour.

Thursdays: 11 A.M. and 3 P.M., General Tours.

Fridays: 11 A.M., Minerals and Prehistoric Exhibits; 3 P.M., General Tour.

There are no tours on Saturdays, Sundays, or the July Fourth holiday.

Persons wishing to participate in the tours should apply at the North Entrance. The tours are free, and no gratuities are to be proffered. Guide-lecturers' services for special tours by parties of ten or more are available free of charge by arrangement with the Director a week in advance.

Gifts to the Museum

Following is a list of some of the principal gifts received during the last month:

From Helmuth Bay—15 specimens of woods, Norway; from School of Forestry, Yale University—37 herbarium specimens, Ecuador; from Robert M. Zingg—21 herbarium specimens, Mexico; from Professor Manuel Valerio—25 herbarium specimens, Costa Rica; from Leslie Wheeler—43 owls, 11 hawks, and a vulture; from J. H. Dekker—a fox and a badger, Iraq; from Henry Dybas—6 snakes, Indiana; from Chicago Zoological Society—4 lizards, 2 sand snakes, a caracal, and a desert monitor; from Bruno Schoemann—3 snakes, Brazil; from Dr. W. E. Hoffmann—8 turtles, South China; from H. B. Conover—a mallard duck and a ground dove, Illinois and Brazil; from Sir Charles Belcher—an orange-crested manakin, British Guiana; from General Biological Supply House—2 salamanders, Portugal; from Lincoln Park Zoo—a polar bear skeleton; from Howard Cleaves—a bobwhite, Wisconsin; from Otto Eckert—a porcupine skeleton, Wisconsin; from Frank L. Thomas—a native copper glacial boulder, Indiana.

NEW MEMBERS

The following persons were elected to membership in Field Museum during the period from May 16 to June 15:

Associate Members

Mrs. Clarence A. Burley, Walter L. Cherry, Jr., W. S. Clithero, Miss Elsa W. Junker, Sigmund Kunstadter, Mrs. William P. Martin, Samuel R. Noble, Samuel J. Walker.

Annual Members

Horace White Armstrong, Edward Buker, Miss M. M. Capper, Carroll G. Chase, Samuel T. Chase, James F. Clancy, Mrs. Schuyler M. Coe, R. Cooper, Jr., Leonard S. Florsheim, D. B. Fulton, Mrs. Cora S. Hirsch, Warren C. Horton, Morton D. Hull, Mrs. Franklin Marling, Jr., Jesse L. McLaughlin, Alfred C. Meyer, J. H. Millsaps, Montrose Newman, Mrs. Leslie H. Nichols, W. H. Parker, Dr. William Raim, Mrs. W. W. Rice, Cranston Spray, Mrs. Leslie Berwyn Steven.

Noted Orientalists Visit Museum

Three of the world's most noted authorities on Chinese art and archaeology, sojourning in Chicago recently, visited Field Museum on June 12 to inspect the Oriental collections of this institution. These visitors, all from England, are Mr. George Eumorfopoulos, founder of the famous Eumorfopoulos Collection recently purchased by the British nation for the Victoria and Albert Museum; Mr. Robert Lockhart Hobson, Keeper of the Department of Ceramics and Ethnography in the British Museum, and cataloguer of the Eumorfopoulos Collection; and Mr. Oscar Raphael, a well-known private collector. A fourth member of their party, Sir Percival David, who has published many important catalogues of Oriental art, was unable to accompany the others on the Museum visit.

附　录

劳费尔传略 (K. S. Latourette)

NATIONAL ACADEMY OF SCIENCES

OF THE UNITED STATES OF AMERICA
BIOGRAPHICAL MEMOIRS
VOLUME XVIII — THIRD MEMOIR

BIOGRAPHICAL MEMOIR

OF

BERTHOLD LAUFER

1874–1934

BY

K. S. LATOURETTE

PRESENTED TO THE ACADEMY AT THE AUTUMN MEETING, 1936

Berthold Laufer

BERTHOLD LAUFER
1874–1934

By K. S. Latourette

Berthold Laufer (Oct. 11, 1874–Sept. 13, 1934) spent most of the years of his maturity in the United States. He was, however, German-reared and educated and to the end preserved many of the attitudes and habits of the European savant. In him Europe made one of its most distinguished gifts to American scholarship. He was born in Cologne, the son of Max and Eugenie (Schlesinger) Laufer. His parents were wealthy and gave him every advantage of education and culture. A brother, Dr. Heinrich Laufer (died July 10, 1935), was an honored physician and for many years practised his profession in Cairo, Egypt.

As a child Berthold Laufer was much interested in dramatics, especially in marionettes. He and his brothers and sisters presented complete plays, all of them original. He himself wrote a number of them and they were usually given on his father's and mother's birthdays. He once cherished dreams of becoming a dramatist and throughout his life was an ardent admirer of Shakespeare. For years he studied music, especially the piano. He had a passion for the great masters and for the opera. Beethoven, Mozart, and Liszt were his favorites.

His father wished him to become a lawyer or a physician and predicted failure for him in his chosen profession of archeology. However, the senior Laufer became reconciled to his son's decision and assisted him in the prolonged and exacting preparation which the young man deemed necessary.

Life in the schools included a decade (1884-1893) in the Friedrich Wilhelms Gymnasium in Cologne. The years 1893-1895 were spent in the University of Berlin. During part of that time (1894-1895) work was taken in the Seminar for Oriental Languages in that city. The doctorate of philosophy was from the University of Leipzig in 1897. Laufer's doctoral dissertation, a critical analysis of a Tibetan text, was dedicated

43

"in love and loyalty to my parents on their silver wedding anniversary." Many years later, in 1931, in the city of his adoption, the University of Chicago appropriately added an honorary doctorate of laws.

Laufer decided on Eastern Asia as his special field and took the time to acquire the necessary linguistic and technical tools. He had courses in Persian, Sanskrit, Pâli, Malay, Chinese, Japanese, Manchu, Mongolian, Dravidian, and Tibetan. Among his teachers were some of the greatest scholars of the day. He studied Buddhism under Dr. Franke, Chinese under Professor Wilhelm Grube, Malay under the grammarian Gabelentz, Tibetan under Dr. Huth, and Japanese under Professor Lange.

In 1898, soon after publishing his doctoral dissertation, Laufer came to the United States. The step was taken at the suggestion of Professor F. Boas, himself German-born. Professor Boas obtained for his young fellow-countryman an invitation to the American Museum of Natural History in New York City. The move so made proved decisive. It was in the United States that Laufer henceforth made his home. Here he did the major part of his scholarly work. Here he married (Bertha Hampton), and here he died. Independent and self-reliant, after he came to the United States he no longer drew support from his parents, but made his own way financially.

Majoring as he did on the Far East, it was important that early in his career he should spend some time in that part of the world. In 1898-1899 he led the Jesup North Pacific Expedition to Saghalin and the Amur region to study the ethnology of the native tribes. The interest so developed and the information obtained are reflected in a number of articles from his pen and, indeed, in his studies throughout the rest of his life. In 1901-1904 he led the Jacob H. Schiff Expedition to China for research and investigation in cultural and historical questions and for the formation of ethnological collections.

Returning to the United States, Laufer became Assistant in Ethnology at the American Museum of Natural History, a position he held from 1904 to 1906. During part of this time, in 1905, he was lecturer in anthropology at Columbia

. 44

University and in 1906-1907 he was lecturer in anthropology and Eastern Asiatic languages in that same institution.

In 1908 Laufer went to the staff of the Field Museum of Natural History and there, in spite of repeated invitations to go elsewhere, sometimes at a marked increase in salary, he remained to the end of his days. Officially the positions held were successively Assistant Curator of the East Asiatic Division, Associate Curator of Asiatic Ethnology, and Curator of Anthropology.

Twice more Laufer made prolonged visits to the Far East, both times for scholarly purposes—in 1908-1910 as leader of the Blackstone Expedition to Tibet and China and in 1923 on the Marshall Field Expedition to China.

In the Field Museum Laufer's duties were multifarious. As Curator of Anthropology he had general oversight of new accessions and of the installation, labeling, and cataloging of materials, and upon him fell the direction of his staff. His chief interest, naturally, was in the Chinese exhibits. Most of these were composed of purchases made during his expeditions to the Far East. His especial pride was the jade collection and he preferred always to show it in person to visitors. He considered it and his monograph on jade as among his major contributions. At the time of his death he had just finished a completely new installation of the entire Chinese Collection. As will be seen from a glance through his bibliography, Laufer edited many of the Museum's publications and actually wrote many of them with his own pen. He was a prodigious worker. Famous in the Museum's staff were his two desks, both piled high with accumulated tasks, and with a swivel chair between them so that he could turn from one to the other.

To his heavy burdens on the staff of the Field Museum, Laufer willingly added many others. There was the constant stream of visitors, some of them distinguished scholars, others Chinese students, and still others youthful beginners in Far Eastern subjects. He was enthusiastic in encouraging Chinese students in scholarly investigations of their own culture.

45

Through his later years he was easily the outstanding American sinologist. To him, then, came for advice and criticism many who aspired to a career in that field. To these embryonic scholars he gave unstintedly of his time. Thoroughly frank in his criticism and in pointing out defects in their work, he was also extraordinarily kind and often went through their manuscripts with minute care, suggesting corrections and additions. He was severe in his condemnation of carelessness, incompetence, or superficiality. Accordingly, his praise, when given—as it often was—became an especially high reward.

Laufer was deeply concerned in promoting an increased interest in the United States in the serious study of Far Eastern cultures. He gathered extensive collections of Chinese books and manuscripts for the Newberry and the John Crerar Libraries in Chicago. One work from the Newberry Library, transferred in 1928 to the Library of Congress, contains the lost Sung (1210 A.D.) *Kêng Chih T'u,* of which no other copy is known. He collaborated with the United States Department of Agriculture in its researches in Far Eastern plants and agricultural methods. When the American Council of Learned Societies formed its Committees on the Promotion of Chinese and Japanese Studies he accepted membership on both. He was the first chairman of the Committee on the Promotion of Chinese Studies and brought to the task creative imagination, and an enthusiasm which led him to devote to it an amazing amount of energy and time. To the leadership which he gave in the initial stages of these Committees must be ascribed much of the remarkable progress which Far Eastern studies have made in the United States in the past decade.

The list of committees and scholarly societies to which Laufer belonged is impressive. He seemed to welcome invitations to help with whatever appeared to him to give promise of promoting scholarship in the fields in which he was interested. It may have been a trace of the vanity which is to be found in most of us—a desire for recognition—or it may have been an urge to accomplish as much as possible, the lure of achievement. Whatever the reason, Laufer was forever taking on new

46

tasks and lengthening his list of membership in committees and societies. In several of these he took a very active part. He was a member of the Advisory Board of the China Institute of America; of the American Committee of the National Council of the Chinese Cultural and Economic Institute; of the Committee on the Promotion of Friendship between America and the Far East; of the board of the American Institute of Persian Art and Archaeology, a fellow of the Ethnological Society, a member of the American Association for the Advancement of Science; of the National Research Council; a member (as this biographical sketch attests) of the National Academy of Sciences, a distinguished member and at one time president of the American Oriental Society, honorary vice-president of the New Orient Society of America, a member and successively vice-president and president of the History of Science Society, a member of the American Anthropological Association, of the German Anthropological Society of Tokyo, of the North China Branch of the Royal Asiatic Society, of the Royal Asiatic Society, of the Société Asiatique, of the Hakluyt Society, of the American Folklore Society, of the Linguistic Society of America, of the Illinois Academy, of the Société de Linguistique, of the Société Finno-Ugrienne, of the Society of Friends of Asiatic Art, of the Society of East Asiatic Art (Berlin), of the Orientals (Chicago), and of the Barth Society (Vienna). He was a corresponding member of the Indian Research Society (Calcutta), an honorary member of the Archeological Society of Finland, and an honorary member and secretary of the American Friends of China (Chicago). He was associate editor of the American Journal of Archeology, a special correspondent of the National Library of Peiping, and a member of the Advisory Council of Yenching University.

To these many activities Laufer added an astonishing amount of writing. His hours were long, from nine to five in his office and evenings in writing or study at home. In one letter he speaks of his sixteen hour day. In his zest for work and with his high standards of accomplishment, often he assumed more than he could do. Partially finished manu-

47

scripts lay in his files for years, uncompleted. Highly sensitive and chronically overworked, at times, especially in his later years, he was unwell and subject to moods of depression. At times, too, ill-health and overwrought nerves made him irritable and extravagant in his censure of fellow scholars. Those who saw below the surface, however, readily forgave these idiosyncrasies, for they knew him to be the soul of loyalty and, when his feelings were touched, prodigal of his time and money. During the War of 1914-1918, for instance, though of German parentage, he gave generous financial assistance to the family of a French sinologist who had lost his life in the struggle.

From his labors Laufer found relief in his home life, in music, and especially in motoring. He was an excellent *raconteur*. His stories were usually drawn from Chinese sources and were always to the point. With his musical and artistic temperament, and with his mastery of Chinese, it was not strange that he found diversion in the rich stores of Chinese poetry. He enjoyed Chinese riddles and had an enormous collection of them which he hoped sometime to publish.

Scholar that he was, Laufer took great interest and pride in his personal library. On it he spent much of his salary. At his death it went, by letter of gift, to the Field Museum.

In the midst of his busy life, as we have said, Laufer took time to do an amazing volume of writing. As will be seen from the appended bibliography, his published works were over two hundred in number and ranged all the way from book reviews and articles of two or three pages to substantial monographs. Geographically these covered all of what used to be known as the Chinese Empire—China proper, Manchuria, Mongolia, Chinese Turkestan, and Tibet. They touched as well on Indian subjects, on Eastern Siberia, on Japan, Sakhalin, the Philippines, and the islands of the Pacific.

Yet within this wide geographical area Laufer's interests were fairly well defined. He did not attempt the impossible task of making himself an expert in all phases of Far Eastern life and culture. The record which we have summarized, the list of his society memberships, and his bibliography indicate

48

the range of his specialization. It was partly linguistic, partly artistic, to a less extent religious, but chiefly anthropological, and in the influence of one culture upon another. Laufer had little or no time for political history. Nor did he evince much concern over current political developments in the Far East, or spend many hours in studying the vast transformation wrought in our own day in the cultures of China and Japan. His interests were centered chiefly on these peoples as they were before the destructive irruption of the Occident. It was the understanding of the older phases of their culture which he sought to promote. After all, it was to archeology that he had early given his affections and it was the ancient life of mankind in the East of Asia which captured his imagination.

Languages were of interest to him mainly as tools. He knew and used an appalling number of them, some well and some only slightly. With the facility of one reared and educated on the Continent of Europe, he wrote in English, French, and German. He knew Chinese and Tibetan, and had some familiarity with Japanese and with several of the languages of India and of Central Asia. Much of his linguistic equipment was in fields in which not many other scholars, and especially American scholars, are proficient. Few, therefore, are competent to judge the entire range of his work.

While languages were to him chiefly means to an end, with his inquiring mind he could not fail to be fascinated by them for their own sakes. Among his writings, for instance, are a study of the genitive in the Altaic tongue, a long article on the prefix a- in Indo-Chinese languages, a small, privately printed brochure on the language of the Yüe-chi or Indo-Scythians, and what is really a major monograph on loan-words in Tibetan. He discussed, too, the origins of the Chinese and the Tibetan languages. He had brief notes on the derivation of our word "booze" and on Jurchi and Mongol numerals.

In the light of his love of music, it is not surprising that Laufer was deeply interested in Chinese art. Since so much of his study of the interchange of products, to be noted in a moment, had to do with examining the legends concerning the

49

results of the western journeys of Chang Ch'ien of the Han dynasty, it is not strange that the art of that period attracted him. One of his most important monographs was on the pottery of the Han dynasty. He had a shorter monograph on Chinese grave sculptures of the Han. He prepared a brochure on archaic Chinese bronzes of Shang, Chou, and Han times. He wrote a brief article on some newly discovered bas-reliefs of the Han.

Yet his interest in art ranged over other periods as well. Buddhist and Christian art in China won his attention. He had a long study of a landscape of Wang Wei, and he wrote on T'ang, Sung, and Yüan paintings. More than one art collector called on him to study his Chinese objects. As late as 1932 he identified in one collection four lost albums of pictures on the themes which were painted for K'ang Hsi in 1696 and then persuaded a patron of Chinese art to present them to the Library of Congress.

He had an interest in philosophy and religion. It is perhaps symptomatic of his emotional temperament that in later years he regretted having devoted so large a proportion of his time to the study of the rationalistic, coldly ethical, and politically and socially minded Confucianism at the expense of the more mystical *Tao Tê Ching* and Chuang Tzu. Not many of his writings dealt primarily with religion. On the great organized Chinese faiths he said little. Incidentally, however, he dealt extensively with Chinese popular religion, especially in some of its earlier forms and as it expressed itself in folklore and magic. So his studies in jade had a good deal to say of the use of that semi-precious stone in magic and religion. He wrote on the development of ancestral images in China and on totemic traces among the Indo-Chinese.

As a scholar, as we have suggested, Laufer was very much the anthropologist. It was in this field that a large proportion of his writing was done. He delighted in taking up specific human tools and practices and putting together all that could be discovered about them. Often the subjects studied were amusing, incidental, and curious rather than of very great prominence. He seems here to have found a kind of diversion. Such were the tree-climbing fish, the domestication of the cormorant in China and Japan, the early history of polo, multiple births among the Chinese, finger-prints, the use of human skulls and bones in Tibet, what he called the pre-history of aviation and of television, certain recondite phases of sex, bird divination among the Tibetans, geophagy, the history of felt, coca and betel chewing, and insect musicians and cricket champions. Others had to do with objects or institutions of more obvious importance—such as the monograph on Chinese clay figures, which he called prolegomena on Chinese defensive armor. Such, too, were his studies of the reindeer and its domestication, and of ivory in China.

Probably Laufer's most important group of contributions lay within the realm of the influence of one culture upon another and of the migration of domesticated plants, of mechanical appliances, and of ideas from people to people. Especially did he devote himself to the interpenetration of cultures in Central and Far Eastern Asia. For this kind of study he was exceptionally well equipped. His knowledge of most of the more widely used languages of the area opened to him the literatures and the inscriptions of many of the peoples involved. His archeological and anthropological interest and training gave zest and background. His phenomenal memory made possible comparisons and put at his disposal a wide range of facts, many of them at first sight seemingly incidental.

In this field were written what some scholars consider his most important single monograph, *Sino-Iranica*. Here he described the migration of various specific cultivated plants. In most instances he traced the introduction of these to China and attempted to determine whether they came from Iran or from some other land. He also included some minerals, metals, drugs, textiles, and precious stones. For some he traced not only the migration to China but also contributions of China to Iran. In appendices he discussed Iranian elements in Mongol, Chinese elements in Turki, and Indian elements in Persian. In this single monograph he used various languages of the Far

51

East and of Central Asia, employed Arabic sources, and evinced a knowledge of the pertinent literature in several languages of western Europe.

Again and again in articles and monographs Laufer dealt with phases of this major theme. His great work on jade included not only China but references to the use of the semi-precious stone in other lands. He was interested in the possible spread of culture features and artifacts from the Amur region into other parts of the Far East and to the Americas. He wrote on the wide extension of amber, on the bird-chariot in China and Europe, on the introduction of maize into Eastern Asia, on the Jonah legend in India, on the cycle of the twelve animals (so familiar in the Far East), on an ancient Turkish rug, on Christian art in China, on the coming of vaccination to the Far East, on Chinese pottery in the Philippines, on Arabic and Chinese trade in walrus and narwhal ivory, on the story of the pinna and the Syrian lamb, on burning-lenses in China and India, on asbestos, salamander, and the diamond, on Chinese and Hellenistic folklore, on the coming of tobacco to Asia, Europe, and Africa, and its use there, on the history of ink in China, Japan, Central Asia, India, Egypt, Palestine, Greece, and Italy, on the migration of American plants, and on the lemon in China and elsewhere. It was characteristic of him that one of his longest and most careful reviews was of T. F. Carter, *The Invention of Printing in China,* in which was discussed the migration of paper from China to Europe and the possible debt of Europe to China for the art of printing.

Laufer published so voluminously partly because he relied extensively upon the prodigious Chinese literature and upon what Chinese scholars had written through the ages. Chinese savants had done the spade work and he made their results available to the Occident. This does not mean that he borrowed without giving credit where credit was due. In his scholarly writings where this did not seem pedantry, he was meticulous in his references to his sources. Moreover, of direct, pedestrian, full-length translation he did very little. In his earlier years, when he was trying out his tools, he published a few translations, perhaps partly as self-imposed literary exercises. Later he did little of this kind of translation. Nor were his writings summaries and popularizations in Western languages of the labors of Far Eastern scholars. He employed treatises in Asiatic languages as he used those of the Occident, critically and as mines of information from which came the many facts which he assembled, especially in his descriptions of objects in the various collections which he gathered or utilized, and in tracing the spread of a given custom, plant, or commodity. In a certain sense his great service was one of synthesis, the comparison and interpretation of existing knowledge. In this he made a distinct contribution to scholarship. Relatively few men, either of the Occident or the Orient, have been equipped in so many of the languages of Central and Eastern Asia. To most Occidental scholars the treasures locked in these languages are as though they were not. His was the function of unsealing them and from the rich stores so disclosed to bring forth and to piece together information in such fashion as to show the interrelation of cultures and the contribution of one to the other.

Of what is usually called generalization Laufer did but little. He wrote few articles attempting to set forth the main outlines of Chinese culture. Once in a long while he attempted it. His brief article on "Some Fundamental Ideas of Chinese Culture" (*Journal of Race Development,* Vol. V, Oct., 1914, pp. 160-174) was one of the few of these efforts. An able younger American sinologist declares that he has found it among the most helpful of Laufer's writings, and states that he is carrying out his own research largely on the basis of the ideas there set forth.

In most of his more serious work Laufer wrote with scholarly objectivity. In it he did not allow his emotions, always strong, or his prejudices, sometimes acute, to enter. Only in infrequent lighter articles did his personal idiosyncracies become obvious. He disciplined himself to observe the same high standards of scientific accuracy by which he measured others.

It would be too much to expect infallibility of Laufer. He

53

would have been the first to insist that his writings must be judged primarily not by their finality but by their assistance to other scholars in expanding the borders of human knowledge. Honest, able work on which others could build and build so well that they could discover in it the flaws of which he could not be aware was probably what he would most desire.

That some of his publications are being criticized by younger scholars who have found them useful is to be expected. Thus it is said that of the two Chinese works on which he leaned heavily in his important monograph on jade, one is very faulty. In a recent number of the *Zapiski* of the Russian Institute of Oriental Studies, N. N. Poppe, writing on *Problems in Buriat Mongol Literary History,* points out what he believes to be deficiencies in Laufer's *Skizze der mongolischen Literatur* (*Revue orientale*, Vol. VIII, 1907, pp. 165-261). Another Russian has recently endeavored to refute something of what Laufer said about the Giliaks. A younger Chinese scholar has recently asserted that in his discussion of the introduction of spectacles to China, Laufer was misled by mistakes in the Chinese sources upon which he relied. (See Ch'iu K'ai-ming, *The Introduction of Spectacles into China,* in *Harvard Journal of Asiatic Studies,* Vol. I, July, 1936, pp. 186-193.) Yet it must be said at once that the work of few if any scholars escape this fate. It must also be added that Laufer was engaged in a revision of his *Jade* which, unhappily, was left unfinished by his untimely death.

In a certain sense Laufer was never completely adjusted to his American environment. In one important respect—in his interest and achievements in anthropology and in the field of culture contacts—he was at home in the atmosphere of American scholarship. Because of the American interest in the social sciences, Laufer could here find congenial spirits who could talk with him as equals and by whom he could be helped. However, in some other ways he remained an alien. In a letter written in his later years he declined to preside at an important meeting on the ground that "a Yank" could do it better. He was, to be sure, loyal to the land of his adoption. However, in East Asiatic studies he stood alone and must often have felt

54

his isolation. This was partly because, until very recently, the United States has had so few experts in that field. Significantly, however, another element entered. Most Americans who specialize in the Far East do so with the purpose of understanding the current situation in that part of the world. They realize that the United States faces the East of Asia across a rapidly narrowing ocean and must be prepared to deal with its peoples successfully and, if possible, amicably. If they are not to make tragic blunders, Americans, so these scholars hold, must understand these peoples and to do so must know their history and culture. American specialists give themselves to Far Eastern studies, partly because they become interested in them for their own sake, but chiefly from the utilitarian purpose of making their country at home in an age in which it must live on terms of intimacy with Eastern Asia. The Far Eastern scholarship of the United States has tended to devote itself to diplomatic and commercial relations, to economic problems, to contacts between the Far Orient and the Occident, and to current changes in the cultures of the Far East.

The most distinguished European savants who have majored in the Far East, on the other hand, have devoted themselves almost exclusively to the older history and cultures of this region. They have not really understood the current situation. Nor have they cared to do so. That the results of their scholarship should be useful in facilitating the intercourse between the West and the East has seemed to them to threaten its objectivity.

In that European atmosphere Laufer received his training and he could never quite adjust himself to the American outlook nor free himself of a certain impatient disdain for it. This attitude was reënforced by the fact that during most of his life America had no sinologists who could begin to equal him in his acquaintance with the languages and in his prodigious learning in the pre-nineteenth century culture. However, in at least his later years Laufer came to see that in dealing with the Far East the United States must develop its own particular type of scholarship adapted to its interests and needs. He recognized that this might attain as high standards of scientific accuracy as had that of Europe. Indeed, he insisted that America must conform to its own patterns and not to those of Europe. Yet probably he never felt entirely reconciled to this phase of the intellectual climate of his adopted land.

Perhaps in this very maladjustment was Laufer's greatest contribution to American scholarship. By representing in the United States in so eminently worthy a fashion and for a generation this European tradition, he enriched American Far Eastern scholarship as he could not have done had he been completely in accord with it.

BIBLIOGRAPHICAL NOTE

The sources for the biographical sketch given above are many—partly the note, based on information given by Dr. Laufer, in *Who's Who in America, 1934-1935* (Vol. 18, p. 1417), partly information kindly provided by Mrs. Laufer and by a former associate and intimate friend, Miss Lucy Driscoll, partly material from Dr. Mortimer Graves of the American Council of Learned Societies, and to a less extent comments by various friends of Dr. Laufer, biographical notices which have appeared since Dr. Laufer's death, and the author's own personal acquaintance, never intimate, but of many years' standing. Among the more important articles on Dr. Laufer are the ones in the *American Anthropologist,* Vol. XXXVIII, pp. 101 ff., *Artibus Asiae,* Vol. IV, pp. 265-270, and *Monumenta Serica,* Vol. I, fasc. 2, pp. 487 ff.

The appended bibliography is the most nearly complete and accurate which has been published. It is based largely upon one compiled by Dr. Laufer himself and which appeared in the *Journal of the American Oriental Society,* Vol. LIV, pp. 352-362, but it has been checked with three other bibliographies and, where feasible, from the articles and monographs themselves. At least some of the inaccuracies appearing in other bibliographies have been eliminated and a number of titles have been discovered and added.

The photograph here reproduced comes through the courtesy of the Field Museum of Natural History.

56

BIBLIOGRAPHY OF BERTHOLD LAUFER, 1895-1934

Japanische Märchen (translated from the Japanese). *Cologne Gazette,* 1895, nos. 98, 120, 144.

Indisches Rezept zur Herstellung von Räucherwerk. Translated from the Tibetan with Tibetan text. *Verhandlungen der Berliner anthropologischen Gesellschaft,* July 18, 1896, pp. 394-398.

Zur Geschichte des Schminkens in Tibet. *Globus,* 1896, vol. LXX, no. 4, pp. 63-65.

Blumen, die unter den Tritten von Menschen hervorsprossen. *Der Urquell,* New Series, 1898, vol. II, part ¾, pp. 86-88.

Eine verkürtze Version des Werkes von den hunderttausend Naga's. Ein Beitrag zur Kenntnis der tibetischen Volkereligion. Tibetan text, translation and introduction. Helsingfors, 1898. 129 pp. *Mémoires de la Société Finno-Ougrienne.*

Einige linguistische Bemerkungen zu Grabowsky's Giljakischen Studien. *Internationales Archiv für Ethnographie,* 1898, Vol. XI, pp. 19-23.

Fünf indische Fabeln aus dem Mongolischen. *Zeitschrift der deutschen morgenländischen Gesellschaft,* 1898, Vol. LII, pp. 283-288. Leipzig.

Neue Materialen und Studien zur buddhistischen Kunst. *Globus,* 1898, vol. LXXIII, No. 2, pp. 27-32. Illustrations.

Studien zur Sprachwissenschaft der Tibeter. *Sitzungsberichte des philos.-philol. und der histor. Classe der k. bayer. Akad. d. Wissenschaften,* München, 1898, part III, pp. 519-594.

Über eine Gattung mongolischer Volkslieder und ihre Verwandtschaft mit türkischen Liedern. *Der Urquell,* New Series, 1898, Vol. II, part ⅞, pp. 145-157.

Ethnological work on the island of Saghalin. *Science,* May 26, 1899, pp. 732-734.

Hohläxte der Japaner und der Südsee-Insulaner. *Globus,* 1899, Vol. 76, no. 2, p. 36.

Die angeblichen Urvölker von Yezo und Sachalin. *Centralblatt für Anthropologie,* 1900, no. 6, pp. 321-330.

Petroglyphs on the Amoor. *American Anthropologist,* N.S., 1899, Vol. I, pp. 746-750.

Preliminary notes on explorations among the Amoor tribes. *American Anthropologist,* N.S., April, 1900, Vol. II, pp. 297-338. Illustrations.

Beiträge zur Kenntnis der tibetischen Medizin. 2 parts. 90 pp. Leipzig, Otto Harrassowitz, 1900. In collaboration with Heinrich Laufer.

Ein Sühngedicht der Bonpo. Aus einer Handschrift der oxforder *Bodleiana. Denkschriften der kaiserlichen Akademie der Wissenschaften in Wien, phil.-hist. Classe,* Vol. XLVI, pp. 1-60. VII. Abhandlung.

Reviews of Grünwedel, *Mythologie des Buddhismus and S. Tajima's selected relics of Japanese art,* vols. I and II. *Globus,* 1900, Vol. LXXVIII, nos. 8 and 19, pp. 129, 310, 311.

57

Über das Va-Zur. Ein Beitrag zur Phonetik der tibetischen Sprache. *Wiener Zeitschrift für die Kunde des Morgenlandes.* Wien, 1898-99, Vol. XII, pp. 289-307; Vol. XIII, pp. 95-109, 199-226.

Felszeichnungen vom Ussuri. *Globus,* 1901, Vol. LXXIX, no. 5, pp. 69-72.

Review of H. Francke's *Der Frülingsmythus der Kesarsage, ein Beitrag zur Kenntnis der vorbuddhistischen Religion Tibets;* and of his *Ladakhi Songs. Wiener Zeitschrift für die Kunde der Morgenlandes,* 1901, Vol. XV, pp. 77-107.

The Decorative Art of the Amur Tribes. *Jesup Expedition Publications, American Museum Memoirs, New York,* Vol. VII, 1902, 86 pp. 4°. 33 plates.

Über ein tibetisches Geschichtswerk der Bonpo. *T'oung Pao,* 2nd Series, Leiden, 1901, Vol. II, no. 1, pp. 24-44.

Zum Märchen von der Tiersprache. *Revue Orientale (Keleti Szemle),* Budapest, 1901, Vol. II, no. 1, pp. 45-52.

Zur Entstehung des Genitivs in den altaischen Sprachen. *Revue Orientale (Keleti Szemle),* Budapest, 1901, Vol. II, no. 2, pp. 133-138.

Verzeichnis der tibetischen Handschriften der königlichen Biblothek zu Dresden. *Zeitschrift der deutschen morgenländischen Gesellschaft.* 1901, Vol. LV, pp. 99-123.

Zwei Legenden des Milaraspa. Translated from the Tibetan, with Tibetan text. *Archiv für Religionswissenschaft,* Tübingen u. Leipzig, 1901, Vol. IV, pp. 1-44.

Aus den Geschichten und Liedern des Milaraspa. Translated from the Tibetan, with Tibetan text. *Denkschriften der kaiserlichen Akademie der Wissenschaften in Wien, phil.-hist. Classe,* 1902, Vol. XLVIII, II. Abhandlung, pp. 1-62.

Mitteilung über die angebliche Kenntnis der Luftschiffahrt bei den alten Chinesen. *Ostasiatischer Lloyd,* Vol. XVII; s. *Orient Bibl.,* 1904, no. 1489, p. 78.

Review of Jul. Lessing, *Chinesische Bronzegefässe. T'oung Pao,* 2d Series, Vol. IV, pp. 264-267; s. *Orient Bibl.,* 1904, no. 1583, pp. 82.

Religiöse Toleranz in China. *Globus,* 1904, Vol. LXXXVI, pp. 219, 220.

Ein buddhistisches Pilgerbild. *Globus,* 1904, Vol. LXXXVI, pp. 386-388.

Zur Geschichte der chinesischen Juden. *Globus,* 1905, Vol. LXXXVII, no. 14, pp. 245-247.

Chinesische Altertümer in der römischen Epoche der Rheinlande. *Globus,* 1905, Vol. LXXXVIII, No. 3, pp. 45-49.

Zum Bildnis des Pilgers Hsüan Tsang. *Globus,* 1905, Vol. LXXXVIII, pp. 257, 258.

Ein angebliches Chinesisches Christusbild aus der T'ang Zeit. *Globus,* 1905, Vol. LXXXVIII, pp. 281-283. Illustrated.

Anneaux nasaux en Chine. *T'oung Pao,* 2d Series, Vol. VI, 1905, pp. 321-323.

58

Introduction to the book of W. Filchner, *Das Kloster Kumbum in Tibet.* Berlin, 1906, pp. ix-xiv.

Obituary notice of Dr. Georg Huth. *T'oung* Pao, 1906, pp. 702-706.

Historical jottings on amber in Asia. *Memoirs American Anthropological Association,* Vol. I, part 3, pp. 211-244.

The Bird-Chariot in China and Europe. *Boas Anniversary Volume,* pp. 410-424.

Editor, Boas anniversary volume. New York, G. E. Stechert & Co., 1906.

Ancient Chinese bronzes. *Craftsman,* April, Vol. XII, pp. 3-15. Illustrated.

A plea for the study of the history of medicine and natural sciences. *Science,* Vol. XXV, pp. 889-895.

The relations of the Chinese to the Philippine Islands. *Smithsonian Miscellaneous Collections,* Vol. L, part 2, Washington, 1907, pp. 248-284.

Zur Geschichte der Brille. *Mitteilungen des Ges. für Geschichte der Medizin,* Vol. VI, 1907, No. 4, pp. 379-385.

A theory of the origin of Chinese writing. *American Anthropologist,* Vol. IX, 1907, pp. 487-492.

Reviews of F. Karseh, *Forschungen über gleichgeschlechtliche Liebe. American Anthropologist,* vol. IX, pp. 390-397. W. v. Hoerschelmann. *Entwicklung der altchin. Ornamentik,* and H. Beckh, *Tib. Übersetzung Kalidasa's Meghaduta. Monist,* Vol. XVII, pp. 634-636.

W. W. Newell and the lyrics of Li-t'ai-po. *American Anthropologist,* Vol. IX, 1907, pp. 655, 656.

The introduction of maize into eastern Asia. *Congres internat. des Americanistes,* Quebec, Vol. I, 1907, pp. 223-257.

Note on the introduction of the groundnut into China. Ibid., 259-262.

Zur buddhistischen Litteratur der Uiguren. *T'oung Pao,* Ser. II, Vol. VIII, 1907, pp. 391-409.

Ein japanisches Frühlingsbild. *Anthropophyteia,* Vol. IV, 1 plate, pp. 279-284.

Skizze der mongolischen Literatur. *Revue orientale,* Vol. VIII, 1907, pp. 165-261. Russian translation published by the Academy of Sciences, Leningrad.

Origin of our dances of death. *Open Court.* vol. XXII, 1908, pp. 597-604.

Die Bru-ža Sprache und die historische Stellung des Padmasambhava. *T'oung Pao,* Vol. IX, 1908, pp. 1-46.

Die Sage von den goldgrabenden Ameisen. *T'oung Pao,* Vol. IX, 1908, pp. 429-452.

Skizze der manjurischen Literatur. *Revue orientale,* Vol. IX, 1908, pp. 1-53.

59

A Mongol irodalom vazlata. A Manszsu Irodalom Vazlata. Illustrated. Hungarian Translation of Sketch of Mongol and Manchu Literatures.

The Jonah legend in India. *Monist,* Vol. XVIII, 1908, pp. 576-578.

Chinese pigeon whistles. *Scientific American,* May, 1908, p. 394. Illustrated.

Chinese pottery of the Han Dynasty. Publication of the East Asiatic Committee of the American Museum of Natural History. Leyden, 1909. 75 plates and 35 figures.

Die Kanjur Ausgabe des Kaisers K'ang-Hsi. *Bulletin de l'Académie impériale des sciences de St. Pétersbourg,* pp. 567-574. St. Petersburg, 1909.

Der Cyclus der zwolf Tiere auf einem alt-turkistanischen Teppich. *T'oung Pao,* Vol. X, 1909, pp. 71-73 (avec note additionelle par E. Chavannes).

Ein homosexuelles Bild aus China (1 plate). *Anthropophyteia,* Vol. VI, 1909, pp. 162-166.

Kunst und Kultur Chinas im Zeitalter der Han. *Globus,* Vol. XCVI, 1909, pp. 7-9, 21-24.

Christian art in China (20 plates). *Mitteilungen des Seminars für orientalische Sprachen,* Vol. XIII, Berlin, 1910, pp. 100-118.

Die Ausnutzung sexueller Energie zu Arbeitsleistungen, eine Umfrage. *Anthropophyteia,* Vol. VII, 1910, pp. 295, 296.

Zur kulturhistorischen Stellung der chinesischen Provinz Shanshi. Beobachtungen auf einer Reise von Tai-yüan nach Hsi-an im Februar 1909. *Anthropos,* Vol. V, 1910, pp. 181-203.

Der Roman einer tibetischen Königin. Tibetischer Text und Übersetzung. 8 figures and book-ornaments after Tibetan designs drawn by Albert Grünwedel. Leipzig, Otto Harrassowitz, 1911. pp. x, 264.

Contributions to Paul Carus, *The fish as a mystic symbol in China and Japan. Open Court,* Vol. XXV, July, 1911, pp. 385-411. Illustrations and quotations from Laufer.

King Tsing, the author of the Nestorian inscription. *Open Court,* August, 1911, pp. 449-454.

The introduction of vaccination into the Far East. *Open Court,* Sept., 1911, pp. 525-531.

Chinese grave-sculptures of the Han period. 10 plates and 14 text-figures. New York, London, and Paris, 1911. 45 pp.

Modern Chinese collections in historical light. *American Museum Journal,* April, 1912, pp. 135-138. Illustrated.

The Chinese Madonna in the Field Museum (1 plate). *Open Court,* Vol. XXVI, Jan., 1912, pp. 1-6. Also as separate brochure.

Confucius and his portraits. *Open Court,* Vol. XXVI, March and April, 1912, pp. 147-168, 202-218 (1 plate and 25 text-figures). Also as separate book.

60

The discovery of a lost book. *T'oung Pao,* Vol. XIII, No. 1, 1912, pp. 97-106 (1 plate).

Five newly discovered bas-reliefs of the Han period. *T'oung Pao,* Vol. XIII, No. 1, 1912, pp. 107-112 (4 plates).

The Wang Chuan Tu, a landscape of Wang Wei. *Ostasiatische Zeitschrift,* Vol. I, No. 1, Berlin, 1912, pp. 28-55 (18 figures).

The name China. *T'oung Pao,* Vol. XIII, 1912, pp. 719-726.

Foreword to *Catalogue of a selection of art objects from the Freer Collection,* Washigton, 1912.

Jade, A study in Chinese archaeology and religion. 68 plates, 6 of which are colored, and 204 text-figures. *Field Museum Anthropological Series,* Vol. X, Chicago, 1912, 370 pp.

China can take care of herself. *Oriental Review,* Vol. II, 1912, pp. 595, 596.

The stanzas of Bharata. *Journal Royal Asiatic Society,* 1912, pp. 1070-1073.

Chinese Sarcophagi. *Ostasiatische Zeitschrift,* Vol. I, No. 3, Berlin, 1912, pp. 318-334. 5 illustrations.

Fish symbols in China. *Open Court,* Nov., 1912, pp. 673-680. Illustrated. (Reprinted from *Jade.*)

Postscript to Cole, *Chinese Pottery in the Philippines. Field Museum Anthropological Series,* vol XII, part 1, July, 1912, pp. 17-47.

The praying mantis in Chinese folk-lore. *Open Court,* No. 1, 1913, pp. 57-60. 3 illustrations. (Extract from *Jade.*)

The Chinese battle of the fishes. *Open Court,* No. 1, June, 1913, pp. 378-381. 1 illustration.

The development of ancestral images in China. *Journal Religious Psychology,* Vol. VI, 1913, pp. 111-123.

Dokumente der indischen Kunst. I. Heft. *Das Citralakshana nach d. tib. Tanjur herausgegeben und übersetzt.* Leipzig, Harrassowitz, 1913, 194 pp.

Descriptive account of the collection of Chinese, Tibetan, Mongol and Japanese books in the Newberry Library. Chicago, 1913. 4 illustrations.

Notes on turquois in the East. *Field Museum Anthropological Series,* Vol. XIII, No. 1, 1913, pp. 1-72. 6 plates (1 colored).

Arabic and Chinese trade in walrus and narwhal ivory. *T'oung Pao,* 1913, pp. 315-370, with addenda by P. Pelliot.

History of the fingerprint system. *Smithsonian Report for 1912,* pp. 631-652, Washington, 1913. (7 plates.)

In memorial of J. Pierpont Morgan. *Ostasiatische Zeitschrift,* Vol. II, 1913, pp. 222-225.

61

Epigraphische Denkmaler aus China. Teil I: *Lamaistische Klosterin-schriften aus Peking, Jehol und Si-ngan* (with O. Franke). Berlin, 1914, fol. 81 plates, 2 portfolios.

The application of the Tibetan sexagenary cycle. *T'oung Pao,* Vol. XIV, 1913, pp. 569-596.

Der Pfau in Babylonien. *Orientalistische Literaturzeitung,* No. 12, 1913, col. 539, 540.

Catalogue of a collection of ancient Chinese snuff-bottles in the possession of Mrs. George T. Smith. Chicago, privately printed, 1913, 64 pp.

Über den Wert chinesicher Papierabklatsche. *Ostasiatische Zeitschrift,* Vol. II, 1913, pp. 346, 347.

Bird divination among the Tibetans (Notes on Document Pelliot, no. 3530, with a study of Tibetan phonology of the ninth century). *T'oung Pao,* Vol. XV, 1914, pp. 1-110. Leyden, E. J. Brill, 1914.

Discussion on relation of archaeology to ethnology. *American Anthropologist,* Vol. XV, Oct.-Dec., 1913, pp. 573-577.

Was Odoric of Pordenone ever in Tibet? *T'oung Pao,* 1914, pp. 405-418.

Obituary notice of F. H. Chalfant. *T'oung Pao,* 1914, pp. 165, 166.

The sexagenary cycle. Once more. *T'oung Pao,* 1914, pp. 278, 279.

Chinese clay figures. Part I. Prolegomena on Chinese Defensive Armor. 64 plates and 55 text-figures. *Field Museum Anthropological Series,* Vol. XIII, No. 2, 1914, pp. 73-315.

Review of H. Beckh, *Verzeichnis der tibetischen Handschriften. Journal Royal Asiatic Society,* Oct., 1914, pp. 1124-1139.

Some fundamental ideas of Chinese culture. *Journal Race Development,* Vol. V, Oct., 1914, pp. 160-174.

The story of the Pinna and the Syrian Lamb. *Journal American Folklore,* Vol. XXVIII, April-June, 1915, pp. 103-128.

The Eskimo screw as a culture-historical problem. *American Anthropologist,* Vol. XVII, 1915, pp. 396-406.

The Prefix A- in the Indo-Chinese languages. *Journal Royal Asiatic Society,* Oct., 1915, pp. 757-780.

Karajang. *Journal Royal Asiatic Society,* Oct., 1915, pp. 781-784.

Optical lenses. I. *Burning-Lenses in China and India. T'oung Pao,* 1915, pp. 169-228.

Burning-lenses in India. *T'oung Pao,* 1915, pp. 562, 563.

Three Tokharian bagatelles. *T'oung Pao,* 1915, pp. 272-281.

Vidanga and cubebs. *T'oung Pao,* 1915, pp. 282-288.

W. W. Rockhill (Obituary Notice). *T'oung Pao,* 1915, pp. 289, 290.

Asbestos and Salamander. An essay in Chinese and Hellenistic Folk-lore. *T'oung Pao,* 1915, pp. 299-373.

Chinese transcriptions of Tibetan names. *T'oung Pao,* 1915, pp. 420-424.

The Diamond. A study in Chinese and Hellenistic Folk-lore. *Field Museum Anthropological Series,* Vol. XV, 1915, No. 1, pp. 1-75.

62

Two Chinese imperial jades. *Fine Arts Journal,* Chicago, Vol. XXXII, June, 1915, pp. 237-241. 4 illustrations.

The Nichols Mo-So Manuscript. *Geographical Review,* Vol. I, April, 1916, pp. 274-285. 4 figures.

Ethnographische Sagen der Chinesen. *Festschrift Kuhn* (München), 1916, pp. 198-210.

Chinesische Schattenspiele, übersetzt von Wilhelm Grube, herausgegeben und eingeleitet von B. Laufer. *Abhandlungen Königl. Bayer. Akad. d. Wissenschaften,* Vol. XXVII, No. 1, 442 pp. 1915. Preface and Introduction (pp. v-xxiv) written by B. Laufer.

The Si-Hia language, a study in Indo-Chinese philology. *T'oung Pao,* 1916, pp. 1-126.

Supplementary notes on walrus and narwhal ivory. *T'oung Pao,* 1916, pp. 348-389.

Se-Tiao. *T'oung Pao,* 1916, p. 390.

The beginnings of porcelain in China. With a technical report by H. W. Nichols. 12 plates and 2 text-figures. *Field Museum Anthropological Series,* Vol. XV, No. 2, pp. 77-182, 1916. The date was arbitrarily changed by the printer to 1917. 106 pp. 7 plates.

Cardan's suspension in China. 1 plate, 1 text-figure. *Holmes Anniversary Volume,* 1916, pp. 288-291.

Review of R. Garbe, *Indien und das Christentum. American Anthropologist,* Vol. XVIII, 1916, pp. 567-573.

Review of J. E. Pogue, *The Turquois. American Anthropologist,* Vol. XVIII, 1916, pp. 585-590.

Burkhan. *Journal American Oriental Society,* vol. XXXVI, 1917, pp. 390-395. Additional note, vol. XXXVII, 1917, pp. 167, 168.

Concerning the history of finger-prints. *Science,* 1917, May 25, pp. 504, 505.

Moccasins. *American Anthropologist,* 1917, pp. 297-301.

Origin of the word Shaman. *American Anthropologist,* 1917, pp. 361-371.

The reindeer and its domestication. *Memoirs, Anthropological Association,* Vol. IV, No. 2, 1917, pp. 91-147.

The vigesimal and decimal systems in the Ainu numerals. *With some remarks on Ainu Phonology. Journal American Oriental Society,* Vol. XXXVII, 1917, pp. 192-208.

The language of the Yüe-Chi or Indo-Scythians. Chicago, R. R. Donnelley and Sons Co., 1917, 14 pp. Privately printed in 50 copies.

Collection of ivory seals in the possession of Mrs. George T. Smith. Privately printed, 1917.

Loan-words in Tibetan. *T'oung Pao,* 1916, pp. 403-552. Leyden, E. J. Brill, 1918.

Religious and artistic thought in Ancient China. *Art and Archaeology*, Dec., 1917, pp. 295-310. 16 illustrations.

Reviews of Diels, *Antike Technik*; Mookerji, *Indian Shipping*; Parmentier, *Guide au Musée de l'Ecole française*; Maspero, *Grammaire de la langue khmere*. *American Anthropologist*, 1917, pp. 71-80, 280-285.

Totemic traces among the Indo-Chinese. *Journal American Folk-lore*, Vol. XXX, 1917, pp. 415-426.

Origin of Tibetan writing. *Journal American Oriental Society*, Vol. XXXVIII, 1918, pp. 34-46.

Edouard Chavannes (Obituary Notice). *Journal American Oriental Society*, Vol. XXXVIII, 1918, pp. 202-205.

Reviews of Lowie, *Culture and ethnology, American Anthropologist*, Vol. XX, 1918, pp. 87-91; Foote, *Foote Coll. of Indian antiquities*, and Rea, *Cat. Prehistoric ant. from Adichanallur*, ibid., pp. 104-106.

La Mandragore (in French). *T'oung Pao*, 1917, pp. 1-30; published 1918.

Malabathron (in French). *Journal asiatique*, Paris, 1918, 50 pp., 12 figures.

The Chinese exhibition. *Bulletin, Art Institute of Chicago*, Dec., 1918, pp. 144-147.

Sino-Iranica. Chinese contributions to the history of civilization in Ancient Iran, with special reference to the history of cultivated plants and products. *Field Museum Anthropological Series*, Vol. XV, 1919, No. 3, 446 pp.

Reviews of Tallgren, *Collection tovestine des antiquités préhist. de Minoussinsk*; Sarkar, *Hindu achievements in exact science*; Starr, *Korean Buddhism*; Mauger, *Quelques considérations sur les jeux en Chine*; Couling, *Encyclopaedia Sinica*. *American Anthropologist*, Vol. XXI, 1919, pp. 78-89. Also, M. Czaplicka, *Turks of Central Asia*; G. Ferrand, *Malaka, le Malaya et Malayur*, and *Apropos d'une carte javannaise du XVI° siècle*; R. Torii, *Études archeologiques. Les Ainu des Iles Kouriles, op. cit.*, pp. 198, 301-308, 459.

Coca and betel-chewing: A query. *American Anthropologist*, Vol. XXI, 1919, pp. 335, 336.

Sanskrit Karketana (in French). *Mémoires de la Société de Linguistique*, Vol. XXII, 1922, pp. 43-46.

Multiple births among the Chinese. *American Journal Physical Anthropology*, Vol. III, 1920, No. 1, pp. 83-96; and *New China Review*, Shanghai, Vol. II, April, 1920, pp. 109-136.

Sex transformation and hermaphrodites in China. *American Journal Physical Anthropology*, Vol. III, 1920, No. 2, pp. 259-262.

The reindeer once more. *American Anthropologist*, Vol. XXII, 1920, pp. 192-197.

Twelve articles and frontispiece contributed to H. Cordier, *Ser Marco Polo*, Murray, London, 1920.

64

Review of L. Wiener, *Africa and the Discovery of America*. *Literary Review* of the *New York Evening Post*, Feb. 5, 1921.

Review of H. Cordier, *Ser Marco Polo: Notes and addenda to Sir Henry Yule's Edition, 1920*. *American Historical Review*, Vol. XXVI, No. 3, April, 1921, pp. 499-501.

Milaraspa. *Tibetische Texte in auswahl übertragen*. Folkwang-Verlag, Hagen, 1922.

A bird's-eye view of Chinese art. *Journal American Institute of Architects*, Vol. X, No. 6, 1922, pp. 183-198. 23 text-figures.

The Chinese gateway. *Field Museum Anthropological leaflet* No. 1, 1922, pp. 1-7. 1 plate.

Archaic Chinese bronzes of the Shang, Chou and Han periods. New York, Parish-Watson, 1922, 24 pp., 4°. 10 plates.

Use of human skulls and bones in Tibet. *Field Museum Anthropological Leaflet* No. 10, Chicago, 1923, 16 pp. 1 plate.

Oriental theatricals. *Field Museum, Guide,* Part I, Chicago, 1923. 60 pp. 11 plates in photogravure.

Review of Shirokogoroff, *Social organization of the Manchus*. *American Anthropologist*, Vol. XXVI, 1924, pp. 540-543.

Tobacco and its use in Asia. *Field Museum Anthropological Leaflet* No. 18, 1924, 40 pp. 10 plates in photogravure.

The introduction of tobacco into Europe. *Field Museum Anthropological Leaflet* No. 19, 1924, 66 pp.

Tang, Sung and Yüan paintings, Paris and Brussels. G. van Oest and Company, 1924, 22 pp., folio. 30 plates.

Chinese baskets. *Anthropology Design Series* No. 3, Field Museum, Chicago, 1925. 38 plates in photogravure, 2 pp. text, quarto.

The tree-climbing fish. *China Journal of Science and Arts*, Vol. III, 1925, No. 1, pp. 34-36.

Archaic bronzes of China. *Art in America*, Vol. XIII, No. 6, pp. 291-307. 4 plates.

Jurchi and Mongol numerals. *Korosi Csoma-Archiv*, Vol. I, No. 2, Dec., 1921, pp. 112-115.

Ivory in China. *Field Museum Anthropological Leaflet* No. 21. 1925, 78 pp. 10 plates in photogravure.

Review of H. A. Giles, *Strange stories from a Chinese studio*. *Journal of American Folk-lore*, Vol. XXXIX, 1926, pp. 86-90.

The Jan Kleykamp collection, Chinese and Japanese paintings. New York, 1925, 40 pp., folio. 40 plates.

Ostrich egg-shell cups of Mesopotamia and the ostrich in ancient and modern times. *Field Museum Anthropological Leaflet* No. 23, 1926, 52 pp. 9 plates in photogravure, 10 text-figures, 1 cover design.

History of ink in China, Japan, Central Asia, India, Egypt, Palestine, Greece, and Italy. In F. B. Wiborg, *Printing Ink, a History*. Harper Bros., New York, 1926, pp. 1-76.

65

Archaic Chinese jades collected in China by A. W. Bahr. Now in Field Museum of Natural History of Chicago. 52 pp., 36 plates, 3 of which are colored. New York. Printed privately for A. W. Bahr by R. R. Donnelley and Sons Co., Chicago, 1927.

Methods in the study of domestications. *Scientific Monthly*, Sept., 1927, pp. 251-255.

Review of T. F. Carter, *Invention of printing in China*. *Journal American Oriental Society*, Vol. XLVII, 1927, pp. 71-76.

Agate, archaeology and folk-lore. *Field Museum Geology Leaflet* No. 8, 1927, pp. 20-35. Illustrated.

Insect-musicians and cricket champions of China. *Field Museum Anthropological Leaflet* No. 22, 1927, 28 pp. 12 plates in photogravure.

The giraffe in history and art. *Field Museum Anthropological Leaflet* No. 27, 100 pp. 9 plates in photogravure, 23 text-figures, 1 vignette, 1 colored cover design. Notice in *Quarterly Review of Biology* IV, 1929, p. 138.

Cricket champions of China. *Scientific American*, Jan., 1928, pp. 30-34. Illustrated.

The prehistory of aviation. *Field Museum Anthropological Series*, Vol. XVIII, No. 1, 1928, 100 pp. 12 plates in photogravure and 1 vignette.

The prehistory of television. *Scientific Monthly*, Nov., 1928, pp. 455-459.

Review of H. Maspero, *La Chine antique*. *American Historical Review*, Vol. XXXIII, 1928, pp. 903, 904.

Turtle fossil arouses interest of scientists (Laufer quoted as authority). *Scientific American*, May, 1929, pp. 451- 452.

Review of W. Barthold, *Turkestan down to the Mongol Invasion*. *American Historical Review*, Vol. XXXIV, 1929, pp. 378, 379.

Review of Chi Li, *The Formation of the Chinese People: An Anthropological Inquiry*. *American Historical Review*, Vol. XXXIV, 1929, pp. 650, 651.

The American plant migration. Reprinted from *Scientific Monthly*, March, 1929, Vol. XXVIII, pp. 239-251.

On the possible origin of our word booze. *Journal of American Oriental Society*, Vol. XLIX, 1929, pp. 56-58.

The early history of felt. Reprinted from *American Anthropologist*, Vol. XXXII, No. 1, January-March, 1930, pp. 1-18.

A Chinese-Hebrew manuscript. A new source for the history of the Chinese Jews. Reprinted from *American Journal of Semitic Languages and Literatures*, Vol. XLVI, No. 3, April, 1930, pp. 189-197.

Mission of Chinese students. *Chinese Social and Political Science Review*, Vol. XIII, No. 3, July, 1929, pp. 285-289.

The Gest Chinese research library at McGill University. Montreal, 1929.

66

The Eumorfopoulos Chinese bronzes. *Burlington Magazine*, Vol. LIV, 1929, pp. 330-336.

Catalogue of a collection of Chinese paintings in the possession of Dr. Frederick Peterson. 1930.

Felt. How it was made and used in ancient times and a brief description of modern methods of manufacture and uses. Chicago, Western Felt Works, 1930. John Crerar Library.

Chinese bells, drums and mirrors. *Burlington Magazine*, Vol. LVII, pp. 183-187. London, October, 1930.

Geophagy. *Field Museum Anthropological Series*, Vol. XVIII, No. 2, Chicago, 1930, pp. 101-198.

Tobacco and its use in Africa. *Field Museum Anthropological Leaflet* No. 29, 1930. By B. Laufer, W. D. Hambly, and R. Linton.

The restoration of ancient bronzes and cure of malignant patina. By Henry W. Nichols, with foreword by Berthold Laufer. Aug., 1930. *Museum Technique Series*, No. 3.

Columbus and Cathay, and the meaning of America to the Orientalist. *Journal American Oriental Society*, Vol. LI, 1931, pp. 87-103.

China and the discovery of America. A monograph published by the China Institute in America. New York, 1931.

The domestication of the cormorant in China and Japan. *Field Museum Anthropological Series*, Vol. XVIII, No. 3, 1931, pp. 205-262.

Inspirational dreams in Eastern Asia. *American Folk-lore Journal*, Vol. XLIV, 1931, pp. 208-216.

The prehistory of aviation. *Open Court*, Vol. XLV, 1931, pp. 493-512. Illustrated.

Tobacco in New Guinea. *American Anthropologist*, Vol. XXXIII, 1931, pp. 138-140.

Paper and printing in Ancient China. Printed for the Caxton Club. Chicago, 1931.

A defender of the faith and his miracles. *Open Court*, Vol. XLVI, 1932, pp. 665-667.

The early history of polo. *Polo, The Magazine for Horsemen*, Vol. VII, 5, Apr. 1932, pp. 13, 14, 43, 44.

Sino-American points of contact. *Scientific Monthly*, Vol. XXXIV, 1932. *Open Court*, Vol. XLVII, 1933, pp. 495-499.

East and West. *Open Court*, Vol. XLVII, 1933, pp. 473-478.

The Jehol pagoda model. *Field Museum News*, Vol. IV, 1933.

Turtle Island. *Open Court*, Vol. XLVII, 1933, pp. 500-504.

Prefaces to *Monograph Series of the New Orient Society of America*. Second Series, 1933. 1. China No. 5; 2. Central and Russian Asia No. 2.

Foreword to Henry Field, *Prehistoric Man. Field Museum Anthropology Leaflet* No. 31, pp. 3, 4. Chicago, 1933.

67

A defender of the faith. *Asia*, Vol. XXXIV, May, 1934, pp. 290, 291.

Etruscans. *Field Museum News*, Vol. V, 1934.

The lemon in China and elsewhere. *Journal American Oriental Society*, vol. LIV, 1934, pp. 143-160.

Rare Chinese brush-holder. *Field Museum News*, Vol. V, 1934.

The Chinese imperial gold collection. A Century of Progress, 1934. Parish-Watson & Co., Inc., New York, 1934.

The Swing in China. *Memoires de la Société Finno-Ougrienne*, Vol. LVII, 1934.

The Noria or Persian wheel. *Pavry Memorial Volume*, London, 1933. Oxford University Press, pp. 238-250.

Chinese Muhammedan bronzes. *Ars Islamica*, Vol. I, 1934, pp. 133-147. Illustrated.

Rye in the Far East and the Asiatic origin of our word series "Rye." *T'oung Pao*, Vol. XXXI, 1935, pp. 237-273.

Not included in the above bibliography are the Reports of the Department of Anthropology, Field Museum, published in the Director's Annual Reports, 1915-1933. These, however, do not appear over Laufer's signature.

Dr. Laufer was the editor of all anthropological publications, leaflets, guides, and design series issued by the Field Museum from 1915 to his death.

In addition, Dr. Laufer left the following unfinished manuscripts:

The Buceros and Hornbill carvings.

History of the cultivated plants of America and their distribution over the Old World. 2 vols., ca. 800-900 pp.

Jade, second revised and enlarged edition.

Chinese domestications, pt. 1: Chicken, Cormorant, and Cat. Five Prehistories.

A History of the Game of Polo.

68